REEF FISH
Identification

Tropical Pacific

Gerald Allen
Roger Steene
Paul Humann
Ned DeLoach

NEW WORLD PUBLICATIONS, INC.

Jacksonville, Florida USA

Print Production
D'Print Pte Ltd

Acknowledgments

This project has been a genuine team effort from the moment it was conceived. The late John Jackson of Odyssey Publishing provided the kick start by introducing us to the great team at New World Publications – Paul Humann, Ned DeLoach, and Eric Riesch. They have been tireless and incredibly efficient through all phases of the production. Paul and Ned are also accomplished photographers and many of their excellent photos are included in the book.

Special praise is also due to my co-author and best mate Roger Steene. By some odd quirk of fate Roger was the first person I happened to meet after arriving on Australian shores with my family in 1972. We have shared amazing experiences over the past 43 years and it's no accident that all my best trips have been the ones that included Roger. Although I confess to being biased, I consider Roger Steene as the world's leading underwater photographer. His brilliant coffee table books, including the recent (2014) 3-volume set *Colour of the Reef,* are testimony to his wide range of skills.

Our species coverage has been greatly enriched by contributions from various colleagues, especially Rudie Kuiter and Fenton Walsh from Australia, Dieter Eichler and Helmut Debelius from Germany, Jack Randall from Hawaii, Scott Michael and Robert Myers from mainland USA, the late Takamasa Tonazuka and his wife Miki from Bali, Indonesia. Most of the color scans were prepared by Mark Allen. Many range extension updates were provided by Kreg Martin and Janet Eyre.

Scientific advice and assistance with identifications was received from Bill Eschmeyer (California Academy of Sciences), Tom Fraser (Florida Museum of Natural History, Gainesville), Tony Gill (The Natural History Museum, London), Dave Greenfield (California Academy of Sciences), Leslie Knapp (Smithsonian Institution), Craig Howson (North Star Charters, Australia), Rudie Kuiter (Museum of Victoria, Melbourne), Helen Larson (Northern Territory Museum, Darwin), John McCosker (California Academy of Sciences), Tom Munroe (Smithsonian Institution), Tom Pietsch (University of Washington), Jack Randall (Bishop Museum, Hawaii), Bill Smith-Vaniz (U.S. Geological Survey, Gainesville, Florida), Victor Springer (Smithsonian Institution), Guy Stevens (Manta Trust) and Rick Winterbottom (Royal Ontario Museum, Toronto).

We are especially grateful to the many generous resort owners and charter boat operators, who provided us with wonderful opportunities for fish photography: Max Ammer (Irian Diving, West Papua), Rob Barrel and Cat Holloway (Nai'a Charters, Fiji), the Batuna family and Danny Charlton (Murex Dive Resorts, Sulawesi), Max Benjamin (Walindi Dive Resort, Papua New Guinea), Mark Ecenbarger (Kungkungan Bay Resort, Sulawesi), Ron Holland (Borneo Divers, Sabah), Bruce Moore (Black Sands Resort, Lembeh, Indonesia), Nyoman and Reno Kirtya (Grand Komodo Tours, Bali), Carol Palmer and Sonny Tjandra (Ambon Diver Centre), Alan Raabe (FeBrina Dive Cruises, Papua New Guinea), Anton Saksono (Pulau Purtri Resort, Java), Frans Seda (Sao Wisata Resort, Flores), Patti Seery (Silolona Charters, Indonesia), the late Takamasa Tonozuka (Dive and Dive's, Bali), Rob Vanderloos (Milne Bay Marine Charters, Papua New Guinea), and Wakatobi Divers (Tukang Besi, Indonesia).

Finally GRA expresses his fondest thanks to his wife Connie for her amazing devotion, continued support, and tolerance of frequent overseas travel. He is also grateful for the ongoing and generous support of the Paine Family Foundation and Conservation International, especially dive buddy and scientific colleague Mark Erdmann (Vice President for Asia-Pacific Marine Programs).

Gerald R. Allen
Perth, Australia

CREDITS

Photography Editor: Eric Riesch
Editor: Ken Marks
Field Editors: Janet Eyre, Kreg Martin
Art Director & Drawings: Michael O'Connell
Print Consultant: D'Print Pte Ltd., Singapore
ISBN 978-1-878-348-60-9
First Edition 2003; Second Edition 2015; Second Printing 2018, Third Printing 2019.
Publisher: New World Publications, Inc. 1861 Cornell Road, Jacksonville, FL 32207, (904) 737-6558, fishid.com, orders@fishid.com.

PHOTO CREDITS

The majority of species images were taken by the four authors. However, several underwater photographers added their work to this collection. The authors appreciate their efforts and assistance in making this book as comprehensive as possible. They include: **Jim Abernethy,** 472mr; **Lyn Adrian,** 43tr; **Mark Allen,** 88br, 133tl; **Tim Allen,** 32tr; **Orhan Aytur,** 401mr; **Jim Black,** 450tr; **Mike Bryant,** 470bl; **Neville Coleman,** 39br; **Helmut Debelius,** 114br, 118tr, 243mr, 436br; **JR Earle,** 177br; **Dieter Eichler,** 113tl, 144mr, 160ml, 179tl, 184ml, 208tr &ml, 218br, 230mr, 232br, 241br, 445tl, 461bl; **Mark Erdmann,** 150tl, 317tr, 335bl, 366mr, 469ml; **Janet Eyre,** 31tr, 343ml, 372tl; **Lynn Funkhouser,** 472tr; **Daniel Geary,** 383mr & br; **Howard Hall,** 439bl; **Wolcott Henry,** 447bl; **John Hoover,** 127ml,132br, 453ml & br, 459tr; **Burt Jones & Maurine Shimlock,** 129tl, 142tl, 162bl, 177tl, 394ml, 395ml, 476 ml; **Peter Kraugh,** 313ml; **Rudie Kuiter,** 59ml, 76bl, 110br, 113br, 121ml, 128mr, 133tr, 136ml, 165ml, 158mr & bl, 175br, 177bl & br, 181mr, 179mr &ml, 182br, 184br, 186tr, 189br, 191ml, 192tl, 194tl, 203mr, 206mr, 209br, 227tl, 229mr, 242bl, 243br, 244br, 245tl, 248bl, 274tr, 290bl & br, 295tl & tr, 316ml, 325 mr, 336bl, 339tl, 340mr, 346ml, 387bl, 396bl, 399tl, 400tr, 411br, 418tr, 419mr, 421br, 427tr, 443mr, 446ml & mr, 447ml, 458ml, mr, bl & br; **Ken Marks,** 17bl, 26br, 42br, 110tl, 137ml, 211ml, 235tl, 298br, 300tl, 424bl, 425br, 428tr, 429tl & mr, 437br, 461tl, 470ml, 471mr, 472tl, 473tr; **Andrea Marshall** 475bl; **Hajime Masuda,** 92br; **Scott Michael,** 64tr, 143tl & tr, 149br, 150br, 205ml, 228tr, 233tl, 358ml, 365ml, 379ml, 381mr, 384tr, 454tl, ml, mr & br, 456tr, 464ml, 468tr, ml & mr, 469tl & bl, 471tr, 474mr; **Rob Myers,** 46mr, 47mr, 73br, 81tl, 83mr, 87ml, 120ml, 126bl, 136ml, 149tr, 151tl, 161mr, 171mr & br, 172tl, 173tr, 177ml, 183tr, 185ml, 196tl, 200tl, 208bl, 209tl, 213tr, 222mr, 224tr, 229tl, 233br, 236tl, 237bl & br, 238tl & tr, 253bl, 274tl, 309tr, 358mr, 359bl, 436ml, 454tr; **Christopher Newbert,** 124m; **Nu Parnuponge,** 475ml; **Mike Phelan,** 127mr; **Jack Randall,** 52bl & br, 98bl, 100br, 103ml, 128bl, 136ml, 148bl & br, 151ml, 168mr, 226br, 243ml, 245tr, 261br, 373ml, 391br, 413tl & tr, 417mr & bl, 418tl, 439mr, 459br, 474ml; **Eric Riesch,** 64mr, 167ml, 253tl, 318bl; **Paddy Ryan,** 403mr; **Thomas Scharz,** 383bl; **Marc Sentis,** 475mr; **Richard Smith,** 164tr, 450tl, ml,mr & bl; **Guy Stevens/Manta Trust,** 475br, 476tl,ml & br; **Mark Strickland,** 472br, 473br; **Tane Sinclair Taylor,** 475tr; **William Tan,** 449tl; **Miki Tonozuka,** 462mr; **Takamosa Tonozuka,** 291ml, 315tr, 330tr, 379br, 412tr, 444br, 445bl, 446tr; **Peter Verhoog,** 468tl, 471bl; **Fenton Walsh,** 150tl, 198tl &tr, 233bl, 278tl; the remaining photographs were taken by **Gerald Allen, Roger Steene, Paul Humann** and **Ned DeLoach.**

About the Authors

Gerald R. Allen is the author of more than 300 scientific articles and 28 books. He served as Senior Curator of Fishes at the Western Australian Museum between 1974-1998 and is now a full-time consultant with Conservation International. Dr. Allen is an international authority on both coral reef fishes and Australian freshwater fishes. He received a Ph.D. in marine zoology from the University of Hawaii in 1971 and since then has dived extensively throughout the Indo-Pacific region, logging over 6,000 hours underwater. Underwater photography is his favorite hobby and several thousand of his photos have appeared in a wide variety of publications. Originally from the USA, Dr. Allen and his wife Connie have resided in Perth, Western Australia for the past 30 years.

Roger Steene has lived his entire life at the front doorstep of Australia's Great Barrier Reef. He became interested in underwater photography at an early stage and his enthusiasm has never waned. This is his eleventh book devoted to marine subjects, having accumulated more than 30 years of underwater experience in the process. He has dived and photographed in all the world's tropical seas from the Caribbean and Galapagos to the Red Sea and Mauritius. He lists Indonesia and Papua New Guinea as his favorite destinations due to their incredible marine biodiversity. Concentrating on close-up photography, his meticulous attention to detail conveys a special impression of marine life.

Paul Humann began photographing marine life in 1964. In a bold move in 1972 he left his established law practice in his hometown of Wichita, Kansas to become the owner/operator of the *Cayman Diver,* the Caribbean's first successful live-aboard diving cruiser. He sold the vessel in 1979 to devote more time to travel, photography and writing. His images and articles have appeared in nearly every diving and wildlife magazine. Together with his partner Ned DeLoach, Paul has written 14 marine life field guides including the popular 3-volume Caribbean *Reef Set, Reef Fish Identification – Galapagos and Coastal Fish Identification – California to Alaska.* When not traveling Paul lives in Davie, Florida where, if not writing about fishes, he tends to another passion, his two-acre palm-studded garden/home fondly known as Mango Manor.

After finishing a degree in education in 1967, **Ned DeLoach** moved from his childhood home in West Texas to Florida so that he would be able to do more of what he loves best – dive. In 1971 he completed his first diving guide to the state, *Diving Guide to Underwater Florida,* which was released in its 11th edition in 2004. Through the 1970s and 1980s Ned was active in Florida's cave diving community and dive/travel writing. A mutual friend introduced Paul to Ned in the mid-80s. Two years later the pair reunited as co-editors of *Ocean Realm* magazine. It was during this time that the idea of producing a series of marine life identification books designed for divers was born. After co-authoring a number of marine life field guides the partners published *Reef Fish Behavior* in 2000. Ned and his wife Anna live in Jacksonville, Florida.

Twenty Identification Groups

1. Disk-shaped/Colorful   16-43

Butterflyfishes

Angelfishes

Spadefishes

2. Large Ovals   44-59

Surgeonfishes

Rabbitfishes

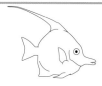

Moorish Idol

3. Small Ovals – Damselfishes   60-101

Damselfishes

Damselfishes/Anemonefishes

4. Sloping Heads/Tapered Bodies   102-121

Snappers

Coral Breams

Emperors

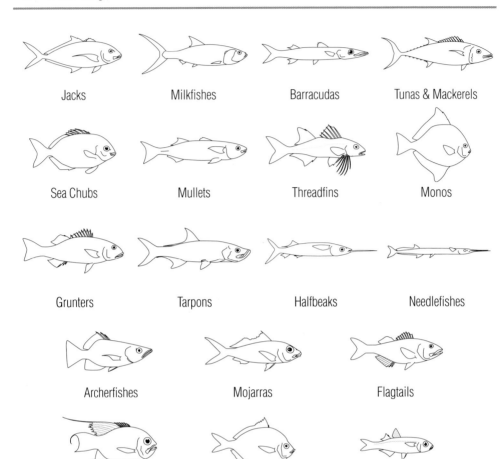

Jacks

Milkfishes

Barracudas

Tunas & Mackerels

Sea Chubs

Mullets

Threadfins

Monos

Grunters

Tarpons

Halfbeaks

Needlefishes

Archerfishes

Mojarras

Flagtails

Pearl Perches

Ponyfishes

Silversides

6. Slender Schoolers/Colorful   138-151

Fusiliers

Anthias

7. Heavy Bodies/Large Lips   152-173

Groupers

Soapfishes

Hawkfishes

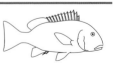

Sweetlips

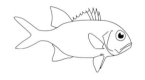
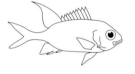
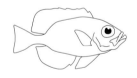
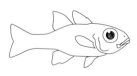

12. Cryptic Crevice Dwellers

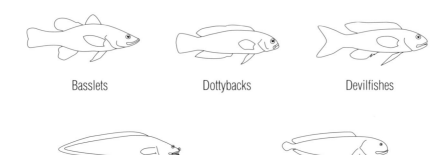

Basslets   Dottybacks   Devilfishes

Brotulas   Cusk-eels

13. Elongate Sand & Burrow Dwellers

Dartfishes   Wormfishes   Sand Divers

Bandfishes   Convict Blennies   Tilefishes   Dragonets

Sandperches   Lizardfishes   Jawfishes   Pearl Fishes

14. Small, Elongate Bottom Dwellers – Gobies

Gobies

15. Small, Elongate Bottom Dwellers – Blennies 352-375

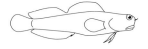

Blennies

Triplefins

16. Odd-Shaped Bottom Dwellers   376-407

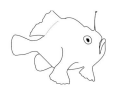

Frogfishes

Batfishes

Sea Moths

Flying Gurnards

Lionfishes

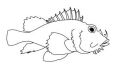

Scorpionfishes

Stonefishes

Waspfishes

Flatheads

Velvetfishes

Coral Crouchers

Stargazers

Flounders

Soles

Boxfishes

Goatfishes

Cornetfishes

Trumpetfishes

Shrimpfishes

Flashlightfishes

Eel-tailed Catfishes

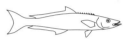

Remoras

Cobias

Sweepers

Triggerfishes

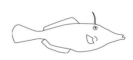

Filefishes

Puffers

Porcupinefishes

Molas

Dolphinfishes

18. Pipefishes & Seahorses

Ghost Pipefishes

Seahorses

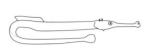

Pipefishes

19. Eels

Morays

Snake Eels

Conger Eels

Garden Eels

20. Sharks & Rays

Wobbegongs

Bamboo Sharks

Cat Sharks

Whale Sharks

Zebra Sharks

Nurse Sharks

Requiem Sharks

Hammerhead Sharks

Wedgefishes

Guitarfishes

Stingrays

Eagle Rays

Cownose Rays

Mantas

How To Use This Book

Identification Groups

Trying to identify a specific fish from the more than 2,000 tropical Pacific species included in this book can be a perplexing task. To help simplify the process, fish families with similar physical or behavioral characteristics have been arranged together into one of 20 color-coded and numbered ID Groups. This approach varies significantly from the traditional system that orders species chronologically by evolutionary development. Although there are a few anomalies, most families, genera and species integrate easily into this visually oriented system.

The ID Groups and their representative families are displayed on the Contents pages. Each group's similar characteristics are listed in italic type at the beginning of its ID Group. It is important for beginning fish watchers to become familiar with the major families that make-up ID Groups, so they can go quickly to the correct section to begin the identification process. Families are scientific groupings based on evolutionary sequence and consequently, typically have similar physical characteristics. An overview of the family's behavioral and physical characteristics (that are observable by divers) is presented at the beginning of each ID Group. The total number of genera and species included in this book, along with diagrams of representative family body shapes, is also given.

Names

Information about each species begins with its common name (that generally used by the English speaking public). Common names are far from standardized and tend to vary from region to region. In some cases there were as many as six different common names applied to a single species. The common names chosen for this text are, in most cases, those names most frequently used in previously published field guides. When this method proved impractical, a name was selected that relates to a readily observable physical feature of the fish, or as a translation of the scientific name. Often where the common name honored an individual, such as the discoverer of the species, the common name was changed to reflect a visually significant feature of the fish in order to help simplify the identification process.

To the right of a species' common name is the species' two-part scientific name printed in italics. These names, rooted in Latin and Greek, are highly standardized and used by scientists throughout the world. The first word (always capitalized) represents the genus. The genus name is given to a group of species, which share a common ancestor, and usually have similar anatomical and physiological characteristics. The second word (never capitalized) is the species. A species includes only animals that are sexually compatible and produce fertile offspring. Each species usually has a combination of visually distinctive features that separates them from all others.

The common and scientific family names follow. Because of its importance in the identification process, the common family name is also printed at the top of left pages where family members appear. Like common species names, common family names also vary between regions. In a few cases, when a distinctive group of fishes within a family are widely known by an alternate name, both names are included together separated by a slash.

The Use of Multiple Photographs for a Species

Many species are presented with more than one photograph. This is necessary to demonstrate differentiations in color, markings and physical features that occur within the same species. Such differentiations are primarily related to one of four categories:

Variations — Species, particularly those from different geographical regions, occasionally exhibit PERMANENT color or marking patterns or physical features distinctly different from the primary species illustrated.

Color and Marking Phases — Often a species may TEMPORARILY alter its color or markings, or, in rare instances, physical features to inhance its camouflage, indicate a change of mood, or for intraspecies communications, such as courtship. Phases can be adapted instantaneously, or, in a few cases, over an extended period of time.

Life Cycle Phases — The juvenile forms (sexually immature individuals) of many species appear distinctly different from adults. In the parrotfish and wrasse families life cycle phases are more complicated: besides juveniles, (denoted in the text as JP [juvenile phase]), adults display two visually distinct phases: the Initial Phase (IP), which generally includes both sexually mature males and females, and the Terminal Phase (TP), which includes only males, that are not only the least abundant, but the largest and most colorful individuals of the species.

Sexes — Males and females of many species display dissimilarity in colors or markings, or differences in body size, or the size and shape of anatomical features, such as fins.

Size

The size, given both in centimeters and inches or feet, represents the maximum size of a species recorded to date.

Description

A species' account is given under the heading ID. Although the visual descriptions in this text might seem redundant to a species' image printed above, this information is often essential when features of an unidentified fish do not exactly match the photograph. In many cases a fish is so distinctive that making a comparison with its photograph easily substantiates its identification. However, because many genera include "similar-appearing" species the identification process is often more complex. Wherever similar-appearing species occur within a genus, every effort has been make to place the species together. Likewise, similar-appearing genera within a family, and similar families within an ID Group have been grouped whenever possible.

To help distinguish between similar-appearing species "distinctive features" that visually differentiate one species from the other have been highlighted with bold text, and where appropriate, an arrow pointing directly to the emphasized characteristic has been superimposed over the photograph.

In some cases the distinctive features emphasized are too small or subtle to establish reliable visual identifications with the naked eye underwater. However, this information, which includes such things as number of scale rows, spine counts, or nostril position, might be relied upon for making identifications from photographic images that can be enlarged and studied in detail.

Behavioral traits that may be observed by a diver and might help in the identification process are also listed under ID following the species' description. This brief information is usually coupled with a species' social organization: solitary, in pairs, form groups, or aggregations, followed by the species' habitat preference and depth range where it typically occurs.

Colors — The colors of many species vary considerably from individual to individual. In such situations, the description might read: "Reddish brown to olive-brown or gray." This means that the fish could be any of the colors or shades between. Many fishes also have an ability to pale, darken, and change colors. Because of this, color alone is rarely relied on for identification.

Markings — The terminology describing markings is defined in the drawings on the following page.

Anatomy — Anatomical features are often referred to as part of the identification process. The features used in this text are pinpointed in the drawings on the following page.

Species Population Distribution

A species' distribution range is presented last. This section begins with a Broad Population Range (highlighted with bold text) followed by a more detailed account.

The range of this book extends east from the far eastern Indian Ocean (the Andaman Sea just west of Thailand, Christmas and Coco-Keeling Islands, far western Indonesia and northwestern Australia) to the Pitcairn Island Group in the southeastern Pacific, then south from the tropical water of south and southwest Japan to the southern end of the Great Barrier Reef. Whenever a species distribution extends outside this vast region, such as to the Red Sea and East Africa in the Indian Ocean or to the Hawaiian Islands in north central Pacific, this information is included;

Markings

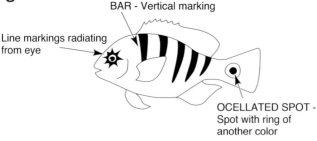

Line markings radiating from eye

BAR - Vertical marking

OCELLATED SPOT - Spot with ring of another color

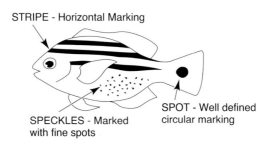

STRIPE - Horizontal Marking

SPECKLES - Marked with fine spots

SPOT - Well defined circular marking

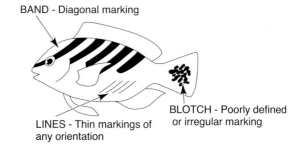

BAND - Diagonal marking

LINES - Thin markings of any orientation

BLOTCH - Poorly defined or irregular marking

however, no attempt has been made to provide a complete inventory of fishes that inhabit areas adjacent to the range of this book.

Because of the random nature of the pelagic dispersal of larval fishes, a species' distribution pattern is, to an extent, in constant flux. This fact coupled with our incomplete knowledge of the general distribution of many fishes often limits the accuracy of information given in this section.

Broad Population Ranges are intended to give readers a quick reference to a species' primary distribution pattern. These ranges are quite arbitrary and do not always follow standard geographic nomenclature. A given fish population might extend completely across a given range or only encompass a significant portion of the area. Some countries within the range are listed, but not all.

Indo-Pacific — A population extending from any point in the western or central Indian Ocean east to the Pacific islands of Hawaii or French Polynesia, and occasionally on to scattered islands east, or to the shore of the Western Hemisphere.

East Indo — The western boundary of this book's range: the Andaman Sea off western Thailand, the Christmas and Coco-Keeling Islands in the eastern Indian Ocean, the shores of western Indonesia and northwestern Australia.

Asian Pacific — A critical region, home to the most bio-diverse population of reef fishes on Earth. In this text, a population bordered in the west by the western shores of Indonesia, then sweeping east well beyond the edge of what is normally considered West Pacific – including the Great Barrier Reef, the Coral Sea, the Solomon Islands, New Caledonia and Vanuatu – and on north to Papua New Guinea, Indonesia, the Philippines, Central and eastern Micronesia, turning northwest to the tropical

Anatomy

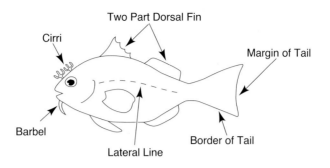

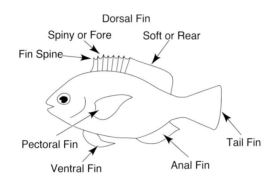

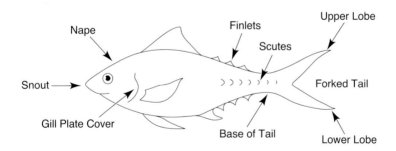

boundaries of Okinawa and the Ryukyu Islands southwest of Japan, and finally completing a crude circle by heading back west across the South China Sea to the Gulf of Thailand.

West Pacific — A population that may include all or part of the Asian Pacific, but also extends eastward into the island nations of Fiji, Tonga and Samoa, or north to the Gilbert and Phoenix Islands of the Central Pacific.

Pacific — A population generally extending from Asian and West Pacific, including Fiji, Tonga and Samoa, east to Hawaii, French Polynesia and occasionally beyond to the Pitcairn Island Group and infrequently on to the far eastern Pacific shores of the Western Hemisphere.

Localized — A limited population center that may include a single island or reef system, but generally representing a somewhat wider area.

Circumtropical — A population extending around the circumference of the world's tropical seas.

Circumglobal — A population extending around the circumference of the world's tropical seas, but also spilling north or south into temperate waters.

IDENTIFICATION GROUP 1

Disk-shaped/Colorful
Butterflyfishes – Angelfishes – Spadefishes

This ID Group consists of thin-bodied fishes with basically round shapes. Generally most are quite colorful.

FAMILY: Butterflyfishes – Chaetodontidae
7 Genera – 73 Species Included

Typical Shape

Typical Shape

Genus *Forcipiger*

Reminiscent of their butterfly namesake, the majority of these small, colorful fishes spend daylight hours flitting about the reef in search of food within rather confined home ranges. Members of the family are easily identified. Only the closely related angelfishes, once classified in the same family, have a similar shape and appearance. However, angelfishes can be easily distinguished by having more robust bodies and by the presence of a sharp spine on the lower edge of their gill covers.

Butterflyfishes typically travel alone, or in pairs, using keen eyesight to spot tiny worms, exposed polyps and other marine invertebrates. Although most species inhabit coral-rich reefs, a few butterflyfishes associate with silty coastal areas, while others gather in huge shoals high above the reef to feed on drifting plankton.

FAMILY: Angelfishes – Pomacanthidae
6 Genera – 51 Species Included

Genus *Pomacanthus*

Genus *Centropyge*

Genus *Genicanthus*

Large, colorful and graceful angelfishes from genus *Pomacanthus* epitomize the classic reef fish for many underwater naturalists. However, in the tropical Indo-Pacific the family is dominated in species numbers by the small elusive members of genus *Centropyge*.

Angelfishes are greatly dependent on the shelter of boulders, caves and coral crevices and so traditionally inhabit areas of heavy coral growth or high profile rock spills. The food of *Centropyge* consists primarily of algae, while *Pomacanthus* consume sponge, algae and benthic invertebrates; some species of *Genicanthus* gather in openwater shoals where they feed on zooplankton, primarily pelagic tunicates.

FAMILY: Spadefishes (Batfishes) – Ephippidae
2 Genera – 6 Species Included

Typical Shape

Although spadefishes, also commonly known as batfishes, are not closely related to butterflyfishes and angelfishes, this small family is placed in ID Group 1 because of the adults' disk-shaped bodies.

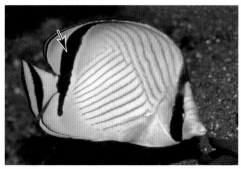

VAGABOND BUTTERFLYFISH *Chaetodon vagabundus*
SIZE: to 23 cm (9 in.)   Butterflyfishes – Chaetodontidae
ID: Yellowish white with chevron pattern of narrow lines; **narrow black band across rear body does not cover entire rear dorsal fin.** Usually solitary or form groups; home ranging. Coastal reefs and outer slopes to 30 m.
Indo-Pacific: E. Africa to Indonesia, Philippines, Micronesia, Hawaii and French Polynesia. – S.W. Japan to E. Australia.

Vagabond Butterflyfish – Juvenile
SIZE: 3-5 cm (1¼ - 2 in.)
ID: Similar to adult but chevron markings faint or absent; black spot on rear dorsal fin, tail clear. Adults and juveniles tolerate wide range of ecological conditions including turbid water and influxes of fresh water near river mouths.

INDIAN VAGABOND BUTTERFLYFISH *Chaetodon decussatus*
SIZE: to 20 cm (8 in.)   Butterflyfishes – Chaetodontidae
ID: Yellowish white with chevron pattern of narrow lines; **wide black band across rear body covers rear dorsal fin.** Usually alone or in pairs; home ranging. Rubble or coral-rich habits to 30 m, often in turbid conditions.
East Indo-Asian Pacific: Maldives and Andaman Sea to Sumatra and E. Timor in S. Indonesia.

THREADFIN BUTTERFLYFISH *Chaetodon auriga*
SIZE: to 23 cm (9 in.)   Butterflyfishes – Chaetodontidae
ID: **Pacific Variation –** White with chevron pattern; rear body and **tail yellow;** spot on rear dorsal fin (except Red Sea), dorsal fin usually trails a thread-like filament. Solitary, in pairs or small groups. Coastal and outer reefs to 40 m.
Indo-Pacific: Red Sea and E. Africa to Indonesia, Philippines, Micronesia, Solomon Is., Hawaii. – S.W. Japan to E. Australia.

CHEVRONED BUTTERFLYFISH *Chaetodon trifascialis*
SIZE: to 18 cm (7 in.)   Butterflyfishes – Chaetodontidae
ID: White with numerous black chevron markings; **black tail with yellow margin.** Territorial; defend coral patches against other butterflyfishes. Coral-rich areas to 12 m.
Indo-Pacific: Red Sea and E. Africa to Indonesia, Philippines, Micronesia, Papua New Guinea, Solomon Is., Hawaii and French Polynesia. – S.W. Japan to N.W. & E. Australia.

CROSSHATCH BUTTERFLYFISH *Chaetodon xanthurus*
SIZE: to 14 cm (5½ in.)   Butterflyfishes – Chaetodontidae
ID: White with **net pattern,** outer tail yellow to orange; broad yellow to orange bar across rear body. Solitary or in pairs. Outer reef slopes and dropoffs, usually among rocks or coral in 12-50 m.
Asian Pacific: Indonesia, Philippines, Micronesia and S.W. Japan.

YELLOWRIMMED BUTTERFLYFISH *Chaetodon guentheri*
SIZE: to 14 cm (5 ½ in.) Butterflyfishes – Chaetodontidae
ID: Silvery white with numerous dark spots; upper and rear body rimmed in yellow. Rocky reefs and coral-rich outer slopes in 5-40 m. Often near cool upwellings.
Asian Pacific: S.W. Japan, N. Sulawesi, Bali and Molucca Is. in Indonesia to S. Papua New Guinea.

ATOLL BUTTERFLYFISH *Chaetodon mertensii*
SIZE: to 13 cm (5 in.) Butterflyfishes – Chaetodontidae
ID: White with chevron markings, outer tail yellow to orange; broad yellow to orange bar across rear body, black eye bar and **dark smudge on nape.** Solitary or in pairs. Lagoons and outer slopes in 10-120 m.
West Pacific: Micronesia, Papua New Guinea, Solomon Is. to French Polynesia. – E. Australia, New Caledonia, Vanuatu to Fiji.

YELLOW-DOTTED BUTTERFLYFISH *Chaetodon selene*
SIZE: to 16 cm (6 ¼ in.) Butterflyfishes – Chaetodontidae
ID: White with numerous faint yellow spots arranged in diagonal rows, fins yellow (except pectorals); blackish rim on rear body. Solitary or in pairs. Rubble or sand bottoms of coastal reefs in 8-50 m.
Asian Pacific: Indonesia, Philippines and N. Papua New Guinea to S.W. Japan.

PANDA BUTTERFLYFISH *Chaetodon adiergastos*
SIZE: to 20 cm (8 in.) Butterflyfishes – Chaetodontidae
ID: Whitish with darker gray bands, fins yellow-orange (except pectoral); oval-shaped black eye bar, and small black spot on nape. Usually in pairs or small groups. Silty inshore reefs and clear outer reefs in 1-30 m.
Asian Pacific: Sabah in Malaysia, Indonesia, Philippines and Micronesia. – S.W. Japan to N.W. Australia.

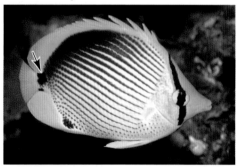

BLACK-BACKED BUTTERFLYFISH *Chaetodon melannotus*
SIZE: to 15 cm (6 in.) Butterflyfishes – Chaetodontidae
ID: White body with many diagonal black lines, black on upper back; **black saddle on base of tail.** Solitary or in pairs; feed on live soft and hard corals. Coral-rich areas of lagoons, reef flats and seaward reefs in 2-20 m.
Indo-West Pacific: Red Sea and E. Africa to Indonesia, Philippines, Micronesia and Papua New Guinea. – S.W. Japan to Australia and Fiji.

SPOT-TAIL BUTTERFLYFISH *Chaetodon ocellicaudus*
SIZE: to 14 cm (5 ½ in.) Butterflyfishes – Chaetodontidae
ID: White body with diagonal black lines; **black spot on base of tail.** Solitary or in pairs. Coral-rich areas of lagoons, reef flats and seaward reefs in 3-50 m. Most common on outer slopes and in reef crest channels.
Asia Pacific: Indonesia, Philippines, Micronesia, Papua New Guinea and Solomon Is. to Great Barrier Reef.

BLACK BUTTERFLYFISH   *Chaetodon flavirostris*
SIZE: to 20 cm (8 in.)   Butterflyfishes – Chaetodontidae
ID: Dark with yellow rim that is most prominent around rear edge of body and tail; dark blotch on nape, large adults develop hump on forehead. Usually in pairs. Coral and rocky reefs in 2-20 m.
Southern Pacific: E. Australia to Rapa and Pitcairn Is. southeast of French Polynesia.

RETICULATED BUTTERFLYFISH   *Chaetodon reticulatus*
SIZE: to 16 cm (6¼ in.)   Butterflyfishes – Chaetodontidae
ID: Black with pale gray scale centers becoming whitish on back and forebody; yellow-edged black eye bar, margin of tail yellow with black-edges. Solitary, in pairs or aggregations. Exposed outer reefs to 30 m.
Pacific: Indonesia, Philippines, Papua New Guinea, Solomon Is. to Hawaii and French Polynesia. – S. W. Japan to Great Barrier Reef.

EASTERN TRIANGULAR BUTTERFLYFISH   *Chaetodon baronessa*
SIZE: to 15 cm (6 in.)   Butterflyfishes – Chaetodontidae
ID: Body roughly triangular; gray with many chevron markings, **yellowish gray tail.** Similar Triangular Butterflyfish [next] has dark triangle marking on tail. In pairs. Near *Acropora* plate coral to 10 m.
East Indo-West Pacific: Cocos-Keeling Is. to Indonesia, Philippines, Micronesia. – S.W. Japan to Great Barrier Reef and Fiji.

TRIANGULAR BUTTERFLYFISH   *Chaetodon triangulum*
SIZE: to 15 cm (6 in.)   Butterflyfishes – Chaetodontidae
ID: Body roughly triangular; gray with many pale yellow chevron markings, **dark triangle on tail.** Similar Eastern Triangular Butterflyfish [previous] lacks this marking. In pairs. Near *Acropora* plate coral to 10 m.
Indian Ocean: E. Africa to Andaman Sea and Java and Bali in Indonesia.

 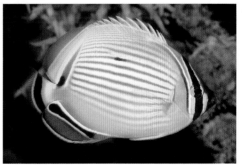

INDIAN REDFIN BUTTERFLYFISH   *Chaetodon trifasciatus*
SIZE: to 15 cm (6 in.)   Butterflyfishes – Chaetodontidae
ID: Pale, becoming bluish toward rear with oblique purplish stripes, anal fin and **base of tail orange;** yellow-edged black band across base of anal fin. Usually in pairs. Coral-rich areas to 20 m.
Indian Ocean: E. Africa to Andaman Sea, Java and Bali in Indonesia.

REDFIN BUTTERFLYFISH   *Chaetodon lunulatus*
SIZE: to 15 cm (6 in.)   Butterflyfishes – Chaetodontidae
ID: Pale with oblique purplish stripes, anal fin red, **base of tail pale;** yellow-edged black band across base of anal fin. Usually in pairs. Coral-rich areas to 20 m.
Pacific: Indonesia, Philippines, Micronesia, Papua New Guinea, Solomon Is., Hawaii to French Polynesia. – S.W. Japan to N.W. Australia and Great Barrier Reef.

Butterflyfishes

BLUE-SPOT BUTTERFLYFISH *Chaetodon plebeius*
SIZE: to 15 cm (6 in.)   Butterflyfishes – Chaetodontidae
ID: Bright yellow with numerous faint stripes; **blue patch on side**, ocellated spot on base of tail and blue-edged eye bar. Solitary or in pairs. Shallow coastal waters including lagoons and seaward reefs to 10 m.
West Pacific: Philippines, Micronesia, Papua New Guinea, Solomon Is. to New Caledonia, Vanuatu and Fiji.

ANDAMAN BUTTERFLYFISH *Chaetodon andamanensis*
SIZE: to 15 cm (6 in.)   Butterflyfishes – Chaetodontidae
ID: Bright yellow with numerous faint stripes; black spot on base of tail, blue-edged eye bar. Solitary or in pairs. Rock and coral reefs near shore and on outer slopes in 10-40 m.
East Indian Ocean: Maldives, Sri Lanka, Andaman Sea to W. Sumatra in Indonesia.

TEARDROP BUTTERFLYFISH *Chaetodon unimaculatus*
SIZE: to 20 cm (8 in.)   Butterflyfishes – Chaetodontidae
ID: White with bright yellow dorsal, anal and ventral fins; black tear-shaped spot on back, adults develop a large bulbous snout. Solitary or form small groups. Lagoons and seaward reefs in 10-60 m.
Pacific: Indonesia, Philippines, Micronesia to Papua New Guinea, Hawaii and French Polynesia. – S.W. Japan to N.W. & E. Australia.

INDIAN TEARDROP BUTTERFLYFISH *Chaetodon interruptus*
SIZE: to 20 cm (8 in.)   Butterflyfishes – Chaetodontidae
ID: Bright yellow; black tear-shaped spot on back, thin black bar from rear dorsal to rear anal fin, adults develop a large bulbous snout. Solitary or form small groups. Lagoon and seaward reefs in areas of hard and soft corals in 10-40 m.
Indian Ocean: E. Africa to Andaman Sea and W. Sumatra in Indonesia.

ECLIPSE BUTTERFLYFISH *Chaetodon bennetti*
SIZE: to 18 cm (7 in.)   Butterflyfishes – Chaetodontidae
ID: Bright yellow; large blue-edged black spot on middle of back, **pair of diagonal blue lines on lower body.** Solitary or in pairs. Coral-rich areas of lagoons and on outer reefs in 5-30 m.
Indo-Pacific: E. Africa to Indonesia, Philippines, Micronesia, New Guinea, Solomon Is. and French Polynesia. – S.W. Japan to Great Barrier Reef.

OVAL-SPOT BUTTERFLYFISH *Chaetodon speculum*
SIZE: to 18 cm (7 in.)   Butterflyfishes – Chaetodontidae
ID: Bright yellow; large black oval-shaped spot on middle of back. Usually solitary or in pairs; shy. Coral-rich reefs in lagoons and on outer reefs in 8-30 m.
West Pacific: Indonesia, Philippines, Micronesia, Papua New Guinea and Solomon Is. – S.W. Japan to N.W. & E. Australia and Fiji.

BLACKCAP BUTTERFLYFISH *Chaetodon wiebeli*
SIZE: to 18 cm (7 in.)   Butterflyfishes – Chaetodontidae
ID: Golden with thin reddish brown bands; black eye bar with prominent white bar behind, **black saddle across upper nape,** tail margin black. Solitary or in pairs. Mixed rock and coral reefs in 4-25 m.
Localized: W. Pacific Rim from Gulf of Thailand and Java Sea in Indonesia to Vietnam, Taiwan and S. W. Japan.

RACCOON BUTTERFLYFISH *Chaetodon lunula*
SIZE: to 21 cm (8¼ in.)   Butterflyfishes – Chaetodontidae
ID: Yellow-orange with dusky back and thin dark diagonal bands; black eye band with white patch behind, connected to **wide diagonal black band running to dorsal fin.** Solitary, in pairs or aggregations. Lagoons and outer reefs to 30 m.
Indo-Pacific: E. Africa to Indonesia, Micronesia, Papua New Guinea, Hawaii and French Polynesia. – S.W. Japan to N.W. & E. Australia.

DOTTED BUTTERFLYFISH *Chaetodon semeion*
SIZE: to 24 cm (9½ in.)   Butterflyfishes – Chaetodontidae
ID: Yellow with horizontal rows of small black spots; dark to pale blue marking on rear snout and nape, dorsal fin trails a thread-like filament. Usually in pairs. Coral-rich areas of lagoons and outer slopes in 2-50 m.
East Indo-Pacific: Sri Lanka to Indonesia, Micronesia, Papua New Guinea and French Polynesia. – S.W. Japan to Great Barrier Reef.

ORIENTAL BUTTERFLYFISH *Chaetodon auripes*
SIZE: to 20 cm (8 in.)   Butterflyfishes – Chaetodontidae
ID: Brown to golden brown with numerous thin dark stripes; prominent black eye bar with white bar behind. Solitary or form aggregations. Rocky reefs with some coral and algal growth to 30 m. Juveniles in tide pools.
Localized: W. Pacific Rim from S. China Sea to S. W. Japan.

BLACKLIP BUTTERFLYFISH *Chaetodon kleinii*
SIZE: to 14 cm (5½ in.)   Butterflyfishes – Chaetodontidae
ID: Light brown with "dirty" white head and diffuse central bar; black lip and ventral fins. Solitary to large aggregations. Rocky reefs and coral-rich areas of lagoons, channels, outer reef slopes in 2-61 m, usually below 10 m.
Indo-West Pacific: E. Africa to Indonesia, Philippines, Micronesia, Solomon Is. and Hawaii. – S.W. Japan to N.W. & E. Australia and Fiji.

LATTICED BUTTERFLYFISH *Chaetodon rafflesii*
SIZE: to 15 cm (6 in.)   Butterflyfishes – Chaetodontidae
ID: Yellow with network of gray lines; broad dark submarginal band on soft dorsal fin and tail. Solitary or in pairs. Primarily coral-rich areas of sheltered coastal reefs, lagoons and outer slopes to 15 m.
East Indo-Pacific: Sri Lanka to Indonesia, Micronesia, Solomon Is., and French Polynesia. – S.W. Japan to Great Barrier Reef and Fiji.

ASIAN BUTTERFLYFISH *Chaetodon argentatus*
SIZE: to 20 cm (8 in.)   Butterflyfishes – Chaetodontidae
ID: White with net pattern; broad black bars or saddle markings on rear head and midbody, black bar across rear body. In pairs or small aggregations. Rock or coral reefs in 5-20 m.
North Asian Pacific: W. Pacific Rim from S. Japan including Ryukyu and Izu Is., Philippines, S. China and Taiwan.

BLACK & WHITE BUTTERFLYFISH *Chaetodon burgessi*
SIZE: to 14 cm (5½ in.)   Butterflyfishes – Chaetodontidae
ID: White; black band from nape to pectoral region, broad black diagonal area across rear of body. Solitary or in pairs. Vertical or undercut dropoffs on outer reefs in 20-80 m, usually below 40 m.
West Pacific: Sabah in Malaysia, E. Indonesia, Philippines, N. New Guinea to Micronesia, Solomon Is. and Fiji.

TINKER'S BUTTERFLYFISH *Chaetodon tinkeri*
SIZE: to 15 cm (6 in.)   Butterflyfishes – Chaetodontidae
ID: White with numerous dark dots, yellow tail; broad black diagonal area across rear of body. Solitary, in pairs or small aggregations. Usually shelter in black corals and sea fans on steep slopes in 27-160 m.
Localized: Micronesia to Hawaii.

YELLOW-CROWNED BUTTERFLYFISH *Chaetodon flavocoronatus*
SIZE: to 12 cm (4¾ in.)   Butterflyfishes – Chaetodontidae
ID: White with numerous dark dots, yellow tail; broad black diagonal area across rear of body, **yellow to orange band on nape.** Solitary or in pairs. Shelter in black corals and sea fans on steep slopes in 35-75 m.
Localized: Known only from Guam (Orote Peninsula) in the Mariana Is., Micronesia.

MARQUESAN BUTTERFLYFISH *Chaetodon declivis*
SIZE: to 15 cm (6 in.)   Butterflyfishes – Chaetodontidae
ID: White with dark spots, fins yellowish; broad diagonal area across rear body gradating orange to black, orange eye bar and snout tip. Rocky reef slopes and steep walls adjacent to sand bottoms below 20 m.
Localized: Line Is. in eastern Central Pacific and Marquesas Is. in French Polynesia.

BLACK & YELLOW BUTTERFLYFISH *Chaetodon smithi*
SIZE: to 17 cm (6¾ in.)   Butterflyfishes – Chaetodontidae
ID: Dark brown head and forebody, yellow rear body and tail. Commonly in large openwater feeding aggregations. Rocky, algal-covered reefs with scattered coral in 10-30 m.
Localized: Rapa I., Ilots de Bass I. (Marotiri) and Pitcairn Is. in S. and S.E. French Polynesia.

TAHITI BUTTERFLYFISH *Chaetodon trichrous*
SIZE: to 12 cm (4 3/4 in.)   Butterflyfishes – Chaetodontidae
ID: Head and forebody white gradating to dark brown with pale scale centers forming grid pattern, yellow tail; black eye bar. Solitary, in pairs or small groups. Primarily in sheltered lagoons in 3-25 m.
Localized: Society, Marquesas and Tuamotu Is. in French Polynesia.

JAPANESE BUTTERFLYFISH *Chaetodon nippon*
SIZE: to 15 cm (6 in.)   Butterflyfishes – Chaetodontidae
ID: Yellowish or brassy with dark gray to brownish head and rear body; black spot on rear dorsal fin. Solitary, in pairs or groups. Rocky coastal reefs in 5-30 m.
North Asian Pacific: W. Pacific Rim from S. Korea and S. Japan to N. Philippines.

GOLDBARRED BUTTERFLYFISH *Chaetodon rainfordi*
SIZE: to 15 cm (6 in.)   Butterflyfishes – Chaetodontidae
ID: Yellow; pair of broad bluish gray bars with orange margins and orange eye bar. Solitary or in pairs; easy to approach. Coastal and offshore reefs areas of sparse coral growth to 15 m.
Localized: S. Papua New Guinea to Great Barrier Reef.

GOLDEN-STRIPED BUTTERFLYFISH *Chaetodon aureofasciatus*
SIZE: to 13 cm (5 in.)   Butterflyfishes – Chaetodontidae
ID: Grayish body rimmed with bright yellow; orange band through eye. Solitary or in pairs; hybridizes with Goldbarred Butterflyfish [previous]. Most common on silty coastal reefs, often near river mouths in 5-15 m.
Localized: N. Australia and S. Papua New Guinea.

SPECKLED BUTTERFLYFISH *Chaetodon citrinellus*
SIZE: to 13 cm (5 in.)   Butterflyfishes – Chaetodontidae
ID: Yellow to whitish with many rows of faint bluish spots; black edge on anal fin. Solitary, in pairs or small groups. Reef flats with some surge and seaward reefs usually in 1-3 m; rarely to 30 m.
Indo-Pacific: E. Africa to Indonesia, Philippines, Micronesia, Hawaii and French Polynesia. – S.W. Japan to N. W. & E. Australia.

SPOTTED BUTTERFLYFISH *Chaetodon guttatissimus*
SIZE: to 12 cm (4 3/4 in.)   Butterflyfishes – Chaetodontidae
ID: Beige with distinctive brown to purplish speckling; broad yellow margin on dorsal fin extends across tail base. Solitary or in pairs. Coral reefs, lagoons and seaward slopes in 5-30 m.
Indian Ocean: E. Africa to Maldives, Andaman Sea, Christmas I. and Bali in Indonesia.

23

MEYER'S BUTTERFLYFISH *Chaetodon meyeri*
SIZE: to 18 cm (7 in.)   Butterflyfishes – Chaetodontidae
ID: White to bluish white; **curving black bands** converging near pectoral fin, yellowish rim encircles body. Solitary or in pairs. Coral-rich areas of clear lagoons and seaward reefs in 2-25 m.
Indo-Pacific: E. Africa, Maldives and Bay of Bengal to Indonesia, Philippines, Micronesia, Papua New Guinea and Line Is. – S. W. Japan to Great Barrier Reef.

ORNATE BUTTERFLYFISH *Chaetodon ornatissimus*
SIZE: to 18 cm (7 in.)   Butterflyfishes – Chaetodontidae
ID: Bluish white with **orange bands**; narrow yellow margin on dorsal and anal fins, yellow-edged black eye bar. Usually in pairs. Coral-rich areas in clear water lagoons and seaward reefs to 36 m.
Indo-Pacific: Maldives to Indonesia, Philippines, Micronesia, Solomon Is. and French Polynesia. – S.W. Japan to Australia.

SADDLED BUTTERFLYFISH *Chaetodon ephippium*
SIZE: to 23 cm (9 in.)   Butterflyfishes – Chaetodontidae
ID: Blue-gray with blue lines on lower body; large white-bordered black patch upper rear body, orange area from snout to ventral fins. Solitary or in pairs. Coral-rich areas in lagoons and seaward reefs to 30 m.
Indo-Pacific: Sri Lanka to Indonesia, Philippines, Micronesia, Papua New Guinea and French Polynesia. – S.W. Japan to Australia.

Saddled Butterflyfish – Juvenile
SIZE: 3-5 cm (1 ¼ - 2 in.)
ID: White, rimmed with yellow; large black teardrop-shaped patch on rear body, dark eyebar and black spot on base of tail. Feed on coral polyps, algae, sponges, fish eggs and assorted benthic invertebrates.

SPOT-BANDED BUTTERFLYFISH *Chaetodon punctatofasciatus*
SIZE: to 12 cm (4 ¾ in.)   Butterflyfishes – Chaetodontidae
ID: Yellowish tan to yellow; **7 gray bars** on upper body and rows of dark spots below, orange eye bar and black spot on nape. Usually in pairs, sometimes with Dot & Dash Butterflyfish [next]. Lagoons and outer reefs to 45 m.
Pacific: Indonesia, Philippines, Micronesia, Papua New Guinea, Solomon Is. to Line Is. – S.W. Japan.to Great Barrier Reef.

DOT & DASH BUTTERFLYFISH *Chaetodon pelewensis*
SIZE: to 13 cm (5 in.)   Butterflyfishes – Chaetodontidae
ID: Yellowish tan with **diagonal rows of dark spots becoming solid bands on upper body**; orange eye bar, black spot on nape. Usually in pairs; sometimes with Spot-banded Butterflyfish (occasionally hybridize). Outer reefs to 30 m.
Pacific: West Papua in Indonesia and Papua New Guinea, Great Barrier Reef to French Polynesia.

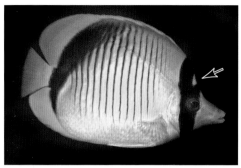

LINED BUTTERFLYFISH *Chaetodon lineolatus*
SIZE: to 30 cm (12 in.) Butterflyfishes – Chaetodontidae
ID: White with vertical black lines; broad black band rear body, **wide black eye bar encloses a white spot on nape.** The largest butterflyfish. Solitary or in pairs. Lagoons and seaward reefs from shallows to 171 m.
Indo-Pacific: E. Africa to Indonesia, Philippines, Micronesia, Hawaii and Line Is. – S.W. Japan to Great Barrier Reef and Fiji.

Lined Butterflyfish – Juvenile
SIZE: 3-5 cm (1¼-2 in.)
ID: Similar to adults, but vertical dark lines on side much fainter; black strip on rear part of body poorly developed, and has large black spot on tail base. Feeds primarily on coral polyps and anemones.

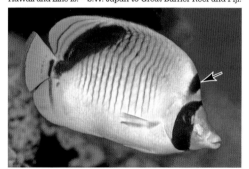

SPOT-NAPE BUTTERFLYFISH *Chaetodon oxycephalus*
SIZE: to 25 cm (10 in.) Butterflyfishes – Chaetodontidae
ID: White with vertical black lines; broad black band rear body, **black eye bar with black patch above on nape.** Usually solitary or in pairs. Coral-rich areas, including both coastal and seaward reefs in 10-40 m.
East Indo-Pacific: Maldives to Indonesia, Philippines, Micronesia, Solomon Is. – S. Japan to Great Barrier Reef and Fiji.

PACIFIC DOUBLE-SADDLE BUTTERFLYFISH *Chaetodon ulietensis*
SIZE: to 15 cm (6 in.) Butterflyfishes – Chaetodontidae
ID: White fore and midbody, bright yellow rear; pair of diffuse dark saddles across back, **black spot on tail base.** Solitary, in pairs or groups. Coral-rich lagoons and seaward reefs to 30 m.
Pacific: E. Australia, Indonesia, Philippines, Micronesia, Papua New Guinea, Solomon Is. to Line Is. – S.W. Japan to Great Barrier Reef.

SADDLEBACK BUTTERFLYFISH *Chaetodon falcula*
SIZE: to 20 cm (8 in.) Butterflyfishes – Chaetodontidae
ID: White becoming **bright yellow on back with 2 black saddles** and rear body; black band on tail base. Solitary, in pairs or groups. Coral-rich areas of lagoons and seaward reefs to 15 m.
Indian Ocean: E. Africa to Andaman Is., Sumatra and W. Java in Indonesia.

FOURSPOT BUTTERFLYFISH *Chaetodon quadrimaculatus*
SIZE: to 16 cm (6¼ in.) Butterflyfishes – Chaetodontidae
ID: Yellow-orange with brown upper body; pair of white spots on back. Solitary or in pairs; feed almost exclusively on *Pocillopora* corals. Exposed rocky reefs with scattered coral growth in 2-15 m.
Northern Pacific: Taiwan and S.W. Japan to Hawaii and French Polynesia.

EIGHT-BANDED BUTTERFLYFISH   *Chaetodon octofasciatus*
SIZE: to 12 cm (4 ³/₄ in.)   Butterflyfishes – Chaetodontidae
ID: Yellow Variation – Yellow to orange with 8 narrow black bars.
Shelters among branching corals; feed exclusively on coral
polyps. Shallow protected lagoons (often turbid/silty) and
inner reefs with good coral cover in 3-20 m.
East Indo-Asian Pacific: Sri Lanka to Indonesia, Philippines,
Micronesia and Solomon Is. – S.W. Japan to Australia.

Eight-banded Butterflyfish – White Variation
ID: White with 8 narrow black bars, and dark spot at base
of tail (occasionally faint).
Asian Pacific: Primarily Micronesia and N. New Guinea.

WHITE COLLAR BUTTERFLYFISH   *Chaetodon collare*
SIZE: to 16 cm (6 ¹/₄ in.)   Butterflyfishes – Chaetodontidae
ID: Dark gray overall with pale scale centers and red tail; white
"collar" marking behind eye. Often in pairs, but may form large
aggregations. Rocky shores and coral-rich outer reefs to 20 m.
East Indo-Asian Pacific: Sri Lanka, Maldives, Andaman Sea
to Malaysia, Brunei and Philippines.

GRAY BUTTERFLYFISH   *Hemitaurichthys thompsoni*
SIZE: to 18 cm (7 in.)   Butterflyfishes – Chaetodontidae
ID: Uniform dark gray; lacks distinguishing marks. Solitary,
in pairs or groups; feed above bottom on plankton. Deep
outer reefs in 10-300 m, occasionally shallow coastal reefs
and lagoons near deep water.
Central Pacific: Hawaii, Phoenix Is. and Johnston Atoll to
Fiji and French Polynesia.

PYRAMID BUTTERFLYFISH   *Hemitaurichthys polylepis*
SIZE: to 18 cm (7 in.)   Butterflyfishes – Chaetodontidae
ID: White pyramid shape formed by yellow triangular patch
on upper forebody and rear dorsal fin; head brown, anal fin
yellow. Form large aggregations; feed on plankton high in
water column. Outer slopes in 3-60 m.
Pacific: Indonesia, Philippines, Micronesia, Solomon Is., Hawaii
and Pitcairn Is. – S. W. Japan to Great Barrier Reef.

BLACK PYRAMID BUTTERFLYFISH   *Hemitaurichthys zoster*
SIZE: to 16 cm (6 ¹/₄ in.)   Butterflyfishes – Chaetodontidae
ID: White pyramid shape formed by dark brown to blackish
head and rear body. Form large aggregations; feed on plankton
high in water column. Outer slopes to 40 m.
Indian Ocean: E. Africa, Maldives and Bay of Bengal to Andaman
Sea and W. Sumatra in Indonesia.

TWO-EYED CORALFISH *Coradion melanopus*
SIZE: to 15 cm (6 in.)   Butterflyfishes – Chaetodontidae
ID: White; **ocellated spot on anal** and rear dorsal fins, pair of closely spaced brown bars behind head, and orange-edged gray bar rear body. In pairs; often near barrel sponges. Coastal and outer reefs in 10-30 m.
Asian Pacific: Bali and Sulawesi in Indonesia to N. & S. New Guinea, Solomon Is., north to the Philippines and S.W. Japan.

ORANGE-BANDED CORALFISH *Coradion chrysozonus*
SIZE: to 15 cm (6 in.)   Butterflyfishes – Chaetodontidae
ID: White with orange bar across rear body; pair of closely spaced brown bars behind head, **ocellated spot on soft dorsal fin and on tail base extending onto tail.** The most common *Coradion* in most areas. Coastal reefs in 3-60 m.
Asian Pacific: Andaman Sea to Indonesia, Philippines, Micronesia, New Guinea and Solomon Is. – S.W. Japan to Australia.

HIGHFIN CORALFISH *Coradion altivelis*
SIZE: to 20 cm (8 in.)   Butterflyfishes – Chaetodontidae
ID: White with wide orange bar across rear body including dorsal and anal fins; pair of closely spaced brown bars behind head. Resembles Orange-banded Coralfish [previous], but adults **lack spot on soft dorsal fin.** Uncommon. Inshore reefs in 3-15 m.
Asian Pacific: Andaman Sea to Indonesia, Philippines, Papua New Guinea and Solomon Is. – S.W. Japan to E. Australia and Vanuatu.

MARGINED CORALFISH *Chelmon marginalis*
SIZE: to 18 cm (7 in.)   Butterflyfishes – Chaetodontidae
ID: Silvery white; black-edged orange forebody bar and eye bar, wide orange bar across rear body including dorsal and anal fins, long snout. Solitary or in pairs. Mainly coastal reefs and near shore islands to 30 m.
Localized: Tropical N.W. Australia to Cape York and N. Great Barrier Reef.

LONG-BEAKED CORALFISH *Chelmon rostratus*
SIZE: to 20 cm (8 in.)   Butterflyfishes – Chaetodontidae
ID: Silvery white; black-edged orange eye bar and two black-edged orange body bars, ocellated spot on rear dorsal fin, long beak-like snout. Solitary or in pairs. Coastal, inner reefs and estuaries, often in turbid water to 25 m.
Indo-Asian Pacific: Andaman Sea to Indonesia, Philippines, Micronesia and Solomon Is. – S.W. Japan to Great Barrier Reef.

BEAKED CORALFISH *Chelmon muelleri*
SIZE: to 18 cm (7 in.)   Butterflyfishes – Chaetodontidae
ID: Silvery white; three orange-brown body bars and narrow orange eye bar, large ocellated spot on rear dorsal fin, **short beak-like snout.** Solitary or in pairs. Shallow coastal reefs with sandy silt or mud bottoms in 2-10 m.
Localized: Tropical N. Australia coast to Great Barrier Reef.

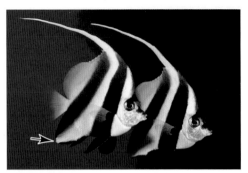

SCHOOLING BANNERFISH *Heniochus diphreutes*
SIZE: to 21 cm (8 ¼ in.)   Butterflyfishes – Chaetodontidae
ID: White with pair of black bands, the **second band ending at tip of anal fin;** elongate dorsal fin filament. **Form large aggregations;** solitary or groups. Outer reef slopes in 5-21 m.
Indo-Southwest Pacific: Red Sea and E. Africa, Andaman Sea, Indonesia. – N.W. & E. Australia to Vanuatu, Kermadec Is. and Fiji. Also 2nd narrow range from Japan to Hawaii.

LONGFIN BANNERFISH *Heniochus acuminatus*
SIZE: to 25 cm (10 in.)   Butterflyfishes – Chaetodontidae
ID: White with pair of black bands, the **second band ending above the tip of the anal fin;** dark rectangle between eyes, long snout compared to similar Schooling Bannerfish [previous]. Lagoons and outer reef slopes in 2-75 m.
Indo-Pacific: E. Africa to Indonesia, Philippines, Micronesia, Solomon Is., French Polynesia. – S. W. Japan to Great Barrier Reef.

MASKED BANNERFISH *Heniochus monoceros*
SIZE: to 23 cm (9 in.)   Butterflyfishes – Chaetodontidae
ID: Rear body and fins yellow; white band runs from dorsal fin filament to lower head with dark bar behind, bump on nape. Solitary, in pairs or small groups. Lagoons and outer reefs with rich coral growth in 2-25 m.
Indo-Pacific: E. Africa to Indonesia, Philippines, Micronesia, New Guinea and French Polynesia. – S.W. Japan to Australia.

PENNANT BANNERFISH *Heniochus chrysostomus*
SIZE: to 18 cm (7 in.)   Butterflyfishes – Chaetodontidae
ID: White with brown to black band on head, midbody and upper rear body, yellow upper snout; tallest dorsal spine trails pennant. Solitary or in pairs. Inshore and outer reefs in 3-45 m.
Pacific: Indonesia, Philippines, Micronesia, Papua New Guinea, Solomon Is. to French Polynesia. – S.W. Japan to Australia and Fiji.

PHANTOM BANNERFISH *Heniochus pleurotaenia*
SIZE: to 17 cm (6 ¾ in.)   Butterflyfishes – Chaetodontidae
ID: White central bar bordered by pair of wide blackish bands that become brown and converge on back; pair of horns just above eyes, bump on nape. Solitary, in pairs or groups. Coral-rich areas to 25 m.
Indian Ocean: Maldives, Sri Lanka, Bay of Bengal and Andaman Sea to W. Java in Indonesia.

HUMPHEAD BANNERFISH *Heniochus varius*
SIZE: to 19 cm (7 ½ in.)   Butterflyfishes – Chaetodontidae
ID: Large brown to black triangular area on body; pair of horns just above eyes, bump on nape. Solitary, in pairs or form groups; often under ledges. Coral-rich areas of lagoons and seaward reef slopes in 2-30 m.
Pacific: Indonesia, Philippines, Solomon Is., French Polynesia. – S. W. Japan to Australia, New Caledonia and Fiji.

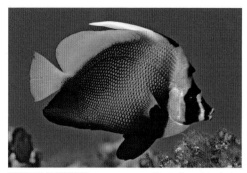

SINGULAR BANNERFISH *Heniochus singularius*
SIZE: to 23 cm (9 in.)   Butterflyfishes – Chaetodontidae
ID: Body and anal fin black, dorsal fin and tail yellow; long white dorsal fin filament, bump on nape, white bar behind eye, ring around snout. Solitary or in pairs. Lagoons and outer reefs, often around shipwrecks in 2-250 m.
East Indo-West Pacific: Maldives to Indonesia, Philippines, Micronesia and Solomon Is. – S.W. Japan to Australia and Fiji.

OCELLATED CORALFISH *Parachaetodon ocellatus*
SIZE: to 18 cm (7 in.)   Butterflyfishes – Chaetodontidae
ID: Silvery white; 5 brown-orange bars across head and body, ocellated spot on tail base and black spot on elevated triangular dorsal fin. Solitary or in pairs. Coastal and inner reefs littered with sponges in 5-40 m.
West Pacific: Indonesia, Philippines, Papua New Guinea and Solomon Is. – S.W. Japan to Australia and Fiji.

BIG LONGNOSE BUTTERFLYFISH *Forcipiger longirostris*
SIZE: to 22 cm (8 ³/₄ in.)   Butterflyfishes – Chaetodontidae
ID: Yellow with black upper head and silvery white below; extremely elongate snout, **black spots on breast**, spot on anal fin below tail base. Solitary or in pairs. Outer reefs in 5-60 m.
Indo-Pacific: Red Sea and E. Africa to Indonesia, Philippines, Micronesia, Papua New Guinea, Solomon Is., Hawaii and French Polynesia. – S.W. Japan to Great Barrier Reef.

Big Longnose Butterflyfish – Dark Variation
ID: Dark brown to nearly black; intermediates are dusky yellowish or yellowish brown.

LONGNOSE BUTTERFLYFISH *Forcipiger flavissimus*
SIZE: to 22 cm (8 ³/₄ in.)   Butterflyfishes – Chaetodontidae
ID: Yellow with black upper head and silvery white below; elongate snout short compared to Big Longnose Butterflyfish [previous], spot on anal fin below tail base. Solitary or in pairs. Coastal and outer reefs in 2-114 m.
Indo-Pacific: Red Sea and E. Africa to Indonesia, Philippines, Micronesia, Solomon Is. and Hawaii. – S. W. Japan to Australia.

CENDERAWASIH LONGNOSE BUTTERFLYFISH *Forcipiger wanai*
SIZE: to 17 cm (6 ³/₄ in.)   Butterflyfishes – Chaetodontidae
ID: Yellowish brown, black upper head, silvery white below; **yellow bar behind head,** dorsal, ventral and anal fins yellow, elongate snout similar to Big Longnose [above], spot on anal fin below tail base. Solitary or in pairs. Coastal and outer reefs.
Localized: Banda Sea and Raja Ampat to Cenderawasih Bay in N.E. Indonesia.

THREE-SPOT ANGELFISH *Apolemichthys trimaculatus*
SIZE: to 25 cm (10 in.) Angelfishes – Pomacanthidae
ID: Bright yellow with blue lips; black spot on nape, gray spot just behind head, **broad black margin on anal fin.** Solitary or in pairs. Outer reefs, steep slopes or dropoffs in 15-60 m.
Indo-West Pacific: E. Africa to Indonesia, Philippines, Micronesia, New Guinea and Fiji. – S. W. Japan to N. W. Australia and Great Barrier Reef.

Three-spot Angelfish – Dark Variation
ID: Occasionally darkish green-yellow to brownish yellow with bright yellow spots on scales.

GOLDEN-SPOTTED ANGELFISH *Apolemichthys xanthopunctatus*
SIZE: to 25 cm (10 in.) Angelfishes – Pomacanthidae
ID: Brownish with golden scale centers and blue lips, **dorsal, anal and tail fins black;** black spot on nape. Solitary or form small groups. Lagoons, passages and outer slopes in 3-40 m.
North Central Pacific: Micronesia, Gilbert, Phoenix and Line Is. in eastern Central Pacific.

Golden-spotted Angelfish – Juvenile
SIZE: 4-7.5 cm (1 1/2 -3 in.)
ID: Brownish with wavy golden bands; black eyebar extends over nape, large pale-edged black patch on rear back and dorsal fin.

WHITE-BELTED ANGELFISH *Apolemichthys griffisi*
SIZE: to 25 cm (10 in.) Angelfishes – Pomacanthidae
ID: Gray with pale gold spots on sides; white belt broadly bordered in black along back to tail base, black eye bar, large spot on nape. Solitary or in pairs. Steep outer reef slopes in 10-60 m.
Pacific: N.E. Indonesia, E. Papua New Guinea to Solomon Is., Gilbert, Phoenix and Line Is. in eastern Central Pacific.

INDIAN YELLOWTAIL ANGELFISH *Apolemichthys xanthurus*
SIZE: to 15 cm (6 in.) Angelfishes – Pomacanthidae
ID: Shades of gray with dark scale markings; yellow "ear" spot, black rear dorsal and anal fins; bright yellow tail. Solitary or in pairs. Coral reefs and rocky shores in 5-25 m.
East Indian Ocean: Sri Lanka to Maldives and Andaman Sea and Thailand.

PURPLE-MASK ANGELFISH *Centropyge venusta*
SIZE: to 12 cm (4 3/4 in.)   Angelfishes – Pomacanthidae
ID: Blue upper head and body, yellow below with large triangular extension invading blue just behind head. Usually solitary; secretive, often upside down under ledges or in caves. Outer reef slopes in 15-35 m.
Localized: W. Pacific Rim from Japan's Ryukyus and Izu Is. to Taiwan, N. Philippines (N. Luzon) and Micronesia.

BLUE VELVET ANGELFISH *Centropyge deborae*
SIZE: to 9 cm (3 1/2 in.)   Angelfishes – Pomacanthidae
ID: Dark blue to nearly black with no markings. A secretive fish usually alone; hide in crevices and ledges. Coral-rich areas or rubble bottoms, most commonly sheltered in outer reefs in 10-70 m.
Localized: Known only from Fiji.

MIDNIGHT ANGELFISH *Centropyge nox*
SIZE: to 9 cm (3 1/2 in.)   Angelfishes – Pomacanthidae
ID: Entirely black except for a **faint yellowish patch on base of pectoral fin.** Secretive and usually solitary; hide in crevices and ledges. Coral-rich areas or rubble bottoms, most commonly sheltered in outer reefs in 10-70 m.
Asian Pacific: Indonesia and Philippines. – S.W. Japan to Great Barrier Reef, New Caledonia and Vanuatu.

WHITE-TAILED PYGMY ANGELFISH *Centropyge fisheri*
SIZE: to 7.5 cm (3 in.)   Angelfishes – Pomacanthidae
ID: Dark blue with **translucent to white tail;** dorsal, anal and ventral fins edged in bright blue, black marks on rear edge of dorsal and anal fins. Solitary or form loose aggregations. Coral rubble of lagoons, passages and outer slopes in 10-60 m.
Indo-Pacific: E. Africa to Indonesia, Philippines, Micronesia, Hawaii and French Polynesia. – S.W. Japan to E. Australia.

YELLOWFIN PYGMY ANGELFISH *Centropyge flavipectoralis*
SIZE: to 10 cm (4 in.)   Angelfishes – Pomacanthidae
ID: Dark brown with pairs of blue-black bars; **yellow pectoral fins.** Similar Brown Pygmy Angelfish [next] lack yellow pectoral fins. Solitary. Areas of coral rubble and reef margins in 3-20 m.
East Indian Ocean: Maldives, Sri Lanka and Andaman Sea.

BROWN PYGMY ANGELFISH *Centropyge multispinis*
SIZE: to 9 cm (3 1/2 in.)   Angelfishes – Pomacanthidae
ID: Dark brown with narrow black bars; **large black ear patch,** anal and ventral fins edged in bright blue. Most common *Centropyge* in Indian Ocean. Rubble bottoms and coral patches to 30 m.
Indian Ocean: Red Sea and E. Africa, Maldives, Sri Lanka to Andaman Sea and N.W. Sumatra in Indonesia.

MULTI-BARRED ANGELFISH   *Centropyge multifasciata*
SIZE: to 10 cm (4 in.)   Angelfishes – Pomacanthidae
ID: Alternating black and white or yellow bars; white-edged black spot on rear dorsal fin. Solitary or form small groups; usually in caves or under ledges, frequently swimming upside down. Steep outer reef slopes in 20-70 m.
East Indo-Pacific: Cocos-Keeling Is. to Philippines, New Guinea and French Polynesia. – S.W. Japan to Great Barrier Reef.

BLUEBACK PYGMY ANGELFISH   *Centropyge colini*
SIZE: to 9 cm (3 ½ in.)   Angelfishes – Pomacanthidae
ID: Pale yellow except for broad patch of blue on upper body extending from nape to soft dorsal fin. Usually solitary; shy, rarely observed. Caves and ledges of steep outer reef slopes in 25-75 m.
East Indo-West Pacific: Cocos-Keeling Is. to C. Indonesia, S. Philippines, Micronesia, Papua New Guinea and Fiji.

MULTICOLOR PYGMY ANGELFISH   *Centropyge multicolor*
SIZE: to 9 cm (3 ½ in.)   Angelfishes – Pomacanthidae
ID: White upper body, yellow face and lower body, dark blue dorsal and anal fins, yellow tail; blue patch on nape with black barring. Usually solitary; shy. Under ledges on steep outer reef dropoffs in 20-90 m.
Central Pacific: Micronesia to Hawaii, south Fiji, Cook Is. and Society Is. in French Polynesia.

OCELLATED PYGMY ANGELFISH   *Centropyge nigriocella*
SIZE: to 6 cm (2 ¼ in.)   Angelfishes – Pomacanthidae
ID: Pale yellowish or whitish; prominent black spot at base of pectoral fin base, ocellated spot on rear dorsal fin. Shy, rarely observed. Coral rubble areas in lagoons and along outer reefs in 4-15 m.
North Central Pacific: Scattered from Mariana Is. in Micronesia to Admiralty Is. in Papua New Guinea to Fiji and Line Is.

JAPANESE PYGMY ANGELFISH   *Centropyge interrupta*
SIZE: to 15 cm (6 in.)   Angelfishes – Pomacanthidae
ID: Orangish red head and forebody with blue spotting and blue behind, yellow tail. Males have stronger blue spotting on head and heavy blue lines on cheeks. Rocky shores and coral patches in 15-60 m.
Localized: S.W. Japan to Kure, Midway Is. and Hawaii.

ORANGEHEAD PYGMY ANGELFISH   *Centropyge hotumatua*
SIZE: to 8 cm (3 ¼ in.)   Angelfishes – Pomacanthidae
ID: Yellow-orange head and deep blue to nearly black body, yellow-orange tail; blue ring around eye, small to large dark spot or marking above pectoral fin base. Solitary or form small aggregations. Rock or coral reefs in 14-45 m.
Localized: Austral and Rapa Is. in French Polynesia, also Pitcairn and Easter Is.

YELLOW PYGMY ANGELFISH *Centropyge heraldi*
SIZE: to 10 cm (4 in.) Angelfishes – Pomacanthidae
ID: Bright yellow; **dusky brown patch behind eye.** Solitary or form loose aggregations. Lagoons and outer reef slopes in areas of mixed coral and rubble in 8-40 m.
Pacific: Brunei, Philippines, Micronesia, N. Papua New Guinea to French Polynesia. – Taiwan, S.W. Japan to Great Barrier Reef and Fiji.

WOODHEAD'S PYGMY ANGELFISH *Centropyge woodheadi*
SIZE: to 10 cm (4 in.) Angelfishes – Pomacanthidae
ID: Bright yellow; **elongate black patch on rear dorsal fin.** Solitary or form loose aggregations. Lagoons and outer reef slopes in areas of mixed coral and rubble in 8-40 m.
Central Pacific: Great Barrier Reef and Coral Sea to Fiji.

LEMONPEEL ANGELFISH *Centropyge flavissima*
SIZE: to 14 cm (5 ¹/₂ in.) Angelfishes – Pomacanthidae
ID: Bright yellow; **blue edge on gill cover** and usually blue ring around eye. Usually in harems. Mimicked by juvenile Mimic Surgeonfish. Coral-rich lagoons and seaward reefs to 25 m.
East Indo-Pacific: Christmas and Cocos-Keeling Is. to Great Barrier Reef, New Caledonia, Vanuatu and French Polynesia.

Lemonpeel Angelfish – Juvenile
SIZE: 3-5 cm (1 ¹/₄ -2 in.)
ID: Bright yellow; blue-rimmed black spot on center of side, blue edge on gill cover.

BICOLOR ANGELFISH *Centropyge bicolor*
SIZE: to 15 cm (6 in.) Angelfishes – Pomacanthidae
ID: Bright yellow head and forebody, deep blue behind with yellow tail; blue saddle across nape extends to eyes. Solitary, in pairs or small groups. Coral-rich and rubble areas of seaward reefs and lagoons, usually in 10-25 m.
West Pacific: Indonesia, Philippines, Micronesia, Papua New Guinea, Solomon Is. – S.W. Japan to N.W. & E. Australia and Fiji.

KEYHOLE ANGELFISH *Centropyge tibicen*
SIZE: to 18 cm (7 in.) Angelfishes – Pomacanthidae
ID: Dark blue; oval-shaped white spot on side, margin of anal fin broadly yellow. Solitary or form small groups. Coral and rubble areas of lagoons and seaward reefs in 4-35 m.
West Pacific: Malaysian Peninsula, Indonesia, Philippines, Micronesia, Papua New Guinea and Solomon Is. – S.W. Japan to N. & E. Australia, New Caledonia and Fiji.

TWO-SPINED ANGELFISH   *Centropyge bispinosa*
SIZE: to 10 cm (4 in.)   Angelfishes – Pomacanthidae
ID: Red-orange undercolor with narrow blue bars, head and fins deep blue to purple; two spines extend from lower gill cover. Solitary or form small groups; shy, stay near shelter. Lagoons and outer reef slopes in 5-45 m.
Indo-Pacific: E. Africa to Indonesia, Philippines, New Guinea, Micronesia and French Polynesia. – S.W. Japan to E. Australia.

Two-spined Angelfish – Variation
ID: The amount of red, orange and blue of this species is variable and occasionally specimens are almost entirely white or orange.

GOLDEN ANGELFISH   *Centropyge aurantia*
SIZE: to 10 cm (4 in.)   Angelfishes – Pomacanthidae
ID: Orange (brown in Sulawesi, Indonesia and Solomon Is.) with rippled bars; blackish ring around eye. Unusually shy species, seldom venture far from coral shelter. Coral-rich areas in 3-20 m.
West Pacific: N.E. Sulawesi in Indonesia to Papua New Guinea, Solomon Is. – Papua New Guinea to N. Great Barrier Reef and Fiji.

FLAME ANGELFISH   *Centropyge loricula*
SIZE: to 10 cm (4 in.)   Angelfishes – Pomacanthidae
ID: Brilliant red to red-orange with about 5 black bars; blue markings on rear edge of dorsal and anal fins. Solitary or form small groups; shy. Seaward reefs and lagoons in 5-60 m.
Central Pacific: Philippines, Papua New Guinea and Solomon Is., Hawaii to French Polynesia. Scattered populations in West Pacific and Indonesia.

RUSTY ANGELFISH   *Centropyge ferrugata*
SIZE: to 10 cm (4 in.)   Angelfishes – Pomacanthidae
ID: Rusty color with numerous brown spots; dorsal and anal fins edged with bright blue. Solitary or form small groups; usually close to reef crevices. Rocky reefs and rubble in 6-30 m.
Localized: S.W. Japan south through N. Mindanao in Philippines. Common in Okinawa.

ORANGEPEEL ANGELFISH   *Centropyge shepardi*
SIZE: to 12 cm (4 ¾ in.)   Angelfishes – Pomacanthidae
ID: Red-orange with narrow dark bars, rear dorsal and anal fins also black. Exposed outer reefs on mixed live and dead coral bottoms in 10-56 m.
Localized: Mariana Is. and Guam in Micronesia and Bonin Is. north to Japan's Izu National Park.

BLACKTAIL ANGELFISH *Centropyge eibli*
SIZE: to 11 cm (4 ¼ in.) Angelfishes – Pomacanthidae
ID: Pale gray with thin brown to orange bars and black tail with blue margin; orange ring around eye, and orange bar across base of pectoral fin. Solitary, occasionally in pairs. Usually in coral-rich areas in 3-25 m.
East Indo-South Asian Pacific: Maldives, Sri Lanka and Bay of Bengal south through S. Indonesia to N.W. Australia.

PEARL-SCALED ANGELFISH *Centropyge vroliki*
SIZE: to 12 cm (4 ¾ in.) Angelfishes – Pomacanthidae
ID: Pale gray head and forebody gradating to black on rear body and adjacent fins; tail black with fine blue margin. Solitary or loose groups. Sheltered coastal reefs and outer slopes to 25 m.
West Pacific: Indonesia, Philippines, Micronesia and Papua New Guinea. – S. W. Japan to E. Australia and Vanuatu.

REGAL ANGELFISH *Pygoplites diacanthus*
SIZE: to 25 cm (10 in.) Angelfishes – Pomacanthidae
ID: Yellow-orange with 7-8 dark-edged bluish white bars, yellow tail; dark patch around eye, blue and orange bands on anal fin. Solitary or in pairs; feed on sponges and tunicates. Lagoons and outer reefs to 48 m.
Indo-Pacific: Red Sea to Indonesia, Philippines, Micronesia and French Polynesia. – S.W. Japan to N. Australia and New Caledonia.

Regal Angelfish – Juvenile
SIZE: 3-7.5 cm (1¼ -3 in.)
ID: Bright yellow to orange; about 5 white bars with black borders on body and head; large blue spot ringed in white and black on rear dorsal fin. Often inhabit caves and crevices.

YELLOW-MASK ANGELFISH *Pomacanthus xanthometopon*
SIZE: to 38 cm (15 in.) Angelfishes – Pomacanthidae
ID: Yellow with large blue spots on scales, blue head forming network pattern with yellow eye mask; large black spot on rear dorsal. Solitary; feed on sponges and tunicates. Coral-rich areas in 5-30 m.
East Indo-Asian Pacific: Maldives to Indonesia, Philippines, Solomon Is. – S.W. Japan to Great Barrier Reef and Vanuatu.

BLUE-GIRDLED ANGELFISH *Pomacanthus navarchus*
SIZE: to 25 cm (10 in.) Angelfishes – Pomacanthidae
ID: Yellow-orange with blue spots and blue lips; deep blue broad girdle wraps body from nape to ventral fins along belly and curving up to above base of tail. Solitary. Coral reefs in 3-40 m.
Asian Pacific: Indonesia, Philippines, Micronesia, Papua New Guinea and Solomon Is. to N. Great Barrier Reef.

EMPEROR ANGELFISH   *Pomacanthus imperator*
SIZE: to 38 cm (15 in.)   Angelfishes – Pomacanthidae
ID: Vivid alternating blue and yellow stripes and yellow tail; blue-edged black eye mask, broad blue-edged black bar behind head. Solitary; make loud drumming sound when alarmed. Coral reefs in 6-60 m.
Indo-Pacific: Red Sea and E. Africa to Indonesia, Philippines, Hawaii and French Polynesia. – S.W. Japan to N. & E. Australia.

Emperor Angelfish – Subadult
SIZE: 9-15 cm (3 1/2 - 6 in.)
ID: Yellow wavy stripes on body; blue vertical lines on head; developing black eye mask and black bar behind head; pale yellowish tail.

Emperor Angelfish – Juvenile
SIZE: 3-8 cm (1 1/4 - 3 1/4 in.)
ID: Blue-black with striking pattern of blue and white concentric circular markings, tail margin transparent. The pattern gradually fades into adult coloration between 8-12 cm.

Emperor Angelfish – Small Juvenile
SIZE: 3 cm (1 1/4 in.)
ID: Dark navy blue with 2 white curving bars on body, another from nape to behind eye and curving to run along lower body and anal fin; white band above mouth to lower corner; white circle on tail base.

BLUE-RINGED ANGELFISH   *Pomacanthus annularis*
SIZE: to 45 cm (18 in.)   Angelfishes – Pomacanthidae
ID: Orangish brown with upward curving blue bands, tail white; blue ring above pectoral fin, numerous blue markings on head. Solitary or in pairs. Coastal reefs, frequently in murky water, also around caves and wrecks to 60 m.
Indo-Asian Pacific: E. Africa to Indonesia, Philippines and Solomon Is. – S.W. Japan to Australia.

Blue-ringed Angelfish – Juvenile
SIZE: 3-8 cm (1 1/4 - 3 1/4 in.)
ID: Bluish black undercolor with narrow closely-spaced pale blue and white bars slightly arched toward head, 2 or 3 are usually slightly wider and more intense, tail mainly whitish or clear without markings.

SEMICIRCLE ANGELFISH *Pomacanthus semicirculatus*
SIZE: to 35 cm (14 in.) Angelfishes – Pomacanthidae
ID: Greenish forebody with blue spotting, black rear body and tail with pale blue spotting; yellow lips, blue markings on gill cover and margins on dorsal, anal and tail fins. Solitary. Sheltered inshore reefs to 40 m.
West Pacific: Indonesia, Philippines, Micronesia, Papua New Guinea, Solomon Is. – S. W. Japan to N.W. & E. Australia and Fiji.

Semicircle Angelfish – Brown Variation
ID: Dark brownish head, pale brown forebody with blue spotting; dark brown rear body and tail with pale blue spotting.

Semicircle Angelfish – Subadult
SIZE: 9-15 cm (3 ½ - 6 in.)
ID: Pale forebody with some blue spotting; head and rear body retain black undercolor of juvenile with blue and white semicircular line markings.

Semicircle Angelfish – Juvenile
SIZE: 3-7.5 cm (1 ¼ - 3 in.)
ID: Bluish black undercolor with distinctive narrow semicircular blue and white markings, becoming more arched toward rear body. Color transformation from juvenile to adult usually occurs between 8-16 cm.

SIX-BANDED ANGELFISH *Pomacanthus sextriatus*
SIZE: to 46 cm (18 in.) Angelfishes – Pomacanthidae
ID: Tan with blue scale centers and 6 dark bars; head dark blue with white bar behind eye, blue spotting on rear dorsal, anal and tail fins. Solitary or in pairs. Coastal, lagoon and outer reefs in 3-60 m.
Asian Pacific: Indonesia, Philippines, Micronesia and Solomon Is. – S. W. Japan to Australia and New Caledonia.

Six-banded Angelfish – Juvenile
SIZE: 3-7.5 cm (1 ¼ - 3 in.)
ID: Black undercolor with 5-6 prominent white bars interspaced with narrower blue bars. Juveniles usually begin color transformation to adult pattern between 8-15 cm.

Angelfishes

ORANGE-FACED ANGELFISH *Chaetodontoplus chrysocephalus*
SIZE: to 20 cm (8 in.)   Angelfishes – Pomacanthidae
ID: Orangish head shading to brown then black toward rear, tail yellow; maze of blue markings on head, blue stripes on body. Coral and rubble in 15-25 m.
Asian Pacific: E. Indonesia to N. Philippines.

PHILIPPINES ANGELFISH *Chaetodontoplus caeruleopuncatus*
SIZE: to 20 cm (8 in.)   Angelfishes – Pomacanthidae
ID: Male – Bluish gray head and nape, remainder of body blackish to bluish with small dots; white margin on rear dorsal and anal fins; yellow tail. **Female –** Similar, except head and nape dull brown. Usually pairs. Coral and rubble in 15-40 m.
Localized: Philippines.

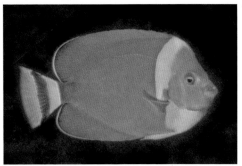

QUEENSLAND YELLOWTAIL ANGELFISH *Chaetodontoplus meredithi*
SIZE: to 25 cm (10 in.)   Angelfishes – Pomacanthidae
ID: Blackish body, head blue (yellow-spotted in male [pictured]) with white bar behind; yellow breast and bright yellow tail, yellow blotch on nape. Solitary or in pairs. Flat bottoms with rocky coral patches in 6-45 m.
Localized: E. Australia from Great Barrier Reef (rare) to Sydney area and Lord Howe I.

WESTERN YELLOWTAIL ANGELFISH *Chaetodontoplus personifer*
SIZE: to 35 cm (14 in.)   Angelfishes – Pomacanthidae
ID: Blackish body, blue head (yellow-spotted in male [pictured]) with white bar behind, orange breast, tail orange with black bar; blotch on nape. Flat bottoms with rocky coral patches in 6-30 m.
Localized: N. W. Australia from Shark Bay to W. edge of Gulf of Carpentaria.

SCRIBBLED ANGELFISH   *Chaetodontoplus duboulayi*
SIZE: to 25 cm (10 in.)   Angelfishes – Pomacanthidae
ID: Dark blue with yellow snout; broad bar on rear head and forebody, yellow stripe under dorsal fin continues to include tail. In pairs or small groups; feed on sponges. Coastal reefs in 5-20 m.
Localized: N. W. & N. E. Australia, Aru Is. in Indonesia to S. New Guinea.

CONSPICUOUS ANGELFISH   *Chaetodontoplus conspicillatus*
SIZE: to 25 cm (10 in.)   Angelfishes – Pomacanthidae
ID: Gray body rimmed in black, orange face with blue eye ring; bluish white margin on dorsal and anal fins, band of orange on pectoral and tail fins. Solitary, in pairs or small groups. Outer reefs in 20-40 m.
Southwestern Pacific: S. Great Barrier Reef to New South Wales, New Caledonia, Lord Howe and Norfolk Is.

PEWTER ANGELFISH *Chaetodontoplus dimidiatus*
SIZE: to 20 cm (8 in.)   Angelfishes – Pomacanthidae
ID: Light gray head and back, black below; dorsal and anal fins black with yellow margin, **tail entirely yellow,** maze of yellow markings on snout and nape. Solitary or in pairs. Inshore and outer rock and coral reefs in 5-30 m.
Localized: Indonesia including Raja Ampat, Molucca Is., Halmahera and N. Sulawesi.

Pewter Angelfish – Juvenile
SIZE: 3-6 cm (1 1/4 - 2 1/4 in.)
ID: Black head and body; pale yellowish bar from between eyes to lips, another from nape to ventral fin; pale yellowish dorsal fin, rear edge of anal fin and tail. Black oblong area on rear half.

BLACK VELVET ANGELFISH *Chaetodontoplus melanosoma*
SIZE: to 20 cm (8 in.)   Angelfishes – Pomacanthidae
ID: Light gray head and back, dark gray below; rear dorsal, **anal and tail fins black with yellow margins,** maze of gold markings on snout and nape. Solitary or in pairs. Inshore and outer rock and coral reefs in 5-30 m.
Asian Pacific: Central to E. Indonesia and Philippines.

GRAYTAIL ANGELFISH *Chaetodontoplus poliourus*
SIZE: to 18 cm (7 in.)   Angelfishes – Pomacanthidae
ID: Purplish gray with white vermiculations; gray tail, white to yellowish patch behind head, black eye bar. Solitary or in pairs. Coral-rich areas of inshore reefs to 20 m.
Asian Pacific: E. Indonesia to Papua New Guinea and Solomon Is.

VERMICULATED ANGELFISH *Chaetodontoplus mesoleucus*
SIZE: to 18 cm (7 in.)   Angelfishes – Pomacanthidae
ID: Purplish gray with white vermiculations; white to yellowish patch behind head, black eye bar. Solitary or in pairs. Coral-rich areas of inshore reefs to 20 m.
Asian Pacific: Sabah in Malaysia, Indonesia, Philippines, Micronesia, Papua New Guinea to Solomon Is. – S.W. Japan to N. Australia.

BALLINA ANGELFISH *Chaetodontoplus ballinae*
SIZE: to 20 cm (8 in.)   Angelfishes – Pomacanthidae
ID: White with black upper body, tail and pectoral fins yellow to orange; dark snout tip and triangular eye mask. Solitary or form small groups. Coral or rock reefs in 10-80 m.
Localized: N. New South Wales (Coffs Harbor, Ballina, Solitary Is.) and Lord Howe I. Generally rare, but common at Ball's Pyramid near Lord Howe I.

ZEBRA ANGELFISH   *Genicanthus caudovittatus*

SIZE: to 20 cm (8 in.)   Angelfishes – Pomacanthidae

ID: Male – White with numerous black bars; **blackened area on middle of dorsal fin.** Solitary or in harems; feed on zooplankton well above bottom; capable of sex reversal. Outer reefs in 15-70 m.

Indian Ocean: Red Sea (common in the Gulf of Aqaba) and E. Africa, Mauritius, Maldives to N. Sumatra in Indonesia.

Zebra Angelfish – Female

SIZE: to 15 cm (6 in.)

ID: Light gray; **black bar above eye,** and black margins on tail.

BLACK-SPOT ANGELFISH   *Genicanthus melanospilos*

SIZE: to 18 cm (7 in.)   Angelfishes – Pomacanthidae

ID: Male – White with numerous black bars; **gold spots on dorsal and tail fins.** Black spot on breast. In pairs or harems; feed on zooplankton above bottom. Outer reef slopes in 20-45 m.

West Pacific: Indonesia, Philippines, Papua New Guinea and Solomon Is. – S.W. Japan to Great Barrier Reef, Vanuatu, New Caledonia and Fiji.

Black-spot Angelfish – Female

SIZE: to 13 cm (5 in.)

ID: Light gray head and lower body becoming yellow on upper half; black tail margins; lack bar above eye (compare female Zebra Angelfish [previous]).

ORNATE ANGELFISH   *Genicanthus bellus*

SIZE: to 18 cm (7 in.)   Angelfishes – Pomacanthidae

ID: Male – Light gray; golden orange stripe along base of dorsal fin and another on midbody, borders of tail blue. Solitary or form groups; feed on plankton above bottom. Outer reefs in 25-110 m; rare above 50 m.

Scattered Pacific: Philippines and Micronesia to Society Is. in French Polynesia.

Ornate Angelfish – Female

SIZE: to 15 cm (6 in.)

ID: Light gray; black eye bar, curved black marking from rear head extends length of dorsal fin, another extends diagonally from rear head onto lower tail border, upper tail border black, blue patch from behind pectoral fin to lower body.

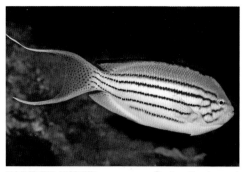

BLACKSTRIPED ANGELFISH   *Genicanthus lamarck*
SIZE: to 23 cm (9 in.)   Angelfishes – Pomacanthidae
ID: Male – White with 3-5 black stripes; broad black stripe on outer edge of dorsal fin, black speckles on tail. Form groups. Outer reefs in 10-50 m.
Asian Pacific: Indonesia, Philippines, Papua New Guinea, Solomon Is. – S.W. Japan to N. Great Barrier Reef to Vanuatu.

Blackstriped Angelfish – Female/Juvenile
SIZE: to 15 cm (6 in.)
ID: Top black stripe broader than adult and angles down to lower tail fin.

PINSTRIPED ANGELFISH   *Genicanthus watanabei*
SIZE: to 15 cm (6 in.)   Angelfishes – Pomacanthidae
ID: Male – Bluish head and back with alternating black and white stripes on remainder of body. Usually form mixed-sex feeding aggregations above reefs. Outer reefs in 12-80 m.
Pacific: S.W. Japan to Micronesia, E. Australia and New Caledonia to French Polynesia. Absent Indonesia, Philippines, and Papua New Guinea.

Pinstriped Angelfish – Female
SIZE: to 12 cm (4 3/4 in.)
ID: Pale blue-gray; short black bar above eye, black margins on dorsal, anal and tail fins.

JAPANESE SWALLOW   *Genicanthus semifasciatus*
SIZE: to 20 cm (8 in.)   Angelfishes – Pomacanthidae
ID: Male – White with numerous close-set wavy bars on upper body, **yellow on head extends into a yellow midbody stripe;** yellow spots on dorsal and tail fins. Usually form same-sex groups. Rock or coral reefs in 15-100 m.
Localized: W. Pacific Rim from S. W. Japan to N. Philippines.

Japanese Swallow – Female
SIZE: to 16 cm (6 1/4 in.)
ID: Dusky brownish gray becoming white on belly; white bar with black borders behind eye; black edged swallowtail.

Spadefishes

SIZE: to 13 cm (5 in.)

ID: Juvenile – Silver; black eye bar and a second bar from dorsal fin onto ventral fins and a third bar across rear body and fins; extremely long dorsal, anal and ventral fins. Occasionally near floating objects. Often solitary, but with increased size form groups.

LONGFIN SPADEFISH   *Platax teira*

SIZE: to 70 cm (28 in.)   Spadefishes – Ephippidae

ID: Silver; dark to faint bar through eye and a second bar from front of dorsal fin to ventral fin, **large dark blotch above rear edge of ventral fin.** Usually form groups. Inshore and outer reefs in 3-25 m.

Indo-Asian Pacific: Red Sea and E. Africa to Solomon Is. – S.W. Japan to Great Barrier Reef and Vanuatu.

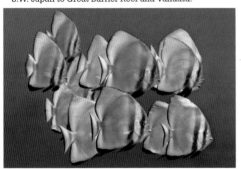

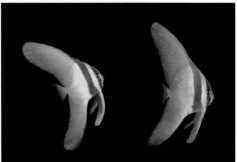

GOLDEN SPADEFISH   *Platax boersii*

SIZE: to 47 cm (19 in.)   Spadefishes – Ephippidae

ID: Silver with yellowish tint, dark to faint bar through eye and a 2nd bar from nape to ventral fins, **head profile is smoothly rounded with virtually no indentation between lower forehead and mouth.** Coastal and outer reef dropoffs in 3-30 m.

Asian Pacific: Indonesia, Philippines, Micronesia, Papua New Guinea and Solomon Is. north to S.W. Japan.

Golden Spadefish – Juvenile

SIZE: to 13 cm (5 in.)

ID: Silver, dusky rear body and fins; black eye bar and a second bar from nape onto ventral fin; elongate, triangular-shaped dorsal and anal fins. Sheltered shoreline reefs.

Circular Spadefish

SIZE: to 7.5 cm (3 in.)

ID: Juvenile – Brown, rear half of body darker brown; brown eye bar and another bar through pectoral and ventral fins. Smaller juveniles orange brown with narrow bar through eye. Occasionally lie on side, or float on surface, mimicking dead leaves.

CIRCULAR SPADEFISH   *Platax orbicularis*

SIZE: to 50 cm (20 in.)   Spadefishes – Ephippidae

ID: Silver with yellowish tint; dark to faint bar through eye and a 2nd bar from nape to ventral fins, **head profile has a small, but distinct indentation between the lower forehead and mouth.** In pairs or groups. Shoreline and outer reefs in 2-35 m.

Indo-Pacific: Red Sea and E. Africa to Indonesia, Micronesia and French Polynesia. – S.W. Japan to N.E. Australia and Fiji.

Batavia Spadefish

SIZE: to 5 cm (2 in.)

ID: Juvenile – Intricate pattern of black and white bands cover small body; long dorsal, anal and ventral fins.

BATAVIA SPADEFISH   *Platax batavianus*
SIZE: to 50 cm (20 in.)   Spadefishes – Ephippidae
ID: Silvery; black eye bar, broad dusky bar from nape to ventral fins, **scattered dark spots often on belly,** older adults develop hump above eyes. Solitary or form groups. Inshore reefs in 5-40 m.
Asian Pacific: Malay Peninsula to Indonesia, Philippines, Papua New Guinea and N. Australia.

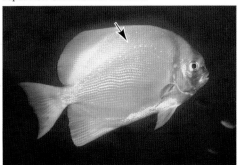

SHORTFIN SPADEFISH   *Zabidius novemaculeatus*
SIZE: to 45 cm (18 in.)   Spadefishes – Ephippidae
ID: Silver; faint eye bar and 2nd faint bar through pectoral fins, **row of pearly white spots just above lateral line,** slight bump between eyes. Usually form groups. Inshore coral reefs, often in turbid water in 2-25 m.
Localized: N. Australia and S. Papua New Guinea.

PINNATE SPADEFISH   *Platax pinnatus*
SIZE: to 37 cm (15 in.)   Spadefishes – Ephippidae
ID: Silver; black eye bar and a second bar through pectoral and onto ventral fin; **distinctive protruding snout.** Young adults have a single wide dusky bar on rear body. Usually solitary. Coastal reefs and seaward slopes in 2-25 m.
West Pacific: Sumatra in Indonesia to Solomon Is. – S.W. Japan to N. Australia, New Caledonia, Vanuatu and Fiji.

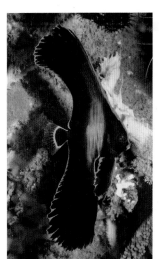

Pinnate Spadefish
SIZE: 13-18 cm (5 - 7 in.)

ID: Older Juvenile – With maturity a vague white bar appears on the side beginning behind the pectoral fin; dorsal and anal fins elongate and become more pointed. Hover in secluded areas.

Pinnate Spadefish
SIZE: to 12 cm (4¾ in.)

ID: Young Juvenile – Black with brilliant orange-red margin around body and fins; fins have scalloped outline, swim with an exaggerated motion, the smaller juveniles believed to mimic similar appearing flatworms.

Large Ovals
Surgeonfishes – Moorish Idol – Rabbitfishes

This ID Group consists of fishes that are fairly large, thin-bodied, have a basic oval shape, and are generally colorful.

FAMILY: Surgeonfishes – Acanthuridae
5 Genera – 55 Species Included

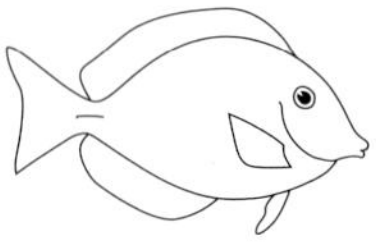

Typical Shape 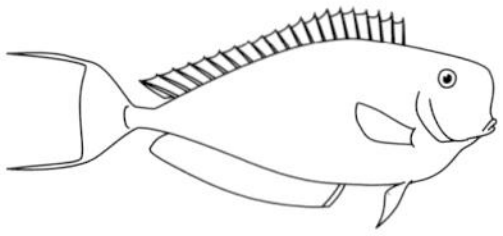   Genus *Naso*    Genus *Naso*

Surgeonfishes have thin, oval bodies with relatively long continuous dorsal and anal fins, small pointed mouths, and crescent tails. Lateral lines are continuous and scales are not conspicuous. A spine (or pair of spines in a few member of genus *Naso*) as sharp as a surgeon's scalpel and located on each side of the body at the base of the tail is the origin of this family's common name. The formidable blades, typically housed within fleshy body grooves, are employed when defending territories, establishing social dominance, or as defense against predators. The tail-base spines, formed by modified scales and attached posteriorly by ligaments to the spinal column, cannot be voluntarily erected. The convex forward-pointing blades extend slightly from their grooves each time the tail flexes. If the exposed point happens to snag into something, the spine extends at a right angle from the body.

Family members are frequently seen reef inhabitants that often mix in loose aggregations. Most species pick algae from the bottom during the day, while others feed on detritus or consume plankton from the water column.

FAMILY: Rabbitfishes – Siganidae
Single Genus – 22 Species Included

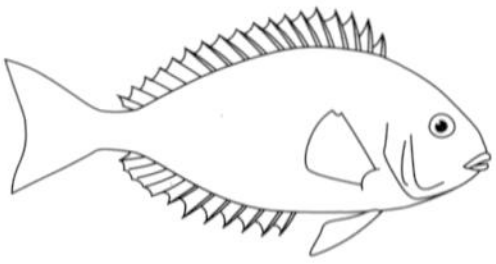

Typical Shape 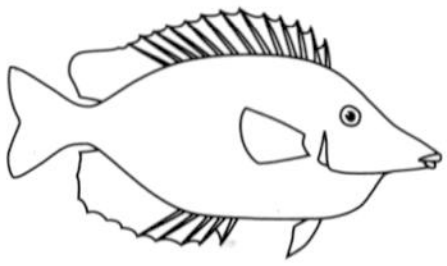   Typical Shape

These moderately sized, oval-shaped fishes have small terminal mouths, continuous lateral lines and venomous dorsal, ventral and anal spines, which can inflict painful wounds. Their common name is derived from a herbaceous diet of sea grasses and algae and a ravenous appetite.

FAMILY: Moorish Idol – Zanclidae
Single Genus – Single Species

Moorish Idol

The Moorish Idol is the lone species in its family. Its bold, black, white and yellow bars and a long filamentous dorsal fin make the thin-bodied omnivore one of the most conspicuous and easily recognizable Indo-Pacific reef species.

ORANGEBAND SURGEONFISH *Acanthurus olivaceus*
SIZE: to 35 cm (14 in.)   Surgeonfishes – Acanthuridae
ID: Light gray head and forebody, dark gray behind; blue-edged elliptical **orange band behind upper gill cover**. Solitary or form groups. Over sand bottoms near reefs in 3-45 m.
Pacific: Indonesia, Philippines, Micronesia, Papua New Guinea, Solomon Is., Hawaii and French Polynesia. – S.W. Japan to Australia.

Orangeband Surgeonfish – Subadult
SIZE: to 7.5 cm (3 in.)
ID: Juvenile – Entirely yellow, with age they gradually develop an elliptical orange band behind the upper gill cover. Individual shown will soon transform into darker adult.

ACHILLES TANG *Acanthurus achilles*
SIZE: to 20 cm (8 in.)   Surgeonfishes – Acanthuridae
ID: Dark to navy blue; large orange teardrop on rear body, white edge on gill cover, and white band at base of dorsal and anal fins, middle of tail orange. Solitary; territorial and aggressive. Surge zone to 4 m.
Central Pacific: Micronesia, New Caledonia and Hawaii to Pitcairn Is. east of French Polynesia.

WHITE-SPOTTED SURGEONFISH *Acanthurus guttatus*
SIZE: to 28 cm (11 in.)   Surgeonfishes – Acanthuridae
ID: Darkish with numerous white spots or streaks on rear body; white bar behind eye and behind pectoral fin, yellowish foretail. Usually form groups. Surge zone to 4 m.
Indo-Pacific: Seychelles to Indonesia, Micronesia, Papua New Guinea, Hawaii and French Polynesia. – S.W. Japan to Great Barrier Reef and New Caledonia.

Surgeonfishes

MIMIC SURGEONFISH   *Acanthurus pyroferus*
SIZE: to 29 cm (11 in.)   Surgeonfishes – Acanthuridae
ID: Brown; curving black band from chin to upper edge of gill cover, **orange patch above pectoral fin base.** Usually solitary. Lagoon and seaward reefs in 4 - 60 m.
Pacific: Indonesia, Philippines, Micronesia, Papua New Guinea and Solomon Is. to French Polynesia. – S.W. Japan to Great Barrier Reef and New Caledonia.

Mimic Surgeonfish – Juvenile
SIZE: to 7 cm (2 ³⁄₄ in.)
ID: Depending on area, mimic one of several *Centropyge* angelfishes to avoid predators. Pictured example mimicking Pearl-scaled Angelfish, but distinguished by lack of gill cover spine found on all angelfish.

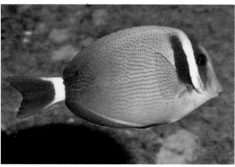

Mimic Surgeonfish – Juvenile
SIZE: to 7 cm (2 ³⁄₄ in.)
ID: Depending on area, mimic one of several *Centropyge* angelfishes to avoid predators. Pictured example mimicking Lemonpeel Angelfish, but distinguished by lack of gill cover spine found on all angelfishes.

WHITEBAR SURGEONFISH   *Acanthurus leucopareius*
SIZE: to 20 cm (8 in.)   Surgeonfishes – Acanthuridae
ID: Tan with thin dark stripes; white bar behind eye bordered by two dark bars. Form schools. Rocky, boulder-strewn areas of surge to 10 m.
Pacific: Two separate populations. North of equator: S. Japan and Micronesia to Hawaii. South of equator: New Caledonia to Pitcairn Is. located east of French Polynesia.

INDIAN MIMIC SURGEONFISH   *Acanthurus tristis*
SIZE: to 25 cm (10 in.)   Surgeonfishes – Acanthuridae
ID: Gray face and tan to dark brown body; curving black band from chin to upper gill cover, black patch above pectoral fin base, white margin on tail. Usually solitary. Lagoon and outer reefs in 2 - 30 m.
Indian Ocean: Seychelles, Chagos, Maldives, Sri Lanka and Andaman Sea to Bali in Indonesia.

Indian Mimic Surgeonfish – Juvenile
SIZE: to 7 cm (2 ³⁄₄ in.)
ID: Depending on area, mimic one of several *Centropyge* angelfishes to avoid predators. Pictured example mimicking Blacktail Angelfish, but distinguished by lack of gill cover spine found on all angelfish.

WHITECHEEK SURGEONFISH   *Acanthurus nigricans*
SIZE: to 21 cm (8 1/4 in.)   Surgeonfishes – Acanthuridae
ID: Dark brown to navy blue or black; white tail with yellow bar; white patch below eye, white ring behind mouth, yellow line at base of dorsal and anal fins, yellow tail spine. Solitary or form groups. Exposed reefs to 40 m.
Pacific: Indonesia, Philippines, Micronesia, Solomon Is., Hawaii, French Polynesia. – S.W. Japan to Great Barrier Reef, New Caledonia.

POWDERBLUE SURGEONFISH   *Acanthurus leucosternon*
SIZE: to 23 cm (9 in.)   Surgeonfishes – Acanthuridae
ID: Black head and blue body; yellow dorsal fin, white ventral anal and tail fins, white patch below mouth extending to pectoral fin. Form large feeding groups. Inshore and outer reefs to 25 m.
Indian Ocean: E. Africa to Andaman Sea, Java, Bali and Komodo in Indonesia.

JAPANESE SURGEONFISH   *Acanthurus japonicus*
SIZE: to 20 cm (8 in.)   Surgeonfishes – Acanthuridae
ID: Yellowish brown to dark blue becomes yellow on extreme rear body; white patch from lips to eye, white tail. Solitary or form small groups. Clear lagoons and outer reef in 2-12 m.
North Asian Pacific: Brunei, Sabah in Malaysia, Sulawesi and Halmahera in Indonesia to Philippines, Micronesia and S.W. Japan.

BLUE-LINED SURGEONFISH   *Acanthurus nigros*
SIZE: to 25 cm (10 in.)   Surgeonfishes – Acanthuridae
ID: Dark to pale bluish brown with numerous fine stripes; dark spot on rear base of dorsal and anal fins, commonly displays white bar on base of tail.. Solitary or form small groups. Clear lagoons and seaward reefs to 90 m.
Pacific: Micronesia to Great Barrier Reef and French Polynesia.

YELLOWMASK SURGEONFISH   *Acanthurus mata*
SIZE: to 40 cm (16 in.)   Surgeonfishes – Acanthuridae
ID: Slender; pale to dark bluish body with numerous blue and dark horizontal lines, **upper lip yellowish**, double yellow bands between eyes. Groups feed on zooplankton. Inshore (often turbid) and outer reefs in 5-25 m.
Indo-Pacific: Red Sea and E. Africa to Indonesia, French Polynesia. – S.W. Japan to Great Barrier Reef and New Caledonia.

DARK SURGEONFISH   *Acanthurus nubilus*
SIZE: to 26 cm (10 1/2 in.)   Surgeonfishes – Acanthuridae
ID: Bluish with numerous blue spots and wavy lines; **lower head profile rounded** (lower head profile is flattened in most surgeonfishes). Solitary or form small groups; feed on zooplankton. Steep outer reef slopes in 20-90 m.
Pacific: Indonesia, Philippines, Micronesia and New Caledonia to French Polynesia.

Surgeonfishes

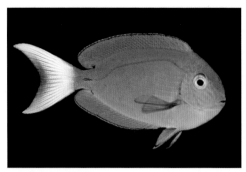

YELLOWFIN SURGEONFISH   *Acanthurus xanthopterus*
SIZE: to 60 cm (24 in.)   Surgeonfishes – Acanthuridae
ID: Blue to grayish brown; yellow to yellowish pectoral fins, yellow band passes through eye, frequently a white ring around tail base. The largest surgeonfish. Solitary or form groups. Sandy areas near reefs in 15-90 m.
Indo-Pacific: E. Africa to Indonesia, Philippines, Micronesia and Mexico. – S.W. Japan to Great Barrier Reef and New Caledonia.

WHITETAIL SURGEONFISH   *Acanthurus thompsoni*
SIZE: to 27 cm (11 in.)   Surgeonfishes – Acanthuridae
ID: Dark brown changeable to pale bluish gray, **white tail;** may display striped pattern, body more slender than most surgeonfishes. Form groups; feed on zooplankton high above the bottom. Seaward reef slopes and steep dropoffs in 4-75 m.
Indo-Pacific: E. Africa to Indonesia, Philippines, Micronesia, Hawaii and French Polynesia. – S.W. Japan to Great Barrier Reef.

EYESTRIPE SURGEONFISH   *Acanthurus dussumieri*
SIZE: to 55 cm (22 in.)   Surgeonfishes – Acanthuridae
ID: Bluish to brownish body with horizontal line markings; yellow stripe through eye, **blue tail with numerous black spots.** Solitary or form small groups. Seaward reefs in 4-100 m.
Indo-Pacific: E. Africa to Indonesia, Philippines, Micronesia, Papua New Guinea, Solomon Is., Hawaii and Line Is. – S.W. Japan to Great Barrier Reef.

ROUNDSPOT SURGEONFISH   *Acanthurus bariene*
SIZE: to 42 cm (17 in.)   Surgeonfishes – Acanthuridae
ID: Brown to yellowish brown with orange dorsal fin; white lips, **orange bar behind gill cover,** round dark spot behind eye. Solitary or in pairs. Clear seaward reefs in 10-50 m.
Indo-West Pacific: Seychelles to Indonesia, Philippines, Micronesia, and Solomon Is. – S.W. Japan to Great Barrier Reef and Fiji.

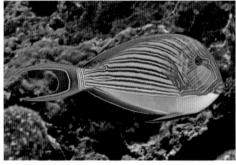

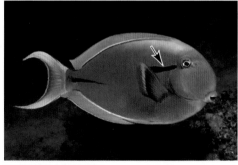

STRIPED SURGEONFISH   *Acanthurus lineatus*
SIZE: to 38 cm (15 in.)   Surgeonfishes – Acanthuridae
ID: Gold undercolor with **numerous black edged blue stripes and bluish belly;** yellow ventral fins, bright blue margins on most fins. Solitary; territorial and aggressive; tail spine venomous. Outer edge of reefs to 6 m.
Indo-Pacific: E. Africa to Indonesia, Micronesia, Solomon Is. and Polynesia. – S.W. Japan to Great Barrier Reef and New Caledonia.

BLACKSTREAK SURGEONFISH   *Acanthurus nigricauda*
SIZE: to 40 cm (16 in.)   Surgeonfishes – Acanthuridae
ID: Pale to dark brown or gray; **long black bar behind eye to above pectoral fin,** thin streak on rear body to tail spine. Solitary or form small groups. Usually over sand near coral or rock outcrops in 3-30 m.
Indo-Pacific: E. Africa to Indonesia, Philippines, Micronesia, Papua New Guinea, French Polynesia. – S.W. Japan to Great Barrier Reef.

BLACKSPINE SURGEONFISH *Acanthurus fowleri*
SIZE: to 45 cm (18 in.)   Surgeonfishes – Acanthuridae
ID: Blue head and brownish body; **arc-shaped marking behind head**, heavy black margin encircles tail spine. Solitary or in pairs; often graze on algae growing on sponges. Outer reef areas in clear water in 10-45 m.
Asian Pacific: Indonesia, Philippines, Papua New Guinea and Solomon Is.

FINE-LINED SURGEONFISH *Acanthurus grammoptilus*
SIZE: to 35 cm (14 in.)   Surgeonfishes – Acanthuridae
ID: Brown body with fine wavy blue lines; yellowish band through eye, black margin on tail spine, **bluish cheek with small orange spots.** Solitary or form groups. Inshore reefs in 2-20 m.
Localized: S. Raja Ampat, Indonesia to N. Australia.

ORANGE SOCKET SURGEONFISH *Acanthurus auranticavus*
SIZE: to 30 cm (12 in.)   Surgeonfishes – Acanthuridae
ID: Head brown without spots; body finely lined with purplish blue and dark brown stripes, orange ring around eye, orange encircles tail spine, **dark band behind eye and above gill cover.** Form small groups. Lagoons and outer reefs to 20 m.
Indo-West Pacific: E. Africa to Indonesia, Philippines, Great Barrier Reef, Solomon Is. and Fiji.

RINGTAIL SURGEONFISH *Acanthurus blochii*
SIZE: to 42 cm (17 in.)   Surgeonfishes – Acanthuridae
ID: Dark blue (almost black) to brown; narrow blue margins on dorsal, anal and tail fins, **small orange spot behind eye,** often display white ring around base of tail. Form schools; graze on algae. Lagoon and outer reef in 2-15 m.
Indo-Pacific: E. Africa to Indonesia, Philippines, Micronesia, Solomon Is., Hawaii and French Polynesia. – S.W. Japan to Great Barrier Reef.

PALE-LIPPED SURGEONFISH *Acanthurus leucocheilus*
SIZE: to 48 cm (19 in.)   Surgeonfishes – Acanthuridae
ID: Dark brown to nearly black; pale lips and pale band on chin; yellow on pectoral fin, **white tail spine.** Solitary. Near dropoffs on seaward clearwater reefs in 4-30 m.
Indo-Pacific: E. Africa to Andaman Sea, Indonesia, Philippines, Micronesia, Papua New Guinea, Solomon Is. and Line Is. – S.W. Japan to Great Barrier Reef.

Pale-lipped Surgeonfish – Juvenile
SIZE: to 5 cm (2 in.)
ID: Grayish brown head and body; bright yellow tail; whitish tail spine.

Surgeonfishes

WHITE-FRECKLED SURGEONFISH *Acanthurus maculiceps*
SIZE: to 40 cm (16 in.)   Surgeonfishes – Acanthuridae
ID: Black head with numerous white spots and grayish body; black streak behind upper edge of gill cover, black margin on tail spine, often a white ring around tail base. Solitary or form groups. Outer reefs to 30 m.
East Indo-West Pacific: Maldives, Andaman Sea to Indonesia, Micronesia and Line Is. – S.W. Japan to Great Barrier Reef and Fiji.

BROWN SURGEONFISH *Acanthurus nigrofuscus*
SIZE: to 21 cm (8 1/4 in.)   Surgeonfishes – Acanthuridae
ID: Brown with numerous orange spots on head; black spot at rear base of last dorsal and anal fin rays. Common on Indo-Pacific reefs. Form large schools; graze on algae growing on rocky surfaces. Inshore and outer reefs to 20 m.
Indo-Pacific: Red Sea and E. Africa to Indonesia, Micronesia, Solomon Is., Hawaii and Polynesia. – S.W. Japan to Great Barrier Reef.

WHITEFIN SURGEONFISH *Acanthurus albipectoralis*
SIZE: to 33 cm (13 in.)   Surgeonfishes – Acanthuridae
ID: Nearly black to light bluish gray, except for white outer third of pectoral fin; occasionally display white lips and pale bar on tail base. Solitary or form small groups; feed on plankton high above bottom. Steep outer reef slopes in 5-20 m.
Southwest Pacific: Great Barrier Reef and Coral Sea to New Caledonia and Fiji.

TENNENT'S SURGEONFISH *Acanthurus tennentii*
SIZE: to 30 cm (12 in.)   Surgeonfishes – Acanthuridae
ID: Brownish gray; **pair of broad black streaks behind upper edge of gill cover,** blue-edged black oval patch around tail spine, margin of tail white. Solitary or form groups. Lagoon and seaward reefs to 40 m.
Indian Ocean: E. Africa to Andaman Sea and Bali in Indonesia.

CONVICT SURGEONFISH *Acanthurus triostegus*
SIZE: to 22 cm (8 3/4 in.)   Surgeonfishes – Acanthuridae
ID: White with 5-6 black bars on head and body. Often feed in large groups. Usually on shallow reefs to 5 m.
Indo-Pacific: E. Africa to Indonesia, Philippines, Micronesia, Papua New Guinea, Solomon Is., Hawaii and French Polynesia. – S.W. Japan to N. & E Australia.

PALETTE SURGEONFISH *Paracanthurus hepatus*
SIZE: to 26 cm (10 in.)   Surgeonfishes – Acanthuridae
ID: Brilliant blue head and body; dramatic black hook-shaped marking, yellow tail with black borders. Solitary or form groups. Young hide in branching corals. Usually on clear, current swept outer reefs to 5 m.
Indo-West Pacific: E. Africa to Indonesia, Philippines, Micronesia and Line Is. – S.W. Japan to Great Barrier Reef and Fiji.

BRUSHTAIL TANG   *Zebrasoma scopas*
SIZE: to 22 cm (8 ³/₄ in.)   Surgeonfishes – Acanthuridae
ID: Yellowish brown shading to nearly black on the tail; dark brush-like patch of bristles in front of white tail spine, tiny pale blue dots or lines on head and body. Solitary or form groups. Lagoons and outer reefs to 50 m.
Indo-Pacific: E. Africa to Indonesia, Philippines, Micronesia, Solomon Is., French Polynesia. – S.W. Japan to Great Barrier Reef.

Brushtail Tang – Juvenile
SIZE: to 5 cm (2 in.)
ID: Forebody pale golden brown with gold spots on head, rear body dark brown to purple; paired dark bars on body, white tail spine. Solitary. Lagoons and reefs to 50 m.

Pacific Sailfin Tang
SIZE: 4-7.5 cm (1 ¹/₂ - 3 in.)
ID: Juvenile – Yellow with alternating dark and pale bars; clear tail, greatly enlarged dorsal and anal fins yellow with thin line markings. Solitary. Rock and coral reefs inside lagoons and on shallow protected reefs.

PACIFIC SAILFIN TANG   *Zebrasoma velifer*
SIZE: to 40 cm (16 in.)   Surgeonfishes – Acanthuridae
ID: White bars alternate with gray to brown bars; **tail white to yellowish to brown without spots,** greatly enlarged dorsal and anal fins are dark gray to brown with pale bands. Solitary or form groups. Lagoon and outer reefs to 45 m.
Pacific: Indonesia, Micronesia, Solomon Is., Hawaii to French Polynesia. – S.W. Japan to Great Barrier Reef and New Caledonia.

INDIAN SAILFIN TANG   *Zebrasoma desjardinii*
SIZE: to 40 cm (16 in.)   Surgeonfishes – Acanthuridae
ID: Generally gray with pattern of bars and spots on sides; enlarged dorsal and anal fins marked with pale lines, **tail dark with blue spots. Juvenile –** Similar, but yellow-gold. Solitary or form groups. Lagoons and outer reefs in 3-30 m.
Indian Ocean: Red Sea and E. Africa to Andaman Sea and N. Sumatra in Indonesia.

LONGNOSE TANG   *Zebrasoma rostratum*
SIZE: to 21 cm (8 ¹/₄ in.)   Surgeonfishes – Acanthuridae
ID: Dark brown to nearly black; white tail spine, protruding snout. Solitary or form groups; feed on filamentous algae. Lagoons and seaward reefs in 3-20 m.
East Central Pacific: Phoenix and Line Is. to French Polynesia and east to Pitcairn Is.

Surgeonfishes

LINED BRISTLETOOTH   *Ctenochaetus striatus*
SIZE: to 26 cm (10 in.)   Surgeonfishes – Acanthuridae
ID: Dark brown with numerous orange spots on head and blue lines on body; may display a small black spot at rear base of dorsal fin. Solitary or form groups; one of most abundant reef fishes. Lagoon and seaward reefs to 35 m.
Indo-Pacific: E. Africa to Indonesia, Philippines, Micronesia, Solomon Is., French Polynesia. – S.W. Japan to Great Barrier Reef.

Lined Bristletooth – Juvenile
SIZE: 3 to 6 cm (1 1/4 to 2 1/4 in.)
ID: Greenish brown head and fins; body brilliantly marked with blue and orange stripes; blue patch below mouth.

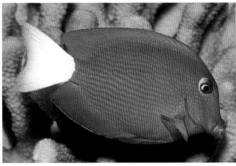

BLUELIPPED BRISTLETOOTH   *Ctenochaetus cyanocheilus*
SIZE: to 18 cm (7 in.)   Surgeonfishes – Acanthuridae
ID: Orange-brown with blue lines on body and small pale yellowish spots on head; blue lips, usually narrow yellow ring around eye. Solitary or form groups; graze on algae. Lagoon and outer reef slopes to 46 m. **Juvenile** – Pale to brownish yellow.
West Pacific: Indonesia, Philippines, Micronesia, Papua New Guinea and Solomon Is. – S. Japan to Australia and Fiji.

PALE-TAILED BRISTLETOOTH   *Ctenochaetus flavicauda*
SIZE: to 13 cm (5 in.)   Surgeonfishes – Acanthuridae
ID: Yellowish brown with thin dark stripes; yellow spots on head; conspicuous white to yellowish tail. Solitary or form groups. Coral-rich areas and rocky reefs to 30 m.
Central Pacific: Phoenix and Line Is., French Polynesia and Pitcairn Is.

HAWAIIAN BRISTLETOOTH   *Ctenochaetus hawaiiensis*
SIZE: to 25 cm (10 in.)   Surgeonfishes – Acanthuridae
ID: Dark olive-brown with numerous thin pale greenish stripes covering head, body, dorsal and anal fins. Solitary. Seaward rock and coral reefs in 10-45 m.
Central Pacific: Micronesia to Hawaii and French Polynesia.

Hawaiian Bristletooth – Juvenile
SIZE: to 6 cm (2 1/2 in.)
ID: Orange with dark gray-blue chevron markings. Solitary. Seaward rock and coral reefs in 10-45 m.

BLUE-SPOTTED BRISTLETOOTH   *Ctenochaetus marginatus*
SIZE: to 22 cm (8 3/4 in.)   Surgeonfishes – Acanthuridae
ID: Head pale brown; body bluish with numerous small white spots, **white margin on base of anal fin,** fins finely striped, except pectorals, which have spots. Solitary or form groups. Seaward reef surge zones in 2-10 m, rarely to 40 m.
Central Pacific: Micronesia to Line Is., French Polynesia and Cocos I. off Central America.

ORANGETIP BRISTLETOOTH   *Ctenochaetus tominiensis*
SIZE: to 18 cm (7 in.)   Surgeonfishes – Acanthuridae
ID: Brown with paler lower head; outer rear portion of dorsal and anal fins yellow orange, white tail. Solitary or small groups. Coral-rich areas of lagoons and outer reef slopes in 5-40 m.
West Pacific: Indonesia, Philippines, Micronesia, Papua New Guinea, Solomon Is., to Great Barrier Reef, Vanuatu and Fiji.

TWOSPOT BRISTLETOOTH   *Ctenochaetus binotatus*
SIZE: to 20 cm (8 in.)   Surgeonfishes – Acanthuridae
ID: Orangish brown with pale blue spots on head and lines on body; black spot at rear base of dorsal and anal fins. **Juvenile –** Dark body and yellow tail. Usually solitary. Rubble areas of lagoons and seaward reefs in 12-53 m.
Indo-Pacific: E. Africa to Indonesia, Philippines, Micronesia, Solomon Is. and French Polynesia. – S.W. Japan to Great Barrier Reef.

YELLOWEYE BRISTLETOOTH   *Ctenochaetus truncatus*
SIZE: to 19 cm (7 1/2 in.)   Surgeonfishes – Acanthuridae
ID: Shades of brown with small pale bluish or yellowish spots on head and body; **bright yellow ring around eye. Juvenile –** Bright yellow with blue edging on dorsal fins. Inhabit outer reef slopes and occasionally lagoons to 21 m.
Indian Ocean: E. Africa to Andaman Sea, Java and Bali in Indonesia.

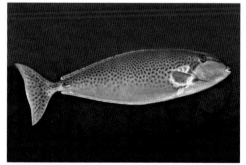

GRAY UNICORNFISH   *Naso caesius*
SIZE: to 45 cm (18 in.)   Surgeonfishes – Acanthuridae
ID: Gray to brownish gray, can rapidly change to pattern of indistinct round to vertically elliptical blotches on upper half of body; no horn and no dark margin on gill cover. Solitary or form aggregations. Outer reefs in 6-60 m.
Pacific: West Papua in Indonesia, Micronesia, Papua New Guinea, Solomon Is., New Caledonia to Hawaii and French Polynesia.

SLENDER UNICORNFISH   *Naso lopezi*
SIZE: to 50 cm (20 in.)   Surgeonfishes – Acanthuridae
ID: Slender body; bluish gray with numerous small round dark gray spots, occasionally displays a large whitish patch around pectoral fin. Form groups; feed on zooplankton along steep outer reef dropoffs in 20-50 m.
Asian Pacific: Indonesia, Philippines, Micronesia, Papua New Guinea, Solomon Is. – S.W. Japan to Great Barrier Reef, Vanuatu.

Surgeonfishes

BLUESPINE UNICORNFISH *Naso unicornis*
SIZE: to 70 cm (28 in.) Surgeonfishes – Acanthuridae
ID: Gray to olive, **tail spines blue;** relatively short forehead horn (does not project past mouth). Solitary or form groups; feed on leafy algae. Lagoons and outer reefs to 80 m.
Indo-Pacific: Red Sea and E. Africa to Indonesia, Philippines, Micronesia, Papua New Guinea, Solomon Is., Hawaii and French Polynesia. – S.W. Japan to Australia.

HUMPBACK UNICORNFISH *Naso brachycentron*
SIZE: to 90 cm (35 in.) Surgeonfishes – Acanthuridae
ID: Gray to brown upper half of body; **unusual profile gives "hump-backed" appearance,** adult males develop long horn on forehead, females have only a slight bump. Often form small groups. Seaward reef slopes in 8-30 m.
Indo-Pacific: E. Africa to Indonesia, Philippines, Micronesia, French Polynesia. – S.W. Japan to Great Barrier Reef and Vanuatu.

PALETAIL UNICORNFISH *Naso brevirostris*
SIZE: to 60 cm (24 in.) Surgeonfishes – Acanthuridae
ID: Varying shades of brown with vertical rows of dark spots or lines (can be faint); **white tail,** white outer edge on gill cover; a broad based tapering horn on forehead. Often in small groups. Open water off lagoons and seaward reefs in 4-46 m.
Indo-Pacific: Red Sea and E. Africa to Indonesia, Philippines, Micronesia and French Polynesia. –S.W. Japan to Great Barrier Reef.

WHITEMARGIN UNICORNFISH *Naso annulatus*
SIZE: to 100 cm (39 in.) Surgeonfishes – Acanthuridae
ID: Olive to brown, but capable of rapid change to pale blue-gray; white lips, tail darkish with white fin rays, adults have long tapering horn on forehead. Form small schools. Steep outer reef slopes in 20-60 m.
Indo-Pacific: E. Africa to Indonesia, Micronesia, Solomon Is., Hawaii and Polynesia. – S.W. Japan to Great Barrier Reef.

HUMPNOSE UNICORNFISH *Naso tonganus*
SIZE: to 60 cm (24 in.) Surgeonfishes – Acanthuridae
ID: Gray gradating to pale yellow on lower side; large bulbous snout, and hump on back. Often form small groups. Seaward reefs in 3-20 m.
Indo-West Pacific: E. Africa to Indonesia, Micronesia, Papua New Guinea and Solomon Is. – S.W. Japan to N. Great Barrier Reef, New Caledonia and Fiji.

SLEEK UNICORNFISH *Naso hexacanthus*
SIZE: to 75 cm (30 in.) Surgeonfishes – Acanthuridae
ID: Brown to bluish gray shading to yellowish lower side (can quickly change to pale blue); black band marking and black margin on gill cover; feed in open water. Dropoffs of seaward reefs in 15-135 m.
Indo-Pacific: Red Sea and E. Africa to Indonesia, Micronesia, Solomon Is., Hawaii and Line Is. – S.W. Japan to Great Barrier Reef.

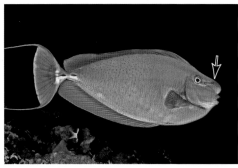

BIGNOSE UNICORNFISH *Naso vlamingii*

SIZE: to 55 cm (22 in.)   Surgeonfishes – Acanthuridae

ID: Head shades of brown, body brown to blue or gray; **blue lips and a blue band in front of eye,** blue vertical lines and spots on sides; can rapidly pale or darken colors and markings. Groups feed in open water along outer reef slopes in 4-50 m.

Indo-Pacific: E. Africa to Indonesia, Micronesia, Solomon Is., Polynesia. – S.W. Japan to S. Great Barrier Reef and New Caledonia.

Bignose Unicornfish – Phase

ID: This species can rapidly pale or darken and change colors from brown to blue or gray in a wide range of phases; blue band between eyes and blue lips displayed in all phases.

BLUETAIL UNICORNFISH *Naso caeruleacauda*

SIZE: to 30 cm (12 in.)   Surgeonfishes – Acanthuridae

ID: Bluish to brownish gray, occasionally yellowish lower body; bluish tail, black margin on gill cover, no horn or bump on forehead. Usually form aggregations. Outer reef slopes in 15-45 m.

Asian Pacific: Indonesia, Philippines, Papua New Guinea, Solomon Is. and Great Barrier Reef.

BLACKSPINE UNICORNFISH *Naso minor*

SIZE: to 23 cm (9 in.)   Surgeonfishes – Acanthuridae

ID: Gray body pales on lower sides; blackish lips, tail spine and its basal plate black, pectoral and tail fins yellowish. Form small to large schools in open water. Lagoons and outer reefs in 12-40 m.

Indo-Asian Pacific: E. Africa to Indonesia, Philippines, Papua New Guinea, Solomon Is. and Great Barrier Reef.

ORANGESPINE UNICORNFISH *Naso lituratus*

SIZE: to 46 cm (18 in.)   Surgeonfishes – Acanthuridae

ID: Brownish gray; yellowish nape, orange tail spines and anal fin, **gray tail with pale yellow margin,** yellow-edged black area from mouth to eye, broad black band on dorsal fin. Solitary or form small groups. Lagoon and outer reefs to 70 m.

Indo-Pacific: Red Sea and E. Africa to Indonesia, Micronesia, Hawaii, Pitcairn Is. – S.W. Japan to Australia and New Caledonia.

ELEGANT UNICORNFISH *Naso elegans*

SIZE: to 45 cm (18 in.)   Surgeonfishes – Acanthuridae

ID: Shades of gray-brown; yellow dorsal fin, **white tail with black submarginal bar,** orange spots around spines on tail base, black band with yellow margins on snout. Solitary or from small groups. Coral reefs and rocky bottoms in 5-30 m.

Indian Ocean: E. Africa to Andaman Sea and Java and Bali in Indonesia.

Surgeonfishes – Moorish Idol – Rabbitfishes

BARRED UNICORNFISH *Naso thynnoides*
SIZE: to 30 cm (12 in.)   Surgeonfishes – Acanthuridae
ID: Pale gray; wide yellowish stripe from eye to tail, numerous dark bluish gray bars on sides, irregular spots on head. Form groups. Lagoons and outer reefs in 2-30 m.
Indo-West Pacific: E. Africa to Indonesia, Philippines, Micronesia, Papua New Guinea and Solomon Is. – S.W. Japan to Great Barrier Reef and Fiji.

MOORISH IDOL *Zanclus cornutus*
SIZE: to 22 cm (8 3/4 in.)   Moorish Idol – Zanclidae
ID: Three wide black and two pale yellow bars; yellow saddle marking on long protruding snout, long trailing dorsal fin filament. Solitary, in pairs or groups; feed primarily on sponges. Lagoons and outer reefs to 180 m.
Indo-Pacific: E. Africa to Indonesia, Philippines, Micronesia, Solomon Is., Hawaii, Easter I. – S.W. Japan to Great Barrier Reef.

VERMICULATE RABBITFISH *Siganus vermiculatus*
SIZE: to 37 cm (15 in.)   Rabbitfishes – Siganidae
ID: Bluish white undercolor with maze of wavy yellowish brown bands; **small dark spots on tail** with straight margin. Usually form groups. Young occur in shallow brackish areas; adults inhabit sheltered reefs to 15 m.
East Indo-West Pacific: India to Indonesia, Philippines, Micronesia, Papua New Guinea, Solomon Is. and Fiji.

RANDALL'S RABBITFISH *Siganus randalli*
SIZE: to 28 cm (11 in.)   Rabbitfishes – Siganidae
ID: Reddish brown to brown with yellowish tint on head; bluish spots cover body shading to maze-like markings on lower rear body and tail. Form schools. Rubble and sandy areas of coral lagoons and bays to 15 m.
Asian Pacific: Micronesia, Papua New Guinea to Solomon Is.

LINED RABBITFISH *Siganus lineatus*
SIZE: to 43 cm (17 in.)   Rabbitfishes – Siganidae
ID: Pale gray with **wavy orangish lines,** occasionally breaking into spots; **large yellow spot below rear base of dorsal fin.** Form large schools. Lagoons, coastal reefs and mangroves to 25 m.
Asian Pacific: Indonesia, Philippines, Micronesia, Papua New Guinea and Solomon Is. – S. Japan to N.W. Australia, S. Great Barrier Reef, New Caledonia and Vanuatu.

GOLDEN RABBITFISH *Siganus guttatus*
SIZE: to 35 cm (14 in.)   Rabbitfishes – Siganidae
ID: Pale gray with **numerous orangish gold spots; large yellow spot below rear base of dorsal fin.** Form small to large schools. Lagoons, coastal reefs and mangroves to 25 m.
Asian Pacific: Andaman Sea to West Papua in E. Indonesia, north to S.W. Japan.

CORAL RABBITFISH *Siganus corallinus*

SIZE: to 25 cm (10 in.)   Rabbitfishes – Siganidae

ID: Yellow-orange undercolor with numerous blue spots; darkish eye bar, may display pattern of dark smudges on back. Adults form pairs in coral-rich areas to 18 m.

Indo-Asian Pacific: Seychelles to Andaman Sea, Indonesia, Philippines, Micronesia, Papua New Guinea and Solomon Is. – S.W. Japan to Great Barrier Reef and New Caledonia.

BLACKEYE RABBITFISH *Siganus puelloides*

SIZE: to 31 cm (12 in.)   Rabbitfishes – Siganidae

ID: Pale blue undercolor with close-set yellow spots; **dark marking under chin** and dark area around eye are remnants of a chin to eye band on juveniles. In pairs; feed on algae, tunicates and sponges. Reef flats and along dropoffs to 20 m.

Indian Ocean: Seychelles to Maldives and Andaman Sea.

JAVA RABBITFISH *Siganus javus*

SIZE: to 40 cm (16 in.)   Rabbitfishes – Siganidae

ID: Pale gray with numerous bluish white spots and wavy gray lines, yellowish head and dorsal and anal fins; usually a large black blotch on tail. Solitary or form small groups. Coastal reefs, occasionally in brackish water mangroves to 15 m.

East Indo-Asian Pacific: Persian Gulf to Indonesia, Philippines, Papua New Guinea. – S. Japan to Great Barrier Reef and Vanuatu.

PEARLSPOTTED RABBITFISH *Siganus margaritiferus*

SIZE: to 30 cm (12 in.)   Rabbitfishes – Siganidae

ID: Pale gray with dense pattern of white spots, often yellowish tints on upper head to spinous dorsal fin; black upper edge on gill cover. Form small to large schools. Coastal reefs, estuaries and harbors to 10 m.

Asian Pacific: Andaman Sea to Indonesia, Philippines, Papua New Guinea and Solomon Is. – S. Japan to Australia.

FINE-SPOTTED RABBITFISH *Siganus punctatissimus*

SIZE: to 30 cm (12 in.)   Rabbitfishes – Siganidae

ID: Purplish brown with numerous close-set pale blue to brown spots, **rear dorsal and anal fins black**; deeply forked yellow to yellowish tail with dark edging. In pairs; feed on algae. Lagoons and along reef slopes in 3-30 m.

West Pacific: Indonesia, Philippines, Micronesia, Papua New Guinea, Solomon Is. – S.W. Japan to N. Great Barrier Reef and Fiji.

GOLD-SPOTTED RABBITFISH *Siganus punctatus*

SIZE: to 40 cm (16 in.)   Rabbitfishes – Siganidae

ID: Pale brown to blue with pattern of close-set dark-edged brown to orange spots; may display pale saddle on tail base, **large dark spot behind gill cover and lack dark margin on tail**. In pairs. Coral areas to 40 m.

West Pacific: Indonesia, Philippines, Micronesia, Papua New Guinea, and Solomon Is. – S.W. Japan to Australia, New Caledonia and Fiji.

Rabbitfishes

ONESPOT RABBITFISH   *Siganus unimaculatus*
SIZE: to 24 cm (9 1/2 in.)   Rabbitfishes – Siganidae
ID: Yellow body usually with elongate black blotch, which can be turned on and off; white head with black band from protruding snout to dorsal fin, wide **black curved bar on gill cover extend to yellow body color.** Solitary or form small groups. Coral-rich areas, often shelter in staghorn corals, to 30 m.
Asian Pacific: S.W. Japan and Philippines to Australia.

FOXFACE RABBITFISH   *Siganus vulpinus*
SIZE: to 25 cm (10 in.)   Rabbitfishes – Siganidae
ID: Yellow body; white head with black band from protruding snout to dorsal fin, broad **black breast marking extend to just above pectoral fin.** Solitary or form groups. Coral-rich areas of lagoons and outer reefs, often shelter in staghorn corals, to 30 m.
West Pacific: Indonesia, Philippines, Micronesia, Papua New Guinea. – Taiwan to Great Barrier Reef, New Caledonia and Vanuatu.

MASKED RABBITFISH   *Siganus puellus*
SIZE: to 38 cm (15 in.)   Rabbitfishes – Siganidae
ID: Yellow with wavy broken blue lines arranged vertically on forebody and horizontal on rear; black bar across eye. In pairs; feed on tunicates and sponges. Rich coral areas in 3-12 m.
Asian Pacific: Indonesia, Philippines, Micronesia and Papua New Guinea. – S.W. Japan to S. Great Barrier Reef, New Caledonia and Vanuatu.

MAGNIFICENT RABBITFISH   *Siganus magnificus*
SIZE: to 23 cm (9 in.)   Rabbitfishes – Siganidae
ID: White head, blackish upper body becomes pale below, yellow pectoral, anal and tail fins; black band from protruding snout to dorsal fin. In pairs. Coral-rich areas in 2-20 m.
East Indian Ocean: Andaman Sea at Similan Is., W. Thailand and Myanmar.

VIRGATE RABBITFISH   *Siganus virgatus*
SIZE: to 30 cm (12 in.)   Rabbitfishes – Siganidae
ID: Yellow upper body with pale blue spots, white lower body; tail yellow, pair of dark bands on head and forebody. Form small to large groups. Usually inshore coastal reefs, often in turbid water to 12 m.
East Indo-Asian Pacific: India to West Papua in E. Indonesia. – S.W. Japan to N. Australia.

BARRED RABBITFISH   *Siganus doliatus*
SIZE: to 25 cm (10 in.)   Rabbitfishes – Siganidae
ID: Light blue to whitish with intricate pattern of thin yellow and blue lines; pair of dark bands on head and forebody. May hybridize with Virgate Rabbitfish [previous]. In pairs or small groups. Inshore and outer reefs in 2-15 m.
West Pacific: Indonesia, Micronesia, Papua New Guinea, Solomon Is. to N. Australia, New Caledonia and Fiji.

BICOLOR RABBITFISH   *Siganus uspi*
SIZE: to 22 cm (8 ³/₄ in.)   Rabbitfishes – Siganidae
ID: Chocolate-brown to rear dorsal and anal fin then abruptly
yellow to tail; yellow pectoral fins, protruding snout. In pairs.
Coral-rich areas of lagoons and seaward reefs to 8 m.
Localized: Known only from New Caledonia and Fiji.

HONEYCOMB RABBITFISH   *Siganus stellatus*
SIZE: to 35 cm (14 in.)   Rabbitfishes – Siganidae
ID: Whitish undercolor with numerous close-set black polygonal
spots forming honeycomb network; **white edging on rear
dorsal, anal and tail fins**. In pairs. Lagoon and seaward reefs
to 30 m.
Indian Ocean: E. Africa to Andaman Sea and Bali in Indonesia.

WHITE-SPOTTED RABBITFISH   *Siganus canaliculatus*
SIZE: to 29 cm (11 in.)   Rabbitfishes – Siganidae
ID: Greenish to yellow-brown with numerous bluish white
spots (change to mottled pattern when resting on bottom);
often dark spot behind upper gill opening. Form aggregations.
Most common on seagrass flats to 4 m.
East Indo-Asian Pacific: Persian Gulf to Andaman Sea, Indonesia,
Philippines, Papua New Guinea. – S. W. Japan to Great Barrier Reef.

AFRICAN WHITESPOTTED RABBITFISH   *Siganus sutor*
SIZE: to 45 cm (18 in.)   Rabbitfishes – Siganidae
ID: Yellowish olive to bronze with bluish white spots. Form
schools. Weedy sand flats, rock and coral reefs to 12 m.
Previously only reported from W. Indian Ocean. Asian Pacific
population currently under study.
Indo-Asian Pacific: E. Africa to E. Andaman Sea and Bali in
Indonesia.

SCRIBBLED RABBITFISH   *Siganus spinus*
SIZE: to 24 cm (9 ¹/₂ in.)   Rabbitfishes – Siganidae
ID: Whitish undercolor with labyrinth brown broken bands;
fins pale with dark mottling. Smallest member of family. Form
small to large schools; graze on algae. Coastal reef flats and
outer reefs to 6 m.
Indo-Pacific: India to Society Is. in French Polynesia. – S.W.
Japan to S. Great Barrier Reef.

FORKTAIL RABBITFISH   *Siganus argenteus*
SIZE: to 30 cm (12 in.)   Rabbitfishes – Siganidae
ID: Blue to bluish gray with numerous small yellow spots and
lines (change to mottled pattern when resting on bottom);
deeply forked tail. Usually form groups. Juveniles inshore,
adults prefer outer reef slopes to 40 m.
Indo-Pacific: Red Sea and E. Africa, Indonesia, Micronesia to French
Polynesia. – S.W. Japan to Great Barrier Reef and New Caledonia.

IDENTIFICATION GROUP 3
Small Ovals – Damselfishes
and Damselfishes/Anemonefishes
This ID Group consists of small fishes with oval, perch-like profiles.

FAMILY: Damselfishes – Pomacentridae
19 Genera – 177 Species Included

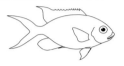

Typical Shape   Genus *Chromis*

These energetic little fishes are an evident part of every coral reef community. Distinctive visual family characteristics include a single nostril on each side of the snout, rather than the usual two, a continuous dorsal fin, an interrupted lateral line, and a forked or lunate tail.

Approximately three-quarters of the 321 known species are found in the Indo-West Pacific. Damselfishes display remarkable diversity of habitat preference, feeding habits, and behavior. Coloration is quite variable, ranging from drab hues of brown, gray, and black to brilliant combinations of orange, yellow, and neon-blue. A number of species have juvenile stages characterized by yellow bodies with bright blue stripes crossing their upper heads and backs.

The species in genera *Plectroglyphidodon*, *Hemiglyphidodon* and *Stegastes* are energetic defenders of tiny filamentous algae plots, which they farm as their primary food source. Trespassers, including divers, are aggressively chased and if necessary given pesky nips. The family's plankton feeders include species of *Chromis*, *Dascyllus*, and *Lepidozygus*. Those species classified in *Abudefduf*, *Chrysiptera*, *Amblyglyphidodon*, *Neopomacentrus* and *Pomachromis* feed on a combination of current-borne plankton, filamentous algae and benthic invertebrates.

All damselfishes are egg layers. Either the male or both partners clear a nest site on the bottom and engage in courtship displays of rapid swimming and fin extension. Males generally turn dark or display broad white blotches during nuptial periods. Courtship and spawning usually occurs around daybreak.

SUBFAMILY: Anemonefishes – Pomacentridae/Amphiprioninae
2 Genera – 17 Species Included

Typical Shape

A subfamily of damselfishes (Amphiprioninae), commonly known as anemonefishes or clownfishes, have developed the ability to live among the stinging tentacles of large sea anemones. About one-third of the anemonefishes inhabit a specific host anemone. However, Clark's Anemonefish, the most wide-ranging member of the subfamily, inhabit all ten known host anemone species.

Anemonefishes live in small social groups with a single large dominant female, a smaller sexually-active male and from two to four even smaller males and juveniles. With the loss of the female, the largest male will change sex and become the harem's new matriarch.

Only three species exhibit a difference in color patterns between the sexes; and juveniles in all but three species resemble adults. Only the Clark's Anemonefish displays a marked difference in color patterns between geographic regions. The diet of most anemonefishes consists of current-borne zooplankton, primarily copepods and larval tunicates, and occasional bits of algae.

BENGAL SERGEANT *Abudefduf bengalensis*

SIZE: to 17 cm (6 ³/₄ in.)   Damselfishes – Pomacentridae

ID: Whitish to pale gray with 6-7 black bars; only member of genus *Abudefduf* to have **rounded tail lobes with blue edging.** Coastal and lagoon reefs to 6 m.

East Indo-Asian Pacific: India and Andaman Sea to Indonesia, Philippines and Papua New Guinea. – S.W. Japan to Great Barrier Reef. Absent Solomon Is.

BLACK-TAIL SERGEANT *Abudefduf lorenzi*

SIZE: to 17 cm (6 ³/₄ in.)   Damselfishes – Pomacentridae

ID: Whitish to yellowish with 5-6 black bars; **large black blotch on base of tail.** Protected coastal areas near shore, frequently next to docks or breakwaters in less than 6 m.

Asian Pacific: Sulawesi and Molucca Is. in Indonesia to Philippines, Papua New Guinea and Solomon Is.

SCISSORTAIL SERGEANT *Abudefduf sexfasciatus*

SIZE: to 19 cm (7 ¹/₂ in.)   Damselfishes – Pomacentridae

ID: Whitish with 5 black bars; **broad dark streak runs length of each tail lobe.** Usually form groups; feed on plankton in open water. Coastal and offshore reefs to 15 m.

Indo-West Pacific: Red Sea and E. Africa to Andaman Sea, Indonesia, Philippines, Micronesia, Papua New Guinea and Solomon Is. – S.W. Japan to Australia and Fiji.

YELLOWTAIL SERGEANT *Abudefduf notatus*

SIZE: to 17 cm (6 ³/₄ in.)   Damselfishes – Pomacentridae

ID: Gray with 5 narrow white bars (including band on tail base); **yellow tail.** Usually form schools, but occasionally solitary. Rocky inshore reefs with moderate to strong wave action to 12 m.

Indo-Asian Pacific: E. Africa to Andaman Sea, Indonesia, Philippines, Micronesia, Papua New Guinea and Solomon Is. – S.W. Japan to Australia.

BANDED SERGEANT *Abudefduf septemfasciatus*

SIZE: to 20 cm (8 in.)   Damselfishes – Pomacentridae

ID: Tan with 6 wide dark brown bars. Form small groups. Rocky inshore reefs with mild to moderate surge conditions to 3 m.

Indo-Pacific: E. Africa to Andaman Sea, Indonesia, Philippines, Micronesia, Papua New Guinea, Solomon Is. and French Polynesia. – S.W. Japan to Great Barrier Reef.

BLACKSPOT SERGEANT *Abudefduf sordidus*

SIZE: to 24 cm (9 ¹/₂ in.)   Damselfishes – Pomacentridae

ID: Tan with 6 wide dark brown bars; **small black saddle marking on upper base of tail.** Usually form small groups. Rocky shoreline reefs with mild to moderate surge to 3 m.

Indo-Pacific: Red Sea and E. Africa to Andaman Sea, Indonesia, Philippines, Micronesia, Papua New Guinea, Solomon Is., Hawaii and Pitcairn Is. – S.W. Japan to E. Australia.

Damselfishes

INDO-PACIFIC SERGEANT *Abudefduf vaigiensis*
SIZE: to 18 cm (7 in.) Damselfishes – Pomacentridae
ID: Gray with 5 black to purple or blue bars (including bar at tail base); frequently with yellow back. Usually form feeding groups in open water or guard nests inside rocky crevices. Shoreline reefs and outer slopes to 12 m.
Indo-Pacific: Red Sea and E. Africa to Indonesia, Micronesia, Solomon Is. to French Polynesia. – S.W. Japan to Great Barrier Reef.

GREEN SERGEANT *Abudefduf whitleyi*
SIZE: to 15 cm (6 in.) Damselfishes – Pomacentridae
ID: Pale green to yellowish green or blue-green with 4-5 narrow black bars; tail and outer edges of dorsal and anal fins blackish. Solitary or form groups. Outer edge of reefs and surge gutters to 5 m.
Southwestern Pacific: Great Barrier Reef, Coral Sea and New Caledonia.

STAGHORN DAMSEL *Amblyglyphidodon curacao*
SIZE: to 11 cm (4 1/4 in.) Damselfishes – Pomacentridae
ID: Pale greenish to whitish with 3 wide, dark green bars; midbody often yellow. Usually form groups. Commonly shelter among branches of staghorn *Acropora* corals on coastal reefs, lagoons and outer slopes to 15 m.
East Indo-West Pacific: Cocos-Keeling Is. to Indonesia and Micronesia. – S.W. Japan to Great Barrier Reef and Fiji.

GOLDEN DAMSEL *Amblyglyphidodon aureus*
SIZE: to 12 cm (4 3/4 in.) Damselfishes – Pomacentridae
ID: Bright yellow to gold including fins; **blue markings around eye,** long pointed rear dorsal and anal fins. Solitary or in pairs. Steep outer reefs slope in 12-35 m.
West Pacific: Andaman Sea to Indonesia, Philippines, Micronesia, Papua New Guinea and Solomon Is. – S.W. Japan to Australia, New Caledonia and Fiji.

BATUNA'S DAMSEL *Amblyglyphidodon batunai*
SIZE: to 8.5 cm (3 1/4 in.) Damselfishes – Pomacentridae
ID: Whitish often with greenish yellow sheen and silvery reflections on forehead and back; upper edge of tail base blackish, **white ventral fins.** Solitary or form groups. Sheltered within branches of *Acropora* corals, often on silty reefs, in 2-10 m.
Asian Pacific: Sabah in Malaysia, Indonesia, Philippines, Micronesia, Papua New Guinea and Solomon Is.

TERNATE DAMSEL *Amblyglyphidodon ternatensis*
SIZE: to 12 cm (4 3/4 in.) Damselfishes – Pomacentridae
ID: Whitish often with yellow hue and silvery reflections on forehead and back; upper edge of tail base blackish, **yellow ventral fins.** Usually form groups. Shelter within branching corals of protected coastal reefs to 12 m.
Asian Pacific: Indonesia, Philippines, Papua New Guinea, Micronesia and Solomon Is. to Vanuatu and New Caledonia.

WHITEBELLY DAMSEL   *Amblyglyphidodon leucogaster*
SIZE: to 13 cm (5 in.)   Damselfishes – Pomacentridae
ID: Gray with pale scale centers, yellow ventral fins; **black edges on dorsal, anal and tail fins.** Lagoons and outer reefs in 2-45 m.
East Indo-Asian Pacific: Cocos-Keeling to Indonesia, Philippines, Micronesia and Solomon Is. – S.W. Japan to N.W. Australia, Great Barrier Reef, New Caledonia and Vanuatu.

ORBICULAR DAMSEL   *Amblyglyphidodon orbicularis*
SIZE: to 13 cm (5 in.)   Damselfishes – Pomacentridae
ID: Pale gray-white with **bright yellow ventral and anal fins.** Formerly considered a variation of Whitebelly Damsel [previous]. Generally solitary, occasionally in pairs. Near shore and outer reefs in 4-40 m.
West Pacific: New Caledonia to Fiji.

CENDERAWASIH DAMSEL   *Amblyglyphidodon flavopurpureus*
SIZE: to 12 cm (4 3/4 in.)   Damselfishes – Pomacentridae
ID: Purplish gray body; yellowish brown nape, blue line markings below eye, yellow tail base extends onto body below rear dorsal fin and yellow borders of tail. Solitary, pairs or small groups. Outer reefs slopes with gorgonian sea fans in 10-30 m.
Localized: Cenderawasih Bay, West Papua in Indonesia.

SILVER-STREAKED DAMSEL   *Altrichthys azurelineatus*
SIZE: to 7.5 cm (3 in.)   Damselfishes – Pomacentridae
ID: Whitish with narrow silver streak on each scale join to form narrow lines; **thin black margin on length of dorsal fin and black borders of tail.** Adults in pairs; guard broods of 20-100 babies; young form groups. Sheltered reefs coral in 2-8 m.
Localized: Calamian Group north of Palawan in Philippines.

GUARDIAN DAMSEL   *Altrichthys curatus*
SIZE: to 6 cm (2 1/4 in.)   Damselfishes – Pomacentridae
ID: Whitish to pale green or olive; narrow silver streak on each scale join to form narrow lines. Adults in pairs; guard broods of 20-100 babies; young form groups. Sheltered reefs in 2-10 m.
Localized: Calamian Group and Cuyo Is. north of Palawan in Philippines.

BLACK-BANDED DAMSEL   *Amblypomacentrus breviceps*
SIZE: to 7 cm (2 3/4 in.)   Damselfishes – Pomacentridae
ID: White; dark eye bar, pair of wide dark bars on upper half of body joined to dark margin on dorsal fin, slender body, adults have filaments on tail lobes, young are yellow to brown on lower half. Sand or silt bottoms of coastal reefs and lagoons in 2-35 m.
East Indo-Asian Pacific: Maldives to Indonesia, Philippines and Papua New Guinea to Solomon Is. and Great Barrier Reef.

Damselfishes

SKUNK ANEMONEFISH　　　*Amphiprion akallopisos*
SIZE: to 11 cm (4 1/4 in.)　　Damselfishes – Pomacentridae
ID: Pinkish orange; white mid-dorsal stripe from head (but not lip) to tail. Can be distinguished from similar Pacific Anemonefish [next] by location. Live with Magnificent and Mertens' Anemones in 3-25 m.
Indian Ocean: E. Africa to Seychelles, Comoro Is., Andaman Sea, Sumatra, Java and Bali in Indonesia.

PACIFIC ANEMONEFISH　　　*Amphiprion pacificus*
SIZE: to 10 cm (4 in.)　　Damselfishes – Pomacentridae
ID: Orange with yellow highlights; white stripe from snout to tail. Nearly identical to Skunk Anemonefish [previous], but distinguished by location. Live with Magnificent Anemones.
West Pacific: New Britain off E. Papua New Guinea to Samoa, Fiji and Tonga.

ORANGE ANEMONEFISH　　　*Amphiprion sandaracinos*
SIZE: to 14 cm (5 1/2 in.)　　Damselfishes – Pomacentridae
ID: Orange; wide white mid-dorsal stripe runs from at least midsnout or upper lip to tail. Live most commonly with Merten's Anemone in 3-20 m.
Asian Pacific: Indonesia, Philippines, Micronesia, Papua New Guinea and Solomon Is. north to S.W. Japan.

PINK ANEMONEFISH　　　*Amphiprion perideraion*
SIZE: to 10 cm (4 in.)　　Damselfishes – Pomacentridae
ID: Pink to orange; **narrow white head bar,** and white mid-dorsal stripe runs from between eyes to tail. Live with 4 anemone species (most commonly the Magnificent Anemone) on reefs in 3-20 m.
East Indo-West Pacific: Cocos-Keeling to Indonesia, Micronesia, Solomon Is., French Polynesia. – S.W. Japan to Australia and Fiji.

WHITE-BONNET ANEMONEFISH　　*Amphiprion leucokranos*
SIZE: to 12 cm (4 3/4 in.)　　Damselfishes – Pomacentridae
ID: Orange to light brown; **broad white patch on nape** tapers toward dorsal fin and smaller patch at mid-dorsal fin, single usually discontinuous bar behind eye. Live with Leathery, Magnificent and Merten's Anemones in 2-12 m.
Localized: West Papua and N.E. Halmahera in Indonesia, N. Papua New Guinea and Solomon Is.

White-bonnet Anemonefish – Variation
ID: Occasionally white bar behind eye extends to join white patch on nape.

CLARK'S ANEMONEFISH   *Amphiprion clarkii*
SIZE: to 14 cm (5 1/2 in.)   Damselfishes – Pomacentridae
ID: Black to entirely orange with pair of white or pale bluish bars, **second bar wide**; tail white or yellow, when body dark usually narrow white bar on tail base, other fins variably black to yellow-orange. Live with 10 anemone species to 55 m.
East Indo-West Pacific: Maldives to Indonesia, Micronesia, Solomon Is. – S.W. Japan to N. Australia, New Caledonia and Fiji.

Clark's Anemonefish – Orange Variation
ID: Exhibit variable amounts of yellow-orange. Some are entirely pale, others, such as the form shown, have a dark patch on rear body. When rear body is orange the narrow white bar on tail base is usually absent. Coloration is influenced by the host anemone species.

ORANGEFIN ANEMONEFISH   *Amphiprion chrysopterus*
SIZE: to 15 cm (6 in.)   Damselfishes – Pomacentridae
ID: Dark body with orange to yellow head, upper back, dorsal and pectoral fins; pair of white or pale bluish bars, **second bar narrow**, tail white or yellow, ventral and anal fins yellow-orange except black in Melanesia. Live with 6 anemone species to 20 m.
Pacific: E. Indonesia, Philippines, Micronesia, Papua New Guinea and Solomon Is. to Coral Sea, Fiji and French Polynesia.

Orangefin Anemonefish – Yellowtail Variation
ID: Occasionally in Micronesia, more numerous in Fiji and Tonga. This variation has yellow tail rather than white. Occasionally body bar can be nearly as wide as the bar on head.

TOMATO ANEMONEFISH   *Amphiprion frenatus*
SIZE: to 7 cm (2 3/4 in.)   Damselfishes – Pomacentridae
ID: Male – Orange to red with a single white or pale bluish head bar; male considerably smaller than female. Live with Bulb-tentacle Anemone on reefs to 12 m.
North Asian Pacific: Brunei, Sabah in Malaysia, Sumatra, Java and Kalimantan in Indonesia and Philippines, north to S.W. Japan.

Tomato Anemonefish – Female
SIZE: to 14 cm (5 1/2 in.)
ID: Primarily black on sides with red snout, breast, belly and fins; white or pale blue head bar. Similar to Red and Black Anemonefish [next page] distinguished by location. **Juvenile –** Have 2-3 white bars.

Damselfishes

RED AND BLACK ANEMONEFISH   *Amphiprion melanopus*
SIZE: to 13 cm (5 in.)   Damselfishes – Pomacentridae
ID: Reddish orange with variable amounts of black on sides; a white to pale bluish bar on head. Young are overall reddish orange with 2-3 narrow white bars. Live with 3 anemone species in lagoons and outer reefs to 10 m.
Pacific: E. Indonesia, Philippines, Micronesia to Solomon Is. – S.E. Philippines to Great Barrier Reef, Vanuatu and New Caledonia.

FIJI ANEMONEFISH   *Amphiprion barberi*
SIZE: to 13 cm (5 in.)   Damselfishes – Pomacentridae
ID: Entirely red-orange with white head bar. Commonly found anemones on shallow reefs. Usually in groups of variable-sized individuals.
Localized: Samoa, Fiji and Tonga.

RED SADDLEBACK ANEMONEFISH   *Amphiprion ephippium*
SIZE: to 12 cm (4 3/4 in.)   Damselfishes – Pomacentridae
ID: Red to reddish orange with variable amount of black on rear body (no bars). Live most commonly with Bulb-tentacle Anemone on shoreline reefs in 2-15 m.
East Indian Ocean: Andaman Sea to Sumatra and Java in Indonesia.

Red Saddleback Anemonefish – Juvenile
ID: Broad white bar on head; dark patch on side becomes larger with maturity.

FALSE CLOWN ANEMONEFISH   *Amphiprion ocellaris*
SIZE: to 9.5 cm (3 3/4 in.)   Damselfishes – Pomacentridae
ID: Orange with 3 white bars, middle bar has forward-projecting bulge; variable amounts of black edging on bars and fins. Shoreline reefs to 15 m.
Asian Pacific: Andaman Sea to N.W. Australia, Central Indonesia and Philippines, S.W. Japan. Rare black variation around Darwin, Australia; brown variation elsewhere.

CLOWN ANEMONEFISH   *Amphiprion percula*
SIZE: to 7.5 cm (3 in.)   Damselfishes – Pomacentridae
ID: Orange with 3 white bars, middle bar has forward-projecting bulge; variable amounts of black edging on bars and fins. Similar False Clown [previous] distinguished by location. Live with 3 anemone species to 15 m.
West Pacific: West Papua in Indonesia, N. Papua New Guinea, Solomon Is., Great Barrier Reef and Vanuatu.

BARRIER REEF ANEMONEFISH *Amphiprion akindynos*
SIZE: to 9 cm (3 1/2 in.) Damselfishes – Pomacentridae
ID: Light to dark brown; pair of white or pale bluish bars, head bar often constricted or discontinuous across top of head. Live with at least 6 anemone species in 3-25 m.
Southwestern Pacific: Great Barrier Reef, Coral Sea, New Caledonia, Loyalty Is., Fiji and Tonga.

SEBAE ANEMONEFISH *Amphiprion sebae*
SIZE: to 16 cm (6 1/4 in.) Damselfishes – Pomacentridae
ID: Varying amounts of black to dark brown and yellow-orange; white head bar and broad somewhat forward slanting midbody bar, all or partial amount of yellow on tail. Usually with Haddon's Anemone on sand bottoms in 2-25 m.
North Indian Ocean: Arabian Peninsula, India, Sri Lanka, Maldives, Andaman Sea to Sumatra and Java in Indonesia.

SADDLEBACK ANEMONEFISH *Amphiprion polymnus*
SIZE: to 12 cm (4 3/4 in.) Damselfishes – Pomacentridae
ID: Varying amounts of black to dark brown and yellow-orange; white head bar and broad somewhat forward slanting midbody bar, white edging on black tail. Usually with Haddon's Anemone on sandy bottoms in 2-35 m.
Asian Pacific: Indonesia, Philippines, Papua New Guinea to Solomon Is. – S.W. Japan to N. Australia.

Saddleback Anemonefish – Variation
ID: Occasionally have orange head that may extend onto forebody and uncommonly to the tail and the white midbody bar extends only onto upper back. When guarding eggs quite aggressive toward divers.

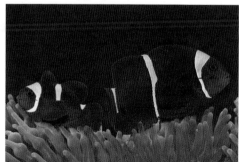

SPINECHEEK ANEMONEFISH *Premnas biaculeatus*
SIZE: to 8 cm (3 1/4 in.) Damselfishes – Pomacentridae
ID: Male (left) & Female (right) – Brilliant red to brownish red with 3 white to gray bars; long spines on cheek. Female several times larger than male. Live in Bulb-tentacle Anemone to 18 m.
Asian Pacific: Andaman Sea, Brunei, Indonesia, Philippines, Papua New Guinea, Solomon Is. to N. Great Barrier Reef and Vanuatu.

Spinecheek Anemonefish – E. Indian Ocean Variation – Female
SIZE: to 16 cm (6 1/4 in.)
ID: In the Indian Ocean and the Andaman Sea the 3 pale bars of the female are yellowish, in the balance of the range the bars are white to pale gray.

Damselfishes

CROSS'S DAMSEL　　　　　*Neoglyphidodon crossi*
SIZE: to 12.5 cm (5 in.)　　　Damselfishes – Pomacentridae
ID: Dark chocolate brown; golden iris, **orange pectoral fin base.** Solitary or loose groups. Rocky shorelines and sheltered coral reefs of bays and lagoons in 2-12 m.
Localized: Bali, Komodo, Flores, Sulawesi, Molucca Is. and W. Papua in Indonesia.

Cross's Damsel – Juvenile
SIZE: 3 - 5 cm (1 1/4 - 2 in.)
ID: Bright red-orange head, upper body and fins, pale mauve lower body; neon-blue stripe extending from snout to rear base of dorsal fin. At a length of about 4 - 6 cm body gradually darkens to adult coloration.

YELLOWTAIL DAMSEL　　　　*Neoglyphidodon nigroris*
SIZE: to 11.5 cm (4 1/2 in.)　　Damselfishes – Pomacentridae
ID: Entirely brown or brown shading to yellow on rear body, tail and adjacent fins; **dark bar on gill cover,** black spot on upper pectoral fin base. Solitary or form loose groups. In passes and along outer reef slopes in 2 - 23 m.
Asian Pacific: Andaman Sea, Indonesia, Philippines, Micronesia to Solomon Is. – S.W. Japan to Great Barrier Reef and Vanuatu.

Yellowtail Damsel – Juvenile
SIZE: 3 - 4 cm (1 1/4 - 1 1/2 in.)
ID: Yellow with pair of black stripes extending from snout to rear dorsal fin base and eye to tail; black spot on upper pectoral fin base.

CARLSON'S DAMSEL　　　　*Neoglyphidodon carlsoni*
SIZE: to 12 cm (4 3/4 in.)　　　Damselfishes – Pomacentridae
ID: Head and body charcoal gray to blue with black scale outlines. **Juvenile –** Similar, but with thin blue streak along back. Solitary. Near entrances of caves and ledges. Fringing reefs and lee side of platform reefs in 1-5 m.
Localized: Known only from Fiji.

OCELLATED DAMSEL　　　　*Neoglyphidodon bonang*
SIZE: to 13.5 cm (5 1/4 in.)　　Damselfishes – Pomacentridae
ID: Dark brown; blue ocellated spot on mid-dorsal fin base and second below last dorsal rays, large adults sometimes lose spots. Among rocks and corals of shoreline reefs, often in turbid conditions to 20 m.
East Indo - Asian Pacific: Sri Lanka and Andaman Sea to Sumatra, Java, Bali and Komodo I. in Indonesia.

BLACK DAMSEL *Neoglyphidodon melas*
SIZE: to 15 cm (6 in.)   Damselfishes – Pomacentridae
ID: Deep blue to black; no distinctive markings. Similar Javanese Damsel [next] has paler coloration. Solitary. Shoreline reefs, lagoons and outer slopes, usually in areas with abundant soft corals to 12 m.
Indo-Asian Pacific: Red Sea and E. Africa to Indonesia, Micronesia and Solomon Is. – S.W. Japan to N. Australia and Vanuatu.

Black Damsel – Juvenile
SIZE: 3-4 cm (1 1/4 - 1 1/2 in.)
ID: Pale blue to white with yellow cap from snout to rear dorsal fin; blue ventral and anal fins with black frontal margin, thin yellow borders on tail. Once thought to be separate species until intermediates were discovered.

JAVANESE DAMSEL *Neoglyphidodon oxyodon*
SIZE: to 14.5 cm (5 3/4 in.)   Damselfishes – Pomacentridae
ID: Gray with black scale outlines and dark fins. Often yellowish patch on upper back. Similar Black Damsel [previous] lacks scale outlines. Solitary or form loose groups. Sheltered shoreline reefs, lagoons and reef flats to 4 m.
Asian Pacific: Indonesia, Philippines and Ashmore Reef (Timor Sea).

Javanese Damsel – Juvenile
SIZE: 3-5 cm (1 1/4 - 2 in.)
ID: Black body; neon-blue lines on head, middle of dorsal fin and upper tail base; white to yellow bar near midbody.

MULTISPINE DAMSEL *Neoglyphidodon polyacanthus*
SIZE: to 14 cm (5 1/2 in.)   Damselfishes – Pomacentridae
ID: Gray-brown with yellowish pectoral fins; subadult [pictured] blue with dark scale outlines, yellow belly. Solitary or form groups. Coral and rocky reefs in 2-30 m.
Southwestern Pacific: S. Great Barrier Reef, New Caledonia, Lord Howe and Norfolk Is.

Multispine Damsel – Juvenile
SIZE: 3-4 cm (1 1/4 - 1 1/2 in.)
ID: Bright yellow; neon-blue line connecting snout with blue ocellated spot on middle of dorsal fin, a second smaller blue ocellated spot at base of last dorsal rays. Solitary or form groups. Coral and rocky reefs in 2-30 m.

Damselfishes

EASTERN BARHEAD DAMSEL *Neoglyphidodon mitratus*

SIZE: to 10 cm (4 in.)　　　Damselfishes – Pomacentridae

ID: Blue-gray; **two brown bars on head with white between,** dark spot on upper pectoral fin base, yellowish pectoral, anal and front of ventral fins. Solitary or form loose groups. Sheltered reef slopes and deeper lagoons in 10 - 45 m.

Asian Pacific: Micronesia, Papua New Guinea and Solomon Is.

Eastern Barhead Damsel – Juvenile

SIZE: 3-4 cm (1 $^1/_4$ - 1 $^1/_2$ in.)

ID: Pale mauve with yellow ventral and anal fins; three orange bars on head, blue ocellated spot on dorsal fin base, dark spot on upper pectoral fin base. Generally remain closer to bottom than adults. Sheltered reef slopes and deeper lagoons in 15 - 45 m.

WESTERN BARHEAD DAMSEL *Neoglyphidodon thoracotaeniatus*

SIZE: to 13.5 cm (5 $^1/_4$ in.)　　　Damselfishes – Pomacentridae

ID: Dark gray gradating to nearly black rear body; **three brown bars on head with white between,** black spot on pectoral fin base; white ventral and gray anal fin. Solitary or loose groups. Sheltered reef slopes and deeper lagoons in 15 - 45 m.

Asian Pacific: Sabah in Malaysia, Indonesia and Philippines.

Western Barhead Damsel – Juvenile

SIZE: 3-4 cm (1 $^1/_4$ - 1 $^1/_2$ in.)

ID: Whitish forebody, dark rear body; yellow ventral fins, three yellowish brown bars on head, blue ocellated spot on dorsal fin base. Generally remain closer to bottom than adults. Sheltered reef slopes and deeper lagoons in 15 - 45 m.

JOHNSTON DAMSEL *Plectroglyphidodon johnstonianus*

SIZE: to 9 cm (3 $^1/_2$ in.)　　　Damselfishes – Pomacentridae

ID: Yellowish tan head often with bluish hue; broad black bar across rear body. Solitary or form loose groups. Often associated with *Acropora* and *Pocillopora* branching corals in passages and outer reef slopes in 2 - 12 m.

Indo - Pacific: E. Africa to Bali and Flores in Indonesia, Hawaii and Pitcairn Is. – S.W. Japan to E. & W. Australia.

Johnston Damsel – Pale Variation

ID: Pale yellowish tan body and bluish head; lacks characteristic dark bar across rear body. Apparently a color variant that is relatively rare and found among normal populations.

WHITEBAND DAMSEL *Plectroglyphidodon leucozonus*
SIZE: to 11.5 cm (4 1/2 in.)   Damselfishes – Pomacentridae
ID: Brown with white bar across midbody; rear edge of gill cover often dark, black spot on upper pectoral fin base. Solitary or form loose groups. Rocky shores and reef flats exposed to wave action to 4 m.
Indo-Pacific: Red Sea and E. Africa to Indonesia, Philippines, Micronesia, Solomon Is. and Pitcairn Is. – S.W. Japan to Australia.

BLACKBAR DAMSEL *Plectroglyphidodon dickii*
SIZE: to 11.5 cm (4 1/2 in.)   Damselfishes – Pomacentridae
ID: Tan with fine dark scale margins, white rear body and tail, pectoral fins have yellow hue; a narrow black bar across rear body. Solitary or form loose groups. Coral-rich areas of lagoons and outer reefs to 12 m.
Indo-Pacific: E. Africa to Indonesia, Micronesia, Papua New Guinea, Solomon Is., French Polynesia. – S.W. Japan to Australia.

BRIGHTEYE DAMSEL *Plectroglyphidodon imparipennis*
SIZE: to 6 cm (2 1/4 in.)   Damselfishes – Pomacentridae
ID: Light blue-gray (nearly white), tail base and tail often yellow, nape may be dusky; black bar across middle of eye, iris silvery. Wave-swept shallows to 3 m.
Indo-Pacific: E. Africa to Hawaii and French Polynesia. – S.W. Japan to E. Australia. Mainly oceanic islands; largely absent from the Asian Pacific.

PHOENIX DAMSEL *Plectroglyphidodon phoenixensis*
SIZE: to 8.5 cm (3 1/4 in.)   Damselfishes – Pomacentridae
ID: Brown with translucent to white tail; four narrow white to yellowish bars and black bar on tail base, may display pale ocellated spot on rear dorsal fin. Solitary or form loose groups. Rocky shores and reef flats exposed to vigorous wave action to 8 m.
Indo-Pacific: E. Africa to Komodo in Indonesia, S.W. Japan to French Polynesia; largely absent from the Asian Pacific.

JEWEL DAMSEL *Plectroglyphidodon lacrymatus*
SIZE: to 11 cm (4 1/4 in.)   Damselfishes – Pomacentridae
ID: Brown with black scale margins, often shading to tan or whitish on rear body and tail; **small blue spots scattered on head and body.** Solitary or form loose groups. Lagoons and outer reefs in 2-12 m.
Indo-Pacific: E. Africa to Indonesia, Philippines, Micronesia, Solomon Is. and French Polynesia. – S.W. Japan to E. Australia.

Jewel Damsel – Juvenile
SIZE: 3-4 cm (1 1/4 - 1 1/2 in.)
ID: Yellowish green with numerous blue spots, which are more pronounced on juveniles.

Damselfishes

INDIAN SURGE DAMSEL   *Chrysiptera brownriggii*
SIZE: to 8 cm (3 1/4 in.)   Damselfishes – Pomacentridae
ID: Light Variation – Two distinct variations: Yellowish with broad blue stripe along top of head and body, and vertical black mark on posterior dorsal fin. Usually found on either side of outer reef crest where wave action is moderate in less than 2 m.
Indian Ocean: E. Africa to Andaman Sea and Sumatra, Java and Bali in Indonesia.

Indian Surge Damsel – Dark Variation
ID: Dark brown with 3 white bars below front and middle parts of dorsal fin and front of tail base; yellow patch on gill cover. This variety usually found on reef crest where wave action is more severe in less than 1 m.

PACIFIC SURGE DAMSEL   *Chrysiptera leucopoma*
SIZE: to 8 cm (3 1/4 in.)   Damselfishes – Pomacentridae
ID: Light Variation – Yellow with prominent blue stripe along top of head and body; oval black spot at end of blue stripe, and **black spot on upper tail base.** Usually found on either side of outer reef crest where wave action is less severe in less than 2 m.
Central Pacific: N. Australia to S. Japan and eastward to Line Is. and French Polynesia.

Pacific Surge Damsel – Dark Variation
ID: Dark brown with complete white bars below front part of dorsal fin and front of tail base; **yellow patch on gill cover and lower forebody.** Usually found on reef crest where heavy wave action in less than 1 m.

BLUELINE DAMSEL   *Chrysiptera caeruleolineata*
SIZE: to 5.5 cm (2 1/4 in.)   Damselfishes – Pomacentridae
ID: Pale yellow to orange often with purplish tints; bright neon-blue stripe from snout to rear dorsal fin, scattered blue spots and lines on head. Usually form small groups. Rubble and rock outcroppings on steep outer slopes in 30-65 m.
West Pacific: Indonesia, Philippines, Micronesia, Papua New Guinea and Solomon Is. – S.W. Japan to W. Australia and Fiji.

KING DAMSEL   *Chrysiptera rex*
SIZE: to 7 cm (2 3/4 in.)   Damselfishes – Pomacentridae
ID: Yellowish white with pale yellow to orange back and grayish head; often blue spots on head including gill cover and scales of back. Solitary or small groups. Near shore rocky substrates just below surge zones in 3-6 m.
Asian Pacific: Indonesia to S.W. Japan and Micronesia.

ONESPOT DAMSEL *Chrysiptera unimaculata*
SIZE: to 8 cm (3 1/4 in.) Damselfishes – Pomacentridae
ID: Usually pale brown to gray forebody with darker rear half of body, occasionally entirely dark, yellowish pectoral fins; usually dark blotch or spot on rear dorsal fin. Solitary or form small groups. Wave-exposed reef flats to 2 m.
Indo-West Pacific: Red Sea and E. Africa to Indonesia, Micronesia and Solomon Is. – S.W. Japan to Australia, Coral Sea and Fiji.

Onespot Damsel – Juvenile
SIZE: 3-4 cm (1 1/4 - 1 1/2 in.)
ID: Yellowish; neon-blue stripe on upper head, blue-ringed black ovate spot below mid-dorsal fin, dark spot at rear dorsal fin. Solitary or form loose groups. Shallow reef flats and shoreline reefs exposed to mild surge to 2 m.

ROLLAND'S DAMSEL *Chrysiptera rollandi*
SIZE: to 5.5 cm (2 1/4 in.) Damselfishes – Pomacentridae
ID: Variable, but most commonly with dark bluish upper head and back to midbody, bluish gray below; **all varieties have long white ventral fins.** Protected shoreline reefs, lagoons and outer slopes in 2-35 m.
Indo-Asian Pacific: Andaman Sea to Indonesia, Philippines, Micronesia, Solomon Is. – S.W. Japan to Australia, New Caledonia.

Rolland's Damsel – Juvenile
SIZE: 3-4 cm (1 1/4 - 1 1/2 in.)
ID: Dusky blue forebody with bright blue "V" extending from snout to dorsal fin; blue-ringed ocellated black spot on dorsal fin.

DUSKYBACK DAMSEL *Chrysiptera caesifrons*
SIZE: to 7 cm (2 3/4 in.) Damselfishes – Pomacentridae
ID: Bluish gray upper head and back; white to yellowish white lower body and tail, blue spots on lower head and small dark "ear" spot. Solitary or small groups. Near shore rocky substrates just below surge zones in 1-6 m.
Asian Pacific: Papua New Guinea, Solomon Is. and Great Barrier Reef to New Caledonia and Vanuatu.

TRACEY'S DAMSEL *Chrysiptera traceyi*
SIZE: to 6 cm (2 1/4 in.) Damselfishes – Pomacentridae
ID: Blue to dark purplish gray, pale yellowish tail base, ventral fins blackish; ovate black blotch at base of dorsal fin. Solitary or form small groups. Lagoons and outer reefs in 5-30 m.
Localized: Micronesia.

Damselfishes

YELLOWFIN DAMSEL   *Chrysiptera flavipinnis*
SIZE: to 7.5 cm (3 in.)   Damselfishes – Pomacentridae
ID: Blue with yellow mid-dorsal stripe running from snout to tail; yellow anal and **dorsal and anal fins yellowish with thin blue line.** Similar Bleeker's Damsel [next] distinguished by location. Rubble and dead coral outcroppings in sandy areas in 3-38 m.
Asian Pacific: S.E. Papua New Guinea, Solomon Is., Great Barrier Reef and Coral Sea to Vanuatu.

BLEEKER'S DAMSEL   *Chrysiptera bleekeri*
SIZE: to 7.5 cm (3 in.)   Damselfishes – Pomacentridae
ID: Blue to purple body with yellow upper head, foreback and dorsal fin; ventral fins often yellow. Similar Yellowfin Damsel [previous] distinguished by location. Solitary or form loose groups close to bottom. Sheltered shoreline reefs in 3-12 m.
Localized: Bali to West Papua in Indonesia and Philippines.

STARCK'S DAMSEL   *Chrysiptera starcki*
SIZE: to 9 cm (3 ¹/₂ in.)   Damselfishes – Pomacentridae
ID: Deep blue with broad area of yellow extending from snout to upper back and most of dorsal fin; tail translucent yellow, **blue ventral fins.** Solitary or form groups. Rocky outcroppings and sand channels on outer slopes in 25-52 m.
West Pacific: Coral Sea and New Caledonia to Fiji.

AZURE DAMSEL   *Chrysiptera hemicyanea*
SIZE: to 5 cm (2 in.)   Damselfishes – Pomacentridae
ID: Brilliant blue over most of head and body; **yellow lower body, ventral, anal and tail fins.** Form small groups. Shelter within coral branches of seaward reefs and lagoons in 3 - 20 m.
Localized: N.W. Australia shelf reefs to S. Sulawesi and West Papua in Indonesia.

GOLDTAIL DAMSEL   *Chrysiptera parasema*
SIZE: to 5 cm (2 in.)   Damselfishes – Pomacentridae
ID: Bright blue head and body with **tail base and tail yellow;** ventral and anal fins blue. Form groups. Shelter within branching corals of protected reefs in 3-15 m.
Asian Pacific: Sabah in Malaysia, Java to Flores in Indonesia and Philippines.

ARNAZ'S DAMSEL   *Chrysiptera arnazae*
SIZE: to 5 cm (2 in.)   Damselfishes – Pomacentridae
ID: Brilliant blue over most of head and body, **yellow rear body and tail including rear dorsal and anal fins;** ventral fin yellow except for 1st spine. Form small groups. Shelter within coral branches on seaward reefs and lagoons in 3-20 m.
Asian Pacific: West Papua, Halmahera, Flores and N. Sulawesi in Indonesia to N.E. Papua New Guinea.

BLUE DEVIL   *Chrysiptera cyanea*
SIZE: to 7.5 cm (3 in.)   Damselfishes – Pomacentridae
ID: Male – Blue with **blue dorsal fin** and scattered yellow to white spots; bright yellow-orange tail (except in some Indonesian localities), lips and ventral fins often yellowish. Lagoons and sheltered shoreline reefs to 10 m.
Asian Pacific: Brunei, Indonesia, Philippines, Micronesia, Solomon Is. – S.W. Japan to Great Barrier Reef.

Blue Devil – Female
SIZE: 7.5 cm (3 in.)
ID: Blue; small black spot at base of rear dorsal fin; **black band from snout through eye** (also present on males). Fiji variation has white to yellowish belly, ventral, anal and tail fins.

SOUTH SEAS DEVIL   *Chrysiptera taupou*
SIZE: to 8 cm (3 1/4 in.)   Damselfishes – Pomacentridae
ID: Blue body and dorsal fin; pale yellow streaked with white underside, anal and tail fins, black band between and through eyes; **often dark spot below base of rear dorsal fin.** Solitary or small groups. Reefs to 5 m.
Southwest Pacific: Great Barrier Reef to Coral Sea, New Caledonia and Fiji.

MILNE BAY DAMSEL   *Chrysiptera cymatilis*
SIZE: to 5.5 cm (2 1/4 in.)   Damselfishes – Pomacentridae
ID: Brilliant blue with lighter blue spotting on head and breast; dark "mask" from snout to eye, **black spot or blotch at base of rear dorsal fin.** Form small groups. Shelter within branching corals of protected seaward reefs and lagoons in 3-20 m.
Localized: Milne Bay Province, E. Papua New Guinea and Solomon Is.

SPRINGER'S DAMSEL   *Chrysiptera springeri*
SIZE: to 5 cm (2 in.)   Damselfishes – Pomacentridae
ID: Brilliant blue occasionally intermixed with black speckling. Similar Milne Bay Damsel [previous] and Sinclair's Damsel [next] distinguished by location. Form small groups. Shelter within branching corals of protected reefs and lagoons in 5-30 m.
Asian Pacific: Sabah in Malaysia to Komodo I., Sulawesi, Halmahera, Molucca Is. and West Papua in Indonesia and Philippines.

SINCLAIR'S DAMSEL   *Chrysiptera sinclairi*
SIZE: to 6 cm (2 1/4 in.)   Damselfishes – Pomacentridae
ID: Blue with blue streaks and small blotches on lower head and breast. Form small groups. Shelter with branching corals of protected seaward reefs and lagoons to 15 m.
Localized: Bismarck Archipelago in Papua New Guinea from Manus I. to N. Solomon Is.

Damselfishes

TWOSPOT DAMSEL
Chrysiptera biocellata

SIZE: to 11 cm (4 1/4 in.)   Damselfishes – Pomacentridae

ID: Dark yellowish brown; wide white midbody saddle and pale tail. **Young Adult** – [pictured] Wide white midbody bar and often dark spot on middle and rear dorsal fin. Rubble and rock outcroppings in lagoons and shore reefs to 5 m.

Indo-West Pacific: E. Africa to Indonesia, Philippines, Micronesia and Solomon Is. – S.W. Japan to Great Barrier Reef and Fiji.

TALBOT'S DAMSEL
Chrysiptera talboti

SIZE: to 6 cm (2 1/4 in.)   Damselfishes – Pomacentridae

ID: Commonly pale mauve but variable (dusky in Fiji); bright yellow area from snout to back, yellow ventral fins, black spot on middle of dorsal fin. Solitary or form small groups. Shoreline reefs, lagoons and outer slopes in 6-35 m.

West Pacific: Andaman Sea to Indonesia, Philippines, Micronesia, Solomon Is., Great Barrier Reef, New Caledonia and Fiji.

BLUESPOT DAMSEL
Chrysiptera oxycephala

SIZE: to 8 cm (3 1/4 in.)   Damselfishes – Pomacentridae

ID: Yellowish gray with numerous tiny blue specks; **dusky "mask" from upper lip to eye. Juvenile –** Blue head and upper body, grayish to yellowish below. Form small to large groups. Sheltered shoreline reefs and lagoons in rich coral areas to 16 m.

Asian Pacific: Indonesia, Philippines, Palau.

GRAY DAMSEL
Chrysiptera glauca

SIZE: to 11.5 cm (4 1/4 in.)   Damselfishes – Pomacentridae

ID: Pale gray with bluish tint and occasional yellowish olive hue on upper half of head and body. Usually form aggregations. Reef flats and near-shore shallows exposed to wave action to 2 m.

Indo-Pacific: E. Africa to Indonesia, Philippines, Micronesia, Solomon Is. and Line Is. – S.W. Japan to Great Barrier Reef.

KUITER'S DAMSEL
Chrysiptera kuiteri

SIZE: to 6 cm (2 1/4 in.)   Damselfishes – Pomacentridae

ID: White with 3 black bars: 1) midhead; 2) forebody behind ventral fins; 3) rear body behind anal fin, **ventral fins white.** Solitary or in pairs. Flat or sloping sand bottoms around rock or coral outcroppings in 15-30 m.

Localized: Sri Lanka, Brunei and N. Bali in Indonesia.

THREEBAND DAMSEL
Chrysiptera tricincta

SIZE: to 6 cm (2 1/4 in.)   Damselfishes – Pomacentridae

ID: White with 3 black bars: 1) midhead; 2) forebody extending onto rear half of ventral fin; 3) rear body extending onto rear half of anal fin, **ventral fin black and white.** Solitary or in pairs. Sand bottoms around rock or coral outcroppings in 10-38 m.

West Pacific: S. Japan to Micronesia, Coral Sea and Fiji.

HUMBUG DASCYLLUS *Dascyllus aruanus*
SIZE: to 7.7 cm (3 in.)   Damselfishes – Pomacentridae
ID: White with 3 black bars; **white tail and solid black ventral fins;** large white spot between eyes. Form groups. Shelter among branching corals when threatened. Inshore and lagoon reefs to 12 m.
Indo-Pacific: Red Sea and E. Africa to Indonesia, Philippines, Micronesia, Solomon Is., French Polynesia. – S.W. Japan to Australia.

BLACK-TAILED DASCYLLUS *Dascyllus melanurus*
SIZE: to 8 cm (3 1/4 in.)   Damselfishes – Pomacentridae
ID: White with 3 black bars; large white spot between eyes extends to lips, **black rear half of tail.** Form groups; shelter within branching corals when threatened. Inshore and lagoon reefs to 12 m.
Asian Pacific: Indonesia, Philippines, Micronesia, Solomon Is., Great Barrier Reef, Coral Sea, New Caledonia and Vanuatu.

YELLOW-TAILED DASCYLLUS *Dascyllus flavicaudus*
SIZE: to 11 cm (4 1/4 in.)   Damselfishes – Pomacentridae
ID: Medium brown to dark brown; **rear dorsal and tail fins slightly yellow to transparent,** black spot on upper pectoral fin base. Form groups near coral heads or anemones in 3-40 m.
Eastern Central Pacific: Society, Tuamotu and Rapa Is. in French Polynesia, also Pitcairn Is.

THREESPOT DASCYLLUS *Dascyllus trimaculatus*
SIZE: to 13.8 cm (5 1/2 in.)   Damselfishes – Pomacentridae
ID: Gray with black scale edges; **fins dark except rear dorsal,** occasionally a suffusion of yellow or orange on head and breast. Usually form groups. Coral and rocky reefs to 55 m.
Indo-Pacific: Red Sea and E. Africa to Indonesia, Philippines, Micronesia, Papua New Guinea, Solomon Is., Line Is. (absent Hawaii). – S.W. Japan to and Great Barrier Reef.

Threespot Dascyllus – Fiji Variation
ID: Orange lower body, ventral, anal and tail fins with black outer margins.

Threespot Dascyllus – Juvenile
SIZE: 3-5 cm (1 1/4 - 2 in.)
ID: Black with large white spot on forehead and another on middle of upper back (both white spots gradually reduced in size with increased growth); occasionally a suffusion of yellow or orange on head and breast. Often near and in anemones, mixing with anemonefishes.

Damselfishes

INDIAN DASCYLLUS   *Dascyllus carneus*
SIZE: to 6 cm (2 1/4 in.)   Damselfishes – Pomacentridae
ID: Tan head, whitish body with diffuse black bar between, often dusky rear body; **numerous small blue spots on head,** wide black margin on dorsal fin. Form groups; shelter within branching corals when threatened. Inshore and outer reefs in 5-35 m.
Indian Ocean: E. Africa to Andaman Sea and Java Sea in Indonesia.

RETICULATED DASCYLLUS   *Dascyllus reticulatus*
SIZE: to 8.5 cm (3 1/4 in.)   Damselfishes – Pomacentridae
ID: Tan to gray head, whitish body with diffuse black bar between, often dusky rear body; wide black margin on dorsal fin. Shelter within branching corals on inshore and outer reefs to 50 m.
East Indo-West Pacific: Cocos-Keeling Is. to Indonesia, Micronesia, Solomon Is. and Line Is. – S.W. Japan to Great Barrier Reef and Fiji.

WHITE DAMSEL   *Dischistodus perspicillatus*
SIZE: to 20 cm (8 in.)   Damselfishes – Pomacentridae
ID: White to pale green; highly variable markings include 2 or 3 black to dusky spots, saddles or bars on forehead, midback and rear back. Solitary or form groups, aggressive. Lagoon and inshore reefs with sand, silt and seagrass bottoms to 10 m.
Asian Pacific: Andaman Sea to Indonesia, Philippines, Micronesia, Solomon Is., Great Barrier Reef and Vanuatu.

White Damsel – Juvenile
SIZE: 3-5 cm (1 1/4 - 2 in.)
ID: Yellowish white with pair of black bars behind head and on midbody; ventral fins yellow. Rocky outcroppings in sandy or weedy areas.

HONEYHEAD DAMSEL   *Dischistodus prosopotaenia*
SIZE: to 18.5 cm (7 1/4 in.)   Damselfishes – Pomacentridae
ID: Golden brown with wide white bar on midbody and white tail; black blotch below front dorsal fin. Solitary or form loose groups. Lagoon and inshore reefs with sand and silt bottoms to 12 m.
Asian Pacific: Andaman Sea to Indonesia, Philippines, Micronesia and Solomon Is. – S.W. Japan to Great Barrier Reef.

Honeyhead Damsel – Juvenile
SIZE: 3-5 cm (1 1/4 - 2 in.)
ID: White with wide brown bar from nape extends over eye and gill cover; second wide brown bar extending from mid-dorsal fin to underside; black spot on rear spinous dorsal fin bordered with yellow ring.

PALE-SPOT DAMSEL *Dischistodus chrysopoecilus*
SIZE: to 15 cm (6 in.)   Damselfishes – Pomacentridae
ID: Dark brown; pale bar across nape, yellowish line and spots on head, **white spot on middle of upper back.** Nest-guarding males (shown here) display light gray head, forebody and tail. Solitary or form loose groups to 5 m.
Asian Pacific: Brunei, Indonesia, Philippines, Micronesia, Papua New Guinea and Solomon Is.

Pale-spot Damsel – Juvenile
SIZE: 3-5 cm (1 1/4 - 2 in.)
ID: Dark brown with **pair of white bars on body end near midbody;** ocellated spot on middle of dorsal fin, blue to yellow streaks and spots on cheek and gill cover. Sand and silt bottoms of sheltered reefs in seagrass or around rocky outcroppings to 5 m.

MONARCH DAMSEL *Dischistodus pseudochrysopoecilus*
SIZE: to 16 cm (6 1/4 in.)   Damselfishes – Pomacentridae
ID: Dark brown (almost black) with blue streak on each scale; **blue lines and spots on head,** white spot on middle of upper back. Coral thickets interspersed with open sand or dead coral to 5 m.
Asian Pacific: Indonesia, Philippines, Papua New Guinea, Great Barrier Reef and Coral Sea.

Monarch Damsel – Juvenile
SIZE: 3-5 cm (1 1/4 - 2 in.)
ID: Brown with pale blue streak on each scale; **pair of white bars on side extends to belly,** large pale ocellated spot on middle of dorsal fin; blue streaks and spots on cheek and gill cover.

BANDED DAMSEL *Dischistodus fasciatus*
SIZE: to 14 cm (5 1/2 in.)   Damselfishes – Pomacentridae
ID: Pale with dark bar through eye with a 2nd from nape to gill cover; **wide brown central body bar,** 4th body bar from rear dorsal to anal fin. Often aggressive toward divers. Shoreline and lagoon reefs with silt and sand bottoms to 8 m.
Asian Pacific: Sabah in Malaysia, Indonesia and Philippines.

BLACKVENT DAMSEL *Dischistodus melanotus*
SIZE: to 15 cm (6 in.)   Damselfishes – Pomacentridae
ID: White with broad brown area on upper head and forebody; dark brown patch on belly, **numerous pale spots on gill cover.** Solitary. Lagoon and inshore coral reefs to 10 m.
Asian Pacific: Sabah in Malaysia, Indonesia, Philippines and Solomon Is. to the Great Barrier Reef and Coral Sea.

Damselfishes

REEF CHROMIS   *Chromis agilis*
SIZE: to 9 cm (3 ½ in.)   Damselfishes – Pomacentridae
ID: Brown with **bluish area on breast and lower head;** no markings except a prominent black spot covering pectoral fin base. Solitary or form groups. Rocky shore or outer reefs in 3-56 m.
Central Pacific: Micronesia to Fiji and French Polynesia, north to Hawaii. Primarily around oceanic islands.

YELLOW-SPOTTED CHROMIS   *Chromis yamakawai*
SIZE: to 15 cm (6 in.)   Damselfishes – Pomacentridae
ID: Gray to brown; prominent black spot covering pectoral fin base, yellow to pale yellow spot at rear base of dorsal fin, **fine blue margins on fins.** Form large schools. Coastal reefs, lagoons and outer slopes in 6-40 m.
Asian Pacific: S.W. Japan to Philippines.

KENN REEF CHROMIS   *Chromis kennensis*
SIZE: to 15 cm (6 in.)   Damselfishes – Pomacentridae
ID: Medium brown with prominent black spot on base of pectoral fin; fine blue margin on anal fin, fine black margin on dorsal fin; **wide yellow borders on forked tail.**
West Pacific: Great Barrier Reef, Coral Sea, New Caledonia and Fiji.

PURA CHROMIS   *Chromis pura*
SIZE: to 12 cm (5 in.)   Damselfishes – Pomacentridae
ID: Gray head and body, yellow rear dorsal fin, tail base and tail; black spot on upper pectoral fin base. Solitary or from small groups. Steep dropoffs near shore in 35-65 m.
Localized: Bali to Alor in S.E. Indonesia.

YELLOW-SPECKLED CHROMIS   *Chromis alpha*
SIZE: to 12.5 cm (5 in.)   Damselfishes – Pomacentridae
ID: Pale blue-gray with bluish or yellow spots on head; **curved whitish lateral line,** small black spot on upper pectoral fin base. Solitary or form groups. Commonly on steep outer reef slopes in 18-95 m.
Pacific: Indonesia, Philippines, Micronesia and Solomon Is. to French Polynesia. – N.W. and N.E. Australia to New Caledonia.

HOWSON'S CHROMIS   *Chromis howsoni*
SIZE: to 5.5 cm (2 ¼ in.)   Damselfishes – Pomacentridae
ID: Cinnamon brown with yellow-orange pectoral fin base; yellow ventral fins and rear edges on dorsal and anal fins, dark margins on dorsal, anal and tail fins. Usually in loose groups. Sheltered coastal reefs and outer reef slopes 3-40 m.
Localized: Currently known only from Milne Bay and Oro Provinces in Papua New Guinea.

PHILIPPINES CHROMIS *Chromis scotochiloptera*
SIZE: to 15 cm (6 in.) Damselfishes – Pomacentridae
ID: Yellowish brown shading to bluish gray on lower body; **black margin on front of anal fin** and borders of tail. Usually form groups. Upper edge of dropoffs in 5-20 m.
Asian Pacific: Sabah in Malaysia, Indonesia and Philippines.

TERNATE CHROMIS *Chromis ternatensis*
SIZE: to 10.5 cm (4 1/4 in.) Damselfishes – Pomacentridae
ID: Golden brown shading to silvery white lower parts; black borders on tail. Indian Ocean variation has yellow hue on back. Large schools in areas of *Acropora* corals in 2-15 m.
Indo-West Pacific: Red Sea and E. Africa to Indonesia, Philippines, Micronesia, Papua New Guinea and Solomon Is. – S.W. Japan to Great Barrier Reef, New Caledonia and Fiji.

GREEN CHROMIS *Chromis cinerascens*
SIZE: to 13 cm (5 in.) Damselfishes – Pomacentridae
ID: Generally olive green, often with yellow hue; ventral fins white to yellow. Blue-green and Black-axil Chromis [below] similar, but larger and on turbid inshore reefs rather than coral-rich areas. Form large schools. Silty coastal reefs in 3-15 m.
East Indo-Asian Pacific: Sri Lanka and Andaman Sea to Brunei, Indonesia and Philippines.

AMBON CHROMIS *Chromis amboinensis*
SIZE: to 10 cm (4 in.) Damselfishes – Pomacentridae
ID: Light gray-brown; **orange spot on pectoral fin base,** dark borders on rear dorsal and anal fins and tail. Solitary or loose groups. Lagoons and outer reefs in 5-65 m.
East Indo-West Pacific: Cocos-Keeling Is. to Indonesia, Philippines, Micronesia, Papua New Guinea, Solomon Is., Great Barrier Reef and Fiji.

BLUE-GREEN CHROMIS *Chromis viridis*
SIZE: to 10 cm (4 in.) Damselfishes – Pomacentridae
ID: Blue to pale green with no markings. Form large schools above coral thickets. Coastal reefs and lagoons in 2-20 m.
Indo-Pacific: Red Sea and E. Africa to Indonesia, Philippines, Micronesia, Papua New Guinea, Solomon Is. and French Polynesia (absent Hawaii). – S.W. Japan to Great Barrier Reef and W. Australia.

BLACK-AXIL CHROMIS *Chromis atripectoralis*
SIZE: to 10 cm (4 in.) Damselfishes – Pomacentridae
ID: Blue to pale green; **black spot on axil of pectoral fin.** Form large schools; feed in open water above coral thickets. Lagoons and outer reefs in 2-15 m.
Indo-Pacific: Madagascar to Andaman Sea, Indonesia, Philippines, Micronesia, Papua New Guinea and Solomon Is. to French Polynesia. – S.W. Japan to Great Barrier Reef.

Damselfishes

YELLOW-AXIL CHROMIS   *Chromis xanthochira*
SIZE: to 14 cm (5 1/2 in.)   Damselfishes – Pomacentridae
ID: Bluish to greenish or olive with dark scale margins; blackish borders on tail, **yellow blotch at base of pectoral fin.** Solitary or form groups. Outer reef slopes in 10-48 m.
West Pacific: Indonesia, Philippines, Micronesia, Papua New Guinea, Solomon Is. to Great Barrier Reef, Vanuatu, New Caledonia and Fiji.

DARKFIN CHROMIS   *Chromis atripes*
SIZE: to 8 cm (3 1/4 in.)   Damselfishes – Pomacentridae
ID: Light brown with **dark edge on anal and rear dorsal fins;** dark bar across eye, small dark spot on upper pectoral fin base. Solitary or form loose aggregations. Steep outer reef slopes and deeper patch reefs in 10-35 m.
West Pacific: Indonesia, Philippines, Micronesia, Papua New Guinea, Solomon Is. – S.W. Japan to Great Barrier Reef and Fiji.

MALAYAN CHROMIS   *Chromis flavipectoralis*
SIZE: to 7.5 cm (3 in.)   Damselfishes – Pomacentridae
ID: Golden brown with white rear body including tail and rear dorsal and anal fins; white ventral fin, **yellow-orange base of pectoral fin.** Solitary or form loose aggregations. Shoreline reefs and seaward reef slopes in 2-16 m.
Indian Ocean: Maldives to Andaman Sea and Java Sea in Indonesia.

STOUT-BODY CHROMIS   *Chromis chrysura*
SIZE: to 16 cm (6 1/4 in.)   Damselfishes – Pomacentridae
ID: Brown with darker spot on each scale, white rear body including tail and rear dorsal and anal fins; **pearly stripe below eye.** Groups usually swim well above bottom. Outer coral and rocky reefs in 6-30 m.
Localized: Three populations: Mauritius and Réunion Is. in W. Indian Ocean; Taiwan and S.W. Japan; Coral Sea and Fiji.

PACIFIC HALF-AND-HALF CHROMIS   *Chromis iomelas*
SIZE: to 9 cm (3 1/2 in.)   Damselfishes – Pomacentridae
ID: Dark brown to black head and forebody, white rear body and tail. Distinguished from Indian Half-and-half Chromis [next] by location. Solitary or form groups. Outer reef areas in 3-35 m.
Pacific: Papua New Guinea, Great Barrier Reef, Coral Sea, Vanuatu, New Caledonia and Fiji and French Polynesia.

INDIAN HALF-AND-HALF CHROMIS   *Chromis fieldi*
SIZE: to 9 cm (3 1/2 in.)   Damselfishes – Pomacentridae
ID: Dark brown to black head and forebody, white rear body and tail. Distinguished from Pacific Half-and-half Chromis [previous] by location. Solitary or form groups. Common near shore and on seaward reef slopes in 2-20 m.
Indian Ocean: Red Sea and E. Africa, Christmas I. and Andaman Sea to Sumatra and Java in Indonesia.

BLUE-AXIL CHROMIS *Chromis caudalis*
SIZE: to 10 cm (4 in.)   Damselfishes – Pomacentridae
ID: Dark gray to brown with white tail including base; tips of rear dorsal and anal fins whitish or clear, **blue spot on pectoral fin base.** Solitary or form small groups. Steep outer reef slopes in 20-50 m.
Pacific: Brunei, Indonesia, Philippines, Micronesia, Papua New Guinea, Solomon Is. to Line Is.

Blue-axil Chromis – Juvenile
SIZE: 3-5 cm (1¼ -2 in.)
ID: Light blue with white extreme rear body and tail base; translucent tail, blue spot covering pectoral fin base.

BICOLOR CHROMIS *Chromis margaritifer*
SIZE: to 9 cm (3½ in.)   Damselfishes – Pomacentridae
ID: Dark brown to black with white rear body including tail and rear dorsal and anal fins; **black spot covers pectoral fin base.** Solitary or form groups. Coastal and offshore reefs in 2-20 m.
East Indo-Pacific: Cocos-Keeling Is. to Indonesia, Philippines, Micronesia and Solomon Is. to French Polynesia. – S.W. Japan to Great Barrier Reef and New Caledonia.

DEEP REEF CHROMIS *Chromis delta*
SIZE: to 7 cm (2¾ in.)   Damselfishes – Pomacentridae
ID: **Gray;** white band around base of tail, tail pale to translucent, black spot covers pectoral fin base. Solitary or form small groups. Steep outer reef slopes in 10-80 m.
East Indo-West Pacific: Cocos-Keeling Is. to Micronesia. – Taiwan to N. Great Barrier Reef, Vanuatu and Fiji.

BLACKBAR CHROMIS *Chromis retrofasciata*
SIZE: to 5.5 cm (2¼ in.)   Damselfishes – Pomacentridae
ID: Yellowish tan with whitish tail; dark band above eye, black bar across rear edge of body. Solitary or form groups. Stay near bottom of lagoons and on outer reefs in 5-65 m.
West Pacific: Indonesia, Philippines, Micronesia, Papua New Guinea and Solomon Is. – S.W. Japan to Great Barrier Reef, New Caledonia, Vanuatu and Fiji.

MIDGET CHROMIS *Chromis acares*
SIZE: to 5.5 cm (2¼ in.)   Damselfishes – Pomacentridae
ID: Blue-gray to light purple; yellow area from eye to pectoral fin, **black front half of anal fin with blue margin,** and broad yellow borders on tail. Form aggregations above coral heads. Clear lagoons and channels and exposed seaward reefs in 2-37 m.
West and Central Pacific: S. Japan, Micronesia to Hawaii, Fiji and French Polynesia.

Damselfishes

PALETAIL CHROMIS   *Chromis xanthura*
SIZE: to 17 cm (6 ³/₄ in.)   Damselfishes – Pomacentridae
ID: Dark charcoal-gray to blackish with **white tail including base, but not rear dorsal and anal fins.** Usually form groups. Outer reef slopes in 3-40 m.
Pacific: Indonesia, Philippines, Micronesia, Papua New Guinea, Solomon Is. to French Polynesia. – S.W. Japan to Great Barrier Reef and New Caledonia.

Paletail Chromis – Variation
ID: Occasionally has black tail. Both variations usually display a dark band behind eye and another from nape to base of pectoral fin. Similar Doublebar Chromis [next] the rear bar starts on the rear gill cover, not the nape.

DOUBLEBAR CHROMIS   *Chromis opercularis*
SIZE: to 17 cm (6 ³/₄ in.)   Damselfishes – Pomacentridae
ID: Charcoal-gray with random pale scale centers; **narrow black bar on gill cover, and broad black bar from rear upper corner of gill cover to pectoral fin base,** often yellow to yellowish tail base. Form groups. Outer reef slopes in 10-40 m.
Indian Ocean: E. Africa to Andaman Sea and Seribu Is. off Java and Bali in Indonesia.

WEBER'S CHROMIS   *Chromis weberi*
SIZE: to 13.5 cm (5 ¹/₄ in.)   Damselfishes – Pomacentridae
ID: Brown with dark margins on scales; dark brown bar on gill cover, and another from upper edge of gill cover to pectoral fin base, **black tips on tail lobes.** Coastal and outer reefs in 3-25 m.
Indo-Pacific: Red Sea and E. Africa to Andaman Sea, Indonesia, Philippines, Micronesia, Papua New Guinea and Solomon Is., to French Polynesia. – S.W. Japan to Australia and Fiji.

TWINSPOT CHROMIS   *Chromis elerae*
SIZE: to 6.5 cm (2 ¹/₂ in.)   Damselfishes – Pomacentridae
ID: Brown to gray; **two white spots, one on base of dorsal fin the other on base of anal fin.** Form small groups. Caves, ledges and black coral thickets, usually on steep slopes in 12-70 m.
East Indo-West Pacific: Maldives to Indonesia, Philippines, Micronesia, Papua New Guinea and Solomon Is. – Taiwan to Great Barrier Reef and Fiji.

SCALY CHROMIS   *Chromis lepidolepis*
SIZE: to 8 cm (3 ¹/₄ in.)   Damselfishes – Pomacentridae
ID: Gray to brown; dark tips on tail fin lobes, **dark blotch on middle edge of anal fin,** black tips on dorsal fin rays. Usually form groups. Shoreline reefs, lagoons and outer slopes in 2-20 m.
Indo-Pacific: Red Sea and E. Africa to Indonesia, Micronesia and Line Is. – S.W. Japan to Great Barrier Reef, New Caledonia and Fiji.

LINED CHROMIS *Chromis lineata*
SIZE: to 5 cm (2 in.)   Damselfishes – Pomacentridae
ID: Yellowish brown undercolor with blue scale rows forming stripes on body; fine blue margins on dorsal, anal and ventral fins. Usually form groups or aggregations above outer reef slopes in 2-10 m.
Asian Pacific: Indonesia, Micronesia, Papua New Guinea and Solomon Is. – Philippines to N. Great Barrier Reef and Vanuatu.

VANDERBILT'S CHROMIS *Chromis vanderbilti*
SIZE: to 6 cm (2 1/4 in.)   Damselfishes – Pomacentridae
ID: Yellowish brown undercolor with blue scale rows forming stripes on body; **lower border of tail and anal fin black.** Usually form aggregations above prominent coral heads on exposed seaward reefs in 2-20 m.
Pacific: Micronesia, Hawaii and French Polynesia. – S.W. Japan to Great Barrier Reef.

YELLOW CHROMIS *Chromis analis*
SIZE: to 14 cm (5 1/2 in.)   Damselfishes – Pomacentridae
ID: Bright yellow to brownish yellow; **white ventral fins,** often dusky band at base of dorsal fin. Solitary or form large groups. Steep outer reef slopes and rocky bottoms in 10-70 m.
West Pacific: Indonesia, Philippines and Micronesia. – S.W. Japan to Great Barrier Reef, New Caledonia and Fiji.

KOMODO CHROMIS *Chromis albicauda*
SIZE: to 17.5 cm (7 in.)   Damselfishes – Pomacentridae
ID: Yellow with brownish to grayish tints on upper head and body; **white tail.** Similar Yellow Chromis [previous] has yellow tail and white ventral fins. Steep outer reef slopes in areas of cool (18-25 C) upwelling in 10-50 m.
Localized: Known only from Lesser Sunda Is. at Bali, Alor and S. Japan.

OVATE CHROMIS *Chromis ovatiformes*
SIZE: to 10 cm (4 in.)   Damselfishes – Pomacentridae
ID: Pale brown with white tail, tail base, rear anal and dorsal fins. Solitary or small groups; remain close to shelter. Outer reefs in 10-30 m.
North Asian Pacific: Brunei, Sabah in Malaysia, Philippines and S.W. Japan.

WHITE-TAILED CHROMIS *Chromis leucura*
SIZE: to 8.5 cm (3 1/4 in.)   Damselfishes – Pomacentridae
ID: Dark blue gray with yellow ventral fins and white tail; black edge marking extends from rear dorsal fin to rear anal fin, black spot covers pectoral fin base. Solitary or form small groups. Around deep boulders or scattered coral in 16-119 m.
Indo-Pacific: Mauritius and Andaman Is. to N. Sumatra in Indonesia, S.W. Japan, New Caledonia, Hawaii and French Polynesia.

Damselfishes

SPINY CHROMIS *Acanthochromis polyacanthus*
SIZE: to 14 cm (5 1/2 in.) Damselfishes – Pomacentridae
ID: Visual ID nearly impossible due to highly variable color and markings depending on geographic location. Similar in shape to *Chromis* damselfishes, but much larger than most. Usually in pairs. Shoreline, lagoon and outer reefs to 65 m.
Asian Pacific: Indonesia, Philippines, Papua New Guinea, and Solomon Is. to Vanuatu.

Spiny Chromis – Variation
ID: One of only a few reef fish lacking a pelagic larval stage, which accounts for the localized range of many variations. Pairs guard babies as well as eggs. Two large adults tending many young can be a clue to identification. This entirely dark variation and similar forms are widespread throughout the Philippines and Indonesia.

Spiny Chromis – Variation
ID: With over 15 different color and marking variations, sometimes separated by only 20-30 km, positive ID can only be established by dorsal fin spines, 17 instead of the 12-14 in other *Chromis*.

Spiny Chromis – Variation

Spiny Chromis – Juvenile Variation
ID: This juvenile variation was photographed in northern Sulawesi and Raja Ampat.

Spiny Chromis – Juvenile Variation
ID: Juveniles tend to cluster in small groups just inside openings of reef pockets. This juvenile variation was photographed in Raja Ampat in vicinity where previous juvenile variation was photographed.

SMOKY CHROMIS *Chromis fumea*
SIZE: to 13 cm (5 in.) Damselfishes – Pomacentridae
ID: Light to dark brown upper head and body, blue to gray below; white spot rear base of dorsal fin, broad dark borders on tail, black spot on pectoral fin base. Form openwater feeding aggregations over rock or coral reefs in 3-25 m.
Asian Pacific: Malaysian Peninsula to Komodo I. in Indonesia. – S.W. Japan to W. and E. Australia.

BARRIER REEF CHROMIS *Chromis nitida*
SIZE: to 9 cm (3 1/2 in.) Damselfishes – Pomacentridae
ID: White; **black bordered brown "skull-cap" runs from snout to mid-dorsal fin,** black borders on tail and anal fin. Usually form groups. Lagoons and outer reefs in coral areas in 5-25 m.
Southwestern Pacific: Central and S. Great Barrier Reef, south to Sydney and Lord Howe I.

GUAM REEF-DAMSEL *Pomachromis guamensis*
SIZE: to 5.5 cm (2 1/4 in.) Damselfishes – Pomacentridae
ID: Grayish head and belly, light blue-green body shading to yellow rear body and tail; a small black spot on upper pectoral fin base. Feed in groups on plankton 1-2 m above bottom. Exposed seaward reefs in 3-33 m.
Localized: Mariana Is. in Micronesia.

SLENDER REEF-DAMSEL *Pomachromis richardsoni*
SIZE: to 9.5 cm (3 3/4 in.) Damselfishes – Pomacentridae
ID: Light brown upper head shading to light blue-gray with dark scale margins on upper body; **black streak on upper tail base, black tail borders.** Form groups. Exposed reefs in 10-20 m.
Indo-West Pacific: Reunion and Mauritius Is. to Solomon Is. – S.W. Japan to Great Barrier Reef, New Caledonia and Fiji.

FUSILIER DAMSEL *Lepidozygus tapeinosoma*
SIZE: to 10 cm (4 in.) Damselfishes – Pomacentridae
ID: Elongate body; greenish brown to yellowish upper body, bluish to pink below, usually a yellow spot at base of last dorsal fin rays. Form feeding aggregations high above bottom. Often in areas of strong current in 5-25 m.
Indo-Pacific: E. Africa to Indonesia, Micronesia and French Polynesia. – S.W. Japan to Great Barrier Reef and Coral Sea.

CHINESE DEMOISELLE *Neopomacentrus bankieri*
SIZE: to 7 cm (2 3/4 in.) Damselfishes – Pomacentridae
ID: Gray with yellow rear dorsal and tail fins; small black spot on upper pectoral fin base. Form schools. Coral and rock outcroppings or debris on sand and silt bottoms in 3-12 m.
Asian Pacific: Vietnam to Java Sea and West Papua in Indonesia and Great Barrier Reef.

Damselfishes

SILVER DEMOISELLE   *Neopomacentrus anabatoides*
SIZE: to 10 cm (4 in.)   Damselfishes – Pomacentridae
ID: Greenish metallic with blue spot on each scale; blue "ear" spot, dark submarginal stripe on each lobe of tail. Form openwater feeding aggregations. Shelter around coral and outcroppings in near shore reefs in 2-15 m.
North Asian Pacific: Andaman Sea to W. Indonesia, South China Sea and Philippines.

CORAL DEMOISELLE   *Neopomacentrus nemurus*
SIZE: to 8 cm (3 ¼ in.)   Damselfishes – Pomacentridae
ID: Blue-gray with yellow rear dorsal, anal and tail fins; **anal fin yellowish entire length,** small dark "ear" spot, and another on upper pectoral fin base, blue dot on each scale of tail base. Silty inshore reefs and lagoons to 10 m.
Asian Pacific: Indonesia, Philippines, Micronesia, New Guinea, Solomon Is. and Vanuatu.

FRESHWATER DEMOISELLE   *Neopomacentrus taeniurus*
SIZE: to 10 cm (4 in.)   Damselfishes – Pomacentridae
ID: Brown with dark scale margins and **yellow tail with dark borders;** yellow edge on rear edge of dorsal and anal fins. Form small groups. Mangroves, lower reaches of freshwater streams and bays with freshwater discharge to 3 m.
Indo-Asian Pacific: E. Africa and Andaman Sea to Micronesia, Papua New Guinea. – S.W. Japan to Australia and Vanuatu.

VIOLET DEMOISELLE   *Neopomacentrus violascens*
SIZE: to 6.7 cm (2 ½ in.)   Damselfishes – Pomacentridae
ID: Brown with **yellow tail, tail base and rear dorsal fin;** large black "ear" spot, and another at base of pectoral fin. Inshore reefs on soft bottoms around coral or rock outcroppings, wharf pilings and wreckage in 5-25 m.
Asian Pacific: Andaman Sea, Indonesia, Philippines, Papua New Guinea, Solomon Is., Vanuatu and Fiji.

REGAL DEMOISELLE   *Neopomacentrus cyanomos*
SIZE: to 9 cm (3 ½ in.)   Damselfishes – Pomacentridae
ID: Dark brown; **rear dorsal fin yellow with yellow or white spot at base,** rear tail yellowish; black "ear" spot. Form groups. More common on sheltered inshore reefs, but also found on outer reefs in 5-18 m.
Indo-Asian Pacific: Red Sea and E. Africa to Indonesia, Micronesia and Solomon Is. – S.W. Japan to Great Barrier Reef and Vanuatu.

YELLOWTAIL DEMOISELLE   *Neopomacentrus azysron*
SIZE: to 7.8 cm (3 in.)   Damselfishes – Pomacentridae
ID: Blue-gray with yellow rear dorsal and tail fins; dark "ear" spot, black spot on upper pectoral fin base. Form schools. Deeper surge channels or near ledges to 12 m.
Asian Pacific: Taiwan to Indonesia, Philippines, Papua New Guinea and Solomon Is. – S.W. Japan to Australia and Vanuatu.

BROWN DEMOISELLE   *Neopomacentrus filamentosus*
SIZE: to 7.5 cm (3 in.)   Damselfishes – Pomacentridae
ID: Gray-brown; **rear dorsal and tail fins clear,** narrow blue margin on most fins. Soft bottoms of lagoons and inshore reefs also around coral and rock outcroppings or wreckage in 5-12 m.
Asian Pacific: Andaman Sea to Indonesia, Philippines, Micronesia, Papua New Guinea and Solomon Is.

IMITATOR DAMSEL   *Pomacentrus imitator*
SIZE: to 11 cm (4 1/4 in.)   Damselfishes – Pomacentridae
ID: Shades of blue to gray with yellow or white tail; **large black spot around base of pectoral fin;** usually dark lower anal fin. Inhabit reef and outer reef slopes in 2 - 15 m.
Central Pacific: Coral Sea to New Caledonia to Fiji and Tonga.

WHITESPOT DAMSEL   *Pomacentrus albimaculus*
SIZE: to 8.5 cm (3 1/4 in.)   Damselfishes – Pomacentridae
ID: Gray to brown with dark scale outlines; **white saddle on upper tail base.** Solitary or form groups. Rocky outcroppings and debris of sandy inshore areas in 10-20 m.
Localized: Cenderawasih Bay in West Papua, Indonesia, Papua New Guinea and Solomon Is.

COLIN'S DAMSEL   *Pomacentrus colini*
SIZE: to 8.6 cm (3 1/4 in.)   Damselfishes – Pomacentridae
ID: Gray with darker scale outlines and **white patch on upper rear body below dorsal fin.** Solitary or form loose groups. Shoreline reefs and lagoons in 10 -18 m.
Localized: S.E. Papua New Guinea from Port Moresby to Milne Bay.

AMBON DAMSEL   *Pomacentrus amboinensis*
SIZE: to 10.5 cm (4 in.)   Damselfishes – Pomacentridae
ID: Body color variable according to locality but most commonly yellow; pale pink to blue blotches and markings on lower head. **Juvenile –** Black spot on rear dorsal fin persists on adults in Andaman Sea. Sandy areas in 2- 40 m.
West Pacific: Andaman Sea to Indonesia, Philippines, Micronesia and Solomon Is. – S.W. Japan to Great Barrier Reef and Fiji.

KOMODO DAMSEL   *Pomacentrus komodoensis*
SIZE: to 10 cm (4 in.)   Damselfishes – Pomacentridae
ID: Brown with darker scale margins; **blue iris with thin inner yellow ring. Juvenile –** Reddish tinge on back and blue-edged black spot on dorsal fin. Solitary or form groups. Rock or boulder-strewn shorelines to 5 m.
Localized: Known only from Komodo I., Indonesia.

Damselfishes

BROWN DAMSEL *Pomacentrus opisthostigma*
SIZE: to 8 cm (3¹/₄ in.) Damselfishes – Pomacentridae
ID: **E. Papua New Guinea Variation –** Color variable from dark brown to yellowish tan; **small yellowish brown mark on upper pectoral fin base.** Solitary. Turbid shores with coral in 3-12 m.
Asian Pacific: Halmahera and West Papua in Indonesia, Philippines and Papua New Guinea. This variation from Milne Bay, Papua New Guinea.

Brown Damsel – West Papua Variation
ID: Brown; **whitish blotch on rear dorsal fin,** small wedge-shaped dark mark on upper pectoral fin base.
Localized: Misool I. in West Papua, Indonesia.

Brown Damsel – Philippines Variation
ID: Purplish brown with whitish tail; wedge-shape mark is extended to form bar across entire pectoral fin base.
Localized: Calamian Is. north of Palawan, Philippines.

BURROUGH'S DAMSEL *Pomacentrus burroughi*
SIZE: to 8 cm (3¹/₄ in.) Damselfishes – Pomacentridae
ID: Reddish brown; **pale yellow blotch on base of rear dorsal fin,** tiny black "ear" spot. Solitary or form loose groups. Sheltered shoreline and lagoon reefs, usually in silty areas in 2-16 m.
Asian Pacific: E. Indonesia, Philippines, Micronesia to New Guinea and Solomon Is.

Burrough's Damsel – E. Sulawesi, Indonesia Variation
ID: Same as typical variation, except lack pale blotch on base of rear dorsal fin.
Localized: Togean and Banggai Is. off E. Sulawesi, Indonesia.

Burrough's Damsel – Juvenile
SIZE: 3-4.5 cm (1¹/₄ - 1³/₄ in.)
ID: Dark gray, nearly black with royal blue spots on head and scale streaks on back, tail pale to translucent; ocellated spot on rear dorsal fin.

Small Ovals – Damselfishes

COLOMBO DAMSEL *Pomacentrus proteus*
SIZE: to 9 cm (3 1/2 in.) Damselfishes – Pomacentridae
ID: Yellowish brown with blue spots or streaks on most scales; **blue line markings extend from snout to foreback,** blue-edged black spot on rear dorsal fin. Solitary or form loose groups. Silty shorelines on mixed rubble and coral reefs in 2-10 m.
East Indian Ocean: Sri Lanka and Andaman Sea.

Colombo Damsel – Juvenile
SIZE: 3-6 cm (1 1/4 - 2 1/4 in.)
ID: Blue on head and back gradating to yellow below; neon-blue line markings on upper head and blue spotting below, black spot edged in neon-blue on rear dorsal fin.

YELLOWEYE DAMSEL *Pomacentrus flavioculus*
SIZE: to 10 cm (4 in.) Damselfishes – Pomacentridae
ID: Gray to blue head and body with black scale margins; yellow to tan or pale gray tail, **gold iris,** large black spot covering pectoral fin base. Solitary or form groups. Usually seen on outer reef slopes in 4-30 m.
Localized: Fiji and Tonga.

ALEXANDER'S DAMSEL *Pomacentrus alexanderae*
SIZE: to 8.6 cm (3 1/4 in.) Damselfishes – Pomacentridae
ID: Gray; black spot covering pectoral fin base, **tips of dorsal spines black.** Usually form groups; a common species. Shoreline, lagoon and outer reefs in 5-30 m.
Asian Pacific: Brunei to Sulawesi in Indonesia and Philippines north to Taiwan and S.W. Japan.

BORNEO DAMSEL *Pomacentrus armillatus*
SIZE: to 8 cm (3 1/4 in.) Damselfishes – Pomacentridae
ID: Purplish brown with pale tail; **thin dark bar across pectoral fin base,** black "ear" spot. Solitary or form small groups. Sheltered shoreline reefs in silty areas with scattered corals in 8-12 m.
Localized: Malaysia to N. Kalimantan (Borneo) in Indonesia and Philippines.

THAI DAMSEL *Pomacentrus polyspinus*
SIZE: to 8.6 cm (3 1/4 in.) Damselfishes – Pomacentridae
ID: Gray with dark scale outlines; **blue lines on snout and forehead,** ocellated spot on rear dorsal fin and may display small black "ear" spot. Solitary or form groups. Shoreline and lagoon reefs in 3-10 m.
Localized: E. Andaman Sea.

Damselfishes

NEON DAMSEL *Pomacentrus coelestis*
SIZE: to 8 cm (3 1/4 in.)   Damselfishes – Pomacentridae
ID: Neon-blue, ventral and anal fins either yellow or blue; **blue margin on anal and tail fin.** Usually form groups. Rubble areas to 12 m.
Pacific: Andaman Sea to Indonesia, Philippines, Micronesia and Solomon Is. to French Polynesia. – S.W. Japan to Great Barrier Reef.

ANDAMAN DAMSEL *Pomacentrus alleni*
SIZE: to 6.5 cm (2 1/2 in.)   Damselfishes – Pomacentridae
ID: Neon-blue with yellow anal fin; **black streak along lower edge of tail.** Usually in groups. Rubble and dead reef areas, both near shore and on outer slopes in 3-15 m.
East Indian Ocean: Andaman Sea and Indian Ocean coasts of Sumatra, Java and Bali in Indonesia.

SIMILAR DAMSEL *Pomacentrus similis*
SIZE: to 7 cm (2 3/4 in.)   Damselfishes – Pomacentridae
ID: Deep blue (generally lacking brightness of other neon-blue damsels); pale yellowish tail and blackish ventral fins. Solitary or form groups. Rubble or sand and silt areas around rock outcroppings or debris to 15 m.
East Indian Ocean: Sri Lanka and Andaman Sea.

GOLDBELLY DAMSEL *Pomacentrus auriventris*
SIZE: to 7 cm (2 3/4 in.)   Damselfishes – Pomacentridae
ID: Neon-blue head and upper body, yellow lower and rear body, anal and tail fins. Solitary or form groups; often mix with Neon Damsels [previous]. Mainly on rubble slopes in 2-15 m.
Asian Pacific: Christmas I. to Indonesia and Micronesia.

SAKSONO'S DAMSEL *Pomacentrus saksonoi*
SIZE: to 10 cm (4 in.)   Damselfishes – Pomacentridae
ID: Pale gray to nearly white; wedge-shaped brownish mark on upper pectoral fin base. Silty coral reefs in 8-15 m.
Localized: Seribu Is. in Java Sea and Sangihe Is. in N. Sulawesi, Indonesia.

BLACKSPOT DAMSEL *Pomacentrus stigma*
SIZE: to 12 cm (4 3/4 in.)   Damselfishes – Pomacentridae
ID: Light gray with **black blotch on rear anal fin.** Solitary or loose groups. Shoreline, lagoon and outer reefs in 2-10 m.
Localized: Brunei and Sabah in Malaysia to Philippines.

SMITH'S DAMSEL   *Pomacentrus smithi*
SIZE: to 6.8 cm (2 3/4 in.)   Damselfishes – Pomacentridae
ID: Light gray with pale yellow wash over nape and upper back and a few blue spots; **bright yellow iris.** Form large busy schools over patches of branching corals. Silty coastal reefs and lagoons in 2-14 m.
Asian Pacific: Indonesia and Philippines.

AUSTRALIAN DAMSEL   *Pomacentrus australis*
SIZE: to 8 cm (3 1/4 in.)   Damselfishes – Pomacentridae
ID: Bluish gray upper head and back, remainder light gray to white; vertical blue streak on scales of back. Large adults have blue on upper body and white below. Solitary or form groups. Coral rock outcroppings in sand or rubble areas in 5-35 m.
Localized: Great Barrier Reef to Sydney, Australia.

BLUE-SPOTTED DAMSEL   *Pomacentrus azuremaculatus*
SIZE: to 9.3 cm (3 3/4 in.)   Damselfishes – Pomacentridae
ID: Light blue-gray (nearly white) with **blue spot or streak on scales of upper forebody.** Solitary or form loose groups. Mainly offshore coral reefs in 5-30 m.
Localized: E. Andaman Sea and Seribu Is. in Java Sea, Indonesia.

REID'S DAMSEL   *Pomacentrus reidi*
SIZE: to 12 cm (4 3/4 in.)   Damselfishes – Pomacentridae
ID: Light gray to brown; **yellowish tail with darkish bar on tail base,** blue spots and streaks on cheek and gill cover, black spot on pectoral fin base. Solitary and form groups. Outer reef slopes and deep lagoons in 12-70 m.
Asian Pacific: Indonesia, Philippines, Micronesia, Solomon Is. to Great Barrier Reef, Coral Sea, Vanuatu and New Caledonia.

TWINSPOT DAMSEL   *Pomacentrus geminospilus*
SIZE: to 8 cm (3 1/4 in.)   Damselfishes – Pomacentridae
ID: Light blue-gray with darker scale outlines; tiny blue spots and blotches on cheek and gill cover. Solitary or form small groups. Sheltered shoreline reefs and lagoons on sloping silty bottoms around coral and rock outcroppings in 3-15 m.
Localized: Sabah in Malaysia and Palawan Province in the Philippines.

Twinspot Damsel – Juvenile
SIZE: 3 - 4.5 cm (1 1/4 - 1 3/4 in.)
ID: Light blue-gray with pair of ocellated spots; one on rear dorsal fin and another across upper tail base.

Damselfishes

BLUE DAMSEL *Pomacentrus pavo*
SIZE: to 11 cm (4 1/4 in.) Damselfishes – Pomacentridae
ID: Shades of blue to light green with **vertical dark streaks on scales;** rear tail yellowish, scattered blue spots on head, dusky "ear" spot. Form groups. Shoreline and lagoon reef coral patches surrounded by sand to 16 m.
Indo-Pacific: E. Africa to Indonesia, Philippines, Micronesia, Solomon Is. and French Polynesia. – Taiwan to Great Barrier Reef.

SCALY DAMSEL *Pomacentrus lepidogenys*
SIZE: to 8.6 cm (3 1/2 in.) Damselfishes – Pomacentridae
ID: Pale blue-gray with **yellow rear dorsal fin and tail base.** Variation from Melanesia and Tonga are light gray with little or no yellow. Shoreline, lagoon and outer reefs to 12 m.
West Pacific: Andaman Sea to Indonesia, Philippines, Micronesia, Papua New Guinea, Solomon Is., Great Barrier Reef, Coral Sea, Fiji and Tonga.

PRINCESS DAMSEL *Pomacentrus vaiuli*
SIZE: to 10 cm (4 in.) Damselfishes – Pomacentridae
ID: Blue with yellow-orange area on upper head and back; blue lines extends from snout onto back and dorsal fin, ocellated spot rear dorsal fin, black "ear" spot. Solitary or form loose aggregations. Lagoons and outer reef slopes in 3-45 m.
West Pacific: Indonesia, Philippines, Micronesia and Solomon Is. – S.W. Japan to Great Barrier Reef and Fiji.

SPECKLED DAMSEL *Pomacentrus bankanensis*
SIZE: to 8 cm (3 1/4 in.) Damselfishes – Pomacentridae
ID: Variable, but commonly orange-brown with blue lines from snout onto back; **white band behind black bar on translucent tail,** ocellated spot on rear dorsal fin, black "ear" spot and black spot on base of pectoral fin. Shoreline and outer reefs to 12 m.
West Pacific: Indonesia, Philippines, Micronesia and Solomon Is. – S.W. Japan to Great Barrier Reef and Fiji.

Speckled Damsel – Variation
ID: Brown with only faint blue lines from snout onto back; all variations have white band on front of translucent tail behind black bar, ocellated spot on rear dorsal fin, black "ear" spot and black spot on base of pectoral fin.

Speckled Damsel – Variation
ID: Yellow-brown body only faint blue lines from snout onto back; all variations have white band on front of translucent tail behind black bar, ocellated spot on rear dorsal fin, black "ear" spot and black spot on base of pectoral fin.

BLUE-GREEN DAMSEL *Pomacentrus callainus*
SIZE: to 9.5 cm (3 ³/₄ in.) Damselfishes – Pomacentridae
ID: Dull blue-green; **wedge-shaped black spot on upper half of pectoral fin base**, translucent moderately forked tail. Shelter in shallow coral reefs in 2-15 m.
Localized: Fiji and Tonga.

ORANGE SPOTTED DAMSEL *Pomacentrus spilotoceps*
SIZE: to 8 cm (3 ¹/₄ in.) Damselfishes – Pomacentridae
ID: Yellowish brown with darkish scale margins; **light blue margin on dorsal and anal fins**, "ear" spot behind eye and dark spot at base of pectoral fin, pale orangish spots on gill cover and around base of pectoral fin. Solitary or form loose aggregations. Lagoons and inshore reefs to 3 m.
Localized: Fiji and Tonga.

OBSCURE DAMSEL *Pomacentrus adelus*
SIZE: to 8 cm (3 ¹/₄ in.) Damselfishes – Pomacentridae
ID: Dark brown with dark "ear" spot; black spot on rear dorsal of younger adults and juveniles. Inshore and outer reefs to 8 m.
Asian Pacific: Andaman Sea to Indonesia, Philippines, Micronesia and Solomon Is. to Great Barrier Reef, Vanuatu and New Caledonia.

FIJI DAMSEL *Pomacentrus microspilus*
SIZE: to 8.5 cm (3 ¹/₄ in.) Damselfishes – Pomacentridae
ID: Charcoal gray with blackish anal fin; whitish half circle above eye, large **diamond-shaped dark marking on pectoral fin base**. Solitary or form loose groups. Lagoons and offshore reefs with silty conditions in 1-35 m.
Localized: Fiji.

WHITETAIL DAMSEL *Pomacentrus chrysurus*
SIZE: to 9 cm (3 ¹/₂ in.) Damselfishes – Pomacentridae
ID: Gray to brown with white tail, often lighter gray on upper back and dorsal fin; upper rim of eye orange. Solitary or form loose groups. Sandy areas of shoreline reefs and lagoons to 3 m.
East Indo-West Pacific: Maldives and Sri Lanka to Indonesia, Philippines, Micronesia, Papua New Guinea and Solomon Is. – S.W. Japan to Great Barrier Reef and New Caledonia.

Whitetail Damsel – Juvenile
SIZE: 3 - 4.5 cm (1 ¹/₄ - 1 ³/₄ in.)
ID: Gray with translucent tail; broad dorsal stripe of orange from snout to rear dorsal fin, ocellated spot on rear dorsal fin.

95

Damselfishes

PHILIPPINE DAMSEL   *Pomacentrus philippinus*
SIZE: to 11 cm (4 1/2 in.)   Damselfishes – Pomacentridae
ID: Dark gray to brown to purplish with **black scale margins;** black spot with yellow margin on pectoral fin base, yellowish tail, rear dorsal and anal fins. Solitary or loose groups. Reef slopes, often near ledges in 3-12 m.
Asian Pacific: Japan to Indonesia, Philippines and N.W. Australia.

Philippine Damsel – Juvenile
SIZE: 3-5 cm (1 1/4 -2 in.)
ID: Blue with black scale margins; black spot on pectoral fin base; yellow tail, rear dorsal and anal fins.

BLACK-MARGINED DAMSEL   *Pomacentrus nigromarginatus*
SIZE: to 9 cm (3 1/2 in.)   Damselfishes – Pomacentridae
ID: Pale to dark gray, usually yellow on rear body, dorsal, anal and tail fins; **black tail margin,** black spot covering pectoral fin base. Solitary. Steep outer reef slopes around coral and rock outcroppings in 20-50 m.
West Pacific: Indonesia, Philippines, Micronesia and Solomon Is. – S.W. Japan to Great Barrier Reef, Coral Sea and Fiji.

GOLDBACK DAMSEL   *Pomacentrus nigromanus*
SIZE: to 8.5 cm (3 1/4 in.)   Damselfishes – Pomacentridae
ID: Gray head and forebody shading to bright yellow on rear body, dorsal and tail fins; large black spot on pectoral fin base, **wide black margin on anal fin.** Solitary or form groups. Lagoons and outer slopes in 6-60 m.
West Pacific: E. Indonesia, Philippines, Micronesia, Papua New Guinea, Solomon Is. and Vanuatu.

NAGASAKI DAMSEL   *Pomacentrus nagasakiensis*
SIZE: to 12 cm (4 3/4 in.)   Damselfishes – Pomacentridae
ID: Gray or occasionally blue with black scale margins; numerous blue lines and spots on head, black dorsal spine tips, large black spot on pectoral fin base, **tail, posterior dorsal and anal fins whitish with faint wavy lines.** Sandy areas around rock in 5-30 m.
East Indo-Asian Pacific: Maldives to Indonesia, Micronesia and Solomon Is. – S.W. Japan to Great Barrier Reef and Vanuatu.

Nagasaki Damsel – Juvenile
SIZE: 3-4.5 cm (1 1/4 -1 3/4 in.)
ID: Dark gray to gray with blue streak on each scale; blue lines and spots on head, black ocellated spot with blue margin on rear dorsal fin, black spot on base of pectoral fin.

WEDGESPOT DAMSEL   *Pomacentrus cuneatus*
SIZE: to 9 cm (3 1/2 in.)   Damselfishes – Pomacentridae
ID: Gray with darker scale margins; wedge-shaped black mark at upper pectoral fin base, black "ear" spot, **blue tips on dorsal spines.** Solitary or form loose groups. Shoreline reefs, often in turbid silty areas to 6 m.
Localized: Singapore, Java Sea, Halmahera, Raja Ampat and Komodo I. in Indonesia.

Wedgespot Damsel – Juvenile
SIZE: 3-4.5 cm (1 1/4 - 1 3/4 in.)
ID: Yellow to light blue with blue lines and bands from snout to rear dorsal fin ending at an ocellated spot.

AZURESPOT DAMSEL   *Pomacentrus grammorhynchus*
SIZE: to 11 cm (4 1/4 in.)   Damselfishes – Pomacentridae
ID: Dark brown to yellow-tan; blue to lavender spots and markings on cheek and gill cover, may have dark "ear" spot, **neon-blue saddle spot over upper tail base.** Tail orange in Java Sea. Shoreline and lagoon reefs in 2-12 m.
Asian Pacific: Indonesia, Philippines, Micronesia and Solomon Is. – Taiwan to Great Barrier Reef and New Caledonia.

Azurespot Damsel – Juvenile
SIZE: 3-4.5 cm (1 1/4 - 1 3/4 in.)
ID: Blue-gray shading to yellow over most of lower and rear body and adjacent fins; neon-blue lines on upper head and upper back, blue spot on upper tail base. Adults and juveniles usually around branching corals.

CHARCOAL DAMSEL   *Pomacentrus brachialis*
SIZE: to 10.5 cm (4 in.)   Damselfishes – Pomacentridae
ID: Charcoal-gray to black; large black spot covering base of pectoral fin. Usually feed in groups in open water a short distance above reef. Passages and outer reef slopes in 6-40 m.
West Pacific: Sabah in Malaysia, Indonesia, Philippines, Micronesia, Papua New Guinea and Solomon Is. to Great Barrier Reef, Coral Sea, Vanuatu, New Caledonia and Fiji.

Charcoal Damsel – Juvenile
SIZE: 3-4.5 cm (1 1/4 - 1 3/4 in.)
ID: Shades of blue except translucent tail; large black spot on base of pectoral fin.

Damselfishes

LEMON DAMSEL *Pomacentrus moluccensis*
SIZE: to 8 cm (3 1/4 in.) Damselfishes – Pomacentridae
ID: Bright yellow; usually display small black spot at upper pectoral fin base, tiny dark "ear" spot, **fine black to blue margin on anal fin.** Usually form groups around live coral patches. Shoreline, lagoon and outer reefs to 14 m.
West Pacific: Andaman Sea to Indonesia, Philippines, Micronesia. – S.W. Japan to Great Barrier Reef to Vanuatu and Fiji.

CORAL DAMSEL *Pomacentrus maafu*
SIZE: to 7 cm (2 3/4 in.) Damselfishes – Pomacentridae
ID: Purplish brown, yellowish on lower head, breast and tail. Small black spot at upper pectoral-fin base and tiny dark "ear" spot. Usually form groups around live coral patches. Shoreline, lagoon and outer reefs in 1-12 m.
Localized: Fiji and Tonga.

BLUEBACK DAMSEL *Pomacentrus simsiang*
SIZE: to 9 cm (3 1/2 in.) Damselfishes – Pomacentridae
ID: Yellow with translucent white tail and tail base; pale spots on cheek and gill cover, usually blue lines on forehead, **small blue or pale spot on upper tail base.** Solitary or form groups. Silty shoreline reefs and lagoons around outcroppings to 10 m.
Asian Pacific: Indonesia, Philippines, Micronesia, Papua New Guinea, Solomon Is. and Vanuatu.

Blueback Damsel – Juvenile
SIZE: 3-4 cm (1 1/4 - 1 1/2 in.)
ID: Yellow with blue upper head and back; neon-blue lines extend from snout to back, a large blue-edged black spot on rear dorsal fin.

BRACKISH DAMSEL *Pomacentrus taeniometopon*
SIZE: to 11.5 cm (4 1/2 in.) Damselfishes – Pomacentridae
ID: Dark brown to nearly black with yellowish tail; often with neon-blue markings on upper head and rear dorsal fin. Solitary or form loose groups. Mangrove areas and lower reaches of freshwater streams to 4 m.
Asian Pacific: Indonesia, Philippines, S.W. Japan, Micronesia, Papua New Guinea and Solomon Is.

Brackish Damsel – Juvenile
SIZE: 3-4.5 cm (1 1/4 - 1 3/4 in.)
ID: Dark brown to nearly black undercolor with brilliant neon-blue lines and dot rows and orange tail; ocellated spot on rear dorsal fin.

SMOKY DAMSEL *Pomacentrus littoralis*

SIZE: to 10.5 cm (4 1/4 in.) Damselfishes – Pomacentridae

ID: Dark charcoal-gray; **green "ear" spot.** Variations from Indonesia and Philippines usually have orange upper rim on eye. Solitary or form loose groups. Silty shoreline reefs to 4 m.

Asian Pacific: Singapore, Brunei, Indonesia, Philippines and N. Australia.

WARD'S DAMSEL *Pomacentrus wardi*

SIZE: to 10 cm (4 in.) Damselfishes – Pomacentridae

ID: Brown with no distinctive markings. **Juvenile –** Yellow with blue lines on upper head, large ocellated spot on dorsal fin. Solitary or form loose groups. Coastal and offshore reefs to 20 m.

Localized: Great Barrier Reef and E. Australian coast south to Sydney area.

THREESPOT DAMSEL *Pomacentrus tripunctatus*

SIZE: to 9.2 cm (3 3/4 in.) Damselfishes – Pomacentridae

ID: Dark gray-brown; **black spot or saddle on upper tail base.** Dead shoreline reefs to 3 m, but typically less than 1 m.

East Indo-Asian Pacific: Sri Lanka and Andaman Sea to Indonesia, Philippines, Micronesia, Papua New Guinea and Solomon Is. – S.W. Japan to Great Barrier Reef, Vanuatu and New Caledonia.

Threespot Damsel – Juvenile

SIZE: 3-4.5 cm (1 1/4 - 1 3/4 in.)

ID: Bluish gray; bright blue lines extend from snout and eyes to foredorsal fin, large black spot ringed with bright blue on middorsal fin, small black saddle ring with bright blue on tail base.

INDONESIAN DAMSEL *Pomacentrus melanochir*

SIZE: to 6.5 cm (2 1/2 in.) Damselfishes – Pomacentridae

ID: Dark charcoal-gray with variable amounts of blue on scales; **small orange spot on upper pectoral fin base.** Usually form groups. Shoreline reefs on rubble slopes, typically where currents are periodically strong to 8 m.

Asian Pacific: Bali, Flores, Timor, Molucca Is., West Papua in Indonesia and Anilao, Luzon in Philippines.

GOLDHEAD DAMSEL *Pomacentrus aurifrons*

SIZE: to 7.5 cm (3 in.) Damselfishes – Pomacentridae

ID: **Yellow on snout, forehead and tips of anterior dorsal spines.** Form groups among sponge, soft corals and branching hard corals. Coastal reefs in 2-14 m.

East Asian Pacific: Papua New Guinea, Solomon Is., Vanuatu, and New Caledonia.

Damselfishes

WESTERN GREGORY   *Stegastes obreptus*
SIZE: to 14.5 cm (6 in.)   Damselfishes – Pomacentridae
ID: Brown with dark scale margins; **black blotch on front dorsal fin.** Solitary or form loose groups. Shoreline and lagoon reefs in 2-6 m.
East Indo-Asian Pacific: Sri Lanka and Andaman Sea to West Papua in Indonesia. – S.W. Japan to N. Australia.

Western Gregory – Juvenile
SIZE: 3-4.5 cm (1 1/4 - 1 3/4 in.)
ID: Yellowish gold with narrow dark scale margins; blue spots on head, large black spot with yellow edging on front of dorsal fin.

PACIFIC GREGORY   *Stegastes fasciolatus*
SIZE: to 16 cm (6 1/4 in.)   Damselfishes – Pomacentridae
ID: Dark brown with blackish scale margins; dusky stripe below lower lip, **violet streak below eye** and scattered violet spots on lower head and body. Rock and coral reefs exposed to surge to 5 m.
Pacific: Indonesia, Philippines, Micronesia to Hawaii and Easter I. – S.W. Japan to Great Barrier Reef.

AUSTRALIAN GREGORY   *Stegastes apicalis*
SIZE: to 13 cm (5 in.)   Damselfishes – Pomacentridae
ID: Brown with darker scale margins; yellow iris, **red-orange margin on rear dorsal fin and upper lobe of tail.** Usually solitary. Coastal reefs and inner parts of Great Barrier Reef to 5 m.
Localized: Great Barrier Reef and E. Australian coast to Sydney area.

CORAL SEA GREGORY   *Stegastes gascoynei*
SIZE: to 15 cm (6 in.)   Damselfishes – Pomacentridae
ID: Brown; **dark scale edges form vertical lines on side,** yellow lower body and fins, golden iris, small black spot on upper pectoral fin base. Solitary or form loose groups. Coral and rocky reefs in 2-30 m.
Southwestern Pacific: Great Barrier Reef and Coral Sea to N. New Zealand.

GOLDEN GREGORY   *Stegastes aureus*
SIZE: to 11 cm (4 1/4 in.)   Damselfishes – Pomacentridae
ID: **Bright yellowish-orange with dusky blue lips;** tiny black saddle on tail base, small black spot on upper pectoral fin base. Solitary or form loose groups. Reef flats, lagoons and outer reefs to 5 m.
Central Pacific: New Caledonia, Gilbert, Fiji, Phoenix, Line and Marquesas Is. in French Polynesia.

BLUNTSNOUT GREGORY *Stegastes punctatus*
SIZE: to 15 cm (6 in.) Damselfishes – Pomacentridae
ID: Brown; **large black blotch on base of rear dorsal fin and back** (pale-edged black spot in Indian Ocean). Form groups; "farm" algae; aggressive toward intruders. Shelter in dead staghorn corals to 5 m.
Indo - West Pacific: Red Sea and E. Africa to Indonesia, Philippines, Solomon Is. and French Polynesia. – S.W. Japan to Australia and Fiji.

DUSKY GREGORY *Stegastes nigricans*
SIZE: to 14 cm (5 1/2 in.) Damselfishes – Pomacentridae
ID: Variable from light brown or gray to nearly black; dark spot at base of rear dorsal fin, **purplish streak below eye.** Form groups; "Farm" algae; aggressive toward intruders. Shelter in branching corals to 10 m.
Indo - Pacific: Red Sea and E. Africa to Indonesia, Micronesia, Solomon Is., French Polynesia. – S.W. Japan to Great Barrier Reef.

WHITEBAR GREGORY *Stegastes albifasciatus*
SIZE: to 11 cm (4 1/4 in.) Damselfishes – Pomacentridae
ID: Dark gray to brown, often white bar across rear body; **large black and white spot on rear base of dorsal fin,** black spot on base of pectoral fin. Solitary or form loose groups. Rubble and boulder areas exposed to wave action to 2 m.
Indo-Pacific: E. Africa to Micronesia and French Polynesia. – S.W. Japan to Australia and New Caledonia. Rare in Asian Pacific.

BIG-LIP DAMSEL *Cheiloprion labiatus*
SIZE: to 7.5 cm (3 in.) Damselfishes – Pomacentridae
ID: Brown; **large swollen lips.** Solitary or loose groups. Associated with branching *Acropora* corals where they feed. on polyps. Sheltered shoreline reefs to 3 m.
Asian Pacific: Andaman Sea, Indonesia, Philippines, Micronesia, Papua New Guinea and Solomon Is. – S.W. Japan to Great Barrier Reef and Vanuatu.

LAGOON DAMSEL *Hemiglyphidodon plagiometopon*
SIZE: to 18 cm (7 in.) Damselfishes – Pomacentridae
ID: Brown; head occasionally pale shading to dark rear body, no distinctive markings. "Farm" algae; aggressive toward intruders. Sheltered shoreline reefs and lagoons often in areas of silting and turbid water to 20 m.
Asian Pacific: Andaman Sea, Indonesia, Philippines, Papua New Guinea and Solomon Is. – S.W. Japan to Great Barrier Reef.

GULF DAMSEL *Pristotis obtusirostris*
SIZE: to 13 cm (5 in.) Damselfishes – Pomacentridae
ID: Light gray to whitish undercolor with blue spot on each scale; occasionally a yellowish to greenish wash on upper head and back. Form groups. Flat sand or rubble bottoms around coral or rock outcroppings in 5-80 m.
East Indo - West Pacific: Sri Lanka to Indonesia, Papua New Guinea and Solomon Is. – S.W. Japan to Great Barrier Reef and Vanuatu.

Sloping Heads/Tapered Bodies
Snappers – Coral Breams – Emperors

This ID Group consists of fishes that have what can best be described as a basic "fish-like" shape with relatively large mouths and notched tails.

FAMILY: Snappers – Lutjanidae
8 Genera – 37 Species Included

Typical Shape

Snappers are medium-sized, oblong fishes with triangular heads. All have shallow, notched tails, and a single, continuous dorsal fin that is often higher in the front. They also have slightly upturned snouts, large mouths, and prominent canine teeth near the front of both jaws. Most species are active, nocturnal predators that primarily feed on fishes, but they also consume cephalopods, gastropods and crustaceans. The two smaller family members in genus *Pinjalo*, pick drifting zooplankton from the currents. Most snappers inhabit shallow to medium depths; however, a few species live along ledges several hundred feet below the surface.

FAMILY: Coral Breams – Nemipteridae
2 Genera – 23 Species Included

Typical Shape

Coral Breams are small- to medium-sized inhabitants of reefs and surrounding sand and rubble areas. All have small terminal mouths that never extend past the eyes, a single unnotched dorsal fin, and indented or forked tails. Although Coral Breams occasionally form groups, most are solitary. The alert, opportunistic bottom-feeders hover just above the sand between short, aggressive dashes in search of food.

FAMILY: Emperors – Lethrinidae
4 Genera – 24 Species Included

Typical Shape

Emperors, close kin to snappers and sweetlips, are common, medium- to large-sized fishes. Most species inhabit the reef's fringes where they feed primarily on sand-dwelling invertebrates. A few species are nocturnal predators and larger species occasionally feed on fishes. Most members of genus *Lethrinus* have the ability to rapidly switch on and off dark mottled patterns, bars or spots.

HUMPBACK SNAPPER   *Lutjanus gibbus*
SIZE: to 50 cm (20 in.)   Snappers – Lutjanidae
ID: Shades of red to gray; maroon forked tail with rounded lobes, high arching nape and foreback, orange around base of pectoral fin. Solitary or form schools. Lagoons, passages and outer reef slopes in 1 - 150 m.
Indo-Pacific: Red Sea and E. Africa to Micronesia and French Polynesia. – S.W. Japan to Australia and New Caledonia.

RED EMPEROR SNAPPER   *Lutjanus sebae*
SIZE: to 80 cm (2 ½ ft.)   Snappers – Lutjanidae
ID: Juvenile/Subadult – White with red-brown midbody bar; band from lip to nape, 2nd band mid-dorsal fin to lower tail. Juveniles from 3-13 cm associate with sea urchins. **Adult –** Solid red. Solitary. Typically deep, sandy bottoms to 100 m.
Indo-Asian Pacific: Red Sea and E. Africa to Indonesia, Philippines. – S.W. Japan to Australia to Vanuatu and New Caledonia.

MALABAR SNAPPER   *Lutjanus malabaricus*
SIZE: to 100 cm (3 ¼ ft.)   Snappers – Lutjanidae
ID: Juvenile – Reddish with narrow dusky stripes; black band from lip through eye to dorsal fin, white-edged black spot covering tail base. **Adult –** Solid red in deep water, not on reefs. Solitary. Coastal and outer reefs in 10 - 40 m.
Indo-West Pacific: Arabian Gulf to Micronesia, Papua New Guinea. – S.W. Japan to Great Barrier Reef, New Caledonia and Fiji.

TIMOR SNAPPER   *Lutjanus timorensis*
SIZE: to 50 cm (20 in.)   Snappers – Lutjanidae
ID: Red; black spot on base of pectoral fin. Young have diagonal dark bar through eye and pearl-white spot on upper tail base [pictured]. Coastal reef slopes, often on sand with scattered reef or log debris in 10-130 m.
East Indo-West Pacific: Sri Lanka and Andaman Sea to Indonesia, Philippines, Solomon Is. and Fiji.

Snappers

LONGSPOT SNAPPER
Lutjanus fulviflamma
SIZE: to 35 cm (14 in.) Snappers – Lutjanidae
ID: Whitish undercolor with **yellow stripes of equal width below lateral line;** yellow tail and anal fin, long oval to rectangular spot on lateral line below mid-dorsal fin. Usually form groups. Estuaries, coastal reefs and outer slopes in 3 - 35 m.
Indo - West Pacific: Red Sea and E. Africa to Indonesia, Micronesia and Solomon Is. – Taiwan to Great Barrier Reef and Fiji.

BLACKSPOT SNAPPER
Lutjanus ehrenbergii
SIZE: to 35 cm (14 in.) Snappers – Lutjanidae
ID: Whitish with **5 thin yellow stripes below lateral line;** yellowish tail and anal fins, large round black spot on lateral line below mid-dorsal fin. Usually form groups. Coastal reefs and estuaries in 1 - 20 m.
Indo - Asian Pacific: Red Sea and E. Africa to Indonesia, Philippines, Micronesia and Solomon Is. – S.W. Japan to N. Great Barrier Reef.

BLUESTRIPE SNAPPER
Lutjanus kasmira
SIZE: to 35 cm (14 in.) Snappers – Lutjanidae
ID: Yellow upper body with 4 blue stripes; **white belly with thin gray to yellow stripes.** Usually congregate around coral outcroppings. Coastal reefs and outer slopes to 35 m.
Indo-Pacific: Red Sea and E. Africa to Micronesia, Hawaii and French Polynesia. – S.W. Japan to S.E. Australia.

BENGAL SNAPPER
Lutjanus bengalensis
SIZE: to 30 cm (12 in.) Snappers – Lutjanidae
ID: Yellow upper body with 4 blue stripes; **white belly with no markings.** Form small groups to large schools around reef outcroppings to 30 m..
Indo - West Pacific: Red Sea and E. Africa to Indonesia, Philippines and Solomon Is.

FIVE-LINED SNAPPER
Lutjanus quinquelineatus
SIZE: to 30 cm (12 in.) Snappers – Lutjanidae
ID: Yellow with **5 blue stripes,** bottom stripe on yellow belly; usually black spot or smudge on rear back. Frequently form groups. Coastal reefs, lagoons and outer reef slopes in 2 - 40 m.
East Indo-West Pacific: Persian Gulf to Andaman Sea, Indonesia, Philippines, Micronesia, Papua New Guinea and Solomon Is. – S.W. Japan to Australia and Fiji.

SPANISH FLAG SNAPPER
Lutjanus carponotatus
SIZE: to 40 cm (16 in.) Snappers – Lutjanidae
ID: Grayish to white undercolor with 5-9 yellow to golden brown stripes; black spot on base of pectoral fin. Solitary or form groups. Turbid coastal reefs, lagoons and outer slopes to 35 m.
East Indo - Asian Pacific: India to Indonesia, Philippines and Papua New Guinea. – S. China to Great Barrier Reef.

GOLDEN-LINED SNAPPER   *Lutjanus rufolineatus*
SIZE: to 28 cm (11 in.)   Snappers – Lutjanidae
ID: Pale reddish with narrow yellow stripes on side; yellow tail, often a small dark spot on rear upper back. Usually form large aggregations; hover above reefs. Steep coastal slopes and outer reefs in 10-50 m.
West Pacific: Andaman Sea, Indonesia, Micronesia, Papua New Guinea and Solomon Is. – S.W. Japan to Great Barrier Reef and Fiji.

YELLOWFIN SNAPPER   *Lutjanus xanthopinnis*
SIZE: to 30 cm (12 in.)   Snappers – Lutjanidae
ID: Silvery with thin yellow stripes, slightly wider stripe from eye to tail; **nearly straight diagonal lines on back above lateral line.** Form small to large aggregations. Congregate around coral or rock outcroppings on coastal reefs and outer slopes in 5-90 m.
Indo-Asian Pacific: Sri Lanka to Andaman Sea, Indonesia and Philippines and Japan.

BIGEYE SNAPPER   *Lutjanus lutjanus*
SIZE: to 30 cm (12 in.)   Snappers – Lutjanidae
ID: Silvery with yellow or yellow-brown stripe from eye to yellow tail with several narrower stripes below; **wavy diagonal lines on back above lateral line.** Form large drifting aggregations. Coastal reefs and outer slopes in 10-90 m.
Indo-Asian Pacific: Red Sea and E. Africa to Indonesia, Philippines and Solomon Is. – S.W. Japan to Australia and Vanuatu.

BUTTON SNAPPER   *Lutjanus boutton*
SIZE: to 28 cm (11 in.)   Snappers – Lutjanidae
ID: Dusky reddish brown, often yellow-orange lower body; occasionally dark eye-sized spot at midback just below dorsal fin. Solitary or form loose aggregations. Coastal reefs, lagoons and outer slopes in 3-20 m.
West Pacific: Indonesia, Philippines, Micronesia, Papua New Guinea and Solomon Is. – S.W. Japan to Australia and Fiji.

MIZENKO'S SNAPPER   *Lutjanus mizenkoi*
SIZE: to 25 cm (10 in.)   Snappers – Lutjanidae
ID: Red to pinkish back becoming grayish below; several faint yellow horizontal lines, yellow anal and tail fins, **two white spots on back.** Coral reefs in 10-70 m.
West Pacific: Indonesia, Milne Bay Province in Papua New Guinea to N. W. Australia and Fiji.

TWOSPOT SNAPPER   *Lutjanus biguttatus*
SIZE: to 20 cm (8 in.)   Snappers – Lutjanidae
ID: Slender; brownish gray back with **2-3 white spots,** reddish brown belly, wide reddish brown stripe from snout to tail and wide white stripe from mouth to tail. Form small to large groups. Coastal reefs, lagoons and seaward slopes in 5-30 m.
East Indo-West Pacific: Maldives to Indonesia, Micronesia and Solomon Is. – Philippines to Great Barrier Reef and Fiji.

Snappers

PAPUAN SNAPPER *Lutjanus papuensis*
SIZE: to 35 cm (14 in.) Snappers – Lutjanidae
ID: Reddish to yellowish head, bluish or gray on upper body, dorsal fin, and tail fin; yellow-orange on sides and lower body, bright yellow anal and ventral fins. Solitary or small groups. Mainly coastal reefs in 6-15 m.
Localized: West Papua in Indonesia, Papua New Guinea and Solomon Is.

ONESPOT SNAPPER *Lutjanus monostigma*
SIZE: to 55 cm (22 in.) Snappers – Lutjanidae
ID: Silver to reddish or yellowish silver with **yellow fins;** may display a horizontally elongate black spot on rear back. Solitary or form small groups. Outer reef areas in 5-60 m.
Indo-Pacific: Red Sea and E. Africa to Andaman Sea, Indonesia, Philippines, Micronesia, Solomon Is. and French Polynesia. – S.W. Japan to Australia and New Caledonia.

MANGROVE SNAPPER *Lutjanus argentimaculatus*
SIZE: to 120 cm (4 ft.) Snappers – Lutjanidae
ID: Grayish with tints of red, green and brown; **darkish fins, darkish scale centers. Juvenile –** Eight white bars and pair of blue streaks across cheek. Solitary or form loose aggregations. Mangrove coasts to steep outer reefs to 120 m.
Indo-Pacific: Red Sea and E. Africa to Indonesia, Micronesia, Solomon Is. and Line Is. – S.W. Japan to S.E. Australia and Fiji.

BLACKTAIL SNAPPER *Lutjanus fulvus*
SIZE: to 40 cm (16 in.) Snappers – Lutjanidae
ID: Silvery white to pale yellow to yellow or tan with dark tail, **yellow pectoral, ventral and anal fins.** Solitary or form loose aggregations. Coastal reefs, lagoons and outer slopes to 75 m.
Indo-Pacific: Red Sea and E. Africa to Andaman Sea, Indonesia, Micronesia, Papua New Guinea, Solomon Is., Hawaii and French Polynesia. – S.W. Japan to Great Barrier Reef.

RUSSELL'S SNAPPER *Lutjanus russellii*
SIZE: to 45 cm (18 in.) Snappers – Lutjanidae
ID: Silvery white with pinkish hue; occasionally have yellow ventral and anal fins, dark tail, large black blotch below rear dorsal fin turned on and off; **often black spot on base of pectoral fin.** Solitary or form groups. Coastal and offshore reefs in 3-80 m.
West Pacific: Indonesia, Philippines, S.W. Japan, Micronesia, Solomon Is. and Fiji.

Russell's Snapper – Juvenile
SIZE: 5-10 cm (2-4 in.)
ID: White with 4 dark brown stripes from snout to rear body; often elongate black spot with pale edge on upper rear body. Similar species in Indian Ocean including Andaman Sea, Striped Snapper, *L. indicus,* distinguished by 7 brown to yellow stripes from rear head to body.

RED SNAPPER *Lutjanus bohar*
SIZE: to 75 cm (2 1/2 ft.) Snappers – Lutjanidae
ID: Red to reddish gray; large robust body, **pronounced groove in front of eye,** upper edge of pectoral fin dark. Solitary or form groups. Lagoons and outer reefs in 5-150 m.

Indo-Pacific: Red Sea and E. Africa to Indonesia, Philippines, Micronesia, Papua New Guinea, Solomon Is. and French Polynesia. – S.W. Japan to Australia.

Red Snapper – Subadult
SIZE: 5-15 cm (2-6 in.)
ID: Grayish brown with two white spots on back just below start and end of soft dorsal fin; translucent tail with dark borders.

BLUBBERLIP SNAPPER *Lutjanus rivulatus*
SIZE: to 80 cm (2 1/2 ft.) Snappers – Lutjanidae
ID: Grayish green with wavy yellow lines on head and yellow fins and outer edge of tail. Solitary. Near shore and outer reefs in 2-100 m.

Indo-Pacific: Red Sea and E. Africa to Indonesia, Philippines, Micronesia, Papua New Guinea, Solomon Is. and French Polynesia. – S.W. Japan to Australia and Fiji.

LUNARTAIL SNAPPER *Lutjanus lunulatus*
SIZE: to 35 cm (14 in.) Snappers – Lutjanidae
ID: Reddish pink with **yellow belly, pectoral, ventral and anal fins;** wide dark crescent on foretail. Solitary or form small groups. Coastal reefs and seaward slopes in 10-30 m.

East Indo-Asian Pacific: Pakistan to Andaman Sea, Indonesia, Philippines, Papua New Guinea and Vanuatu.

DARKTAIL SNAPPER *Lutjanus lemniscatus*
SIZE: to 65 cm (2 1/4 ft.) Snappers – Lutjanidae
ID: Gray-brown to olive, silvery or reddish with black tail; **pectoral, ventral and anal fins match color of body** (never yellow). Solitary or form groups. Coastal reefs, lagoons and outer slopes in 2-80 m.

East Indo-Asian Pacific: Sri Lanka to Indonesia, Philippines, S. Papua New Guinea and Great Barrier Reef.

Darktail Snapper – Juvenile
ID: Gray with dark stripe from snout through eye to tail base with white borders from behind eye to tail base.

107

Snappers

BLACK-BANDED SNAPPER   *Lutjanus semicinctus*
SIZE: to 35 cm (14 in.)   Snappers – Lutjanidae
ID: Olive to yellowish back, white below; **7 black bars on upper body come to points,** large black spot covering tail base. Solitary or form small groups. Lagoons and seaward reefs in 5-35 m.

West Pacific: Indonesia, Philippines, Micronesia, Papua New Guinea, Solomon Is., Great Barrier Reef (rare), New Caledonia, Vanuatu and Fiji.

CHECKERED SNAPPER   *Lutjanus decussatus*
SIZE: to 30 cm (12 in.)   Snappers – Lutjanidae
ID: White with 6 brown stripes on body; **5-6 dark bars across back forming netted pattern on upper body,** black spot on base of tail. Solitary or form groups. Coastal reefs, lagoons and outer slopes in 5-35 m.

Indo-Asian Pacific: India to Indonesia, Philippines and Micronesia. – S.W. Japan to W. Australia.

BROWNSTRIPE SNAPPER   *Lutjanus vitta*
SIZE: to 40 cm (16 in.)   Snappers – Lutjanidae
ID: Whitish with diagonal brownish lines above lateral line and horizontal below; **yellow to brown or black stripe from eye to base of tail.** Solitary or form groups. Coastal and offshore reefs in 10-72 m.

Indo-Asian Pacific: Seychelles to Indonesia, Philippines, Micronesia and Solomon Is. – S.W. Japan to Australia and New Caledonia.

YELLOWTAIL FALSE FUSILIER   *Paracaesio xanthura*
SIZE: to 40 cm (16 in.)   Snappers – Lutjanidae
ID: Blue with yellow area on back extending from forehead to tail. Form schools. Below diver depths in tropics, but in 10-50 m in subtropics.

Indo-West Pacific: E. Africa to S.W. Japan, Papua New Guinea and Fiji. Rarely sighted in Asian Pacific, often near cool upwellings.

PINJALO SNAPPER   *Pinjalo pinjalo*
SIZE: to 50 cm (20 in.)   Snappers – Lutjanidae
ID: Variable shades from reddish gray to red that can quickly intensify or pale; **yellow ventral fins,** more robust than similar Slender Pinjalo [next]. Form schools. Coastal reefs and outer slopes in 15-100 m.

East Indo-Asian Pacific: Persian Gulf to Taiwan, Indonesia, Philippines, Micronesia, Papua New Guinea and Solomon Is.

SLENDER PINJALO   *Pinjalo lewisi*
SIZE: to 50 cm (20 in.)   Snappers – Lutjanidae
ID: Variable shades from grayish red to bright red that can quickly intensify or pale; may display pale spot on upper base of tail. Similar Pinjalo Snapper [previous] distinguished by yellow ventral fins. Form schools. Outer reefs in 20-100 m.

East Indo-West Pacific: Persian Gulf to Sabah in Malaysia, Indonesia, Philippines, Papua New Guinea, Solomon Is. and Fiji.

FALSE FUSILIER   *Paracaesio sordida*
SIZE: to 35 cm (14 in.)   Snappers – Lutjanidae
ID: Purplish brown to blue with faint dark stripes; lower lobe of tail red, slender elongate body with deeply forked tail similar to fusilier family (Caesionidae). Form schools. Deep steep outer reefs in 30-200 m.
Indo-West Pacific: Red Sea to S.W. Japan, Micronesia, Papua New Guinea and Fiji.

BLACK SNAPPER   *Macolor niger*
SIZE: to 60 cm (2 ft.)   Snappers – Lutjanidae
ID: Gray to gray-brown with numerous indistinct blotches; **no blue lines or spots on head,** large eye with dull gold iris. Solitary or form schools. Steep slopes of lagoons, passes and outer reefs in 3-90 m.
Indo-West Pacific: Red Sea and E. Africa to Indonesia, Micronesia and Solomon Is. – S.W. Japan to Great Barrier Reef and Fiji.

Black Snapper – Juvenile
SIZE: 4-15 cm (1 ½ - 6 in.)
ID: Distinctive black and white pattern; **tips of tail lobes white.** Solitary; swim with jerky motion. Upper edge of steep slopes in 5-15 m.

MIDNIGHT SNAPPER   *Macolor macularis*
SIZE: to 60 cm (2 ft.)   Snappers – Lutjanidae
ID: Black with pale line markings on scales; **blue line and spot markings on head,** gold gill cover, lower head and forebody, eye with gold iris. Solitary or form groups. Edge of steep slopes of lagoons, passes and outer reefs in 5-50 m.
East Indo-West Pacific: Maldives to Indonesia, Philippines, Micronesia, Solomon Is. – S.W. Japan to Great Barrier Reef and Fiji.

Midnight Snapper – Intermediate
SIZE: 15-25 cm (6-10 in.)
ID: Body color and head markings similar to adults, but commonly have several white spots on back and midlateral; white stripe from behind pectoral fin onto tail. Often have gold undercolor on lower head that extends below pectoral fins.

Midnight Snapper – Juvenile
SIZE: 4-15 cm (1½-6 in.)
ID: Distinctive black and white pattern, **tips of tail lobes clear;** white stripe through center of tail, very long ventral fins. Solitary; swims with jerky motion. Upper edge of steep slopes in 5-15 m.

109

CHINAMANFISH *Symphorus nematophorus*
SIZE: to 80 cm (2½ ft.)   Snappers – Lutjanidae
ID: Reddish to yellowish brown with faint to distinct irregular bars and numerous faint bluish stripes on head and body; occasionally filaments extend from upper rear of dorsal fin. Solitary or form large schools. Coastal reefs to 50 m.
Asian Pacific: Brunei, Indonesia, Philippines, Papua New Guinea and Solomon Is. – S.W. Japan to Great Barrier Reef and Vanuatu.

Chinamanfish – Young and Subadults
ID: On young and subadults the blue head and body stripes are much brighter and obvious; filaments more commonly extend from the upper rear dorsal fin.

SAILFIN SNAPPER *Symphorichthys spilurus*
SIZE: to 60 cm (2 ft.)   Snappers – Lutjanidae
ID: Undercolor shades of yellow with numerous blue stripes from head to tail; black ocellated spot on tail base, dark saddle bar behind head. Form schools. Mixed sand and coral to 60 m.
Asian Pacific: Indonesia, Philippines, Micronesia and Solomon Is. – S.W. Japan to Great Barrier Reef and New Caledonia.

Sailfin Snapper – Subadult
ID: Subadults trail long filaments from dorsal and anal fins. With age adults lose the trailing fin filaments and develop a steep snout (almost squared-off head profile). **Juvenile –** Broad black stripe bordered with pale blue extending from eye to tail-base spot.

SMALLTOOTH JOBFISH *Aphareus furca*
SIZE: to 55 cm (22 in.)   Snappers – Lutjanidae
ID: Blue-gray; slender body with large mouth and strongly forked tail, pectoral fins long, dark outline on rear edge and thin **bar on gill cover.** Solitary or form small groups. Lagoons and seaward reefs in 5-100 m.
Indo-Pacific: Red Sea and E. Africa to Indonesia, Philippines, Micronesia, Solomon Is., French Polynesia. – S.W. Japan to Australia.

GREEN JOBFISH *Aprion virescens*
SIZE: to 110 cm (3½ ft.)   Snappers – Lutjanidae
ID: Dark green to blue to bluish gray; slender cylindrical body with strongly forked tail, pectoral fins short, no distinctive markings. Usually solitary. Lagoons, reef passes and outer slopes in 5-150 m.
Indo-Pacific: Red Sea and E. Africa to Micronesia, Solomon Is., Hawaii, French Polynesia. – S.W. Japan to Great Barrier Reef.

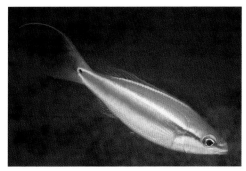

PARADISE WHIPTAIL   *Pentapodus paradiseus*
SIZE: to 25 cm (10 in.)   Coral Breams – Nemipteridae
ID: Gray; **white to yellow stripe bordered by dark gray stripe from eye to tail joins white V-shaped mark and small black spot on tail base.** Solitary or form small groups. Similar Butterfly Whiptail [next] distinguished by location. Coastal reefs in 10-70 m.
Asian Pacific: N. & E. Papua New Guinea, Solomon Is. and Great Barrier Reef.

Paradise Whiptail – Juvenile
SIZE: 3-5 cm (1¼-2 in.)
ID: Dark back and white lower body; broad yellow stripe on side bordered by black stripe below. Solitary. Coastal reefs in 10-70 m.

BUTTERFLY WHIPTAIL   *Pentapodus setosus*
SIZE: to 25 cm (10 in.)   Coral Breams – Nemipteridae
ID: Light gray; gold to yellow stripe from eye to upper base of tail joins a pale lower stripe to form a V with central dark spot; adults have filament extending from upper tail lobe. Solitary or form small groups. Coastal reefs in 5-50 m.
North Asian Pacific: Sabah in Malaysia, Singapore, Indonesia and Philippines.

NORTHWEST AUSTRALIAN WHIPTAIL   *Pentapodus porosus*
SIZE: to 25 cm (10 in.)   Coral Breams – Nemipteridae
ID: Light gray; **yellow stripe from eye extends onto forebody,** blue V-shaped mark and black spot on tail base, thin blue stripe along base of dorsal fin. Solitary or form small groups. Coastal reefs in 5-50 m.
Localized: N.W. Australia, West Papua in Indonesia and S. Papua New Guinea.

SMALL-TOOTHED WHIPTAIL   *Pentapodus caninus*
SIZE: to 25 cm (10 in.)   Coral Breams – Nemipteridae
ID: Greenish gray to bluish gray back shading to whitish lower body; broad yellowish stripe from eye to base of tail, narrow stripe on back, no dark band between eyes. Solitary or form small groups. Silty coastal reefs and deep lagoons in 2-35 m.
Asian Pacific: Brunei to Indonesia, Philippines, Micronesia. – S.W. Japan to New Caledonia.

JAPANESE WHIPTAIL   *Pentapodus nagasakiensis*
SIZE: to 20 cm (8 in.)   Coral Breams – Nemipteridae
ID: Brownish back shading to whitish lower body; white stripe bordered with wide yellow to gold diffuse stripes extends from snout through lower eye to tail, slender with pointed snout. Solitary or form small groups. Coastal reefs in 20-100 m.
Asian Pacific: S.W. Japan, Indonesia and Philippines to N.W. Australia.

Coral Breams

DOUBLE WHIPTAIL *Pentapodus emeryii*
SIZE: to 35 cm (14 in.) Coral Breams – Nemipteridae
ID: Blue back shading to whitish lower body with thin yellowish stripe on back and wider midlateral stripe; slender with pointed snout and elongate filament on each tail lobe. Solitary or form small groups. Silty coastal reefs in 2-35 m.
Asian Pacific: Brunei, Sabah in Malaysia, E. Indonesia and Philippines to N.W. Australia.

Double Whiptail – Juvenile
SIZE: 4-8 cm (1 1/2 - 3 1/4 in.)
ID: Deep blue to purple; **yellow to gold midlateral stripe somewhat wider than back stripe.**

YELLOW-STRIPED WHIPTAIL *Pentapodus aureofasciatus*
SIZE: to 25 cm (10 in.) Coral Breams – Nemipteridae
ID: Bluish back and white belly; creamy yellow midlateral stripe and yellowish streak on back. Solitary or in pairs. Open water over sand near reefs in 5-20 m.
West Pacific: Brunei, Sabah in Malaysia, Indonesia, Philippines, Papua New Guinea, Solomon Is. to New Caledonia and Fiji.

Yellow-striped Whiptail – Juvenile
SIZE: 4-7 cm (1 1/2 - 2 3/4 in.)
ID: Blue body; **yellow to gold midlateral stripe same width as yellow to gold stripe on back.** Usually solitary. Rubble areas.

PAPUAN WHIPTAIL *Pentapodus numberii*
SIZE: to 25 cm (10 in.) Coral Breams – Nemipteridae
ID: Dark gray back, shading to yellowish or bluish white on lower body; bluish stripe on side of snout runs to gill cover, yellow midbody stripe can be switched on or off. Solitary or small groups. Coastal reefs from 12-60 m.
Asian Pacific: Halmahera, Raja Ampat and West Papua in E. Indonesia to Papau New Guinea and Solomon Is.

THREE-STRIPED WHIPTAIL *Pentapodus trivittatus*
SIZE: to 28 cm (11 in.) Coral Breams – Nemipteridae
ID: Brown to gray, underside of head and body white; pale saddles and stripe on back, white midbody blotch shading into stripe toward tail. Solitary or in pairs. Coastal reefs and lagoons in 2-30 m.
Asian Pacific: Malaysian Peninsula to Indonesia, Philippines, Micronesia and Solomon Is.

WHITE-SHOULDERED WHIPTAIL *Pentapodus bifasciatus*
SIZE: to 20 cm (8 in.)   Coral Breams – Nemipteridae
ID: Brown to dark gray on upper body with 3 white stripes; white marking on upper rear edge of gill cover; slender with pointed snout. Solitary or groups. Silty coastal reefs in 2-20 m.
Asian Pacific: Malaysian Peninsula, Singapore, W. Indonesia and Philippines.

RAINBOW MONOCLE BREAM *Scolopsis temporalis*
SIZE: to 40 cm (16 in.)   Coral Breams – Nemipteridae
ID: Gray back with pale blue to white stripe and wider bright stripe below; pair of blue bands between eyes, darkish spot behind upper rear eye, narrow orange line on cheek. Solitary or form groups. Mixed sand of coastal and lagoon reefs to 35 m.
West Pacific: Bali and E. Sulawesi in Indonesia, Papua New Guinea and Solomon Is. to Vanuatu and Fiji.

BRIDLED MONOCLE BREAM *Scolopsis bilineata*
SIZE: to 25 cm (10 in.)   Coral Breams – Nemipteridae
ID: Dark gray to yellow upper body, white below; black-edged white band runs from below eye to rear of dorsal fin, 3 yellow to white stripes on upper head. Solitary or form small groups. Sand and rubble fringe of reefs to 25 m.
East Indo-West Pacific: Maldives to Indonesia, Philippines, Micronesia and Solomon Is. – S.W. Japan to Australia and Fiji.

Bridled Monocle Bream – Juvenile
SIZE: 3-5 cm (1 1/4 - 2 in.)
ID: Alternating black and yellow stripes on upper body, whitish below; large black spot on foredorsal fin. Juveniles in Fiji entirely yellow with small black spot on foredorsal fin; similar to and believed to mimic the Canary Fangblenny. Sand and rubble fringe of reefs to 25 m.

THREE-LINED MONOCLE BREAM *Scolopsis trilineata*
SIZE: to 25 cm (10 in.)   Coral Breams – Nemipteridae
ID: Shades of gray; 3 white to yellowish curving lines extend from eye to back and base of dorsal fin, dark band between eyes with bluish band above. Solitary or form groups. Sheltered coastal reefs and lagoons in 1-10 m.
West Pacific: Indonesia, Philippines, Micronesia, Papua New Guinea and Solomon Is. – Taiwan to Great Barrier Reef and Fiji.

LATTICE MONOCLE BREAM *Scolopsis taenioptera*
SIZE: to 25 cm (10 in.)   Coral Breams – Nemipteridae
ID: Yellowish gray upper body and white below; dark midlateral stripe, red spot on pectoral fin base, bluish stripe or patch on snout extends between eyes. Solitary or form groups; shy. Sandy fringe of reefs or muddy estuaries in 2-25 m.
Asian Pacific: Andaman Sea to Brunei, Indonesia, Philippines, S. Papua New Guinea. – Taiwan to Australia.

Coral Breams

PEARLY MONOCLE BREAM   *Scolopsis margaritifera*
SIZE: to 25 cm (10 in.)   Coral Breams – Nemipteridae
ID: Pearly shades of gray; frequently 2-3 longitudinal rows of
yellow spots on side and yellow pectoral fin base, two narrow
whitish bars below rear dorsal fin. Solitary or form groups. Sand
and rubble fringe of coastal reefs and lagoons in 2-25 m.
Asian Pacific: Andaman Sea, Indonesia, Philippines, Micronesia
and Solomon Is. – Taiwan to Great Barrier Reef and Vanuatu.

Pearly Monocle Bream – Juvenile
SIZE: 4-6 cm (1 1/2 - 2 1/4 in.)
ID: White upper body bordered with pair of black stripes and
yellow below; black spot on front of dorsal fin. Solitary or
form groups. Sand and rubble fringe of coastal reefs and
lagoons in 2-25 m.

WHITESTREAK MONOCLE BREAM   *Scolopsis ciliata*
SIZE: to 26 cm (10 1/2 in.)   Coral Breams – Nemipteridae
ID: Pale gray body; 2-4 longitudinal rows of orange spots on
side, white streak below base of dorsal fin. Solitary or form
groups. Sandy fringe of coastal and lagoon reefs in 2-25 m.
Frequent silty areas.
Asian Pacific: Andaman Sea, Indonesia, Philippines and
Micronesia. – S.W. Japan to Great Barrier Reef and Vanuatu.

Whitestreak Monocle Bream – Juvenile
SIZE: 4-6 cm (1 1/2 - 2 1/4 in.)
ID: Dark gray above and lighter below with white stripe below
dorsal fin base and white midlateral stripe of varying width to
tail base. Solitary or form groups. Coastal reefs in 8-25 m.

YELLOWSTRIPE MONOCLE BREAM   *Scolopsis aurata*
SIZE: to 30 cm (12 in.)   Coral Breams – Nemipteridae
ID: Grayish back with narrow pale stripes; broad yellow to
orange-brown midlateral stripe, dark gray snout with bluish
band connecting eyes. Solitary or form small groups. Sandy
bottoms of lagoons and coastal reefs to 30 m.
East Indian Ocean: Maldives, Sri Lanka, and Andaman Sea
to W. Sumatra, S. Java and Bali in Indonesia.

JAPANESE MONOCLE BREAM   *Scolopsis japonicus*
SIZE: to 25 cm (10 in.)   Coral Breams – Nemipteridae
ID: Brown with dark scale outlines; whitish bar on head
behind eye and **yellow dorsal, anal and tail fins.** Similar
Vosmer's Monocle Bream [next] has whitish tail. **Juvenile –** Similar,
but with white belly. Sand and rubble bottoms in 5-30 m.
Asian Pacific: Indonesia, Philippines. – S.W. Japan to Australia.

VOSMER'S MONOCLE BREAM *Scolopsis vosmeri*
SIZE: to 20 cm (8 in.)   Coral Breams – Nemipteridae
ID: Purplish brown with dark scale centers; wide white bar on head behind eye and **whitish tail and base.** Similar Japanese Monocle Bream [previous] distinguished by yellow dorsal, anal, and tail fins. Solitary or small groups. Sand and rubble bottoms in 2-25 m.
W. Asian Pacific: E. Andaman Sea, Brunei and W. Indonesia.

Vosmer's Monocle Bream – Juvenile
SIZE: 4-8 cm (1½ - 3 in.)
ID: Silver gray with white midlateral stripe bordered with black. Inshore waters in 3-8 m, often in turbid conditions.

PALE MONOCLE BREAM *Scolopsis affinis*
SIZE: to 25 cm (10 in.)   Coral Breams – Nemipteridae
ID: Pale gray body with yellowish tail; 3-4 rows of black dots on upper back. Solitary or form small groups. Sandy bottoms of lagoons and coastal reefs in 5-35 m.
Asian Pacific: Andaman Sea to Indonesia, Philippines, Micronesia, Papua New Guinea and Solomon Is. – S.W. Japan to Great Barrier Reef.

Pale Monocle Bream – Juvenile
SIZE: 4-7 cm (1½ - 2¾ in.)
ID: White with two dusky stripes on back and wide black midlateral stripe. Solitary or form small groups. Sandy bottoms of lagoons and coastal reefs in 5-35 m.

STRIPED MONOCLE BREAM *Scolopsis lineata*
SIZE: to 25 cm (10 in.)   Coral Breams – Nemipteridae
ID: Dark gray to blackish upper body with 2-3 white stripes and irregular whitish bars, white below. Solitary or form groups. Sandy fringe of coastal reefs, lagoons and seaward slopes in 1-20 m.
Asian Pacific: Andaman Sea to Indonesia, Philippines, Micronesia and Solomon Is. – S.W. Japan to Great Barrier Reef and Vanuatu.

PEARL-STREAKED MONOCLE BREAM *Scolopsis xenochrous*
SIZE: to 22 cm (8¾ in.)   Coral Breams – Nemipteridae
ID: Brownish body with pale underside; brown-edged diagonal blue streak behind head followed by a series of brown spots and elongate pearly streak. Solitary or form groups. Rubble areas of coastal reefs, lagoons and outer slopes in 5-50 m.
East Indo-Asian Pacific: Maldives to Indonesia, Philippines, Papua New Guinea, Solomon Is. – S.W. Japan to Great Barrier Reef.

MONOGRAM MONOCLE BREAM *Scolopsis monogramma*
SIZE: to 38 cm (15 in.)   Coral Breams – Nemipteridae
ID: Pale gray to whitish body; blue band between eyes and margin on yellowish tail, often display elongate brown blotch on side with adjacent yellowish areas. Solitary or form small groups. Sandy fringe of coastal reefs and lagoons in 2-50 m.
Asian Pacific: Andaman Sea to Indonesia, Philippines, Papua New Guinea. – S.W. Japan to Great Barrier Reef, New Caledonia.

Monogram Monocle Bream – Pale Phase
ID: Pale grayish when brown blotch and yellowish areas are not displayed; blue band between eyes and blue margin on yellowish tail; older adults have a long filament extending from upper tail lobe.

STRIPED LARGE-EYE BREAM *Gnathodentex aureolineatus*
SIZE: to 30 cm (12 in.)   Emperors – Lethrinidae
ID: Silvery gray to brown; dark scale rows on back, 4-5 brown to gold stripes on sides, **yellow-orange blotch below rear dorsal fin.** Solitary to large aggregations. Shallow coastal reefs, lagoons and outer slopes in 3-20 m.
Indo-Pacific: E. Africa to Indonesia, Philippines, Micronesia, Solomon Is., French Polynesia. – S.W. Japan to Great Barrier Reef.

HUMPNOSE BIGEYE BREAM *Monotaxis grandoculis*
SIZE: to 60 cm (2 ft.)   Emperors – Lethrinidae
ID: Silvery gray with dark scale margins; black spot on base of pectoral fin, dark borders and margin of tail, often yellowish upper lip and bar on rear gill cover. Solitary or form groups. Coastal reefs, lagoons and outer slopes to 100 m.
Indo-Pacific: Red Sea and E. Africa to Micronesia, Hawaii and French Polynesia. – S.W. Japan to Australia and New Caledonia.

Humpnose Bigeye Bream – Variation
ID: Black back with two white bars; silvery below lateral line; dark to black borders and margin on tail, often yellowish upper lip and bar on rear gill cover.

Humpnose Bigeye Bream – Juvenile
SIZE: 4-10 cm (1 1/2 -4 in.)
ID: White with narrow dark bar through eye; wide dark bars extend from nape, spiny and soft dorsal fins onto midbody, yellowish forked tail often with dark stripe down middle of each lobe.

REDFIN BREAM *Monotaxis heterodon*

SIZE: to 35 cm (14 in.) Emperors – Lethrinidae

ID: Gray or brown on back with 3 narrow white bars; black spot on axil of pectoral fin, reddish to yellowish forked tail. Solitary or form groups. Coastal reefs, lagoons and outer slopes to 100 m.

Indo-West Pacific: Seychelles to Indonesia, Philippines, Micronesia, Papua New Guinea. – S.W. Japan to Great Barrier Reef and Fiji.

Redfin Bream – Subadult

SIZE: 4-10 cm (1 1/2 -4 in.)

ID: Dark to nearly black upper body divided by 3 narrow white bars; **black spot on base of pectoral fin,** gray lower body; reddish to yellowish forked tail.

EYEBROWED LARGE-EYE BREAM *Gymnocranius superciliosus*

SIZE: to 45 cm (18 in.) Emperors – Lethrinidae

ID: Silvery gray with **dark "eyebrow" marking above eye;** eye bar extends to lower head; black spot on each scale of back form rows; forked tail. Solitary or form small groups. Sand or rubble bottoms, adjacent rocks and reefs in 15-40 m.

West Pacific: Raja Ampat, Indonesia to Great Barrier Reef, New Caledonia and Fiji.

JAPANESE LARGE-EYE BREAM *Gymnocranius euanus*

SIZE: to 45 cm (18 in.) Emperors – Lethrinidae

ID: Silvery gray with **scattered small black blotches.** Solitary or form small groups. Open sand or rubble bottoms near reefs in lagoons and on outer slopes in 15-50 m.

West Pacific: Sabah in Malaysia, Indonesia, Philippines, Micronesia, Papua New Guinea and Solomon Is. – S.W. Japan to Great Barrier Reef and Fiji.

GRAY LARGE-EYE BREAM *Gymnocranius griseus*

SIZE: to 35 cm (14 in.) Emperors – Lethrinidae

ID: Silvery gray with several irregular dark bars on head and body. Young have heavily blotched pattern with eye bar, and bar across pectoral fins. Solitary or form small groups. Sand or rubble bottoms of coastal reefs in 15-80 m.

East Indo-Asian Pacific: India and Andaman Sea to Indonesia, Micronesia and Papua New Guinea to S.W. Japan.

BLUE-LINED LARGE-EYE BREAM *Gymnocranius grandoculis*

SIZE: to 80 cm (2 1/2 ft.) Emperors – Lethrinidae

ID: Adult – Silvery gray with **wavy blue lines on cheek;** no yellow tint on snout, tail fin moderately forked with pointed tips. **Juvenile –** Silvery gray; several thin bars on side; eye bar. Solitary or form small groups. Sand or rubble bottoms in 15-100 m.

Indo-Pacific: Red Sea and E. Africa to Indonesia, Micronesia, Solomon Is. and French Polynesia. – S.W. Japan to Great Barrier Reef.

Emperors

LONGFACE EMPEROR *Lethrinus olivaceus*
SIZE: to 100 cm (3 ¼ ft.) Emperors – Lethrinidae
ID: **Elongate body with long pointed snout;** gray to olive with no distinctive markings, often display mottled pattern. Largest species in family. Solitary or form groups; highly active and fast swimming. Sand bottoms of lagoons and outer slopes to 185 m.
Indo-West Pacific: Red Sea and E. Africa to Indonesia, Philippines, Micronesia, and Solomon Is. – S.W. Japan to Australia and Fiji.

SPANGLED EMPEROR *Lethrinus nebulosus*
SIZE: to 80 cm (2 ½ ft.) Emperors – Lethrinidae
ID: Elongate pointed snout; pale gray with blue to white scale centers, **blue streaks on cheek.** Solitary to large groups. Flat sand bottoms in the vicinity of reefs to 75 m; also frequent seagrass beds and mangrove areas.
Indo-West Pacific: Red Sea and E. Africa to Indonesia, Philippines and Solomon Is. – S.W. Japan to Australia and Fiji.

SPOTCHEEK EMPEROR *Lethrinus rubrioperculatus*
SIZE: to 50 cm (20 in.) Emperors – Lethrinidae
ID: Bright silver to brownish silver with diffuse stripe and bar markings on lower body; reddish brown edge on rear gill cover. Solitary or form small groups. Sand and rubble bottoms of coastal reefs and outer slopes to 40 m.
Indo-Asian Pacific: East Africa to Indonesia, Philippines, Micronesi and Solomon Is. – S.W. Japan to Australia and New Caledonia.

YELLOWLIP EMPEROR *Lethrinus xanthochilus*
SIZE: to 60 cm (2 ft.) Emperors – Lethrinidae
ID: Elongate body; unmarked silvery pale gray to olive, also mottled and blotched pattern, **yellow upper lip,** yellow to orange spot on base of pectoral fin. Solitary or form small groups. Sand and rubble bottoms near reefs in 5-30 m.
Indo-Pacific: Red Sea and E. Africa to Micronesia, Solomon Is., French Polynesia. – S. Japan to Great Barrier Reef, New Caledonia.

SMALLTOOTH EMPEROR *Lethrinus microdon*
SIZE: to 70 cm (2 ¼ ft.) Emperors – Lethrinidae
ID: Long pointed snout; silvery gray, **streaks radiate from front lower quarter of eye.** Solitary or form groups. Coastal reefs, lagoons and outer slopes in 10-80 m.
Indo-Asian Pacific: Red Sea and E. Africa to Indonesia, Philippines, Micronesia and Papua New Guinea. – S.W. Japan to Great Barrier Reef.

Smalltooth Emperor – Mottled Phase
ID: May camouflage by changing to mottled and blotched shades of brown, especially at night, but still distinguished by dark streaks radiating from fore lower quarter of eye. Most species of emperors can quickly change to a similar mottled pattern primarily when near bottom.

THUMBPRINT EMPEROR — *Lethrinus harak*
SIZE: to 50 cm (20 in.) Emperors – Lethrinidae
ID: Pale gray with **dark elongate blotch on middle of side, occasionally some yellow to orange near or around blotch.** Solitary or form groups. Sandy shallows next to shore, coastal reefs and lagoons to 20 m.
Indo-West Pacific: Red Sea and E. Africa to Indonesia, Philippines, Micronesia and Solomon Is. – S.W. Japan to N. E. Australia and Fiji.

Thumbprint Emperor – Phase
ID: Has the ability to rapidly fade and intensify dark midbody spot. **Two narrow lines below eye are distinctive when spot is faded.**

PINKEAR EMPEROR — *Lethrinus lentjan*
SIZE: to 50 cm (20 in.) Emperors – Lethrinidae
ID: Pale silvery gray; **bright red streak on rear edge of gill cover.** Solitary or form groups. Sandy areas of coastal reefs, lagoons and outer slopes in 10-50 m.
Indo-West Pacific: Red Sea and E. Africa to Indonesia, Philippines, Micronesia, Papua New Guinea and Solomon Is. – S.W. Japan to N. Australia, New Caledonia and Fiji.

Pinkear Emperor – Mottled Phase
ID: May camouflage by changing to mottled and blotched shades of brown especially when resting on bottom at night, but still distinguished by bright red streak on rear edge of gill cover. Most species of emperors can quickly change to a similar mottled pattern.

ORANGE-STRIPED EMPEROR — *Lethrinus obsoletus*
SIZE: to 50 cm (20 in.) Emperors – Lethrinidae
ID: Pale gray; **yellow stripe from base of pectoral fin to tail.** Solitary or form small groups. Seagrass beds and sand and rubble areas of coastal reefs, lagoons and outer slopes to 30 m.
Indo-West Pacific: Red Sea and E. Africa to Indonesia, Philippines, Micronesia, Papua New Guinea and Solomon Is. – S.W. Japan to Australia, New Caledonia and Fiji.

Orange-striped Emperor – Blotched Phase
ID: May quickly camouflage to mottled and blotched shades of brown especially when sheltering in seagrass, branching gorgonians or corals also at night, but still distinguished by yellow stripe from base of pectoral fin to tail. Most species of emperors can quickly change to a similar mottled pattern.

Emperors

YELLOWFIN EMPEROR *Lethrinus erythracanthus*
SIZE: to 70 cm (2 1/4 ft.) Emperors – Lethrinidae
ID: Dark bluish head and dark gray body with yellow or occasionally reddish fins. Easily distinguished from most emperors by large size. Solitary. Deep lagoons and outer reef slopes in 15-120 m.
Indo-Pacific: E. Africa to Indonesia, Philippines, Micronesia, Solomon Is., French Polynesia. – S.W. Japan to Great Barrier Reef.

Yellowfin Emperor – Juvenile
SIZE: 5-10 cm (2-4 in.)
ID: Gray head, yellowish body with several narrow white stripes or broken lines on side. Solitary. Deep lagoons and outer reef slopes in 15-120 m.

LONGSPINE EMPEROR *Lethrinus genivittatus*
SIZE: to 25 cm (10 in.) Emperors – Lethrinidae
ID: Silvery brown with brown bars, spots and mottling; **second spine of dorsal fin longest** (often much longer). Solitary. Grass beds, estuaries, mangroves and shallow areas of sand around coastal reefs.
Asian Pacific: Indonesia, Philippines, Micronesia. Solomon Is. – S.W. Japan to Australia (absent Great Barrier Reef) and Vanuatu.

BLACKBLOTCH EMPEROR *Lethrinus semicinctus*
SIZE: to 35 cm (14 in.) Emperors – Lethrinidae
ID: Brown forehead and back, white lower body; yellowish head and midside patch, **short dusky band behind gill cover;** dark patch on posterior can be faded or darkened. Solitary. Sandy reef flats and lagoon adjacent to coral reefs in 4-35 m.
East Indo-West Pacific: Sri Lanka to Indonesia, Philippines, Micronesia, Solomon Is. – S.W. Japan to Great Barrier Reef and Fiji.

GRASS EMPEROR *Lethrinus laticaudis*
SIZE: to 56 cm (22 in.) Emperors – Lethrinidae
ID: Pale silvery gray; often marked with diffuse dark blotches and irregular bars, **blue line markings radiate from lower edge of eye and pale blue spots on cheek.** Solitary or form groups. Sandy areas near reefs in 5-35 m.
Asian Pacific: Lesser Sunda Is. in Indonesia, N. Australia, Papua New Guinea, Solomon Is. to Vanuatu and New Caledonia.

SWEETLIP EMPEROR *Lethrinus miniatus*
SIZE: to 90 cm (3 ft.) Emperors – Lethrinidae
ID: Long pointed snout; gray, dark centers on scales of back, reddish brown upper head, lips often red; **base of pectoral fins bright red,** may display alternating light and dark bars. Solitary or form groups. Sand and rubble between reefs in 5-35 m.
Southwest Pacific: N. Australia, Coral Sea and New Caledonia.

AMBON EMPEROR *Lethrinus amboinensis*

SIZE: to 70 cm (2¼ ft.) Emperors – Lethrinidae

ID: Pale gray with narrow brown scale margins; occasionally small dark spots on a few to many scales, **snout from lip to eye slightly convex** (most members of genus have slightly concave snout). Solitary. Sand and rubble near coral reefs in 5-30 m.

West Pacific: Indonesia, Philippines, Micronesia, Papua New Guinea and Solomon Is. – S.W. Japan to N.W. Australia and Fiji.

LONGFIN EMPEROR *Lethrinus erythropterus*

SIZE: to 50 cm (20 in.) Emperors – Lethrinidae

ID: Red to yellow-brown, often with faint bars; red fins, pair of pale to bright white bars on base of tail. Solitary or form small groups. Coastal reefs, lagoons and outer slopes in 2-25 m.

Indo-Asian Pacific: E. Africa and Andaman Sea to Philippines, Micronesia, Papua New Guinea and Solomon Is.

SLENDER EMPEROR *Lethrinus variegatus*

SIZE: to 20 cm (8 in.) Emperors – Lethrinidae

ID: Small elongate body; mottled shades of green to brown mixed with white, wide darkish midlateral stripe, pale spots on fins. Usually form groups. Seagrass beds and sandy areas near coral reefs to 15 m.

Indo-West Pacific: Red Sea and E. Africa to Indonesia, Micronesia, Solomon Is. – S.W. Japan to Great Barrier Reef and New Caledonia.

ORNATE EMPEROR *Lethrinus ornatus*

SIZE: to 40 cm (16 in.) Emperors – Lethrinidae

ID: Silvery; **narrow red bar on cheek and gill cover**, broad yellowish stripes on sides. Solitary or form small groups. Seagrass beds and sand and rubble areas of coastal reefs and lagoons to 30 m.

East Indo-Asian Pacific: Sri Lanka to Indonesia, Philippines, Micronesia, Papua New Guinea. – S.W. Japan to Great Barrier Reef.

YELLOWTAIL EMPEROR *Lethrinus atkinsoni*

SIZE: to 41 cm (16 in.) Emperors – Lethrinidae

ID: Blue gray to yellowish tan, shading to white lower body with thin dark scale margins; yellow tail base and tail. Solitary or form groups. Sand, rubble, and seagrass areas adjacent to coral reefs in 2-30 m.

West Pacific: Indonesia, Philippines, Micronesia, Papua New Guinea and Solomon Is. – S.W. Japan to Australia and Fiji.

Yellowtail Emperor – Variation

ID: Capable of displaying yellowish head, broad yellow midlateral stripe on side and yellowish tail.

Silvery

Jacks – Barracudas – Tunas & Mackerels – Others

This ID Group consists of fishes that are silver to gray in color, and are generally unpatterned; however, several species have bluish, yellowish or greenish tints and occasional markings. All have forked tails.

FAMILY: Jacks (Trevallies) – Carangidae
16 Genera – 31 Species Included

Typical Shape

Jacks, also commonly known as trevallys, are strong open-water swimmers that on occasion form large schools that roam for great distances. Although primarily pelagic, solitary jacks often feed along the fringes of outer reef slopes. These voracious predators of fishes, and in a few cases crustaceans, are generally silvery, have laterally compressed or torpedo-shaped bodies with deeply sloping heads, large eyes and mouths, slender tail bases and widely forked tails. On many species, scales at the rear of their single, continuous lateral line form a series of short spiny structures called scutes. Jacks vary greatly in size from the small, aggregation-oriented, plankton-feeding scads, to the Giant Trevally, which can reach a length of 165 cm (5 ½ ft.). The wide geographic distribution of most species indicates a lengthy pelagic larval stage.

FAMILY: Barracudas – Sphyraenidae
Single Genus – 7 Species Included

Typical Shape

Barracudas are relatively large, silvery, elongate fishes with long jaws filled with an awesome array of pointed teeth. They have two, low, widely separated dorsal fins and widely forked tails indicative of their typically pelagic existence. These rapacious fish predators appear quite frightening; however, they present little or no threat to divers. The few substantiated attacks on humans involved spearfishing or fish feeding activities. A few species form large spiraling schools that occasionally allow a cautious diver to enter their midst. The Great Barracuda, the largest family member reaching a length of 180 cm (6 ft.), are typically solitary reef inhabitants that often approach divers out of curiosity not menace.

FAMILY: Tunas & Mackerels – Scombridae
5 Genera – 5 Species Included

Typical Shape

Tunas are streamlined, spindle-shaped open-water fishes with two dorsal fins that fold into grooves and finlets between the second dorsal and their deeply forked or lunate tails. At least two small keels extend from each side of the narrow tail base. Their upper bodies vary between shades of silvery iridescent blues and greens with countershadings of white on the belly. Pelagic species, built for speed and endurance are rapacious predators of squids and fishes in the near-surface zone of the open ocean.

FAMILY: Others

Milkfishes – Chanidae

Sea Chubs – Kyphosidae

Mullets – Mugilidae

Threadfins – Polynemidae

Monos – Monodactylidae

Grunters – Terapontidae

Tarpons – Megalopidae

Halfbeaks – Hemiramphidae

Needlefishes – Belonidae

Archerfishes – Toxotidae

Mojarras – Gerreidae

Flagtails – Kuhliidae

Pearl Perches – Glaucosomatidae

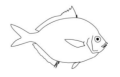

Ponyfishes – Leiognathidae

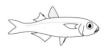

Silversides – Atherinidae

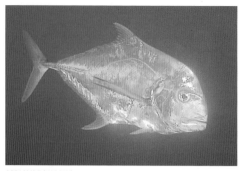

AFRICAN POMPANO   *Alectis ciliaris*
SIZE: to 130 cm (4 1/4 ft.)   Jacks – Carangidae
ID: Silver, often have bluish or greenish tints; deep body with steep head profile and deeply forked tail, scales not obvious. Young form schools, large adults often solitary. Pelagic; often near dropoffs to 100 m.
Circumtropical.

African Pompano – Young Adult
SIZE: to 90 cm (3 ft.)
ID: Front lobes of dorsal and anal fins of young adults trail long filamentous rays that are lost with age.

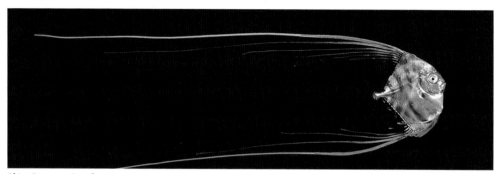

African Pompano – Juvenile
SIZE: body to 13 cm (5 in.)
ID: Diamond-shaped juveniles trail long thread-like filaments from dorsal and anal fins many times the length of their bodies.

GOLD-SPOTTED TREVALLY   *Carangoides fulvoguttatus*
SIZE: to 103 cm (3 1/4 ft.)   Jacks – Carangidae
ID: Slivery with many small golden or brassy spots; often 5 faint darkish bars on side, large adults have 3 or 4 dark blotches along middle of side. Solitary or form schools. Coastal reefs, lagoons and outer slopes to 100 m.
Indo-Asian Pacific: Red Sea and E. Africa to Indonesia, Micronesia, Solomon Is. – S.W. Japan to Great Barrier Reef and New Caledonia.

BLUDGER TREVALLY   *Carangoides gymnostethus*
SIZE: to 90 cm (3 ft.)   Jacks – Carangidae
ID: Silvery with a few brown or yellow spots scattered on sides. Juveniles and young adults form schools, adults usually solitary. Sheltered coasts and over deeper offshore reefs to at least 70 m.
Indo-Asian Pacific: Red Sea and E. Africa to Indonesia, Micronesia, Solomon Is. – S.W. Japan to Great Barrier Reef and New Caledonia.

BARCHEEK TREVALLY   *Carangoides plagiotaenia*
SIZE: to 46 cm (18 in.)   Jacks – Carangidae
ID: Silvery; **narrow dark bar on gill cover**. Solitary or form small groups. Most common along edge of steep outer reef slopes in 2-200 m.

Indo-West Pacific: Red Sea and E. Africa to Indonesia, Philippines, Micronesia, Papua New Guinea and Solomon Is. – S.W. Japan to Australia to Fiji.

ORANGE-SPOTTED TREVALLY   *Carangoides bajad*
SIZE: to 55 cm (22 in.)   Jacks – Carangidae
ID: Brassy silver to yellow-orange with many small orange spots on sides. Solitary or form small groups. Coastal reefs and outer slopes to 70 m.

Indo-Asian Pacific: Red Sea and E. Africa to Indonesia, Philippines, Micronesia, Papua New Guinea and Solomon Is. – S.W. Japan to Australia.

Orange-spotted Trevally – Variation
ID: Silver head and body with scattered orange spots; orange dorsal, anal and tail fins.

Orange-spotted Trevally – Variation
ID: Mixed silver and yellow-orange head, body and fins.

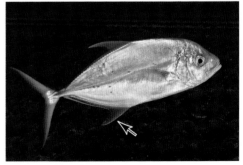

COACHWHIP TREVALLY   *Carangoides oblongus*
SIZE: to 46 cm (18 in.)   Jacks – Carangidae
ID: Silvery; lower lobe of tail, anal and pectoral fins occasionally yellowish, **fore lobe of rear dorsal fin elongate**. Solitary or form small schools. Coastal reefs and lagoons, usually over sandy bottoms to 50 m.

Indo-West Pacific: E. Africa to Indonesia, Philippines, Papua New Guinea and Solomon Is. – S.W. Japan to Australia and Fiji.

YELLOW-SPOTTED TREVALLY   *Carangoides orthogrammus*
SIZE: to 70 cm (2 1/4 ft.)   Jacks – Carangidae
ID: Silvery with elliptical yellow spots or blotches on sides; **bluish fins,** occasionally faint darkish bars on side. Solitary or form small schools. Commonly over sand in lagoons, passes and around outer reefs in 3-160 m.

Indo-Pacific: E. Africa to Indonesia, Micronesia, Solomon Is., Hawaii and French Polynesia. – S.W. Japan to Australia.

COASTAL TREVALLY *Carangoides coeruleopinnatus*
SIZE: to 41 cm (16 in.) Jacks – Carangidae
ID: Silvery; 5 darkish bars and scattered yellow spots on body, **high arching lateral line**, blackish spot on upper gill cover, occasionally yellowish fins. Often form small groups. Over sandy bottoms near reefs to 25m.

Indo-Pacific: Red Sea and E. Africa to Hawaii – S. W. Japan to Australia, Fiji and Samoa.

BLUE TREVALLY *Carangoides ferdau*
SIZE: to 70 cm (2 1/4 ft.) Jacks – Carangidae
ID: Silvery; 7-10 darkish bars and scattered yellow spots on body, occasionally yellowish fins. Commonly form schools. Over sand bottoms near reefs to 60m.

Indo-Pacific: Red Sea and E. Africa to Hawaii - S. W. Japan to Australia, Fiji and French Polynesia.

BLACK JACK *Caranx lugubris*
SIZE: to 80 cm (2 1/2 ft.) Jacks – Carangidae
ID: Silvery gray to brown to nearly black; small dark spot at upper end of gill cover, scutes, rear dorsal, anal and rear of tail fins dark. Solitary or form small schools. Mainly offshore reefs in 20-70 m.

Circumtropical.

BLUEFIN TREVALLY *Caranx melampygus*
SIZE: to 100 cm (3 1/4 ft.) Jacks – Carangidae
ID: Silvery iridescent blue to green with dense spotting on upper two-thirds of body, blue to bluish fins. Solitary or form schools. Inhabit a variety of reef habitats, but most common on outer reefs to 190 m; can be locally abundant.

Indo-Pacific: Red Sea and E. Africa to Indonesia, Micronesia, Hawaii and far eastern Pacific. – S.W. Japan to Australia.

BRASSY TREVALLY *Caranx papuensis*
SIZE: to 68 cm (2 1/4 ft.) Jacks – Carangidae
ID: Silver, often with bronze to greenish tints; **lower tail lobe yellowish with white margin**, dark speckles above lateral line, white spot behind upper edge of gill cover. Solitary or form schools. Over lagoon to seaward reefs to 30 m.

Indo-Pacific: E. Africa to Indonesia, Micronesia, Solomon Is. and French Polynesia. – S.W. Japan to Great Barrier Reef and Fiji.

BIGEYE TREVALLY *Caranx sexfasciatus*
SIZE: to 85 cm (2 3/4 ft.) Jacks – Carangidae
ID: Silver (males turn black when courting); small black spot on upper end of gill cover, **white tip on fore lobe of rear dorsal fin**, relatively large eye. Form large schools. Most common on clear outer reefs to 50 m.

Indo-Pacific: Red Sea and E. Africa to Indonesia, Philippines, Micronesia, Solomon Is., French Polynesia. – S.W. Japan to Australia.

GIANT TREVALLY *Caranx ignobilis*
SIZE: to 165 cm (5 1/2 ft.) Jacks – Carangidae
ID: Large; silvery with numerous scattered small black spots, **small black area on upper base of pectoral fin,** steep forehead profile. Usually solitary. Most common on seaward reef slopes to 80 m.
Indo-Pacific: Red Sea and E. Africa to Indonesia, Philippines, Micronesia, Solomon Is., French Polynesia. – S.W. Japan to Australia.

SILVER TREVALLY *Pseudocaranx dentex*
SIZE: to 94 cm (3 1/4 ft.) Jacks – Carangidae
ID: Bluish silver with midbody yellowish stripe to tail and yellowish edge below dorsal and anal fins to tail; black spot on rear gill cover. Pairs or form small schools. Deep bottom feeders to 200 m; shallower in cool water.
Pacific: Indonesia to Hawaii, French Polynesia and Easter I.

YELLOWTAIL AMBERJACK *Seriola lalandi*
SIZE: to 193 cm (6 1/2 ft.) Jacks – Carangidae
ID: Elongate; silvery with a bronze to yellow stripe along middle of side to tail, **yellowish to yellow tail.** Form large schools. Usually offshore, but occasionally over reefs in 5-60 m.
Circumglobal: Primarily in subtropical seas.

GREATER AMBERJACK *Seriola dumerili*
SIZE: to 188 cm (6 1/4 ft.) Jacks – Carangidae
ID: Elongate; silvery; dark band runs from lip, across eye, to origin of dorsal fin, occasionally an amber stripe from gill cover to tail. Usually solitary, occasionally form small schools. Pelagic, but occasionally over reefs to 360 m.
Circumtropical.

ALMACO JACK *Seriola rivoliana*
SIZE: to 120 cm (4 ft.) Jacks – Carangidae
ID: Silvery; high back profile; **dark band runs from eye to front of dorsal fin.** Form schools. Pelagic, but occasionally over reefs to 245 m; often around floating objects.
Circumglobal: In tropical and subtropical seas.

RAINBOW RUNNER *Elagatis bipinnulata*
SIZE: to 120 cm (4 ft.) Jacks – Carangidae
ID: Silvery olive-blue; long slender body with pair of light blue stripes on sides, often with broader olive or yellowish stripe between. Form schools. Most common on outer reefs to 150 m.
Circumtropical.

127

Jacks

SMALL-SPOTTED DART   *Trachinotus baillonii*
SIZE: to 54 cm (22 in.)   Jacks – Carangidae
ID: Silver; blunt broadly rounded snout, forelobes of rear dorsal and anal fins elongate, long widely forked tail lobes, 1-5 small black spots on middle of side. Coastal waters; often in surge zones along sandy beaches to 3 m; form schools.
Indo-Pacific: Red Sea and E. Africa to Indonesia, Philippines, Solomon Is. and French Polynesia. – S.W. Japan to Australia.

SNUBNOSE POMPANO   *Trachinotus blochii*
SIZE: to 70 cm (2¼ ft.)   Jacks – Carangidae
ID: Silver; ventral and anal fins frequently yellow or orange, blunt broadly rounded snout; forelobes of rear dorsal and anal fins elongate, long widely forked tail lobes. Solitary or form schools. Coral and rocky reefs in 10-50 m.
Indo-Pacific: Red Sea and E. Africa to Indonesia, Micronesia, Solomon Is. and French Polynesia. – S.W. Japan to Australia.

DOUBLE-SPOTTED QUEENFISH   *Scomberoides lysan*
SIZE: to 70 cm (2¼ ft.)   Jacks – Carangidae
ID: Silver with double row of 6-8 dusky round blotches on side; black spot on tip of fore lobe of rear dorsal fin. Often form small schools. Coastal reefs, lagoons and outer slopes to 100 m.
Indo-Pacific: Red Sea and E. Africa to Indonesia, Philippines, Micronesia, Solomon Is., Hawaii and French Polynesia. – S.W. Japan to Australia.

TALANG QUEENFISH   *Scomberoides commersonnianus*
SIZE: to 120 cm (4 ft.)   Jacks – Carangidae
ID: Silver with single row of large dark spots on upper body. Solitary or form groups. Coastal reefs, lagoons and outer slopes to 25 m.
Indo-Asian Pacific: Red Sea and E. Africa to Indonesia, Philippines, Papua New Guinea and Solomon Is. – S.W. Japan to Australia.

Golden Trevally
ID: Juvenile – Bright yellow with 7-11 black bars on side. (Difficult to see smaller bars.) Occasionally confused with Pilotfish (following page), which are silver and display 6-7 black body bars. Tiny juveniles often with jellyfishes. Small groups of individuals to 5 cm accompany large pelagic fishes.

GOLDEN TREVALLY   *Gnathanodon speciosus*
SIZE: to 120 cm (4 ft.)   Jacks – Carangidae
ID: Adult – Generally silvery gray occasionally display vague bars, yellow highlights and/or scattered blotches. Coastal, lagoons and outer reefs. Often forage in sand for invertebrates.
Indo-Pacific: Red Sea and E. Africa to Indonesia, Philippines, Micronesia, Papua New Guinea, Solomon Is., Hawaii and French Polynesia. – S.W. Japan to Australia.

LONGRAKERED TREVALLY *Ulua mentalis*
SIZE: to 85 cm (2 ³/₄ ft.) Jacks – Carangidae
ID: Silvery blue-green back with silvery sides; dusky to blackish 1st dorsal fin and forked tail, large protruding lower jaw. Form schools. Shallow coastal waters to 50 m.

Indo-Asian Pacific: Red Sea and E. Africa to Indonesia, Philippines, Micronesia, Papua New Guinea and Great Barrier Reef. Absent Solomon Is.

BIGEYE SCAD *Selar crumenophthalmus*
SIZE: to 30 cm (12 in.) Jacks – Carangidae
ID: Silvery, often with yellow stripe on side; deep-body, large eye (diameter greater than snout length), **scutes only on rear third of lateral line**. Form schools. Coastal reef and lagoons to 170 m.

Circumtropical.

OXEYE SCAD *Selar boops*
SIZE: to 26 cm (10 in.) Jacks – Carangidae
ID: Silvery with broad pale to bright yellow stripe from eye to tail; **dark spot confined to rear edge of gill cover; wide row of scutes from midbody to tail, large eye.** Form large schools. Coastal reefs, lagoons and outer slopes to 170 m.

Asian Pacific: Indonesia, Philippines, Papua New Guinea, Solomon Is. to Australia, New Caledonia and Vanuatu.

YELLOWSTRIPE SCAD *Selaroides leptolepis*
SIZE: to 24 cm (9 ¹/₂ in.) Jacks – Carangidae
ID: Silvery with broad pale to bright yellow stripe from eye to tail; **dark blotch on rear edge of gill cover extends onto forbody,** narrow row of scutes from midbody to tail. Form large schools. Often around wharf pilings or reef edges to 20 m.

East Indo-Asian Pacific: Persian Gulf to Indonesia, Philippines, Papua New Guinea, Solomon Is. – S.W. Japan and N. Australia.

HERRING SCAD *Alepes vari*
SIZE: to 55 cm (22 in.) Jacks – Carangidae
ID: Silvery with dark blotch on rear gill cover and dusky forked tail. Form schools. Often near surface in coastal waters and over reefs to 40 m.

Indo-Asian Pacific: Red Sea and Persian Gulf to Indonesia and Philippines. – S.W. Japan to Australia.

YELLOWTAIL SCAD *Atule mate*
SIZE: to 30 cm (12 in.) Jacks – Carangidae
ID: Silvery often with numerous narrow dusky bars on side; yellow forked tail, dark spot on edge of gill cover behind eye. Form schools. Coastal waters, including mangroves and over reefs to 50 m.

Indo-Pacific: Red Sea and E. Africa to Indonesia, Philippines, Solomon Is., Hawaii and French Polynesia – S.W. Japan to Australia.

PILOTFISH *Naucrates ductor*
SIZE: to 75 cm (2 1/2 ft.) Jacks – Carangidae
ID: Silver with **5-7 wide black bars**. Closely associated with pelagic sharks and rays. Form schools near front of host. Juveniles occasionally associate with jellyfishes.
Circumtropical.

BRASS STRIPED BARRACUDA *Sphyraena helleri*
SIZE: to 80 cm (2 1/2 ft.) Barracudas – Sphyraenidae
ID: Silvery; long cylindrical body and large underslung jaw with pointed teeth, pair of thin brassy stripes on sides. Form large daytime schools; disperse to feed at night. Coastal, lagoon and outer reefs to 60 m.
Indo-Pacific: E. Africa to Micronesia, Hawaii and French Polynesia. – S.W. Japan to Australia. Rare much of Asian Pacific.

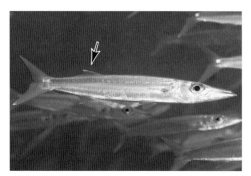

BIGEYE BARRACUDA *Sphyraena forsteri*
SIZE: to 65 cm (2 1/4 ft.) Barracudas – Sphyraenidae
ID: Silvery; long cylindrical body and large underslung jaw with pointed teeth, **rear dorsal fin dusky with white tip,** blackish spot behind base of pectoral fin. Form schools. Reefs and outer reef slopes to 300 m.
Indo-Pacific: E. Africa to S.W. Japan, Micronesia and French Polynesia. – S.W. Japan to New Caledonia.

GREAT BARRACUDA *Sphyraena barracuda*
SIZE: to 170 cm (5 1/2 ft.) Barracudas – Sphyraenidae
ID: Silvery; long cylindrical body and large underslung jaw with pointed teeth, usually has a few scattered dark blotches, **large black patches on dorsal and tail fin lobes.** Can display barred or mottled pattern when resting near bottom. Solitary or form small groups. Reefs, usually in relatively shallow water to 15 m.
Circumtropical.

PICKHANDLE BARRACUDA *Sphyraena jello*
SIZE: to 140 cm (4 1/2 ft.) Barracudas – Sphyraenidae
ID: Silvery with **yellowish tail**; long cylindrical body and large underslung jaw with pointed teeth, about 20 wavy bars primarily on upper half of the body. Form schools. Coastal, lagoons and outer reefs to about 60 m.
Indo-West Pacific: E. Africa to Indonesia, Philippines, Micronesia and Solomon Is. – S.W. Japan to Australia to Fiji.

BLACKFIN BARRACUDA *Sphyraena qenie*
SIZE: to 140 cm (4 1/2 ft.) Barracudas – Sphyraenidae
ID: Silvery with **dusky to dark tail with dark margin**; long cylindrical body and large underslung jaw with pointed teeth, about 18-22 chevron-shaped dark markings on sides. Form large schools. Seaward reefs to 50 m.
Indo-Pacific: Red Sea and E. Africa to Indonesia, Micronesia, Solomon Is. and Panama. – S.W. Japan to Great Barrier Reef.

YELLOWTAIL BARRACUDA   *Sphyraena flavicauda*

SIZE: to 40 cm (16 in.)   Barracudas – Sphyraenidae

ID: Silvery; long cylindrical body and large underslung jaw with pointed teeth, **dark midbody stripe with a second on lower body,** yellowish tail, often short bars on back. Form schools. Coastal and offshore reefs to 20 m.

Indo-West Pacific: E. Africa to Indonesia, Philippines, Micronesia and Solomon Is. – S.W. Japan to Australia to Fiji.

OBTUSE BARRACUDA   *Sphyraena obtusata*

SIZE: to 30 cm (12 in.)   Barracudas – Sphyraenidae

ID: Small size, silvery; yellowish to yellow tail, long cylindrical body with underslung jaw and pointed teeth, pair of diffuse brownish stripes on side. Form schools. Coastal reefs, lagoons and outer slopes to 20 m.

Indo-West Pacific: Red Sea and E. Africa to Indonesia, Micronesia and Solomon Is. – S.W. Japan to Australia and Fiji.

DOUBLE-LINED MACKEREL   *Grammatorcynus bilineatus*

SIZE: to 65 cm (3 1/4 ft.)   Tunas & Mackerels – Scombridae

ID: Silver; long slender unmarked body with dorsal and anal finlets; **2 lateral lines, one on upper the other on lower side.** Lagoon, outer reefs and around dropoffs and steep pinnacles to 15 m.

Indo-West Pacific: Red Sea to Indonesia, Philippines, Micronesia, Papua New Guinea, Solomon Is. – S.W. Japan to Australia and Fiji.

DOGTOOTH TUNA   *Gymnosarda unicolor*

SIZE: to 220 cm (7 1/4 ft.)   Tunas & Mackerels – Scombridae

ID: Silver; long somewhat stocky body with dorsal and anal finlets, **pale tips on rear dorsal and anal fins,** single lateral line. Most common tuna on coral reefs. Deeper lagoons, passes and outer reef slopes to 60 m.

Indo-West Pacific: Red Sea and E. Africa to Indonesia, Philippines, Micronesia and Solomon Is. – S.W. Japan to Australia and Fiji.

WAHOO   *Acanthocybium solandri*

SIZE: to 210 cm (7 ft.)   Tunas & Mackerels – Scombridae

ID: Silver; long slender body with dorsal and anal finlets and elongate pointed snout, occasionally display wavy-bar pattern. Solitary or in pairs. Pelagic, but occasionally pass near reefs to 20 m.

Circumtropical.

NARROW-BARRED SPANISH MACKEREL *Scomberomorus commerson*

SIZE: to 245 cm (8 ft.)   Tunas & Mackerels – Scombridae

ID: Silvery; long slender body with dorsal and anal finlets, **display numerous thin wavy-bars, white tip on anal fin.** Solitary. Pelagic, but occasionally pass near or over reefs to 70 m.

Indo-West Pacific: Red Sea to Indonesia, Philippines, Micronesia – S.W. Japan to Australia, New Caledonia and Fiji.

LONG-JAWED MACKEREL *Rastrelliger kanagurta*

SIZE: to 38 cm (15 in.) Tunas & Mackerels – Scombridae

ID: Silvery; faint spotting on upper back with narrow stripes below, black spot under pectoral fin. Coastal, lagoons and seaward reefs, often in turbid water to 90 m.

Indo-West Pacific: Red Sea and E. Africa to Indonesia, Philippines, Micronesia, Papua New Guinea and Solomon Is. – S.W. Japan to Australia and Fiji.

Long-jawed Mackerel

ID: Form tightly bunched schools; feed on plankton with widely opened mouths.

BLUESPOT MULLET *Moolgarda seheli*

SIZE: to 30 cm (12 in.) Mullets – Mugilidae

ID: Silvery with tint of brown to green on back and whitish sides with dark stripes along scale rows; yellow pectoral fins with blue spot on base; bluish dorsal, anal and tail fins. Form groups. Sheltered coasts and coral lagoons, occasionally in tidal creeks.

Indo-Pacific: Red Sea and E. Africa to Micronesia, Solomon Is., Hawaii and French Polynesia. – S.W. Japan to Australia and Fiji.

FRINGELIP MULLET *Crenimugil crenilabis*

SIZE: to 26 cm (10 1/4 in.) Mullets – Mugilidae

ID: Olive back, silvery side and belly; pectoral fins yellowish with dark spot at base. Coastal reef flats and lagoons to 20 m. Occasionally form large schools.

Indo-Pacific: E. Africa to Indonesia, Philippines, Micronesia, Papua New Guinea, Solomon Is. and French Polynesia. – S.W. Japan to Australia.

SQUARETAIL MULLET *Ellochelon vaigiensis*

SIZE: to 60 cm (2 ft.) Mullets – Mugilidae

ID: Silvery with large diamond-shaped scales, **black pectoral fins**, yellowish tan square-cut tail. Form schools. Protected sandy shorelines of lagoons and reef flats to 10 m.

Indo-Pacific: E. Africa to Indonesia, Philippines, Micronesia, Papua New Guinea, Solomon Is. and French Polynesia. – S.W. Japan to Great Barrier Reef and New Caledonia.

ACUTE-JAWED MULLET *Neomyxus leuciscus*

SIZE: to 46 cm (18 in.) Mullets – Mugilidae

ID: Silvery with **yellow spot on base of pectoral fin**. Form schools. Lagoons and around seaward reefs to 10 m.

Pacific: S. Japan, Micronesia, Hawaii, Line Is. and French Polynesia.

LONGJAW BONEFISH   *Albula argentea*
SIZE: to 70 cm (2 1/4 in.)   Bonefishes – Albulidae
ID: Silvery with rounded head profile; blackish mark on snout tip. Form groups in sheltered, shallow lagoons to 90 m.
Pacific: Indonesia to Philippines, Papua New Guinea, Solomon Is., Hawaii and French Polynesia. – Japan to Great Barrier Reef.

MILKFISH   *Chanos chanos*
SIZE: to 120 cm (4 ft.)   Milkfishes – Chanidae
ID: Silvery shading to bluish green on back; single dorsal fin, large dark deeply forked tail, **ventral fins at midbody below dorsal fin.** Form schools. Near surface of lagoons and seaward reefs to 30 m.
Indo-Pacific: Red Sea and E. Africa to Indonesia, Philippines, Solomon Is. and Central America. – S.W. Japan to Australia.

INDO-PACIFIC TARPON   *Megalops cyprinoides*
SIZE: to 90 cm (3 ft.)   Tarpons – Megalopidae
ID: Silver; large prominent scales, pectoral fins low on body, ventral fins on belly below dorsal fin, deeply forked tail. Solitary or form small groups. Inner bays, river mouths and mangroves to 50 m.
Indo-Pacific: Red Sea to N. Micronesia and French Polynesia. – S. Korea to S.E. Australia.

SILVER MONO   *Monodactylus argenteus*
SIZE: to 27 cm (10 3/4 in.)   Monos – Monodactylidae
ID: Silver; dorsal and tail fins yellowish to yellow, round laterally compressed body with prominent triangular dorsal and anal fins. Form schools. Estuaries, harbors and silty inshore reefs to 10 m.
Indo-Asian Pacific: Red Sea and E. Africa to Indonesia, Philippines, Micronesia, Solomon Is. – S.W. Japan to Australia.

ORANGEFIN PONYFISH   *Photopectoralis bindus*
SIZE: to 11 cm (4 1/4 in.)   Ponyfishes – Leiognathidae
ID: Silvery gray with subtle maze-like pattern on back; deep body; orange spot at front of dorsal fin. Usually form schools. Coastal rock and coral reefs mixed with sand bottoms to 40 m.
Indo-Asian Pacific: Red Sea to Indonesia, Philippines, Papua New Guinea and Solomon Is. – S.W. Japan to Australia and New Caledonia.

TOOTHPONY   *Gazza minuta*
SIZE: to 14 cm (5 1/2 in.)   Ponyfishes – Leiognathidae
ID: Silvery gray; upper body with thin wavy yellow lines; edge of dorsal fin dark. front of anal fin yellow. Coastal waters with silty bottoms to 75 m.
Indo-Asian Pacific: Red Sea, India and Sri Lanka to Indonesia, Philippines, Papua New Guinea and Solomon Is. – S.W. Japan to Australia and Vanuatu.

SILVER GRUNTER *Mesopristes argenteus*
SIZE: to 30 cm (12 in.) Grunters – Terapontidae
ID: Silvery gray with white belly; white 1st spine on ventral and anal fins, yellow iris. **Small Juvenile –** White with 4 or 5 black stripes and yellow fins. Solitary or form groups. Estuaries and stream mouths, occasionally near reefs to 3 m.
Asian Pacific: Indonesia, Philippines, Papua New Guinea and Solomon Is. to N.E. Australia.

CRESCENT-BANDED GRUNTER *Terapon jarbua*
SIZE: to 30 cm (12 in.) Grunters – Terapontidae
ID: Silvery with pattern of curved darkish bands on body; striped tail. Form schools. Estuaries, stream mouths and along sandy beaches in areas of brackish water in 20-290 m.
Indo-West Pacific: Red Sea and E. Africa to Indonesia, Philippines, Papua New Guinea and Solomon Is. – S.W. Japan to Australia and Fiji.

SAND BASS *Psammoperca waigiensis*
SIZE: to 47 cm (19 in.) River Perches – Latidae
ID: Slivery gray to brown; eyes have glassy appearance. Solitary or small groups; nocturnal. Often in dark recesses around weedy areas in 3-12 m.
Asian Pacific: Indonesia, Philippines and S. Papua New Guinea. – S.W. Japan to N. Australia.

BANDED ARCHERFISH *Toxotes jaculatrix*
SIZE: to 20 cm (8 in.) Archerfishes – Toxotidae
ID: Silvery white with 4 or 5 wedge-shaped black bars on upper half of sides; dorsal fin well back on rear body. Swim near surface; "shoot" down insect prey with jet of water from mouth. Associate with mangroves.
East Indo-Asian Pacific: India to Indonesia, Papua New Guinea, Solomon Is. Australia, and Vanuatu.

PHARAO FLYINGFISH *Cypselurus naresii*
SIZE: to 25 cm (10 in.) Flyingfishes – Exocoetidae
ID: Juvenile – Gold back, silver below, chin barbel; red margin on dorsal fin. Often associate with floating debris. **Adult –** Reddish brown back and silvery below with irregular reddish tan bars; large red spot on tip on lower lobe of tail; wide fleshy tab extends from lower jaw. Oceanic, surface waters to 20 m.
Indo-West Pacific: E. Africa to Fiji. – S.W. Japan to Australia.

SAILOR FLYINGFISH *Prognichthys sealei*
SIZE: to 25 cm (10 in.) Flyingfishes – Exocoetidae
ID: Juvenile – Yellowish back; no chin barbel; 2 to 5 cm ($^3/_4$ to 2 in.). Often associate with floating debris. **Adult –** Blue gray upper head, silvery below; short blunt snout with small mouth; large pectoral and ventral fins. Oceanic, surface waters to 20 m.
Indo-Pacific: E. Africa to Central and South America in the Eastern Pacific.

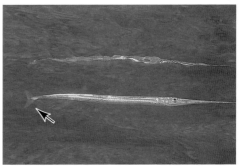

CROCODILE NEEDLEFISH   *Tylosurus crocodilus*
SIZE: to 135 cm (4 ¹/₂ ft.)   Needlefishes – Belonidae
ID: Silvery; extremely slender with elongate upper and lower jaws and numerous needle-like teeth, **forked tail with larger lower lobe** and black keel on base. Solitary or form small groups. Near surface of lagoons and inshore reefs.
Circumtropical.

KEELTAIL NEEDLEFISH   *Platybelone argalus*
SIZE: to 45 cm (17 ³/₄ in.)   Needlefishes – Belonidae
ID: Silvery; extremely slender with elongate upper and lower jaws, **forked tail with lobes of nearly equal size** and black keel on base. Solitary or small form groups. Near surface of lagoons and inshore reefs.
Circumtropical.

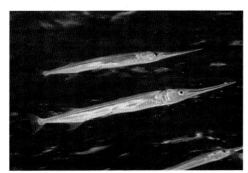

REEF NEEDLEFISH   *Strongylura incisa*
SIZE: to 75 cm (2 ¹/₂ ft.)   Needlefishes – Belonidae
ID: Silvery; extremely slender with elongate upper and lower jaws and numerous needle-like teeth, **tail margin wavy with larger lower lobe**, no keel on tail base. Solitary or form small groups. Near surface of lagoons and inshore reefs.
West Pacific: Andaman Sea to Indonesia, Micronesia, Papua New Guinea, Solomon Is. – S.W. Japan to Great Barrier Reef and Fiji.

ESTUARINE HALFBEAK   *Zenarchopterus dispar*
SIZE: to 16 cm (6 ¹/₄ in.)   Halfbeaks – Hemiramphidae
ID: Silvery; slender body with very short upper jaw and elongate sword-like lower jaw, **margin of tail straight (not forked).** Common in estuaries, but occasionally at surface above coral reefs and near mangroves.
Indo-West Pacific: E. Africa to Indonesia, Philippines, Micronesia, Papua New Guinea, Solomon Is., New Caledonia and Fiji.

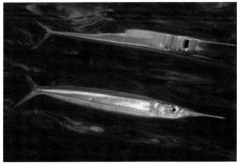

DUSSUMIER'S HALFBEAK   *Hyporhamphus dussumieri*
SIZE: to 30 cm (12 in.)   Halfbeaks – Hemiramphidae
ID: Silvery; slender body with very short upper jaw and elongate sword-like lower jaw, tail forked with lower lobe longer than upper. Form schools. Near surface of lagoons and seaward reefs.
Indo-Pacific: Seychelles and Maldives to Indonesia, Micronesia, Solomon Is. and French Polynesia. – S.W. Japan to Australia.

BLACKBARRED HALFBEAK   *Hemiramphus far*
SIZE: to 44 cm (17 in.)   Halfbeaks – Hemiramphidae
ID: Dark blue back, silvery below; 3-9 dark bars on upper side; yellowish dorsal fin, tail has blue lower lobe and yellowish upper lobe, red tip on lower jaw. Coastal areas with seagrass and algae.
Indo-West Pacific: E. Africa to Indonesia, Philippines, Micronesia, Solomon Is. – S.W. Japan to Australia and Fiji.

DEEP-BODIED SILVERBIDDY
Gerres erythrourus
SIZE: to 30 cm (12 in.) Mojarras – Gerreidae
ID: Silver with dusky bars and darkish tail, **yellow ventral fins.** Form loose groups to small schools. Sand flats and slopes to 40 m.
Indo-Asian Pacific: India to Andaman Sea, Indonesia, Philippines, Micronesia, Papua New Guinea and Solomon Is. – S.W. Japan to Australia.

LONGTAIL SILVERBIDDY
Gerres longirostris
SIZE: to 44 cm (17 in.) Mojarras – Gerreidae
ID: Silver with obvious scales; several rows of dusky (usually oval) markings on sides; young have 7-8 dusky bars. Solitary or form loose groups. Sandy areas near reefs to 12 m.
Indo-Asian Pacific: Red Sea and E. Africa to Andaman Sea, Indonesia, Philippines, Micronesia, Papua New Guinea and Solomon Is. – S.W. Japan to Australia.

OBLONG SILVERBIDDY
Gerres oblongus
SIZE: to 33 cm (13 in.) Mojarras – Gerreidae
ID: Silver with **oval spots forming broken stripes.** Solitary or form loose groups. Sandy areas near reefs to 10 m.
Indo-Asian Pacific: Red Sea and E. Africa to Andaman Sea, Indonesia, Philippines, Micronesia, Papua New Guinea and Solomon Is. – S.W. Japan to Great Barrier Reef.

BLACKTIP SILVERBIDDY
Gerres oyena
SIZE: to 24 cm (10 in.) Mojarras – Gerreidae
ID: Silver with obvious scales; unmarked except for a **blackish margin on dorsal fin** (more intense at front). Solitary or form small groups. Common along sandy beaches and sand bottoms near reefs to 10 m.
Indo-West Pacific: Red Sea and E. Africa to Indonesia, Micronesia and Solomon Is. – S.W. Japan to Australia and Fiji.

BARRED FLAGTAIL
Kuhlia mugil
SIZE: to 24 cm (9 1/2 in.) Flagtails – Kuhliidae
ID: Silvery; perch-like fish with **striped pattern on tail.** Frequently form schools. Rocky surf zones to 18 m. Juveniles often inhabit rocky tide pools.
Indo-Pacific: Red Sea and E. Africa to Central America. – S.W. Japan to Australia. Tend to be uncommon in Asian Pacific.

SIXFEELER THREADFIN
Polydactylus sexfilis
SIZE: to 60 cm (2 ft.) Threadfins – Polynemidae
ID: Silver; under-slung jaw, 6 elongate rays extend from lower pectoral fins. Rays extend to scour sand when feeding. Solitary or form small groups. Sand, mudflats and mangroves to 50 m.
Indo-Pacific: E. Africa to Andaman Sea, Indonesia, Philippines, Micronesia, Papua New Guinea, Solomon Is., Hawaii and French Polynesia. – S. Japan to Australia.

THREADFIN PEARL-PERCH *Glaucosoma magnificum*
SIZE: to 32 cm (13 in.)   Pearl Perches – Glaucosomatidae
ID: Silvery with greenish to yellowish mottling on body; **wide dark bar through eye and 2 more behind on head,** white spot under rear dorsal fin; long filament extends from dorsal and occasionally anal and tail fins. Form schools. Open water and reefs in 10-30 m.
Localized: Great Barrier Reef, N. Australia and S. Papua New Guinea.

Threadfin Pearl-perch – Variation
ID: Change to reddish brown upper body gradating to silvery below when near bottom.

LOWFIN CHUB *Kyphosus vaigiensis*
SIZE: to 45 cm (18 in.)   Sea Chubs – Kyphosidae
ID: Silvery gray with narrow bronzy stripes; rear dorsal fin not elevated, **outer edge of anal fin aligns with outer edge of upper tail lobe.** Form small to large groups. Rocky shores, reef flats, lagoons and outer reefs to 25 m.
Circumtropical.

TOPSAIL CHUB *Kyphosus cinerascens*
SIZE: to 45 cm (18 in.)   Sea Chubs – Kyphosidae
ID: Silvery gray with thin dark horizontal lines on side; **rear dorsal fin distinctly elevated** (higher than tallest dorsal spines). Form small to large groups. Rocky shores, reef flats, lagoons and outer reefs to 25 m.
Circumtropical.

STRIPEY *Microcanthus strigatus*
SIZE: to 16 cm (6 1/4 in.)   Sea Chubs – Kyphosidae
ID: White to yellow with 5-6 black stripes. Solitary or form small or dense aggregations. Lagoons, rocky areas and shallow coral reefs to 140 m.
Localized: Isolated populations in E. and W. Australia, N. New Caledonia, Taiwan, S.W. Japan and Hawaii.

ROBUST SILVERSIDE *Atherinomorus lacunosus*
SIZE: to 14 cm (5 1/2 in.)   Silversides – Atherinidae
ID: Robust body; greenish silver often with blue reflections on gill cover, two dorsal fins with ventrals below and behind pectorals. Form large tightly packed schools in sheltered waters coastal reefs. The numerous members are difficult to distinguish to species.
Indo-West Pacific: Red Sea and E. Africa to Indonesia, Micronesia and Solomon Is. – S.W. Japan to Great Barrier Reef and Fiji.

Slender Schoolers/Colorful
Fusiliers – Anthias

This ID Group consists of slender bodied, fast-moving fishes that gather in large numbers in openwater to feed on current-borne zooplankton.

FAMILY: Fusiliers – Caesionidae
4 Genera – 17 Species Included

Typical Shape

Fusiliers, close relatives of snappers, are a small family of fishes confined to the tropical waters of the Indo-Pacific. Family members typically have slender torpedo-shaped bodies, small terminal mouths with protrusible upper jaws, and deeply forked tails. Fusiliers typically congregate in large, fast-swimming zooplankton-feeding aggregations in openwater along outer reef slopes. Such schools often consist of mixed species. During the day small assemblies from aggregations approach the reef where they mill about in loose groups and often attend cleaning stations. While near the reef several species develop rusty brown complexions. Fusiliers also seek the reefs' protection to sleep at night.

SUBFAMILY: Anthias – Serranidae/Anthiinae
5 Genera – 37 Species Included

Typical Shape   Typical Shape

Anthias, small sea basses classified in the subfamily Anthiinae, spend much of the day feeding on zooplankton just above the reef's protection. On many current-swept outer reef slopes in the tropical Pacific pulsating clouds of the small, brightly colored plankton-pickers represent the largest and most visually dramatic concentrations of fish life. The huge feeding aggregations, often numbering in the hundreds or even thousands, are made up of many small coexisting social units known as harems. Each harem consists of a single dominant male, a few lesser males and a cluster of females and juveniles.

Like other sea basses, anthias are sequential hermaphrodites that begin life as females, and later, influenced by social or environmental cues change into males. A few males eventually rise through a strictly controlled pecking order to gain exclusive dominance of a harem. These typically larger, more brilliantly colored individuals, which often display filamentous fin streamers, constantly defend their social position and bevy of females by aggressively challenging neighboring males and lesser males under their control. At dusk a colony's dominant males begin frenzied up and down, zigzagging courtship dances culminating in side-by-side spawning rushes with individual females in their harems.

BLUESTREAK FUSILIER   *Pterocaesio tile*
SIZE: to 25 cm (10 in.)   Fusiliers – Caesionidae
ID: Silvery blue; several dark scale row stripes on back, wider dark stripe below, **wide iridescent blue stripe from gill cover to tail,** black streak on tail lobes. Form aggregations, often mix with other fusiliers. Clear water slopes and reefs to 60 m.
Indo-Pacific: E. Africa to Indonesia, Micronesia and French Polynesia. – S.W. Japan to Great Barrier Reef and New Caledonia.

Bluestreak Fusilier – Red Phase
ID: Most fusiliers have the ability to change their typical open-water colors to reddish brown shades, especially on the lower body, when they associate with reefs for sanctuary to attend cleaning stations or to sleep at night.

RUDDY FUSILIER   *Pterocaesio pisang*
SIZE: to 21 cm (8 1/4 in.)   Fusiliers – Caesionidae
ID: Vary from a solid color to combinations of blue to blue-green to silvery red; red to black tail tips, **yellow snout** and iris. Form aggregations, often mix with other fusiliers. Steep slopes and shallow coastal and seaward reefs to 30 m.
Indo-West Pacific: E. Africa to Andaman Sea, Indonesia, Philippines and Micronesia.– S. Japan to Australia and Fiji.

TWINSTRIPE FUSILIER   *Pterocaesio marri*
SIZE: to 35 cm (14 in.)   Fusiliers – Caesionidae
ID: Silvery blue to blue-green; **yellow stripe on side covers lateral line except on tail base,** a second stripe on back, dark tail tips. Form aggregations, often mix with other fusiliers. Steep slopes and coastal, lagoon and seaward reefs to 30 m.
Indo-Pacific: E. Africa to Andaman Sea, Micronesia and French Polynesia. – S. Japan to Great Barrier Reef and Fiji.

THREESTRIPE FUSILIER   *Pterocaesio trilineata*
SIZE: to 20 cm (8 in.)   Fusiliers – Caesionidae
ID: Silvery; 3 yellowish to brownish stripes alternate with three pale stripes on back, dark tail tips. Form aggregations, often mix with other fusiliers. Steep slopes, shallow coastal, lagoon and seaward reefs to 30 m.
West Pacific: Sabah in Malaysia, to Indonesia and Micronesia. – Philippines to E. Australia and Fiji.

DOUBLE-LINED FUSILIER   *Pterocaesio digramma*
SIZE: to 21 cm (8 1/4 in.)   Fusiliers – Caesionidae
ID: Silvery blue to blue-green; **thin yellow stripe on side below lateral line except above on tail base** and another on back, dark tail tips. Form aggregations, often mix with other fusiliers. Steep slopes, patch reefs and seaward reefs to 30 m.
West Pacific: E. Malay Peninsula to Indonesia and Solomon Is. – S.W. Japan to Great Barrier Reef, New Caledonia and Fiji.

Fusiliers

RANDALL'S FUSILIER
Pterocaesio randalli

SIZE: to 25 cm (10 in.)
Fusiliers – Caesionidae

ID: Silvery blue to reddish blue or blue-green; **large elongate yellow blotch on forebody**, black to reddish tail tips. Form aggregations, often mix with other fusiliers. Steep slopes and coastal, lagoon and seaward reefs to 30 m.

Asian Pacific: E. Andaman Sea to Sabah in Malaysia, Indonesia and Philippines.

GOLDBAND FUSILIER
Pterocaesio chrysozona

SIZE: to 21 cm (8 1/4 in.)
Fusiliers – Caesionidae

ID: Silvery brownish blue to green back and pale lower body; **wide yellow stripe** from eye tapers toward tail base, dark tail tips. Form aggregations. Coastal, lagoons and outer reefs in 2-25 m.

Indo-Asian Pacific: Red Sea and E. Africa to Indonesia, Philippines and Papua New Guinea. – S.W. Japan to Australia.

NARROWSTRIPE FUSILIER
Pterocaesio tessellata

SIZE: to 25 cm (10 in.)
Fusiliers – Caesionidae

ID: Bluish green upper half and reddish to white below; **narrow yellow stripe** covers lateral line, dark tail tips. Form aggregations. Most commonly on steep slopes, also coastal and seaward reefs to 30 m.

East Indo-Asian Pacific: Maldives to Indonesia, Philippines, Micronesia and Papua New Guinea. – S.W. Japan to Australia and Vanuatu.

WIDE-BAND FUSILIER
Pterocaesio lativittata

SIZE: to 23 cm (9 in.)
Fusiliers – Caesionidae

ID: Silvery blue; **yellow stripe from eye bulges above pectoral fin and tapers toward tail base**, dark tail tips. Form aggregations, often mix with other fusiliers. Steep slopes and deep passes, also coastal and seaward reefs in 10-50 m.

East Indo-Pacific: Maldives to West Papua in Indonesia, Micronesia, Papua New Guinea and Line Is.

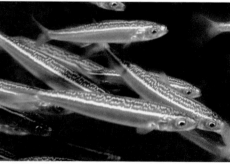

SLENDER FUSILIER
Gymnocaesio gymnoptera

SIZE: to 18 cm (7 in.)
Fusiliers – Caesionidae

ID: Bluish green back, silvery pink to reddish below; 2-3 diffuse wavy stripes on back and narrow yellow to brown stripe below, dark tail tips. Form aggregations, often mix with other fusiliers. Coral reefs in 2-20 m.

Indo-West Pacific: Red Sea and E. Africa to Indonesia, Micronesia and Solomon Is. – Philippines to Great Barrier Reef and Fiji.

MOTTLED FUSILIER
Dipterygonotus balteatus

SIZE: to 14 cm (5 1/2 in.)
Fusiliers – Caesionidae

ID: Brownish to bronze back, silvery below; pair of thin dark wavy stripes on back and pale tan stripe below, dark tail tips. Form aggregations, often mix with other fusiliers. Coastal, lagoon and seaward reefs to 20 m.

East Indo-Asian Pacific: Gulf of Aden to Indonesia, Philippines, Micronesia and Solomon Is. – Taiwan to Great Barrier Reef.

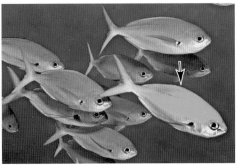

YELLOWTAIL FUSILIER — *Caesio cuning*

SIZE: to 50 cm (20 in.) Fusiliers – Caesionidae

ID: Silvery blue to blue-green; **yellow on rear upper back to tail base and tail,** deeper body than other fusiliers. Form large aggregations. Upper edge of steep slopes and around coastal patch reefs, often in silty areas with reduced visibility to 30 m.

East Indo-West Pacific: Sri Lanka to Indonesia, Micronesia. – S.W. Japan to Great Barrier Reef, New Caledonia and Vanuatu.

BLUE AND YELLOW FUSILIER — *Caesio teres*

SIZE: to 40 cm (16 in.) Fusiliers – Caesionidae

ID: Silvery blue; **yellow on center of upper back to lower tail base and tail,** black blotch on pectoral fin base. Form large aggregations, often mix with other fusiliers. Upper edge of steep slopes and around coastal patch reefs and seaward reefs to 30 m.

Indo-Pacific: E. Africa to Indonesia, Philippines, Micronesia, Solomon Is. and Line Is. – S.W. Japan to Great Barrier Reef and Fiji.

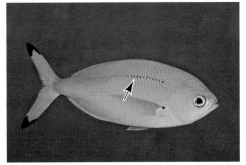

YELLOWBACK FUSILIER — *Caesio xanthonota*

SIZE: to 40 cm (16 in.) Fusiliers – Caesionidae

ID: Silvery blue with **yellow nape, back and tail.** Form large aggregations often mix with other fusiliers. Upper edge of steep slopes and around coastal patch reefs and seaward reefs to 30 m.

Indo-Asian Pacific: Red Sea and E. Africa to Indonesia.

LUNAR FUSILIER — *Caesio lunaris*

SIZE: to 40 cm (16 in.) Fusiliers – Caesionidae

ID: Silvery blue; black tip on tail lobes, **row of black dashes on arching lateral line. Young –** Yellow tail with black tips. Form large aggregations, often mix with other fusiliers. Upper edge of steep slopes, coastal patch and seaward reefs to 30 m.

Indo-West Pacific: Red Sea and E. Africa to Indonesia, Micronesia and Solomon Is. – S.W. Japan to Great Barrier Reef and Fiji.

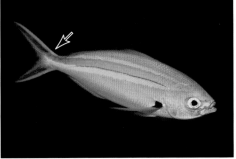

VARIABLE-LINED FUSILIER — *Caesio varilineata*

SIZE: to 40 cm (16 in.) Fusiliers – Caesionidae

ID: Silvery blue with **2-6 yellow stripes;** black streak near tip of each tail lobe. Form aggregations, often mix with other fusiliers. Upper edge of steep slopes and around coastal patch reefs and outer reefs in 2-25 m.

Indian Ocean: Red Sea and E. Africa to Andaman Sea and N.W. Sumatra in Indonesia.

SCISSORTAIL FUSILIER — *Caesio caerulaurea*

SIZE: to 35 cm (14 in.) Fusiliers – Caesionidae

ID: Silvery blue to green; **yellow stripe on side from head to tail joins dark streak on upper lobe of forked tail.** Form large aggregations, often mix with other fusiliers. Edge of slopes and around coastal patch reefs and seaward reefs to 30 m.

Indo-West Pacific: Red Sea and E. Africa to Indonesia, Micronesia and Solomon Is. – S.W. Japan to Great Barrier Reef and Fiji.

141

Anthias

SAILFIN ANTHIAS *Rabaulichthys altipinnis*
SIZE: to 6 cm (2 1/2 in.) Anthias – Serranidae/Anthiinae
ID: Nuptial Male – Reddish back shading to orange below; 3-4 maroon body bars, **tall, sail-like dorsal fin,** tail blue to blue green. **Male –** Red orange head and forebody, becoming lavender behind. **Female –** Red orange with red tips on tail lobes; white edge on low dorsal fin. Small groups to 30-40 m.
Localized: West Papua in Indonesia, Papua New Guinea, Coral Sea.

MAGENTA SLENDER ANTHIAS *Luzonichthys waitei*
SIZE: to 5.8 cm (2 1/4 in.) Anthias – Serranidae/Anthiinae
ID: Magenta with yellow to orange upper head and back; **deep violet to yellow on tail borders,** 2 separate dorsal fins. Form huge schools, often mix with other anthias species. Steep outer reef slopes to 35 m.
Indo-West Pacific: Seychelles to Indonesia and Solomon Is. – S.W. Japan to Australia, New Caledonia and Fiji.

YELLOWNOSE SLENDER ANTHIAS *Luzonichthys whitleyi*
SIZE: to 5.8 cm (2 1/4 in.) Anthias – Serranidae/Anthiinae
ID: Yellow to orange upper head, back and tail base; pink to lavender over remainder of body including tail, 2 separate dorsal fins. Form plankton-feeding aggregations 1-3 m above substrate, often mix with other anthias species. Steep slopes of outer reefs in 15-50 m.
Pacific: Christmas I. to Micronesia, Solomon Is. and Line Is.

EARLE'S SLENDER ANTHIAS *Luzonichthys earlei*
SIZE: to 5.5 cm (2 1/4 in.) Anthias – Serranidae/Anthiinae
ID: Male – Yellow to orange on upper head and back (not including tail base), pink to magenta over remainder of body including tail, often silvery reflection on cheek and midside. Outer reefs in 15-165 m.
East Indo-Asian Pacific: Maldives, Micronesia and Hawaii.

YELLOWBACK ANTHIAS *Pseudanthias evansi*
SIZE: to 10 cm (4 in.) Anthias – Serranidae/Anthiinae
ID: Male – Violet with bright yellow back that angles to lower tail; scattered yellow spots on body, **pale band through eye.** Form large plankton-feeding aggregations 1-3 m above substrate of outer reefs slopes to 40 m.
Indian Ocean: E. Africa and Mauritius to Christmas I. and Andaman Sea.

Yellowback Anthias – Female
ID: Violet with bright yellow back from nape diagonally to lower tail; orange band through eye. Aggregations may number in the hundreds and contain far more females than males.

LONGFIN ANTHIAS   *Pseudanthias ventralis*
SIZE: to 7 cm (2 ³/₄ in.)   Anthias – Serranidae/Anthiinae
ID: Male – Lavender with variable amounts of yellow on upper head, back and tail base; yellow fins, except red often on dorsal fin, random magenta stripes and spots on back; long ventral fins. Form small groups on outer reefs in 26-68 m.
Pacific: Great Barrier Reef to New Caledonia, French Polynesia, north to Mariana Is. in Micronesia.

Longfin Anthias – Female
ID: Shades of lavender with yellow back and fins.

TWOSPOT ANTHIAS   *Pseudanthias bimaculatus*
SIZE: to 7 cm (2 ³/₄ in.)   Anthias – Serranidae/Anthiinae
ID: Male – Shades of red or orange with lavender to purple scribble markings; 2 orange to yellow bands bordered with lavender to purple extend from eye to pectoral fin, tips of tail lobes bluish. Form small plankton-feeding aggregations 1-3 m above reefs and steep slopes in 20-60 m.
Indian Ocean: E. Africa to Java and Bali in Indonesia.

Twospot Anthias – Female
ID: Red to orange or magenta head and forebody shading to bright yellow back, tail base and dorsal, anal and tail fins. Turbid coastal reefs and clear outer reefs.

SQUARESPOT ANTHIAS   *Pseudanthias pleurotaenia*
SIZE: to 20 cm (8 in.)   Anthias – Serranidae/Anthiinae
ID: Male – Orangish red to magenta; **large square violet spot on side,** long 3rd dorsal spine. Form loose groups containing far more females than males. Steep seaward slopes in 10-180 m, usually below 25 m.
West Pacific: Indonesia, Philippines, Micronesia, Papua New Guinea and Solomon Is. – S.W. Japan to Great Barrier Reef and Fiji.

Squarespot Anthias – Female
ID: Orange with yellow fins; **pair of violet stripes run from eye to tail base** (compare similar female Scalefin Anthias p. 147). Form harems with dominant males. Socially dominant female capable of sex change over period of 2-3 weeks.

CHARLENE'S ANTHIAS   *Pseudanthias charleneae*

SIZE: to 10 cm (4 in.)   Anthias – Serranidae/Anthiinae

ID: Male – Reddish orange upper head, body shading to pale lavender undercolor with large orange scale spots; orange band from snout to lower eye extending to edge of gill cover, red bar below middle of spinous dorsal fin. Form aggregations. Outer reef slopes in 30-55 m.

Localized: Bali and West Papua in Indonesia.

REDSTRIPE ANTHIAS   *Pseudanthias fasciatus*

SIZE: to 20 cm (8 in.)   Anthias – Serranidae/Anthiinae

ID: Female – Orange to peach; **red stripe from gill cover to tail. Male –** Distinguished by long tail lobes. Solitary or form small groups. Most commonly in caves (may orient upside down to ceiling) of deep outer reefs and steep slopes in 20-68 m.

Indo-Asian Pacific: Red Sea to Indonesia, Philippines, Micronesia, and Solomon Is. – S.W. Japan to Great Barrier Reef and Fiji.

BARTLETT'S ANTHIAS   *Pseudanthias bartlettorum*

SIZE: to 6.2 cm (2 1/2 in.)   Anthias – Serranidae/Anthiinae

ID: Male – Lavender with yellow upper head, back, dorsal and tail fins; **violet tail borders,** long 2nd dorsal fin spine. **Female –** Similar, but lack long fin spine. Form plankton-feeding aggregations above substrate. Steep outer reef slopes in 4-30 m.

Pacific: N.E. Papua New Guinea and Micronesia to Line Is.

CAVE ANTHIAS   *Pseudanthias* sp. cf. *flavicauda*

SIZE: to 8 cm (3 1/4 in.)   Anthias – Serranidae/Anthiinae

ID: Male – Red to lavender; **yellow ventral fins,** tail yellow becoming reddish in large males [pictured], other fins red to lavender, elongate 3rd dorsal spine. Possibly same as Yellowtail Anthias *P. flavicauda* from Fiji to Tonga. Inhabit caves and overhangs of steep seaward reefs in 25 to over 45 m.

Localized: Micronesia, likely more widespread.

BARRIER REEF ANTHIAS   *Pseudanthias engelhardi*

SIZE: to 10 cm (4 in.)   Anthias – Serranidae/Anthiinae

ID: Male – Pale lavender-pink with **pale yellowish to orangish tint on head** and forebody; red bar on side, pale band below eye, 3rd dorsal spine slightly elongate. Form plankton-feeding aggregations 1-3 m above outer reefs in 50-70 m.

Southwest Pacific: N. Great Barrier Reef, E. Papua New Guinea and Solomon Is. to Fiji.

Barrier Reef Anthias – Female

ID: Pale lavender pink with yellow tint on head.

BICOLOR ANTHIAS
Pseudanthias bicolor

SIZE: to 13 cm (5 in.) Anthias – Serranidae/Anthiinae

ID: Male – Orange upper body, lavender to pale pink below; long 2nd and 3rd dorsal fin spines with yellow tips. **Female –** Similar, but spines shorter and lack yellow tips. Form schools. Ledges and outcroppings of lagoon and outer reefs in 5-68 m.

Indo-Pacific: Mauritius to Indonesia, Micronesia, Solomon Is. to Hawaii and Line Is. – S.W. Japan to N. Great Barrier Reef.

FLAME ANTHIAS
Pseudanthias ignitus

SIZE: to 8 cm (3¼ in.) Anthias – Serranidae/Anthiinae

ID: Male – Orange to yellow body, lavender to pink head and bright red dorsal fin; **red tail.** Form plankton-feeding aggregations above upper edges of steep slopes in 10-30 m.

Indian Ocean: Maldives and Andaman Sea to N. Sumatra in Indonesia.

REDFIN ANTHIAS
Pseudanthias dispar

SIZE: to 9.5 cm (3¾ in.) Anthias – Serranidae/Anthiinae

ID: Male – Orange to yellow body, lavender to pink head and bright red dorsal fin; translucent tail. Form plankton-feeding aggregations 1-3 m above substrate; males erect dorsal fin during courtship. Upper edge of steep slopes in 15 m.

Pacific: Indonesia, Philippines, Micronesia, Solomon Is. and Line Is. – S.W. Japan to Great Barrier Reef to Fiji.

Redfin Anthias – Female

ID: Orange to peach upper head, body and fins, lower head pale; 2 narrow violet to lavender or yellowish bars extend from eye to pectoral fin base. Form schools, females far outnumber males.

RANDALL'S ANTHIAS
Pseudanthias randalli

SIZE: to 9.5 cm (3¾ in.) Anthias – Serranidae/Anthiinae

ID: Male – Yellowish orange back changing to pale or deep magenta below; red dorsal fin, **wide red band on anal fin. Nuptial Male –** Wide red stripe from behind eye to lower tail base. Form small groups. Steep dropoffs in 15-120 m.

Asian Pacific: Sabah in Malaysia to Indonesia, Philippines, Micronesia, north to S.W. Japan.

Randall's Anthias – Female

ID: Yellowish snout and upper back shading to pink below; yellow tail. Form small groups.

Anthias

PURPLE ANTHIAS   *Pseudanthias tuka*

SIZE: to 12 cm (4 ³/₄ in.)   Anthias – Serranidae/Anthiinae

ID: Male – Purple; yellow to yellowish chin and **dark purple blotch on rear dorsal fin.** Form plankton feeding-aggregations 1-3 m above outer reef slopes, but also coastal reefs in 2-40 m.

Asian Pacific: Indonesia to Micronesia and Solomon Is. – S.W. Japan to N.W. Australia, Great Barrier Reef, New Caledonia and Vanuatu.

Purple Anthias – Female

ID: Purple to lavender; yellow stripe along back to tip of upper tail lobe, yellow border lower tail lobe. Aggregations contain far more females than males.

PURPLE QUEEN   *Pseudanthias pascalus*

SIZE: to 17 cm (6 ³/₄ in.)   Anthias – Serranidae/Anthiinae

ID: Male – Purple with numerous small dark blue to orange spots; snout forms fleshy protuberance, pale chin and often orange stripe from snout through eye. Form plankton-feeding aggregations high above outer reef slopes in 5-60 m.

Pacific: Sabah in Malaysia, Sulawesi in Indonesia, Micronesia to French Polynesia. – S.W. Japan to Great Barrier Reef and Fiji.

Purple Queen – Female

ID: Purple to purplish red but no spotting; red to orange stripe from eye to edge of gill cover; dorsal, anal and ventral fins much shorter than male. Aggregations contain far more females than males.

STOCKY ANTHIAS   *Pseudanthias hypselosoma*

SIZE: to 19 cm (7 ¹/₂ in.)   Anthias – Serranidae/Anthiinae

ID: Male – Pinkish with extensive red area on upper head and foreback, **red blotch on dorsal fin;** rounded tail, elongate ventral fins and large anal fin. Form schools above coral outcroppings on sheltered coastal reefs and lagoons to 35 m.

East Indo-West Pacific: Maldives to Indonesia, Philippines, Micronesia and Solomon Is. – S.W. Japan to Great Barrier Reef and Fiji.

Stocky Anthias – Female

ID: Orange with pinkish lavender tint; slightly forked **tail with red tips on lobes,** pale lavender line from eye to pectoral fin lobe. In groups females far outnumber males.

SCALEFIN ANTHIAS   *Pseudanthias squamipinnis*
SIZE: to 15 cm (6 in.)   Anthias – Serranidae/Anthiinae
ID: Male/Red Variation – Shades of red with yellow spots on body scales; **purple blotch on outer pectoral fin,** long 3rd dorsal spine. Form small to huge plankton-feeding aggregations above shallow coastal, lagoon and outer reefs in 2-20 m.
Indo-West Pacific: Red Sea and E. Africa to Indonesia, Philippines, Micronesia and Solomon Is. – S.W. Japan to E. Australia and Fiji.

Scalefin Anthias – Male/Purple Variation
ID: Shades of purple, body may have tints of yellow or green; purple blotch on outer pectoral fin on all male variations, long 3rd dorsal spine. Feed on plankton in strong currents high above the substrate. Often the most common anthias species on shallow reefs.

Scalefin Anthias – Female
ID: Orange; violet-edged orange stripe runs from eye to pectoral fin base. In groups or aggregations females greatly outnumber males.

THREADFIN ANTHIAS   *Pseudanthias huchtii*
SIZE: to 12 cm (4 ³/₄ in.)   Anthias – Serranidae/Anthiinae
ID: Male – Variable, most commonly shades of pale yellow to green, occasionally with grayish tints; bright red stripe from eye to pectoral fin base and **red border on ventral fins.** Solitary or form small feeding groups on clear outer reefs in 4-20 m.
Asian Pacific: Indonesia, Philippines, Micronesia, Papua New Guinea, Solomon Is., Vanuatu and Great Barrier Reef.

Threadfin Anthias – Male/Lavender Variation
ID: Occasionally shades of lavender or gray with lavender tints; all males have long 3rd dorsal spine, source of common name. This variation most common from Palau to Vanuatu.

Threadfin Anthias – Female
ID: Dull yellow or greenish yellow with brownish tone on back and nape; bright yellow borders on tail. Groups contain far more females than males.

Anthias

LUZON ANTHIAS　　　　*Pseudanthias luzonensis*
SIZE: to 14.5 cm (5 ³/₄ in.)　　Anthias – Serranidae/Anthiinae
ID: Male – Peach to pinkish with reddish snout; **several yellow to red narrow wavy stripes on side**, large red spot on mid-dorsal fin. Outer reef slopes in 20-60 m.
West Pacific: Indonesia, Philippines, Micronesia and Solomon Is. – S.W. Japan to Great Barrier Reef, Vanuatu and Fiji.

Luzon Anthias – Female
ID: Shades of pinkish orange upper body, shading to white below; narrow orange band extends from lower eye to base of pectoral fin.

WHITESPOTTED ANTHIAS　　　　*Pseudanthias hutomoi*
SIZE: to 12 cm (4 ³/₄ in.)　　Anthias – Serranidae/Anthiinae
ID: Male – Pinkish to lavender; reddish band with violet borders from eye to pectoral fin, **2 white spots on back** and whitish blotch on rear body and upper tail base. Form small schools. Cluster around outcroppings on coastal reefs in 30-70 m.
Asian Pacific: Brunei, Indonesia, Philippines and Papua New Guinea.

Whitespotted Anthias – Female
ID: Pinkish with reddish scale margins and yellowish brown upper head; wide orangish band from eye to pectoral fin, **3-4 small white spots on back** and upper tail base.

OLIVE ANTHIAS　　　　*Pseudanthias olivaceus*
SIZE: to 12 cm (4 ³/₄ in.)　　Anthias – Serranidae/Anthiinae
ID: Male – Olive to dark gray; yellow spots on scales of lower body may align to form bars, yellow stripe behind eye. Form small plankton-feeding aggregations around rock and coral formations of seaward reefs to 34 m.
Central Pacific: Cook, Austral and Phoenix Is. to French Polynesia and Line Is.

Olive Anthias – Female
ID: Purple or reddish gray with bright yellow tail; red border on dorsal fin. Sometimes associated with Fusilier Damselfish which apparently mimics anthias.

SILVERSTREAK ANTHIAS   *Pseudanthias cooperi*
SIZE: to 14 cm (5 1/2 in.)   Anthias – Serranidae/Anthiinae
ID: Male – Red to orange to pale pink with **red tail and tail base**; silver-white streak below eye, red patch or bar on midside (may be intense or faint). Form small plankton-feeding aggregations off current swept outer reefs and dropoffs in 15-60 m.
Indo-Pacific: E. Africa to Indonesia, Philippines, Micronesia, Solomon Is., Line Is. – S.W. Japan to Australia and Fiji.

Silverstreak Anthias – Female
SIZE: to 11 cm (4 1/4 in.)   Anthias – Serranidae/Anthiinae
ID: Olive upper body gradating to reddish head and lavender belly; reddish band from eye to pectoral fin, red dorsal and tail fins.

REDBAR ANTHIAS   *Pseudanthias rubrizonatus*
SIZE: to 10 cm (4 in.)   Anthias – Serranidae/Anthiinae
ID: Male/Pink Variation – Peach on head shading to pale pink on rear body; pale stripe below eye, **red bar on side.** Form clusters around coral outcroppings, sometimes in turbid water. Coastal and seaward reefs in 10-58 m.
Asian Pacific: Andaman Sea to Indonesia, Philippines, Micronesia, Papua New Guinea, Solomon Is. – S.W. Japan to Australia and Fiji.

Redbar Anthias – Male/Yellow Variation
ID: Peach on head shading to yellow on rear body, occasionally entirely yellow; pale stripe below eye and red bar on side. **Female –** Light red with yellow mark on each scale except belly where whitish, a narrow violet band from below eye to lower pectoral base, red tipped tail lobes.

SUNSET ANTHIAS   *Pseudanthias parvirostris*
SIZE: to 8.5 cm (3 1/4 in.)   Anthias – Serranidae/Anthiinae
ID: Male – Yellow to orange upper body, pinkish to magenta lower body; **purple dorsal fin and borders on tail.** Reddish with white dorsal fin and tail borders in Indian Ocean. Form small aggregations near patch reefs and outcroppings of deep outer slopes in 30-65 m.
Indo-Asian Pacific: Mauritius to Indonesia, Philippines, Micronesia, Papua New Guinea, Solomon Is., north to S. Japan.

Sunset Anthias – Female
ID: Pinkish undercolor with yellow spots on scales to solid bright yellow; violet line markings on snout and top of head.

Anthias

PAINTED ANTHIAS　　　　*Pseudanthias pictilis*
SIZE: to 13.5 cm (5 ¼ in.)　　Anthias – Serranidae/Anthiinae
ID: Male – Reddish lavender; violet bar edged with orange below middle of soft dorsal fin, **pale band on tail.** Form openwater aggregations above seaward reef slopes in 20-40 m.
Southwest Pacific: S. Great Barrier Reef, New Caledonia, Lord Howe Is. and Fiji.

Painted Anthias – Female
ID: Violet-pink sides and belly gradating to yellow upper back, dorsal and tail fins. Form openwater aggregations, usually far more abundant than males.

LORI'S ANTHIAS　　　　*Pseudanthias lori*
SIZE: to 12 cm (4 ¾ in.)　　Anthias – Serranidae/Anthiinae
ID: Male – Lavender to red with orange to yellow spots on body scales; 3-5 bright red bars on back, **wide bright red stripe on upper tail base.** Steep outer reef slopes and drop-offs in 25-60 m.
Pacific: Sabah in Malaysia to Indonesia, Philippines, Micronesia, and French Polynesia. – S.W. Japan to Australia and Fiji.

Lori's Anthias – Female
ID: Red bars on back more prominent than those of male.

PRINCESS ANTHIAS　　　　*Pseudanthias smithvanizi*
SIZE: to 9.5 cm (3 ¾ in.)　　Anthias – Serranidae/Anthiinae
ID: Male – Lavender with orangish spots on body scales; red to reddish upper head and back, **purple border on upper tail fin lobe.** Form small groups, sometimes mix with Lori's Anthias [previous]. Steep outer reef slopes in 6-70 m.
East Indo-West Pacific: Cocos-Keeling Is. to Indonesia, Micronesia and Solomon Is. – S.W. Japan to Great Barrier Reef.

Princess Anthias – Female
ID: Lavender undercolor with orangish spots on body scales; white to pale blue stripe on back continues into white to pale blue tail with red borders.

YELLOWSPOTTED ANTHIAS   *Pseudanthias flavoguttatus*
SIZE: to 8.5 cm (3 1/4 in.)   Anthias – Serranidae/Anthiinae
ID: Lavender to red with yellow spots on scales; alternating bright red and white bars on tail base and back, **yellow lower tail lobe.** Form small schools. Lurk near ledges and caves of deep outer reef slopes and dropoffs below 30 m.
Asian Pacific: Christmas I., Indonesia and Micronesia.

HAWK ANTHIAS   *Serranocirrhitus latus*
SIZE: to 13 cm (5 in.)   Anthias – Serranidae/Anthiinae
ID: Deep pink to orangish with yellow spot on scales of upper body; large bright yellow spot on upper gill cover, **yellow bands radiate from eye.** Solitary or form small groups. Swim upside down under ledges on outer reef dropoffs in 15-70 m.
West Pacific: Indonesia, Philippines, Micronesia, Solomon Is. – S.W. Japan to Great Barrier Reef, New Caledonia and Fiji.

DWARF PERCHLET   *Plectranthias nanus*
SIZE: to 3.5 cm (1 1/2 in.)   Anthias – Serranidae/Anthiinae
ID: White to tan undercolor with irregular reddish brown blotches and spots; diagonal bar below eye, **dark spot on upper and lower tail base.** Solitary. Crevices and recesses of coastal, lagoon and outer reefs in 6-55 m.
East Indo-Pacific: Cocos-Keeling Is. to Indonesia, Philippines, Hawaii and French Polynesia. (Only around islands.)

LONGFIN PERCHLET   *Plectranthias longimanus*
SIZE: to 3.6 cm (1 1/2 in.)   Anthias – Serranidae/Anthiinae
ID: White to tan undercolor; irregular reddish brown blotches, diagonal brown bar below eye, **white spot on upper and lower tail base and behind dorsal fin.** Solitary. Crevices and recesses of reefs in 6-73 m.
Indo-West Pacific: E. Africa to Solomon Is. – S.W. Japan to Great Barrier Reef and Fiji. Continental margins and large islands.

CHEQUERED PERCHLET   *Plectranthias inermis*
SIZE: to 4.3 cm (1 3/4 in.)   Anthias – Serranidae/Anthiinae
ID: Whitish undercolor; large square red blotches align to form bars, **first three dorsal spines yellowish,** fins translucent or lightly spotted. Solitary. Base of steep slopes in rubble or in crevices of seaward slopes in 14-65 m.
East Indo-Asian Pacific: Christmas I. to Bali, Molucca Is. and West Papua in Indonesia, Philippines and Papua New Guinea.

REDBLOTCH PERCHLET   *Plectranthias winniensis*
SIZE: to 4.8 cm (1 3/4 in.)   Anthias – Serranidae/Anthiinae
ID: Yellowish orange to yellowish brown; reddish botching on tail base, **small pale spot on back between 1st and 2nd dorsal fins.** Solitary; cryptic crevice dwellers. Outer reef slopes in 23-58 m.
Indo-Pacific: Red Sea to Indonesia, Philippines, Micronesia, Hawaii and French Polynesia. – Great Barrier Reef and New Caledonia to Vanuatu.

Heavy Bodies/Large Lips
Groupers – Soapfishes – Hawkfishes – Sweetlips

This ID Group consists of fishes with heavy, robust bodies.

FAMILY: Sea Basses – Serranidae

As a group members of family Serranidae are difficult to define, but can be distinguished by three spines on their gill cover, a long, continuous dorsal fin, a complete lateral line, large mouths with more than one row of teeth, and typically rounded tail fins. Serranids, which include the species-rich grouper complex, has undergone dramatic alterations in recent years. For example, anthias, presented in ID Group 6, were previously classified in family Anthiidae, but are now included in Serranidae as subfamily Anthiinae; likewise soapfishes once classified in family Grammistidae are now considered members of the sea bass family in subfamily Grammistinae. Also, a small group of diminutive sea basses in subfamily Liopropomatinae are presented in ID Group 11 because of their cryptic nature. Most sea basses are hermaphroditic beginning life as females and later changing into males. A few species, however, develop both male and female gonads simultaneously.

FAMILY: Sea Basses/Groupers - Serranidae
8 Genera – 55 Species Included

Typical Shape

Groupers, the most recognizable members of the sea bass family, are also locally known as rockcods, cods, hinds, and trouts. All have strong, stout bodies and large mouths filled with more than one row of teeth. In the Indo-Pacific, they vary in size from the Giant Grouper reaching a length of 231 cm (7 1/2 ft.) to the Blacktip Grouper attaining less than 40 cm (16 in.). Smaller groupers mature in one year, while larger species take many years to reach sexual maturity. Spawning is seasonal and controlled by moon phase. Many of the larger species travel from miles around, at precise times, to spawn in mass aggregations at traditional sites.

Groupers, subject to regular infestations of external parasites, spend significant amounts of time at preferred cleaning stations within their home ranges where larger individuals establish proprietary claims. Groupers are solitary carnivores that hunt near the bottom. Although awkward in appearance, groupers can cover short distances quickly. Fishes or crustaceans are drawn into their gullets by a powerful suction created when they rapidly open cavernous mouths. Held securely by hundreds of small, rasplike teeth that cover the jaws, tongue and palate, the prey is swallowed whole.

SUBFAMILY: Soapfishes – Serranidae/Grammistinae
6 Genera – 6 Species Included

Typical Shape

The small group of fishes, known as soapfishes, exhibit a mixed bag of body shapes, but all have the unique ability to exude a soapy skin toxin (grammistin), which makes them unpalatable to predators. Soapfishes typically have upturned mouths, protruding lower jaws, and rounded tail fins. They generally inhabit shallow waters, and are solitary night-hunters that tend to lie on the bottom or hide inside crevices during the day.

FAMILY: Hawkfishes – Cirrhitidae
7 Genera – 14 Species Included

Typical Shape

Hawkfishes are a family of small, stout-bodied bottom fishes, without swim bladders, that typically establish territories within the branches of soft and hard corals. Initially curious, the picturesque fishes perch in an exposed position until closely approached before darting into protected areas where their thick lower pectoral fins can be used to wedge themselves in place. A series of cirri attached near the tips of their dorsal fin spines easily and reliably identifies family members. The sequential hermaphrodites live in small harems. All, except the plankton-feeding Lyretail Hawkfish, are lie-in-wait predators of small fishes and crustaceans.

FAMILY: Sweetlips (Grunts) – Haemulidae
2 Genera – 12 Species Included

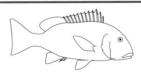

Typical Shape

Sweetlips are closely related to and resemble snappers, but are not quite as large, have smaller mouths, more prominent lips, and lack the snappers' canine teeth. The fishes are known in many regions as grunts because of a "grunting" sound produced by rubbing flat teeth plates together, which is amplified by adjacent air bladders. The nocturnal predators of bottom-dwelling crustaceans spend daylight hours hovering alone or in groups near the reef's structure. Boldly patterned and erratic swimming juveniles only vaguely resemble their adult forms.

Groupers

SLENDER GROUPER *Anyperodon leucogrammicus*
SIZE: to 60 cm (2 ft.) Groupers – Serranidae
ID: Slender body; greenish to brownish gray with red-orange spots covering head and body, usually 3-4 pale stripes of variable intensity depending on mood. Solitary. Sheltered coastal and outer reefs in 1-15 m.
Indo-West Pacific: Red Sea and E. Africa to Indonesia, Philippines, Micronesia, Solomon Is. – S.W. Japan to Great Barrier Reef and Fiji.

Slender Grouper – Juvenile
SIZE: to 12 cm (4 3/4 in.)
ID: Gold with dark edged pale blue stripes; 1-2 blue-edged black spot on base of tail and often another on base of rear dorsal fin.

PEACOCK GROUPER *Cephalopholis argus*
SIZE: to 60 cm (2 ft.) Groupers – Serranidae
ID: Brown covered with small black-edged blue spots, broad blue borders on rear dorsal, anal, pectoral and tail fins. Solitary or form groups of up to 12 including a single dominant male. Variable habitats to outer reef slopes to 15 m.
Indo-Pacific: Red Sea and E. Africa to Indonesia, Micronesia, Solomon Is., French Polynesia. – S.W. Japan to Great Barrier Reef.

Peacock Grouper – Pale Phase
ID: Can pale or darken rapidly. Frequently display 5-6 pale bars on rear body in both the dark and pale phases.

BLUESPOTTED GROUPER *Cephalopholis cyanostigma*
SIZE: to 30 cm (12 in.) Groupers – Serranidae
ID: Brown to reddish brown, occasionally lightly mottled or display bars; covered with numerous dark edged pale bluish spots. Solitary. Coastal, lagoon and seaward reefs to 50 m.
Asian Pacific: E. Malay Peninsula, Indonesia, Philippines, Micronesia, Papua New Guinea and Solomon Is. to Great Barrier Reef.

Bluespotted Grouper – Juvenile
SIZE: 5-13 cm (2-5 in.)
ID: Dark gray to brown with yellow fins; occasionally display bars. It was once described as a separate species, *C. xanthopterus,* but transitional specimens revealed its true identity. Like other genus members feed on crustaceans and fishes.

HARLEQUIN GROUPER *Cephalopholis polleni*
SIZE: to 35 cm (13³/₄ in.) Groupers – Serranidae
ID: Yellow to greenish yellow; bright blue or violet stripes on head, body and fins. Solitary, often inhabit caves Clear water of outer reefs and steep slopes, seldom in less than 30 m.
Indo-Pacific: Comoro Is., Indonesia, Philippines and Micronesia to Line Is., north to S.W. Japan. Known only from scattered location within its range.

Harlequin Grouper – Juvenile
SIZE: 5-13 cm (2-5 in.)
ID: Reddish pink with blue stripe from upper eye to rear gill cover and another from snout to below eye to rear gill cover; few faint blue stripes on body, bright yellow borders on tail.

BLUELINED GROUPER *Cephalopholis formosa*
SIZE: to 34 cm (13¹/₂ in.) Groupers – Serranidae
ID: Dark brown to yellowish brown with dark blue, primarily horizontal, lines on head, body and fins. Solitary. Dead silty reefs in sheltered waters and coastal reefs to 15 m.
East Indo-Asian Pacific: India, Andaman Sea, Indonesia, Philippines, north to Taiwan.

LEOPARD GROUPER *Cephalopholis leopardus*
SIZE: to 18.5 cm (7¹/₄ in.) Groupers – Serranidae
ID: Mottled reddish brown with numerous red-orange to pinkish red spots; **dark brown saddle on upper tail base with smaller saddle behind,** dark brown streak on upper corner of tail. Solitary. Coastal, lagoon and outer reefs to 35 m.
Indo-Pacific: E. Africa to Indonesia, Philippines, Micronesia, Solomon Is. to French Polynesia. – S.W. Japan to Australia.

FLAGTAIL GROUPER *Cephalopholis urodeta*
SIZE: to 27 cm (10³/₄ in.) Groupers – Serranidae
ID: Brown to reddish, darker toward tail; occasionally display faint pale bars, **diagonal white lines across corners of tail.** Indian Ocean variation lacks these lines and has a dark spot on upper gill cover. Solitary. Variable reef habitats to 60 m.
Pacific: Indonesia, Philippines, Micronesia, Solomon Is. to French Polynesia. – S.W. Japan to Great Barrier Reef.

STRAWBERRY GROUPER *Cephalopholis spiloparaea*
SIZE: to 21 cm (8¹/₄ in.) Groupers – Serranidae
ID: Red to pale reddish orange with dark red to brownish red mottling; commonly covered with faint pale spots, **blue to pale submarginal outline on tail.** Solitary. Common on steep outer slopes in 15-108 m.
Indo-Pacific: E. Africa to Sabah in Malaysia, Philippines, Micronesia to French Polynesia. – S.W. Japan to Australia.

Groupers

CORAL GROUPER *Cephalopholis miniata*
SIZE: to 50 cm (20 in.)   Groupers – Serranidae
ID: Orange-red to reddish brown with numerous dark-edged blue spots; a narrow blue margin on all fins except pectorals, occasionally display pale bars. Solitary. Coastal, lagoon and seaward reefs in 1-50 m.
Indo-Pacific: Red Sea and E. Africa to Indonesia, Philippines, Micronesia to Line Is. – S.W. Japan to Great Barrier Reef and Fiji.

Coral Grouper – Small Juvenile
SIZE: 5-13 cm (2-5 in.)
ID: Orange-red with scattering blue-gray spots on head and body; yellow rear dorsal, pectoral, anal and tail fins. Inhabit caves and crevices, rarely in open.

FRECKLED GROUPER *Cephalopholis microprion*
SIZE: to 23 cm (9 in.)   Groupers – Serranidae
ID: Completely dark brown or display whitish area covering much of body; dark-edged blue spots cover head and breast. Solitary. Dead silty reefs in sheltered coastal waters in 2-20 m.
West Pacific: Malay Peninsula to Indonesia, Philippines, Micronesia, Papua New Guinea and Solomon Is. to Great Barrier Reef, Vanuatu and Fiji.

Freckled Grouper – Variation
ID: Occasionally display large light colored area covering most of rear half of body.

TOMATO GROUPER *Cephalopholis sonnerati*
SIZE: to 50 cm (20 in.)   Groupers – Serranidae
ID: Orange-red to reddish brown with **dense covering of red spots on head;** more loosely scattered, fainter spots on body and fins. Solitary; frequent shrimp cleaning stations. Lagoon and outer reefs in 10-150 m.
Indo-Pacific: E. Africa to Indonesia, Philippines, Micronesia and Line Is. – S.W. Japan to Great Barrier Reef and Fiji.

Tomato Grouper – Variation
ID: Blotchy shades of red-orange to brown often forming 6-7 bars; scattering of small white spots on fins. Solitary. Patch reefs below 20 m.

SADDLE GROUPER *Cephalopholis sexmaculata*
SIZE: to 50 cm (20 in.) Groupers – Serranidae
ID: Orange-red with numerous small blue spots; often lines on head, **6-7 pale saddles on back that commonly extend onto side.** Solitary or form groups. Caves on outer reef slopes in 10-150 m.
Indo-Pacific: E. Africa to Indonesia, Philippines, Micronesia, and French Polynesia. – S.W. Japan to Great Barrier Reef.

STARRY GROUPER *Cephalopholis polyspila*
SIZE: to 25 cm (10 in.) Groupers – Serranidae
ID: Brown with numerous small bright blue spots on head, body and fins; bright blue outer edge on ventral, anal and tail fins. Solitary and secretive. Rich coral areas in 3-13 m.
Localized: Andaman Sea, including Myanmar, Thailand, and Andaman Is. and W. Sumatra in Indonesia.

CHOCOLATE GROUPER *Cephalopholis boenak*
SIZE: to 26 cm (10 in.) Groupers – Serranidae
ID: Brown; usually display 7-8 dark bars on side, tail with dark corners edged in blue; dark spot on upper rear grill cover. Solitary; feed primarily on crustaceans. Dead silty reefs in sheltered waters and coastal reefs in 4-30 m.
Indo-Asian Pacific: E. Africa to Andaman Sea, Indonesia, Micronesia, Solomon Is. – S.W. Japan to Great Barrier Reef and New Caledonia.

MASKED GROUPER *Gracila albomarginata*
SIZE: to 40 cm (16 in.) Groupers – Serranidae
ID: Brown head shading to nearly black on body; large white square on midbody, midlateral row of short narrow dark bars, **line markings on head,** dark spot on white tail base. Solitary; hover in open water. Steep outer reef slopes in 15-120 m.
Indo-Pacific: E. Africa to Indonesia, Philippines, Micronesia, Solomon Is., French Polynesia. – S.W. Japan to Great Barrier Reef.

Masked Grouper – Juvenile
SIZE: 5-13 cm (2-5 in.)
ID: Dark violet to lavender with bright red margins on rear dorsal, anal and tail fins. Solitary. Outer reefs in 2-20 m.

Masked Grouper – Subadult
SIZE: 20-30 cm (8-12 in.)
ID: Brown with bright red margins on rear dorsal, anal and tail fins. Body bars and mask stripes of adult are beginning to show. Solitary. Outer reefs in 2-20 m.

Groupers

BLUE AND YELLOW GROUPER   *Epinephelus flavocaeruleus*
SIZE: to 80 cm (2 1/2 ft.)   Groupers – Serranidae
ID: Young - Sky blue with bright yellow fins, tail base and
upper lip. **Adult** - Shades of dark blue with yellow upper lip;
bright yellow fins of young gradually fades to match color of
adult. Solitary. Patch reefs and sandy areas 10-150 m.
Indo-Asian Pacific: East Africa to Andaman Sea, Sumatra, Java
and Bali in Indonesia.

LONGSPINE GROUPER   *Epinephelus longispinis*
SIZE: to 54 cm (21 1/4 in.)   Groupers – Serranidae
ID: Grayish brown with paler blotches; brown spots on head
and diagonally-elongate spots on body, **2 to 4 large dark
blotches on dorsal fins.** Solitary. Coastal reefs to 70 m.
Indo-Asian Pacific: E. Africa to Andaman Sea and Java, Bali
and Lombok in S. Indonesia.

HIGHFIN GROUPER   *Epinephelus maculatus*
SIZE: to 60 cm (2 ft.)   Groupers – Serranidae
ID: Brownish gray to brown covered with dark brown
polygonal spots; **pair of prominent white saddles on
forehead and middle of back.** Solitary. Often on open sand
at base of reefs in coastal, lagoon and outer reefs in 2-100 m.
E. Indo-West Pacific: Cocos-Keeling Is. to Indonesia, Philippines,
Micronesia. – S.W. Japan to E. & N.W. Australia and Fiji.

BLACKSADDLE GROUPER   *Epinephelus howlandi*
SIZE: to 44 cm (17 in.)   Groupers – Serranidae
ID: Blotched pale gray and brownish gray with brownish
black spots on head, body and fins; dark saddles on back and
upper tail base, white margins on dorsal, anal and tail fins.
Solitary. Lagoon and outer reefs in 1-37 m.
West Pacific: Micronesia to Great Barrier Reef to Fiji.

DUSKYTAIL GROUPER   *Epinephelus bleekeri*
SIZE: to 76 cm (2 1/2 ft.)   Groupers – Serranidae
ID: Whitish with numerous orange to reddish or dark brown
spots on head and body, spots on fins less well defined; often
display several faint dark bars on body, **last 2/3 of tail purplish
and without spots.** Solitary. Silty coastal reefs to 50 m.
East Indo-Asian Pacific: Arabian Gulf to Indonesia and
Philippines. – Taiwan to N.W. Australia.

SNUBNOSE GROUPER   *Epinephelus macrospilos*
SIZE: to 43 cm (17 in.)   Groupers – Serranidae
ID: Dusky white with large polygon-shaped brown spots
covering head, body and fins; white margin on dorsal, anal
fins, **rounded tail with rows of dark spots and thin white
margin.** Solitary. Coastal, lagoon and outer reefs in 3-44 m.
Indo-Pacific: E. Africa to Indonesia, Philippines, Micronesia,
Solomon Is., French Polynesia. – S.W. Japan to Great Barrier Reef.

TWINSPOT GROUPER *Epinephelus bilobatus*
SIZE: to 33 cm (13 in.) Groupers – Serranidae
ID: Light gray undercolor with polygonal spots in varying shades of brown; **white stripe on back below dorsal fin,** 2-3 blackish bi-lobed spots at base of dorsal fin. Solitary. Coastal reefs with weed bottoms and scattered coral patches in 4-50 m.
Asian Pacific: N.W. Australia to West Papua in Indonesia.

GREASY GROUPER *Epinephelus tauvina*
SIZE: to 75 cm (2½ ft.) Groupers – Serranidae
ID: Blotchy dusky white to pale brown with reddish brown spots on head, body and fins; darkish saddles on back, **series of black spots form dark margin on rear dorsal, anal and tail fins.** Solitary. Coastal, lagoon and outer reefs to 20 m.
Indo-Pacific: Red Sea to Micronesia, Solomon Is. and French Polynesia – S.W. Japan to Great Barrier Reef.

AREOLATE GROUPER *Epinephelus areolatus*
SIZE: to 40 cm (16 in.) Groupers – Serranidae
ID: Gray to whitish with numerous large close-set brown spots that become smaller and more numerous with maturity; **narrow white straight margin on tail** (tails rounded on most grouper species). Solitary. Fine sediment bottoms to 200 m.
Indo-West Pacific: Red Sea and E. Africa to Indonesia, Philippines, Micronesia. – S.W. Japan to N.W. Australia and Fiji.

SPECKLED GROUPER *Epinephelus cyanopodus*
SIZE: to 100 cm (3¼ ft.) Groupers – Serranidae
ID: Pale bluish gray; profuse small black spots on head, body and fins. Juveniles and subadults [pictured] have black margin on tail and black ventral fin tips. Solitary. Lagoons and outer reefs over sand bottoms near patch reefs to 150 m.
West Pacific: S. China Sea, Indonesia, Philippines, Micronesia, Papua New Guinea to Solomon Is. – S.W. Japan to Australia and Fiji.

LONGFIN GROUPER *Epinephelus quoyanus*
SIZE: to 38 cm (15 in.) Groupers – Serranidae
ID: Whitish with closely set polygonal spots in varying shades of brown; **white tips on foredorsal spines,** broad dark margin on anal fin. Solitary. Silty coastal reefs to 50 m.
Asian Pacific: Andaman Sea to Indonesia, Philippines and Papua New Guinea. – S.W. Japan to Australia.

HONEYCOMB GROUPER *Epinephelus merra*
SIZE: to 32 cm (13 in.) Groupers – Serranidae
ID: **White with light-edged polygonal spots in shades of brown that sometimes join to form dark patches.** Solitary. Coastal, lagoon and sheltered outer reefs to 50 m, but usually less than 20 m.
Indo-Pacific: E. Africa to Indonesia, Philippines, Micronesia, Solomon Is. and French Polynesia. – S.W. Japan to Australia.

Groupers

FOURSADDLE GROUPER *Epinephelus spilotoceps*
SIZE: to 31 cm (12 1/4 in.) Groupers – Serranidae
ID: Whitish with closely packed polygonal spots in varying shades of brown; **4 dark saddles.** Solitary. Coastal, lagoon and outer reefs to 20 m.
Indo-Pacific: E. Africa to Indonesia, Philippines, Micronesia, Papua New Guinea, Solomon Is. and Line Is. – Japan to W. Australia.

HEXAGON GROUPER *Epinephelus hexagonatus*
SIZE: to 26 cm (10 in.) Groupers – Serranidae
ID: Whitish with closely packed polygonal spots in varying shades of brown; spots on upper body often pale and poorly defined, **5 dark saddles.** Solitary. Outer reefs exposed to surge to 30 m.
Indo-Pacific: Red Sea to Indonesia, Papua New Guinea, Micronesia, Solomon Is., French Polynesia. – S.W. Japan to Great Barrier Reef.

CAMOUFLAGE GROUPER *Epinephelus polyphekadion*
SIZE: to 75 cm (2 1/2 ft.) Groupers – Serranidae
ID: Brown with dusky white blotches, especially on upper body; numerous small dark brown spots covering head, body and fins, **dark saddle on tail base.** Solitary. Clear water of lagoons and outer reefs to 46 m.
Indo-Pacific: Red Sea and E. Africa to Indonesia, Philippines, Micronesia, Solomon Is., French Polynesia. – S.W. Japan to Australia.

ONE-BLOTCH GROUPER *Epinephelus melanostigma*
SIZE: to 33 cm (13 in.) Groupers – Serranidae
ID: Bluish white undercolor with closely packed polygonal spots in varying shades of brown; **merging spots on back form dark saddle under mid-dorsal fin.** Solitary. Lagoons, reef flats and seaward slopes to 10 m.
Indo-Pacific: E. Africa to Indonesia, Philippines, Micronesia, Solomon Is. and Line Is. – S.W. Japan to N.W. Australia.

NETFIN GROUPER *Epinephelus miliaris*
SIZE: to 53 cm (21 in.) Groupers – Serranidae
ID: Small brown polygonal spots cover body; much larger dark spots on all fins except spinous dorsal fin. Solitary; young inhabit mangroves and seagrass beds, adults in deeper water. Coastal reefs and outer slopes to 180 m.
Indo-West Pacific: E. Africa to Indonesia, Philippines, Micronesia and Solomon Is. to Fiji, north to S.W. Japan.

CLOUDY GROUPER *Epinephelus erythrurus*
SIZE: to 43 cm (17 in.) Groupers – Serranidae
ID: Dark gray with irregular pale spots and blotches randomly joined to form maze-like pattern. Solitary. Turbid harbors and estuaries with muddy or silty bottoms to 20 m.
East Indo-Asian Pacific: India to Andaman Sea, Malaysia, Java, Kalimantan and Sulawesi in Indonesia.

PALEMARGIN GROUPER *Epinephelus bontoides*
SIZE: to 30 cm (12 in.) Groupers – Serranidae
ID: Gray-brown with scattered reddish brown to black spots; **fins dark gray with narrow pale margins.** Solitary. Sheltered coastal reefs over mud, rock or cobble bottoms in 2-30 m.
Asian Pacific: Taiwan, Indonesia and Philippines to New Britain east of Papua New Guinea.

CORAL ROCK GROUPER *Epinephelus corallicola*
SIZE: to 49 cm (19 in.) Groupers – Serranidae
ID: Pale dusky gray and brown blotches on head and body, small widely spaced black spots on head, body and fins; usually 3-4 dark blotchy saddles on mid-back to tail base.
Asian Pacific: E. Malay Peninsula, Indonesia, Philippines, Micronesia, Papua New Guinea, Solomon Is. – Taiwan to Australia.

WHITESPOTTED GROUPER *Epinephelus coeruleopunctatus*
SIZE: to 60 cm (2 ft.) Groupers – Serranidae
ID: Brownish gray to charcoal with white spots and blotches; **pectoral, anal and convex tail fins black;** only grouper to have underlying netted pattern. Solitary. Inside or near caves of coastal, lagoon and seaward reefs in 4-65 m.
Indo-West Pacific: Arabian Gulf to Indonesia, Philippines, Micronesia, Solomon Is. – S.W. Japan to Great Barrier Reef and Fiji.

SURGE GROUPER *Epinephelus socialis*
SIZE: to 42 cm (17 in.) Groupers – Serranidae
ID: Shades of brown to olive with numerous small close-set brown spots; scattered white spots, **dark margin on tail, rear dorsal and anal fins.** Solitary. Surge areas of reef flats, rocky rubble and deep tide pools to 3 m.
Central Pacific: Insular areas from S. Japan, and Micronesia to Fiji, Line Is. and Rapa I. in French Polynesia.

SPECKLEDFIN GROUPER *Epinephelus ongus*
SIZE: to 31 cm (12 1/4 in.) Groupers – Serranidae
ID: Brown with large white blotches; numerous small pale spots on head, body and fins, the spots join to form wavy stripes on larger individuals. Solitary and cryptic. Near caves and ledges of coastal and lagoon reefs in 5-25 m.
Indo-West Pacific: E. Africa to Indonesia, Philippines, Micronesia and Solomon Is. – S.W. Japan to Australia and Fiji.

Speckledfin Grouper – Juvenile
SIZE: 5-13 cm (2-5 in.)
ID: Dark brown with numerous small white to yellow spots on head, body and fins. Solitary. Shelter under ledges or close to coral outcroppings of coastal, lagoon and outer reef slopes in 1-20 m.

Groupers

MALABAR GROUPER *Epinephelus malabaricus*
SIZE: to 234 cm (7 3/4 ft.)   Groupers – Serranidae
ID: Large; barred or mottled shades of gray to brown to olive with small whitish spots and blotches, covered with numerous small dark spots. Solitary. Protected areas of reefs, lagoons and estuaries to 10 m.
Indo-Pacific: Red Sea to Indonesia, Philippines, Micronesia, Solomon Is., Line Is. – S.W. Japan to New Caledonia and Fiji.

BROWN-MARBLED GROUPER *Epinephelus fuscoguttatus*
SIZE: to 100 cm (3 1/4 ft.)   Groupers – Serranidae
ID: Pale yellowish brown with numerous close-set small brown spots of variable intensity; five vertical series of irregular dark brown blotches, **small black saddle on tail base.** Solitary. Coastal, lagoon and outer reef slopes to 60 m.
Indo-West Pacific: Red Sea and E. Africa to Indonesia, Philippines, Micronesia and Solomon Is. – S.W. Japan to Australia and Fiji.

ORANGE-SPOTTED GROUPER *Epinephelus coioides*
SIZE: to 95 cm (3 1/4 ft.)   Groupers – Serranidae
ID: Tan to dark gray-brown with numerous orangish spots on head, body and fins; **four irregular ladder or H-shaped dark bars** and 3-4 dark saddles on back. Solitary. Turbid coastal reefs and estuaries to 100 m.
Indo-West Pacific: Red Sea and E. Africa to Indonesia, Philippines, Micronesia and Solomon Is. – S.W. Japan to Australia and Fiji.

POTATO GROUPER *Epinephelus tukula*
SIZE: to 200 cm (6 1/2 ft.)   Groupers – Serranidae
ID: Huge; pale grayish with large round to ovate dark gray or blackish blotches on body, dark gray bands and blotches on head, spoke-like markings radiating from eye. Solitary. Coastal, lagoon and seaward reefs in 5-150 m.
Localized: E. Africa to N. Australia and Great Barrier Reef.

GIANT GROUPER *Epinephelus lanceolatus*
SIZE: to 270 cm (8 3/4 ft.)   Groupers – Serranidae
ID: Huge; mottled shades of dark gray to dark brown with small whitish spots and blotches. Largest Indo-Pacific grouper attaining a weight of at least 288 kg (635 lbs.). Solitary. Coastal reefs, lagoons and outer slopes in 3-100 m.
Indo-Pacific: Red Sea and E. Africa to Indonesia, Philippines, Micronesia and Pitcairn Is. – S.W. Japan to Australia.

WAVY-LINED GROUPER *Epinephelus undulosus*
SIZE: to 75 cm (2 1/2 ft.)   Groupers – Serranidae
ID: Pale gray to reddish gray; small dark spots on head, **dark wavy lines** and spots on body, fins bluish on smaller fish. Solitary. Coastal reefs, usually on open sand bottoms near low coral or rock outcroppings in 15-90 m.
Indo-Asian Pacific: E. Africa to Indonesia, Philippines, Micronesia, Papua New Guinea and Solomon Is. – Taiwan to Australia.

BLACKTIP GROUPER *Epinephelus fasciatus*
SIZE: to 36 cm (14 1/4 in.) Groupers – Serranidae
ID: Highly variable from pale to medium greenish gray, reddish yellow or brown and scarlet; often with 5-6 dark bars of variable intensity. Solitary. Coastal, lagoon and seaward reefs in 3-160 m.
Indo-Pacific: Red Sea and E. Africa to Indonesia, Philippines, Micronesia, Solomon Is. and Line Is. – S.W. Japan to Australia.

Blacktip Grouper – Variation
ID: Dark reddish brown with dark body bars; top of head reddish brown with 2 pale bands across nape, **black tips on dorsal fin spines** (source of common name) are not always present.

HALFMOON GROUPER *Epinephelus rivulatus*
SIZE: to 45 cm (18 in.) Groupers – Serranidae
ID: Reddish brown head, mottled pale brown body with 5-6 irregular dark brown bars; **dark brown tail with pale speckling.** Solitary. Coastal reefs, often in weedy areas in 10-150 m.
Indo-West Pacific: E. Africa to S.W. Japan, Australia and New Caledonia.

REDMOUTH GROUPER *Aethaloperca rogaa*
SIZE: to 60 cm (2 ft.) Groupers – Serranidae
ID: Dark gray to black, occasionally with orangish cast; frequently with pale bar across abdomen, inside mouth red. Solitary. Usually inside or near caves or under ledges in coral-rich areas of seaward reefs in 3-50 m.
Indo-West Pacific: Red Sea and E. Africa to Indonesia, Philippines, Micronesia, Solomon Is. – S.W. Japan to Australia and Fiji.

BARRAMUNDI *Cromileptes altivelis*
SIZE: to 66 cm (2 1/4 ft.) Groupers – Serranidae
ID: Pale greenish white to light greenish brown with large widely spaced black spots; compressed body with long sloping nape and **concave profile above eyes.** Solitary and reclusive. Usually on dead, silty reefs to 25 m.
Asian Pacific: Andaman Sea to Indonesia, Philippines, Micronesia and Solomon Is. – S.W. Japan to Great Barrier Reef and Vanuatu.

Barramundi – Juvenile
SIZE: 5 - 13 cm (2 - 5 in.)
ID: White with fewer and larger black spots on head, body and fins than adult. Reclusive, hide near or in reef crevices and caves. Swim with unusual undulating motion similar to juvenile sweetlips.

Groupers – Soapfishes

YELLOW-EDGED LYRETAIL　　　　*Variola louti*
SIZE: to 81 cm (2½ ft.)　　　Groupers – Serranidae
ID: Violet to orange-red to brown with violet to blue spots on head, body and fins; lyre-shaped **yellow margin on rear dorsal, anal and tail fins.** Solitary. Clear waters of lagoons and outer reefs in 3-240 m.
Indo-Pacific: Red Sea and E. Africa to Indonesia, Philippines, Micronesia, Solomon Is. to Pitcairn Is. – S.W. Japan to Australia.

Yellow-edged Lyretail – Juvenile
SIZE: 5-13 cm (2-5 in.)
ID: Reddish brown back and upper head, red to white below with broad black stripe between; small blue spots on back and upper head, white stripe from middle of forehead to snout.

WHITE-EDGED LYRETAIL　　　*Variola albimarginata*
SIZE: to 60 cm (2 ft.)　　　Groupers – Serranidae
ID: Brownish orange to pink or red with violet spots on head, body and fins; lyre-shaped **tail with white margin.** Solitary. Coastal reefs, lagoons and outer reefs in 4-200 m.
Indo-West Pacific: E. Africa to Indonesia, Philippines, Micronesia, Papua New Guinea and Solomon Is. – S.W. Japan to Australia and Fiji.

HIGHFIN CORAL GROUPER　　*Plectropomus oligacanthus*
SIZE: to 75 cm (2½ ft.)　　　Groupers – Serranidae
ID: Red to reddish brown to lavender-brown with series of **oblique blue lines on head;** blue spots on body and fins with some that join to form vertical lines on midside. Solitary. Steep slopes in 5-147 m.
Asian Pacific: Sabah in Malaysia, Indonesia, Philippines, Micronesia, Papua New Guinea, Solomon Is. to E. Australia.

SPOTTED CORAL GROUPER　　*Plectropomus maculatus*
SIZE: to 125 cm (4 ft.)　　　Groupers – Serranidae
ID: Red, pale gray or olive to dark brown with numerous small blue spots on body and fins; **elongate blue spots on head,** blue margin on tail, blue ring around eye. Solitary. Silty coastal reefs in 5-50 m.
Asian Pacific: Malay Peninsula and Philippines to Papua New Guinea, Solomon Is. and Australia.

LEOPARD CORAL GROUPER　　*Plectropomus leopardus*
SIZE: to 75 cm (2½ ft.)　　　Groupers – Serranidae
ID: Red, pale gray or olive to dark brown with **small blue spots on head,** body and fins; thin blue margin on tail, blue ring around eye, may display bars. Solitary. Coastal and lagoon reefs in 3-100 m.
West Pacific: South China Sea, Indonesia, Philippines, Micronesia and Solomon Is. – S.W. Japan to Australia and Fiji.

BLACKSADDLE CORAL GROUPER   *Plectropomus laevis*
SIZE: to 125 cm (4 ft.)   Groupers – Serranidae
ID: **Pale Variation** – Whitish with 4 black saddles on body and black band above eyes, fins yellow. Generally smaller than the dark form and is possibly restricted to juveniles and females. Solitary. Lagoons and seaward reefs in 4-90 m.
Indo-Pacific: E. Africa to Indonesia, Philippines, Micronesia, Solomon Is., French Polynesia. – S.W. Japan to Great Barrier Reef.

Blacksaddle Coral Grouper – Dark Variation
ID: Dark gray to olive head, light gray to olive body with 3-4 dark bars or saddles; small dark-edged blue spots on head body and fins. This variation is possibly restricted to males, but no conclusive evidence. Feeds on a variety of reef fishes.

SQUARETAIL CORAL GROUPER   *Plectropomus areolatus*
SIZE: to 80 cm (2 1/2 ft.)   Groupers – Serranidae
ID: Whitish to pale gray with numerous small dark-edged blue spots on head, body and fins; **dark fins and dark margin on tail,** frequently 4-5 dark blotches forming saddles on back. Solitary. Coastal reefs, lagoons and outer reefs in 2-20 m.
Indo-West Pacific: Red Sea to Indonesia, Philippines, Micronesia, Solomon Is. – S.W. Japan to Australia and Fiji.

ARROWHEAD SOAPFISH   *Belonoperca chabanaudi*
SIZE: to 15 cm (6 in.)   Soapfishes – Serranidae
ID: Long slender body and elongate pointed head; dark bluish gray with black speckling, **yellow spot on tail base,** blue-rimmed black blotches on dorsal and ventral fins. Solitary. Caves and crevices on steep outer reef slopes in 4-50 m.
Indo-Pacific: E. Africa to Indonesia, Philippines, Micronesia, Solomon Is. to French Polynesia. – S.W. Japan to Australia.

DOUBLEBANDED SOAPFISH   *Diploprion bifasciatum*
SIZE: to 25 cm (10 in.)   Soapfishes – Serranidae
ID: Yellow head, body and fins; dark brown to black eye bar and broad bar from mid-dorsal fin to anal fin. Solitary or form small groups. Coastal reefs, often in turbid water to 18 m.
East Indo-Asian Pacific: Maldives to Andaman Sea, Indonesia, Philippines, Micronesia, Papua New Guinea, Solomon Is. – S.W. Japan to Australia, New Caledonia and Vanuatu.

Doublebanded Soapfish – Juvenile
SIZE: 2.5 - 5 cm (1 - 3 in.)
ID: Blue forebody, yellow toward rear. Large black blotch on upper foredorsal fin.

Soapfishes – Hawkfishes

SPOTTED SOAPFISH   *Pogonoperca punctata*
SIZE: to 30 cm (12 in.)   Soapfishes – Serranidae
ID: Thick robust body; brown with dense covering of small white spots, 5 black saddles on head, back and tail base, skin flap on chin. Solitary or in pairs. Clear waters of outer reef slopes in 15-216 m.
Pacific: Indonesia, Philippines, Micronesia, Papua New Guinea, Solomon Is. to French Polynesia. – Japan to Australia.

Spotted Soapfish – Juvenile
SIZE: 2.5 - 7.5 cm (1 - 3 in.)
ID: Dark brown with double row of large yellowish white spots.

OCELLATED SOAPFISH   *Grammistops ocellatus*
SIZE: to 13 cm (5 in.)   Soapfishes – Serranidae
ID: Brown with large pale-edged dark brown to blackish spot on gill cover. Similar Mottled Soapfish [next] has larger scales and is mottled. Solitary or in pairs; cryptic. Deep inside caves and crevices of lagoons and outer slopes in 8 - 30 m.
Indo - Pacific: E. Africa to Indonesia, Philippines, Micronesia, Solomon Is. and French Polynesia. – S.W. Japan to Australia.

MOTTLED SOAPFISH   *Pseudogramma polyacanthum*
SIZE: to 8.6 cm (3¼ in.)   Soapfishes – Serranidae
ID: Mottled brown; large pale-edged dark spot on gill cover. Similar Ocellated Soapfish [previous] has much smaller scales and is not mottled. Solitary and cryptic. Caves and crevices of coral reefs to 61 m.
Indo - Pacific: E. Africa to Micronesia, Hawaii, Pitcairn Is. east of French Polynesia. – S.W. Japan to Australia.

SIXLINED SOAPFISH   *Grammistes sexlineatus*
SIZE: to 27 cm (10¾ in.)   Soapfishes – Serranidae
ID: Black with 6 - 9 white to gold stripes. Solitary. Under ledges or in recesses of lagoons, reef flats or seaward reefs to 20 m.
Indo - Pacific: Red Sea and E. Africa to Indonesia, Philippines, Micronesia, Solomon Is. and French Polynesia. – S.W. Japan to Great Barrier Reef and W. Australia.

LONGNOSE HAWKFISH   *Oxycirrhites typus*
SIZE: to 13 cm (5 in.)   Hawkfishes – Cirrhitidae
ID: Long "needle-nosed" snout; white with red bars and stripes forming netted pattern of squares. Solitary. Perch among branches of black coral and gorgonian sea fans on steep outer reef slopes in 12-100 m, usually below 25 m.
Indo - Pacific: Red Sea to Indonesia, Micronesia, Hawaii and French Polynesia. – S.W. Japan to Great Barrier Reef and Fiji.

TWINSPOT HAWKFISH   *Amblycirrhitus bimacula*
SIZE: to 8.5 cm (3 1/4 in.)   Hawkfishes – Cirrhitidae
ID: Whitish with irregular brown bars; **large pale-edged black spots under rear dorsal fin and on gill cover.** Solitary and cryptic. Caves and crevices of seaward reefs in 2-20 m.
Indo-Pacific: E. Africa to Indonesia, Philippines, Micronesia, Papua New Guinea, Solomon Is., Hawaii and French Polynesia. – S.W. Japan to Australia.

GOLDEN HAWKFISH   *Cirrhitichthys aureus*
SIZE: to 13.8 cm (5 1/2 in.)   Hawkfishes – Cirrhitidae
ID: Yellow to orange; faint dusky bars on side, dark brown spots just above gill cover and across nape, yellow fins. Solitary. Outer reef dropoffs, often on sponges, usually below 20 m.
North Asian Pacific: S.W. Japan to Hong Kong, single sightings from S. Luzon in Philippines and Bali in Indonesia.

THREADFIN HAWKFISH   *Cirrhitichthys aprinus*
SIZE: to 9 cm (3 1/2 in.)   Hawkfishes – Cirrhitidae
ID: White undercolor with irregular red to brown bars and blotches; **pale-edged dark brown spot on gill cover.** Solitary or form small groups. Coastal and seaward reefs in 5-40 m.
Asian Pacific: Malay Peninsula, Indonesia, Philippines, Micronesia to Papua New Guinea. – S.W. Japan to Great Barrier Reef and W. Australia.

DWARF HAWKFISH   *Cirrhitichthys falco*
SIZE: to 7 cm (2 3/4 in.)   Hawkfishes – Cirrhitidae
ID: White with a pair reddish brown saddles on forebody; small red-brown blotches on rear body often align as bars, **pair of reddish bars below eye.** Solitary. Near coral heads on seaward reefs in 4-46 m.
East Indo-West Pacific: Maldives to Indonesia, Philippines, Micronesia, Papua New Guinea. – S.W. Japan to Australia and Fiji.

PIXY HAWKFISH   *Cirrhitichthys oxycephalus*
SIZE: to 8.5 cm (3 1/4 in.)   Hawkfishes – Cirrhitidae
ID: White undercolor with irregular red-brown spots on body and fins. Similar Threadfin Hawkfish [above] has ocellated spot on gill cover. Solitary. Coastal, lagoon and outer reefs to 40 m.
Indo-Pacific: Red Sea and E. Africa to Indonesia, Philippines, Micronesia, Papua New Guinea, Solomon Is. and French Polynesia. – S.W. Japan to Australia.

Pixy Hawkfish – Red Variation
ID: Color pattern variable according to locality and depth. Individuals from deeper water frequently have a pinkish undercolor with more closely-set, irregular red to maroon spots. Pictured individual from N. Sulawesi, Indonesia. All variations lack the bars under eye of similar Threadfin Hawkfish and Dwarf Hawkfish [above].

Hawkfishes

FLAME HAWKFISH *Neocirrhites armatus*
SIZE: to 9 cm (3 ¹/₂ in.) Hawkfishes – Cirrhitidae
ID: Brilliant red; broad zone of black on back, also black ring around eye. Solitary and wary. Shy; perch among branches of *Pocillopora* and *Stylophora* branching corals. Clear seaward reefs to 10 m.
Pacific: Great Barrier Reef to Fiji, Line Is. and French Polynesia.

LYRETAIL HAWKFISH *Cyprinocirrhites polyactis*
SIZE: to 14 cm (5 ¹/₂ in.) Hawkfishes – Cirrhitidae
ID: Orange-brown with faint brown freckling; yellowish forked tail. Solitary or form groups. Unlike other hawkfishes, which feed on the bottom, hover in open water feeding on plankton over steep slopes and coral outcroppings in 10-132 m.
Indo-West Pacific: E. Africa to Indonesia, Philippines, Micronesia, Solomon Is. – S.W. Japan to Great Barrier Reef to Fiji.

WHITESPOTTED HAWKFISH *Cirrhitus pinnulatus*
SIZE: to 28 cm (11 in.) Hawkfishes – Cirrhitidae
ID: Blotchy brown with white spots on body. One of the larger hawkfishes. Solitary. Seaward reefs in wave-affected areas to 3 m.
Indo-Pacific: Red Sea and E. Africa to Indonesia, Philippines, Micronesia, Papua New Guinea, Solomon Is., Hawaii, French Polynesia. – S.W. Japan to Great Barrier Reef and W. Australia.

YELLOW HAWKFISH *Paracirrhites xanthus*
SIZE: to 11 cm (4 ¹/₄ in.) Hawkfishes – Cirrhitidae
ID: Yellow; short diagonal black line behind eye. Solitary. Perch among branches of *Pocillopora* coral on exposed outer reefs in 3-25 m.
Central Pacific: Phoenix Is. and Tuamotu and Society Is. in French Polynesia.

ARC-EYE HAWKFISH *Paracirrhites arcatus*
SIZE: to 14 cm (5 ¹/₂ in.) Hawkfishes – Cirrhitidae
ID: Shades of brown; **arc-shaped marking of orange, red and blue lines behind eye**, 3 orange dashes on lower edge of gill cover, Frequently broad white stripe on side to tail. Solitary. Perch on coral heads of coastal and seaward reefs to 35 m.
Indo-Pacific: E. Africa to Indonesia, Philippines, Micronesia, Solomon Is., Hawaii and French Polynesia. – S.W. Japan to Australia.

NISUS HAWKFISH *Paracirrhites nisus*
SIZE: to 10 cm (4 in.) Hawkfishes – Cirrhitidae
ID: Lower edge of **elliptical marking behind eye extends across cheek to lip;** broad white stripe on rear body bordered with wide black bands. Solitary or in pairs. Among *Pocillopora* coral on exposed outer reefs in 3-25 m.
Central Pacific: Phoenix Is. in Central Pacific and Tuamotu Is. in French Polynesia.

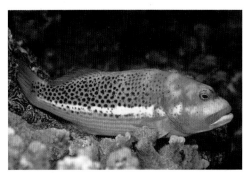

HALFSPOTTED HAWKFISH   *Paracirrhites hemistictus*
SIZE: to 29 cm (11 1/2 in.)   Hawkfishes – Cirrhitidae
ID: Pale Variation – Brown to yellowish brown or gray on head and body; white midlateral stripe with yellow below, numerous dark spots on upper body. Solitary. Perch in corals on clear, exposed seaward reefs to 18 m.
Central Pacific: Great Barrier Reef to Line Is. and Pitcairn Is.

Halfspotted Hawkfish – Dark Variation
ID: Dark brown to dark gray; numerous dark spots cover entire body, single small white spot on midbody.

FRECKLED HAWKFISH   *Paracirrhites forsteri*
SIZE: to 22.5 cm (8 3/4 in.)   Hawkfishes – Cirrhitidae
ID: Colors quite variable, most commonly shades of brown; numerous small spots "freckles" on head and forebody, brown striped area toward rear of body. Solitary. Perch on corals in coastal, lagoon and seaward reefs to 35 m.
Indo-Pacific: Red Sea and E. Africa to Indonesia, Philippines, Solomon Is., Hawaii, French Polynesia. – S.W. Japan to Australia.

Freckled Hawkfish – Variation
ID: Brown head and forebody and at a variable point on rear body becomes yellow continuing onto tail base and tail. Most commonly perch on branching corals including *Acropora, Pocillopora* and *Stylophora*.

Freckled Hawkfish – Juvenile and Subadult
ID: Red to dark brown on upper half of body becoming black on rear with white lower body; yellow stripe on back, **"freckles" on head is key to identification of all variations.**

Freckled Hawkfish – Juvenile and Subadult
ID: Red upper head and body; occasionally tail base and some of lower body grayish to nearly black; red to brownish freckles on head.

Sweetlips

GOLD-SPOTTED SWEETLIPS *Plectorhinchus flavomaculatus*
SIZE: to 60 cm (2 ft.) Sweetlips – Haemulidae
ID: Gray with orange and blue stripes on head and small orange spots on upper body extending onto dorsal and tail fins. Solitary or form small groups. Sheltered coastal reefs and lagoons in 2-25 m.
Indo-Asian Pacific: Red Sea and E. Africa to Papua New Guinea. – S.W. Japan to Australia and New Caledonia.

DIAGONAL-BANDED SWEETLIPS *Plectorhinchus lineatus*
SIZE: to 50 cm (20 in.) Sweetlips – Haemulidae
ID: White with **black diagonal body bands,** white belly; lips and fins yellow, spotted dorsal, anal and tail fins. Solitary or form small groups; generally nocturnal but drift in the open during the day. Coastal reefs, lagoons and seaward reefs to 35 m.
Asian Pacific: Indonesia, Philippines, Micronesia and Solomon Is. – S.W. Japan to N. Great Barrier Reef and New Caledonia.

GOLDSTRIPED SWEETLIPS *Plectorhinchus chrysotaenia*
SIZE: to 51 cm (20 in.) Sweetlips – Haemulidae
ID: Silvery-blue with **numerous thin yellow stripes;** bright yellow fins. Solitary or form groups. Coral-rich areas of coastal reefs and lagoons in 6-70 m. Formerly classified as *P. celebicus.*
Asian Pacific: Malaysia to Indonesia and Solomon Is. – S.W. Japan to N. Great Barrier Reef and New Caledonia.

ORIENTAL SWEETLIPS *Plectorhinchus vittatus*
SIZE: to 60 cm (2 ft.) Sweetlips – Haemulidae
ID: White with **black stripes that extend to belly,** lips and fins yellow, spotted dorsal, anal and tail fins. Solitary or form small groups; nocturnal, but drift in open during the day. Coastal reefs, lagoons and seaward reefs in 2-25 m.
Indo-West Pacific: E. Africa to Indonesia, Philippines and Micronesia. – S.W. Japan to N. Great Barrier Reef and Fiji.

Oriental Sweetlips – Large Juvenile/Subadult
SIZE: 10-15 cm (4-6 in.)
ID: White with yellowish head; irregular pattern of broad black stripes, bands and spots on head, body and fins. Solitary; remain close to shelter.

Oriental Sweetlips – Juvenile
SIZE: 4-8 cm (1 1/2 - 3 1/4 in.)
ID: Dark brown with several large irregular white spots bordered with pale yellow to orange and red. Solitary; remain close to shelter. Swim with rapid undulating motion.

STRIPED SWEETLIPS   *Plectorhinchus lessonii*
SIZE: to 40 cm (16 in.)   Sweetlips – Haemulidae
ID: White with 4-5 brown stripes on upper body, **white unmarked lower body**; yellow lips, broken lines on head, spots on yellowish dorsal, anal and tail fins. Solitary. Coastal reefs, lagoons and seaward reefs to 35 m.
Asian Pacific: Indonesia, Philippines, Micronesia and Solomon Is. – S.W. Japan to Great Barrier Reef and New Caledonia.

Striped Sweetlips – Juvenile
SIZE: 4-8 cm (1 1/2 - 3 1/4 in.)
ID: Orange with white stripe above eye extending to upper rear dorsal fin and upper border of tail; black midlateral stripe from eye extends onto tail, black lower half of pectoral fin.

RIBBON SWEETLIPS   *Plectorhinchus polytaenia*
SIZE: to 40 cm (16 in.)   Sweetlips – Haemulidae
ID: Bright yellow with bold pattern of **black-edged pale blue stripes running from head to tail**. Solitary or may form small resting groups during the day; actively forage at night on small invertebrates. Coastal and seaward reefs in 5-40 m.
Asian Pacific: Sabah in Malaysia to Indonesia, Philippines, Papua New Guinea and Solomon Is. to N.W. Australia.

Ribbon Sweetlips – Juvenile
SIZE: 5-13 cm (2-5 in.)
ID: Orange; **white stripe above eye extends to lower rear dorsal fin and upper portion of tail**, black midlateral stripe from eye extends onto tail; black lower half of pectoral fin.

GIANT SWEETLIPS   *Plectorhinchus albovittatus*
SIZE: to 100 cm (3 1/4 ft.)   Sweetlips – Haemulidae
ID: Dusky gray mottled body; faint wide dark bar behind pectoral fin, black ventral fins and black edges on dorsal, anal and tail fins. Solitary. Lagoon and seaward reefs in 2-50 m.
Indo - West Pacific: Red Sea and E. Africa to Indonesia, Philippines, Micronesia, Solomon Is. – S.W. Japan to Great Barrier Reef and Fiji.

Giant Sweetlips – Juvenile
SIZE: 7.5 - 15 cm (3-6 in.)
ID: Yellow undercolor with black stripe from above eye to base of soft dorsal fin; wide black stripe from eye to tail base; pectoral fins translucent yellow. Similar Ribbon Sweetlips, juvenile [above] pectoral fins dark with pale yellow borders.

Sweetlips

DOTTED SWEETLIPS   *Plectorhinchus picus*
SIZE: to 85 cm (2 ³/₄ ft.)   Sweetlips – Haemulidae
ID: Whitish with profuse small black spots covering head, body and fins, unmarked white belly; **thin black margin on rear edge of gill cover** (similar Many-spotted Sweetlips [below] lack this marking). Solitary. Clear outer reefs and lagoons in 5-50 m.
Indo-Pacific: Seychelles to Indonesia, Philippines, Micronesia, Solomon Is. and French Polynesia. – S.W. Japan to Australia.

Dotted Sweetlips – Juvenile
SIZE: 7.5 - 20 cm (3-10 in.)
ID: Black with broad white area on lower body, white snout; pair of white saddles on back, white pectoral fins.

MANY-SPOTTED SWEETLIPS   *Plectorhinchus chaetodonoides*
SIZE: to 50 cm (20 in.)   Sweetlips – Haemulidae
ID: Whitish to yellowish or greenish with unmarked whitish belly; profuse dark brown spots on head, body and fins. Usually solitary. Often near ledges of coastal reefs, lagoons and seaward slopes in 2-30 m.
East Indo-West Pacific: Maldives to Indonesia, Philippines, Micronesia, Solomon Is. – S.W. Japan to Great Barrier Reef and Fiji.

Many-Spotted Sweetlips – Large Juvenile/Subadult
SIZE: 7.5 - 20 cm (3-8 in.)
ID: White undercolor, occasionally with yellowish or greenish tinting on back; pattern of numerous brown spots that encircle several large white spots. Usually among corals of sheltered coastal reefs and lagoons in 1-12 m.

Many-Spotted Sweetlips – Juvenile
SIZE: 4-10 cm (1¹/₂ - 4 in.)
ID: Brown with large white spots with dark borders. Solitary; remain close to shelter. Swim with peculiar undulating motion. Very small individuals possibly mimic toxic flatworms.

BLUBBERLIP   *Plectorhinchus gibbosus*
SIZE: to 70 cm (2 ¹/₄ ft.)   Sweetlips – Haemulidae
ID: Gray to brown with charcoal-gray head; black margin on cheek and gill cover, dark outlines around scales, may display dusky wide bar on back and forebody, high back profile. Solitary. Coastal reefs, lagoons and outer slopes to 25 m.
Indo-West Pacific: Red Sea and E. Africa to Indonesia, Philippines, Micronesia and Solomon Is. – S.W. Japan to Australia and Fiji.

Blubberlip – Large Juvenile
SIZE: 6-20 cm (2¼ - 8 in.)
ID: Brown with large pale blotches. Small juveniles are solid or nearly solid brown. Often venture into fresh water.

INDONESIAN SWEETLIPS   *Diagramma melanacra*
SIZE: to 50 cm (20 in.)   Sweetlips – Haemulidae
ID: Gray with brown spotting on body, larger blackish spots on **yellow dorsal and tail fins;** black ventral and anal fins. Large juveniles have combination of stripes and longitudinal rows of spots. Solitary or form groups over sand bottoms. Coastal reefs and lagoons in 3-40 m.
Localized: Sabah in Malaysia to Indonesia.

SILVER SWEETLIPS   *Diagramma picta*
SIZE: to 90 cm (3 ft.)   Sweetlips – Haemulidae
ID: Unmarked silvery gray except fine dark spotting on dorsal and tail fins. Solitary or form groups. Sand bottoms of coastal reefs and lagoons in 5-40 m.
Indo-West Pacific: Red Sea to Indonesia, Philippines and Micronesia. – S.W. Japan to Australia, New Caledonia, Vanuatu and Fiji.

Silver Sweetlips – Large sub-adult
SIZE: 60 cm (2 ft.)
ID: Similar to Indonesian Sweetlips [above] but with larger dark spotting on whitish dorsal and tail fins rather than yellow fins.

Silver Sweetlips – Large Juvenile
SIZE: to 25 cm (10 in.)
ID: Yellowish with bold brown to black stripes on back; dark spots on belly and lower head.

Silver Sweetlips – Juvenile
SIZE: 5-10 cm (2-4 in.)
ID: White undercolor with pair of broad black stripes on upper body, lower head and body yellow; large black marking on dorsal fin, with age narrow broken stripe forms between and another below large stripes. Solitary; swim with rapid undulating motion. Sandy areas around rock outcroppings or debris.

173

Swim with Pectoral Fins – Parrotfishes

This ID Group consists of fishes that primarily use their pectoral fins to swim. (Besides the closely related Wrasses [next ID Group], few other fishes swim primarily with their pectoral fins.)

FAMILY: Parrotfishes – Scaridae
7 Genera – 41 Species Included

Terminal Phase – Typical Shape
Initial Phase – Typical Shape

Virtually every reef in the Indo-Pacific is home to a bustling community of parrotfishes. These robust assemblies are typified by foraging herds of nondescript, two- to six-inch females, and large, solitary, brightly colored males, which spend their day methodically taking bite after bite of algae from the bottom. Parrotfishes share many traits with their close relatives and forerunners, the wrasses. Members of both families swim primarily with their pectoral fins, change sex from females to males, generally exhibit two or more color patterns within the same species, and have complex social and mating systems. However, unlike wrasses, which use conspicuous canines to capture hard-shelled crustaceans, the teeth of parrotfishes have fused into powerful beaks capable of rasping filamentous algae from the porous external skeletons of dead coral. In the process, large quantities of the reef's structure – calcium carbonate – are routinely consumed. On average, nearly 75 percent of the gut content of parrotfishes is composed of inorganic sediment. Broad, bony teeth plates, known as the pharyngeal mill grind the grit into tiny bits, and a long, specialized alimentary tract, without a true stomach, extracts food, leaving the remains with nowhere to go but back to the reef as sand. In fact, the family's copious processing system generates much of the sand associated with tropical reefs and beaches. It has been estimated that large parrotfishes deposit more than 5,000 pounds of sediment annually.

Contrary to the territorial nature of many reef fishes, that partition and defend limited food supplies, parrotfishes specialize in a plentiful and rapidly renewable plant food source, allowing several species to share overlapping feeding grounds harmoniously. The average parrotfish diet also includes the tender, uncalcified tips of algal bushes, seagrass blades, occasional crustaceans and, now and again, a bite or two of sponge. A few of the larger Indo-Pacific species feed in part on living coral.

During maturation most parrotfishes go through a series of changes that dramatically alter their colors, markings and body shapes. These transformations are marked by a JUVENILE PHASE (JP), INITIAL PHASE (IP) and TERMINAL PHASE (TP), which represents the largest, brightest, most aggressive and least numerous individuals. IP typically consists of females and, in some species, may also include males. Most parrotfishes are hermaphroditic and go through a sex change to become TP, while the IP males of other species simply mature into TP without changing sex. TP, always males, have reached the apex of sexual maturity and experience the greatest reproductive success.

Most parrotfishes live in harems with a single dominant TP, and from two to seven female IP with which they exclusively mate. All species are pelagic spawners that typically release gametes in paired male/female spawning rushes daily at traditional times and locations. A strict, size-related pecking order governs social rank within harems. Such hierarchies allow the largest IP in a harem to transform into a terminal male after the disappearance of the harem's previous TP. These two- to three-week metamorphoses reorder the gonads from ovaries to testes and confer a bright new coat on the recently transformed females.

BLEEKER'S PARROTFISH   *Chlorurus bleekeri*
SIZE: to 49 cm (19 in.)   Parrotfishes – Scaridae
ID: TP – Green with pink to lavender scale edges; **large pale yellowish to greenish patch on cheek** with green border, dark green margin and borders on tail. Sheltered coastal reefs, lagoons and outer slopes in 3-35 m.
West Pacific: Indonesia, Philippines, Micronesia, Papua New Guinea, Solomon Is. to E. Australia, Vanuatu and Fiji.

Bleeker's Parrotfish – IP
SIZE: to 39 cm (15½ in.)
ID: Dark brown with 3-4 faint pale whitish bars; orange to bright yellow tail base.

REEFCREST PARROTFISH   *Chlorurus frontalis*
SIZE: to 50 cm (20 in.)   Parrotfishes – Scaridae
ID: TP – Green with pink bar on each scale; pink to lavender bands extend from rear eye with patch above, broad band on chin. Form small schools. Reef flats and edge of outer reefs in 10-15 m.
West & Central Pacific: Great Barrier Reef to E. Papua New Guinea, Solomon Is., Micronesia and Line Is.

BOWER'S PARROTFISH   *Chlorurus bowersi*
SIZE: to 30 cm (12 in.)   Parrotfishes – Scaridae
ID: TP – Green with rosy pink streak on scales; broad lavender patch on snout, large triangular orange patch on forebody. Usually solitary. Coral-rich areas of lagoons or upper edge of channels in 2-20 m.
North Asian Pacific: Brunei, Sabah in Malaysia, Indonesia, Philippines, Micronesia, north to S.W. Japan.

INDIAN PARROTFISH   *Chlorurus capistratoides*
SIZE: to 35 cm (13¾ in.)   Parrotfishes – Scaridae
ID: TP – Green back with rosy pink streak on each scale, pale below and **yellowish pectoral region;** green bands around mouth and eye. Form groups. Inner and outer reef crests to 15 m, usually in shallow surge zone.
Indo-Asian Pacific: E. Africa to Seychelles and Andaman Sea to Brunei, Bali and Flores in Indonesia and Palawan in Philippines.

Indian Parrotfish – IP
SIZE: to 30 cm (12 in.)
ID: Dark gray with pale pinkish snout and tail; 4-5 whitish bars and yellowish outer margin on pectoral fin.

Parrotfishes

PACIFIC BULLETHEAD PARROTFISH   *Chlorurus spilurus*
SIZE: to 40 cm (16 in.)   Parrotfishes – Scaridae
ID: TP – Highly variable shades of green with lavender scale
edges; **pale green to white tail base,** pale yellowish cheek;
blue to lavender to pale green patch on snout. Solitary. Coral
reefs and adjacent rubble to 30 m.
Pacific: Indonesia to Philippines, Micronesia, Solomon Is.,
Hawaii and Pitcairn Is. – S.W. Japan to Australia.

Pacific Bullethead Parrotfish – IP
SIZE: to 26 cm (10 in.)
ID: Light reddish brown head and forebody shading to dark rear
body with 3-4 vertical rows of small white spots (spots can
rapidly fade or intensify). Often form small to large groups. **IP/JP –**
Similar head and body to IP, but has a white tail with large
central black spot.

Pacific Bullethead Parrotfish – JP
SIZE: to 10 cm (4 in.)
ID: Alternating brown and white stripes.

JAPANESE PARROTFISH   *Chlorurus japanensis*
SIZE: to 30 cm (12 in.)   Parrotfishes – Scaridae
ID: TP – Light green with broad purple band from forehead to
belly; lavender patch on snout. **IP –** Dark brown; red-orange tail
with dark margin. Solitary. Sheltered seaward and lagoon reefs
to 20 m.
West Pacific: Sulawesi in Indonesia, Philippines and Micronesia.
– S.W. Japan to E. Australia and Fiji.

ROUNDHEAD PARROTFISH   *Chlorurus strongylocephalus*
SIZE: to 70 cm (2 1/4 ft.)   Parrotfishes – Scaridae
ID: TP – Steep forehead; green to greenish blue with lavender-
pink streak on body scales, **extensive yellow area on cheek.**
Solitary or with smaller females. Coastal, lagoon and outer reefs
in 2-35 m.
Indian Ocean: E. Africa to Andaman Sea and Cocos-Keeling Is.
to Sumatra, Java and Bali in S.W. Indonesia.

Roundhead Parrotfish – IP
SIZE: to 50 cm (20 in.)
ID: Yellow-green upper half of head, back and tail, red to dark
brown or greenish below; dark green around mouth, blue-
green margin and borders on tail.

STEEPHEAD PARROTFISH *Chlorurus microrhinos*
SIZE: to 70 cm (2¼ ft.) Parrotfishes – Scaridae
ID: Blunt forehead profile; shades of green to blue-green with lavender-pink scale edges, blue snout and pale lower head. **IP –** Similar to TP. Solitary or in groups. Sheltered reefs to 50 m.
Pacific: Bali in Indonesia, Philippines and Micronesia to Pitcairn Is. – S.W. Japan to Australia.

Steephead Parrotfish – Red Variation
ID: Blunt forehead profile; reddish upper two-thirds of head and body, yellow or whitish below, yellow fins. Red variation is generally rare. Both variations can be either TP or IP.
West Pacific: Great Barrier Reef and S. Pacific Is.

Steephead Parrotfish – JP
SIZE: to 15 cm (6 in.)
ID: Black to dark brown with 3-4 yellow to pale stripes.

BLACK PARROTFISH *Chlorurus oedema*
SIZE: to 35 cm (14 in.) Parrotfishes – Scaridae
ID: Solid black with pronounced hump on forehead; contrasting white dental plates, rounded tail. Form small groups. Coastal reefs to 50 m.
Asian Pacific: Brunei, Sabah in Malaysia, W. Kalimantan and Java in Indonesia, Philippines to S.W. Japan.

MARBLED PARROTFISH *Leptoscarus vaigiensis*
SIZE: to 35 cm (14 in.) Parrotfishes – Scaridae
ID: IP – Well camouflaged; green or olive to greenish brown, often mottled and speckled, **two horizontal rows of widely spaced white spots on upper side. TP –** Similar, but with white stripe on middle of side. Form groups. Seagrass beds to 10 m.
Indo-Pacific: Red Sea and E. Africa to Easter Is. in southeast Pacific. – S.W. Japan to N. New Zealand.

RAGGEDTOOTH PARROTFISH *Calotomus spinidens*
SIZE: to 19 cm (7½ in.) Parrotfishes – Scaridae
ID: TP/IP – Small; brown or greenish brown; **broken stripe of rose spots runs below dorsal fin.** Usually form groups; often camouflaged. Seagrass beds and dense patches of seaweed to 25 m.
Indo-West Pacific: E. Africa, Indonesia, Philippines, Micronesia and Solomon Is. – S.W. Japan to Australia and Fiji.

Parrotfishes

STAREYE PARROTFISH *Calotomus carolinus*
SIZE: to 50 cm (20 in.) Parrotfishes – Scaridae
ID: TP – Blue-green with **orange-pink bands radiating from eye.** Solitary or form groups. Lagoon and seaward reefs on coral, rubble, weed and seagrass bottoms in 2-30 m.

Indo-Pacific: E. Africa to Indonesia, Philippines, Micronesia, Papua New Guinea, Solomon Is. and French Polynesia. – S. Japan to Great Barrier Reef.

Stareye Parrotfish – IP
SIZE: to 39 cm (15 1/2 in.)
ID: Mottled brown with whitish patches on back; spoke-like bands radiate from eye.

SPOTTED PARROTFISH *Cetoscarus ocellatus*
SIZE: to 80 cm (2 1/2 ft.) Parrotfishes – Scaridae
ID: TP – Green with pinkish red spots and lines on head; band from mouth to belly, stripes or bands on fins. Usually with group of IP. Lagoon and seaward reefs to 30 m.
Indo-Pacific: E. Africa to Indonesia, Philippines, Micronesia and Solomon Is. and French Polynesia. – S. Japan to Australia.

Spotted Parrotfish – IP
SIZE: to 60 cm (2 ft.)
ID: Pale yellowish to white back, pale greenish mid and lower body with black scale edges forming netted pattern, gray head with golden iris. Solitary or in harems with other IP.

Spotted Parrotfish – JP
SIZE: 2 1/2 to 10 cm (1 to 4 in.)
ID: White with broad orange bar encircling head from eye to rear gill cover; orange outer rim on tail and outer half of foredorsal fin, with age a black spot develops in the center of the orange area on dorsal fin. Solitary.

LONGNOSE PARROTFISH *Hipposcarus harid*
SIZE: to 60 cm (2 ft.) Parrotfishes – Scaridae
ID: TP – Long snout; pale yellow to green with vertical blue streaks on scales, **blue tail** with elongate lobes. **IP –** Similar, but lack elongate tail lobes. Form aggregations. Sheltered areas, usually on sand bottoms near reefs to 25 m.
Indian Ocean: Red Sea and E. Africa to Andaman Sea and Java in Indonesia.

PACIFIC LONGNOSE PARROTFISH   *Hipposcarus longiceps*
SIZE: to 50 cm (20 in.)   Parrotfishes – Scaridae
ID: IP – Long snout; pale yellowish gray with pale vertical streaks on scales, yellow tail base, short tail lobes. **TP –** Similar, but lack yellow tail base. Lagoons and seaward reefs on sand bottoms near reefs in 2-40 m.
Pacific: Indonesia, Philippines, Micronesia, Solomon Is., French Polynesia. – S.W. Japan to Great Barrier Reef and New Caledonia.

Pacific Longnose Parrotfish – JP
SIZE: to 10 cm (4 in.)
ID: Pale gray upper body and pearly white lower body; wide orange stripe from snout to tail base, black spot on rear tail base.

BARTAIL PARROTFISH   *Scarus caudofasciatus*
SIZE: to 50 cm (20 in.)   Parrotfishes – Scaridae
ID: TP – Green to blue-green with pinkish scale margins; green band on snout extends under and behind eye, green band extends from upper eye to forebody, pink dorsal and anal fins with blue margins, reddish pink streak on tail lobes. Outer reefs in 10-40 m.
Indian Ocean: E. Africa to Andaman Sea including Myanmar and W. Thailand.

EAST INDIES PARROTFISH   *Scarus hypselopterus*
SIZE: to 38 cm (15 in.)   Parrotfishes – Scaridae
ID: TP – Body salmon to green-brown; green tail base and blue tail, green head with **reddish worm markings radiating from eye** and pale greenish to green stripe from eye to behind pectoral fin. Solitary. Coral and rubble areas in 10-30 m.
Asian Pacific: Brunei, Sabah in Malaysia, N. Indonesia, Philippines and Micronesia to S.W. Japan and Taiwan.

East Indies Parrotfish – IP
SIZE: to 30 cm (12 in.)
ID: Pinkish gray; yellow dorsal fin, tail base and tail, greenish ventral fins, large black spot near front of anal fin.

East Indies Parrotfish – JP
SIZE: to 20 cm (8 in.)
ID: Gray to brown with two white stripes and white belly; black spot near front of anal fin. Solitary or form small groups.

Parrotfishes

YELLOWFIN PARROTFISH   *Scarus flavipectoralis*
SIZE: to 30 cm (12 in.)   Parrotfishes – Scaridae
ID: **TP –** Variable bicolor with pale purplish, brown head and forebody and pale green rear body or vise versa; **green band from snout to pectoral fin,** often elongate yellow patch on tail base. Solitary or form groups. Lagoons and outer reefs in 8-40 m.
Asian Pacific: Malay Peninsula, Indonesia, Philippines, Micronesia to Great Barrier Reef and New Caledonia.

Yellowfin Parrotfish – IP
SIZE: to 26 cm (10 in.)
ID: Pale yellowish green to light gray with dark gray tail, bright yellow pectoral fin. Usually form groups.

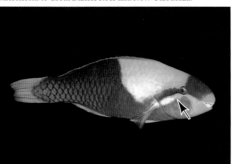

YELLOW-BARRED PARROTFISH   *Scarus dimidiatus*
SIZE: to 30 cm (12 in.)   Parrotfishes – Scaridae
ID: **TP –** Dark bluish to greenish gray with large blue-green to green area over nape and upper forebody; **dark and pale bands from eye to pectoral fin.** Lagoon and seaward reefs to 25 m.
West Pacific: Malay Peninsula to Indonesia, Philippines and Micronesia. – S.W. Japan to Great Barrier Reef, Vanuatu and Fiji.

Yellow-barred Parrotfish – IP
SIZE: to 22 cm (8 ¾ in.)
ID: Yellowish with gray head and 3 gray saddle-bars on back; may have whitish stripe from eye to tail.

DARKCAPPED PARROTFISH   *Scarus oviceps*
SIZE: to 30 cm (12 in.)   Parrotfishes – Scaridae
ID: **TP –** Blue-green with narrow pink scale margins; **lime green and dark blue-green pectoral fin,** upper head and forebody darker shade of purple. Solitary. Reef flats, lagoons and outer slopes to 20 m.
Pacific: Indonesia, Philippines and Micronesia to French Polynesia. – S.W. Japan to Australia.

Darkcapped Parrotfish – IP
SIZE: to 25 cm (10 in.)
ID: Light gray with dark gray "cap" on upper head and forebody; 1-2 yellow bands behind; white or yellowish cheek.

BLUEPATCH PARROTFISH *Scarus forsteni*
SIZE: to 55 cm (22 in.) Parrotfishes – Scaridae
ID: TP – Green with pink scale edges; darkish "cap" on upper head, **wide green band encircles mouth,** occasionally pink area on central body. Solitary or form small groups. Lagoon and seaward reefs in 3-30 m.
Pacific: Indonesia, Philippines and Micronesia to Pitcairn Is. – S.W. Japan to Great Barrier Reef.

Bluepatch Parrotfish – IP
SIZE: to 36 cm (14 1/4 in.)
ID: Reddish brown; dark patch runs from eye to base of pectoral fin, yellow to brown patch runs on mid-side from pectoral fin to tail with blueish spot near middle. Solitary.

CHAMELEON PARROTFISH *Scarus chameleon*
SIZE: to 31 cm (12 in.) Parrotfishes – Scaridae
ID: TP – Blue to green; blue-green band links eyes; narrow green band encircles mouth, **broad area of pale salmon behind pectoral fin.** Can quickly alter or intensify color. Coral-rich areas of lagoon and seaward reefs to 35 m.
West Pacific: Indonesia, Philippines and Micronesia. – S.W. Japan to Australia and Fiji.

Chameleon Parrotfish – IP
SIZE: to 20 cm (8 in.)
ID: Brown, often with highly contrasting white belly. Can quickly change or intensify color.

BRIDLED PARROTFISH *Scarus frenatus*
SIZE: to 50 cm (20 in.) Parrotfishes – Scaridae
ID: TP – Shades of green, **abrupt transition from dark to light green at tail base;** pale green bands around mouth. Usually solitary. Seaward slopes and reef crests to 25 m.
Indo-Pacific: Red Sea and E. Africa to Andaman Sea, Indonesia, Philippines, Micronesia and French Polynesia. – S.W. Japan to Great Barrier Reef and W. Australia.

Bridled Parrotfish – IP
SIZE: to 36 cm (14 in.)
ID: Gray to red with broad blackish stripes (sometimes broken into series of diamond-shaped marks) on sides; dorsal, anal and ventral fins frequently bright red. Solitary or form small groups.

Parrotfishes

FESTIVE PARROTFISH — *Scarus festivus*
SIZE: to 45 cm (1½ ft.)   Parrotfishes – Scaridae
ID: **TP –** Green to blue green with orange to violet streak on each scale; **2 green bands extend from eye to eye,** occasionally yellow-green spot on tail base and green to violet crescent on tail. Lagoon and seaward reefs in 3-30 m.
Indo-Pacific: E. Africa to S.W. Japan, Indonesia, Micronesia, Solomon Is., French Polynesia. Rarely sighted in Asian Pacific.

Festive Parrotfish – IP
SIZE: to 30 cm (12 in.)
ID: Brown with white lower head and body; faint green bar on tail.

BLUEBARRED PARROTFISH — *Scarus ghobban*
SIZE: to 75 cm (2½ ft.)   Parrotfishes – Scaridae
ID: **TP –** Blue to green with salmon to pink scale margins; usually pale salmon area above and behind pectoral fin, **blue to green bands on chin and behind eye.** Solitary. Silty inshore reefs and areas of sand and rubble in 2-30 m.
Indo-Pacific: Red Sea and E. Africa to Indonesia, Philippines, Micronesia, Solomon Is., French Polynesia. – S. Japan to N. Australia.

Bluebarred Parrotfish – IP
SIZE: to 62 cm (2 ft.)
ID: Yellow to yellow green; usually 4-5 diffuse blue bars (sometimes broken into spots) on sides; blue bands around mouth and eye, blue fin margins. Solitary.

GREENTHROAT PARROTFISH — *Scarus prasiognathos*
SIZE: to 70 cm (2¼ ft.)   Parrotfishes – Scaridae
ID: **TP –** Dark green with **yellowish upper head and green below;** broad green band across snout. Often form large aggregations. Reef flats and seaward slopes in 3-25 m.
East Indo-Asian Pacific: Maldives to Indonesia, Philippines, Micronesia, Papua New Guinea. – S.W. Japan to W. Australia.

Greenthroat Parrotfish – IP
SIZE: to 40 cm (16 in.)
ID: Reddish brown with some scattered white speckling. Compare with similar IP Filament-fin Parrotfish.

VIOLETLINED PARROTFISH *Scarus globiceps*
SIZE: to 28 cm (11 in.) Parrotfishes – Scaridae
ID: **TP** – Shades of green and blue with salmon-pink scale margins; numerous green broken lines and spots on upper head and pale lower head, **2-3 violet stripes on belly.** Solitary. Reef flats, lagoons and seaward reefs to 30 m.
Indo-Pacific: E. Africa to Indonesia, Philippines, Micronesia, Solomon Is. and French Polynesia. – S.W. Japan to Great Barrier Reef.

Violetlined Parrotfish – IP
SIZE: to 20 cm (8 in.)
ID: Grayish brown with 3 thin white stripes on belly, may display several pale curved bars on sides and tail base.

PALENOSE PARROTFISH *Scarus psittacus*
SIZE: to 30 cm (12 in.) Parrotfishes – Scaridae
ID: **TP** – Nuptial TP turn yellow. Green with salmon-pink scale margins; lavender-gray snout and bluish tail. Form groups. Reef flats, lagoons and seaward slopes in 2-25 m.
Indo-Pacific: Red Sea to Indonesia, Philippines, Micronesia, Papua New Guinea, Solomon Is., Hawaii and French Polynesia. – S. Japan to Great Barrier Reef and W. Australia.

Palenose Parrotfish – IP
SIZE: to 20 cm (8 in.)
ID: Reddish brown to gray with pale snout. Feed in large mixed-species schools.

SURF PARROTFISH *Scarus rivulatus*
SIZE: to 40 cm (16 in.) Parrotfishes – Scaridae
ID: **TP** – Green to blue body, bright green pectoral fins; **orange patch on gill cover,** wavy green bands/lines on head. Form schools. Silty coastal reefs, lagoons and seaward reefs to 20 m.
Asian Pacific: Andaman Sea to Indonesia, Philippines and Micronesia. – S.W. Japan to Great Barrier Reef, W. Australia and New Caledonia.

Surf Parrotfish – IP
SIZE: to 30 cm (12 in.)
ID: Gray or gray-brown with 2 pale stripes on belly.

Parrotfishes

YELLOWBAR PARROTFISH   *Scarus schlegeli*
SIZE: to 38 cm (15 in.)   Parrotfishes – Scaridae
ID: TP – Dark green to blue except lighter on upper head and foreback; **short bright yellow bar on midback continues as pale bar below.** Solitary or groups. Coral-rich areas of coastal, lagoon and outer reefs to 40 m.
Pacific: E. Indonesia, Philippines and Micronesia to French Polynesia. – S.W. Japan to Australia.

Yellowbar Parrotfish – IP
SIZE: to 15 cm (6 in.)
ID: Grayish brown with 5-6 whitish bars. Sometimes form large feeding aggregations in areas rich with soft and stony corals.

MARQUESAN PARROTFISH   *Scarus koputea*
SIZE: to 31 cm (12 in.)   Parrotfishes – Scaridae
ID: TP – Gray head and body with black scale margins; yellow-green tail base and **white stripe on side of snout. IP –** Shades of red with black scale margins; white patch on lower side and white stripe on anal fin base. Inner bays and outer reef slopes to 18 m.
Localized: Marquesas Is. in French Polynesia.

HIGHFIN PARROTFISH   *Scarus longipinnis*
SIZE: to 40 cm (16 in.)   Parrotfishes – Scaridae
ID: TP – Purple with wide pale bar on forebody; green band runs from eye to pectoral fin, high rounded yellow dorsal fin, elongate tail lobes. **IP –** Light brownish orange, often with dark bars. Clear outer slopes and atoll reefs in 10-55 m.
West & Central Pacific: Great Barrier Reef and Coral Sea to Pitcairn Is.

GREENSNOUT PARROTFISH   *Scarus spinus*
SIZE: to 30 cm (12 in.)   Parrotfishes – Scaridae
ID: TP – Small; green to blue body, green snout and nape and yellow to yellow-green cheek/gill cover. Solitary. Outer reefs in 2-25 m.
West Pacific: Indonesia, Philippines and Micronesia. – S.W. Japan to Great Barrier Reef and Fiji.

Greensnout Parrotfish – IP
SIZE: to 20 cm (8 in.)
ID: Dark brown to nearly black with 4 faint pale bars or vertical rows of whitish spots on side.

TRICOLOR PARROTFISH   *Scarus tricolor*
SIZE: to 40 cm (16 in.)   Parrotfishes – Scaridae
ID: **TP** – Shades of green, often with strong tints of pink or yellow on sides; green bands around mouth, dorsal and anal fins and **long, pointed lobes of tail lavender-pink with dark margins.** Solitary. Outer slopes in 10-40 m.
Indo-Pacific: E. Africa to Indonesia, Philippines, Micronesia, Solomon Is. and Line Is. – S.W. Japan to Great Barrier Reef.

Tricolor Parrotfish – IP
SIZE: to 40 cm (16 in.)
ID: Dark gray to blackish head and upper back shading to blue-green or blue on lower side with blackish margins on scales, red tail, yellow-orange anal fin, dusky orange or yellowish ventral fins.

REDLIP PARROTFISH   *Scarus rubroviolaceus*
SIZE: to 70 cm (2¼ ft.)   Parrotfishes – Scaridae
ID: **TP** – Shades of green, often bicolor with darker forebody; green to blue band on upper lip and **double bands on chin,** numerous blue to blue-green stripes on tail. Solitary or in pairs. Outer reef slopes to 30 m.
Indo-Pacific: E. Africa to Indonesia, Philippines, Micronesia, Solomon Is., Hawaii and French Polynesia. – S.W. Japan to Australia.

Redlip Parrotfish – IP
SIZE: to 48 cm (19 in.)
ID: Shades of red to reddish brown to gray with small black spots and irregular lines on scales; often bicolor with darker forebody, fins and usually lips red. Frequently accompany large males.

RED PARROTFISH   *Scarus xanthopleura*
SIZE: to 54 cm (22 in.)   Parrotfishes – Scaridae
ID: **TP** – Green with pink scale margins; dark green lips, **irregular dark green patch on lower cheek and chin,** faint banding or spotting on head. Solitary. Clearwater lagoons and seaward reefs in 3-30 m.
East Indo-Asian Pacific: Cocos-Keeling Is. to Indonesia and Micronesia. – S.W. Japan to N.W. Australia and Coral Sea.

Red Parrotfish – IP
SIZE: to 33 cm (13 in.)
ID: Bright red with 3-4 faint pale bars on side. Rare. Solitary or form small groups.

Parrotfishes

FILAMENT-FIN PARROTFISH   *Scarus altipinnis*
SIZE: to 50 cm (20 in.)   Parrotfishes – Scaridae
ID: TP – Green with scales edged in salmon-pink; dark bands around mouth, **short filamentous extension from middle of dorsal fin.** Form large aggregations. Shallow protected reefs and outer slopes to 30 m.
Pacific: S.W. Japan to Micronesia, Papua New Guinea, Solomon Is., Great Barrier Reef and Coral Sea to Line Is. and French Polynesia.

Filament-fin Parrotfish – IP
SIZE: to 37 cm (15 in.)
ID: Red-brown with scales edged in darker shade; a few small white spots often align to form 3-4 vertical rows. Compare with similar IP Greenthroat Parrotfish and the IP Greensnout Parrotfish.

Filament-fin Parrotfish – JP
SIZE: to 10 cm (4 in.)
ID: Gray with yellow snout and upper head; white tail base and translucent tail, several dark blotches on dorsal fin and several vertical rows of white spots on rear of body.

SWARTHY PARROTFISH   *Scarus niger*
SIZE: to 35 cm (14 in.)   Parrotfishes – Scaridae
ID: TP – Dark reddish brown becoming purplish green with size; red lips, **yellow to green spot or streak behind eye,** dark bands around mouth. Solitary except during courtship. Coral-rich areas to 20 m.
Indo-Pacific: Red Sea and E. Africa to Indonesia, Micronesia, French Polynesia, Solomon Is. – S.W. Japan to Great Barrier Reef.

Swarthy Parrotfish – IP
SIZE: to 26 cm (10 in.)
ID: Red head, belly and fins except tail; wavy black and white stripes on body to foretail, dark bands around mouth and eye. Coral-rich areas of coastal reefs, lagoons and outer slopes to 20 m.

Swarthy Parrotfish – JP
SIZE: to 9 cm (3 1/2 in.)
ID: Black to red-brown with numerous small white spots, tail translucent; often display white bar across tail base. Usually in small groups on rubble bottoms near reefs.

QUOY'S PARROTFISH
Scarus quoyi

SIZE: to 30 cm (12 in.) — Parrotfishes – Scaridae

ID: TP – Highly variable green to purplish or bluish; occasionally pink patch on pectoral fin region, usually lime-green saddle on tail base, **green "moustache" and green patch on cheek.** Solitary or form small groups. Sheltered coastal reefs in 2-18 m.

East Indo-West Pacific: India to Indonesia, Philippines, Micronesia and Solomon Is. – S.W. Japan to Great Barrier Reef, New Caledonia, Fiji.

Quoy's Parrotfish – IP

SIZE: to 20 cm (8 in.)

ID: Gray back and reddish lower body; 4 wide brown bars on midbody with narrow yellowish bars between and tail base. Compare with similar IP Greenthroat Parrotfish and Greensnout Parrotfish.

GREENLIP PARROTFISH
Scarus viridifucatus

SIZE: to 32 cm (13 in.) — Parrotfishes – Scaridae

ID: TP – Greenish with pink scale margins, prominent blue-green patch on side of snout extends onto chin. **IP –** Dark brown. Closely related to *S. spinus* in the Pacific. Solitary. Coastal, lagoon and outer reefs to 15 m.

Indian Ocean: E. Africa, Aldabra, Seychelles, Maldives, Cocos-Keeling Is. to Sumatra and Bali in Indonesia.

BUMPHEAD PARROTFISH
Bolbometopon muricatum

SIZE: to 120 cm (4 ft.) — Parrotfishes – Scaridae

ID: TP – Largest parrotfish; large hump on forehead, greenish gray. Form feeding groups. Use head to break coral. Lagoons and seaward reefs to 40 m.

Indo-Pacific: Red Sea and E. Africa to Indonesia, Philippines, Micronesia, Papua New Guinea, Solomon Is. and French Polynesia. – S.W. Japan to Great Barrier Reef.

Bumphead Parrotfish – IP

SIZE: to 33 cm (13 in.)

ID: Dark head and body with 5 vertical rows of faint white spots; hump on forehead beginning to develop.

Bumphead Parrotfish – JP

SIZE: to 10 cm (4 in.)

ID: Brown to green with 5 vertical rows of small white spots; nape and snout smoothly sloping to the lip. Usually form small groups.

Swim with Pectoral Fins – Wrasses
Wrasses/Tuskfishes, Hogfishes & Razorfishes

This ID Group consists of fishes that primarily use their pectoral fins to swim. (Besides the closely related parrotfishes [previous ID Group], few other fishes primarily use their pectoral fins to swim.)

FAMILY: Wrasses – Labridae
32 Genera – 190 Species Included

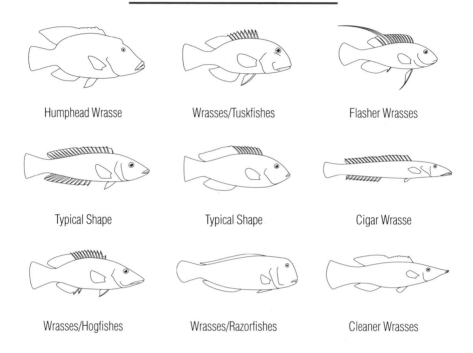

Humphead Wrasse	Wrasses/Tuskfishes	Flasher Wrasses
Typical Shape	Typical Shape	Cigar Wrasse
Wrasses/Hogfishes	Wrasses/Razorfishes	Cleaner Wrasses

Like the closely related parrotfishes, wrasses have large, noticeable scales, swim primarily with their pectoral fins, are often quite colorful, and have the ability to change sex from female to male when certain age, growth or social criteria are met. But unlike parrotfishes, which scrape algae from rocks with fused beaks, most wrasses feed on hard-shelled invertebrates including crabs, shrimps, brittle stars and small gastropods. When discovered, the delicacies are seized or scissored free with a prominent set of protruding canines and crushed with powerful pharyngeal teeth.

In the Indo-Pacific, only gobies (Gobiidae) outnumber wrasses in quantity of species. Wrasses vary greatly in both body shape and size. Whether small or large, slender or deep-bodied, all wrasses have terminal mouths, usually displaying prominent canines, thick lips, and a single, continuous dorsal fin. Because wrasses are relatively small, common, and constantly on the go, underwater observers often fail to take note of their fascinating behaviors and exquisite beauty. By day, most wrasses swim busily in loose, often mixed feeding groups just about the bottom. At dusk, they are the first fishes to bed down and the last to rise after daybreak; smaller species typically dive beneath the sand to sleep, others wedge themselves in reef recesses.

During maturation most wrasses go through a series of changes that dramatically alter their colors, markings and body shapes. These transformations are marked by a JUVENILE PHASE (JP), INITIAL PHASE (IP) and the TERMINAL PHASE (TP), which represents the largest, brightest, most aggressive and least numerous individuals. Some species even display additional color patterns,

INTERMEDIATE, between the primary phases. IP consists of sexually mature females and, in some species, immature but sexually active males. Most wrasses are hermaphroditic and go through a sex change to become TP, while the males of other species simply mature into TP without changing sex. TP, always males, have reached the apex of sexual maturity and generally experience the greatest reproductive success.

Wrasses are all pelagic spawners that typically release gametes daily at traditional times and locations. The TP of many species protect and patrol the boundaries of home ranges encompassing the territories of from three to six feeding herds of IP females with which they attempt to exclusively mate. Chases of encroaching TP and IP males, which occur throughout the day and require costly energy, increase as spawning time approaches. In large populations of certain species young IP males become the dominant reproducers, spawning in mass with passing females while TP, unable to control the competitors, are forced to the periphery of the spawning grounds.

YELLOWTAIL CORIS   *Coris gaimard*
SIZE: to 31 cm (12¹/₄ in.)   Wrasses – Labridae
ID: TP – Variable from shades of blue to green and red with yellow tail; bright yellow or green midbody bar, brilliant blue spots on rear body and tail base. Solitary. Sand and rubble bottoms adjacent to coral reefs in 3-50 m.

East Indo-Pacific: Cocos-Keeling Is. to Indonesia, Micronesia, Hawaii and French Polynesia. – S. Japan to Great Barrier Reef.

Yellowtail Coris – IP
SIZE: to 25 cm (10 in.)
ID: Reddish head, green body with blue rear body and base of tail; red dorsal and anal fins; yellow tail, brilliant blue spots on rear body and base of tail, spike-like 1st dorsal fin spine on both IP and TP. Similar, African Coris, *C. cuvieri*, from E. Africa to Andaman Sea differs only in lacking yellow tail.

Yellowtail Coris – JP
SIZE: to 10 cm (4 in.)
ID: Bright red-orange; 5 black-edged white saddles across head and back. Solitary. Sand and rubble bottoms at base of reefs.

GOLDLINE CORIS   *Coris aurilineata*
SIZE: to 11.5 cm (4¹/₂ in.)   Wrasses – Labridae
ID: TP – Alternating yellow and pale blue-green to green stripes; **ocellated (usually black) spot on upper base of tail. IP –** Similar with large ocellated spot on mid-dorsal fin. Rubble and sand bottoms with algae to 25 m.

Localized: S. Great Barrier Reef and New Caledonia to New South Wales in S.E. Australia.

Wrasses

CLOWN CORIS *Coris aygula*

SIZE: to 60 cm (2 ft.) Wrasses – Labridae

ID: TP – Dark green, appearing nearly black underwater; few faint orangish lines radiate from eye; occasionally a pale bar behind pectoral fin and others on body. Solitary. Sand and rubble patches near reefs in 2-30 m.

Indo-Pacific: Red Sea and E. Africa to Indonesia, Philippines, Micronesia to French Polynesia. – S. Japan to Australia.

Clown Coris – IP

SIZE: to 40 cm (16 in.)

ID: Light gray head with black spots; dark gray rear body separated by wide white bar.

Clown Coris – JP

SIZE: to 7.5 cm (3 in.)

ID: Whitish with black spots on head and forebody; two white-ringed black spots on dorsal fin with orangish area below; black spot near upper tail base.

PIXIE CORIS *Coris pictoides*

SIZE: to 11 cm (4 1/4 in.) Wrasses – Labridae

ID: Black to dark brown upper body, white below, fins pale; **thin white line from top of eye to end of dorsal fin.** Solitary or form small groups. Sand and rubble bottoms near coral reefs in 10-55 m.

Asian Pacific: Indonesia and Philippines to Australia and New Caledonia.

BATU CORIS *Coris batuensis*

SIZE: to 15 cm (6 in.) Wrasses – Labridae

ID: TP – Whitish to pale green; several dark bars and narrow white bars on upper body, **pale-rimmed black spot on mid-dorsal fin.** Solitary. Sand and rubble bottoms surrounding reefs of lagoons and seaward reefs to 30 m.

Indo-West Pacific: E. Africa to Indonesia, Philippines, Micronesia, Solomon Is. – S. Japan to Great Barrier Reef and Fiji.

Batu Coris – JP

SIZE: to 6.5 cm (2 1/2 in.)

ID: White head and forebody changing to green with numerous red spots.

PALE-BARRED CORIS *Coris dorsomacula*
SIZE: to 38 cm (15 1/4 in.) Wrasses – Labridae
ID: TP – Green; darker back and pale lower head and breast, **series of narrow whitish bars on side,** orange stripe from snout, through eye to upper side; lime green tail with coverging reddish bands. Solitary. Sand and rubble in 5-40 m.
West Pacific: South China Sea and Indonesiai. – S. W. Japan to E. Australia and Fiji.

Pale-barred Coris – IP
SIZE: to 23 cm (9 in.)
ID: Shades of red with narrow pale bars on side; yellow-edged black "ear" spot, small black spot on rear dorsal fin.

RUST-BANDED WRASSE *Pseudocoris aurantiofasciata*
SIZE: to 30 cm (12 in.) Wrasses – Labridae
ID: TP – Dark blue-gray; several narrow black bars just behind pectoral fin (may have orange bars between), narrow white bar across rear body. Form groups. Outer reef slopes in 25-50 m.
East Indo-Pacific: Cocos-Keeling and Christmas Is. to Indonesia, Philippines, S. Japan, Micronesia, N.E. Papua New Guinea and French Polynesia.

PHILIPPINE WRASSE *Pseudocoris bleekeri*
SIZE: to 15 cm (6 in.) Wrasses – Labridae
ID: TP – Green with combination of dark bars; elongate spots and saddles, **yellow patch on middle of side. IP –** Light brown with pale belly; black blotch on rear gill cover and upper tail base. Rubble and coral slopes of seaward reefs in 10-30 m.
Asian Pacific: Bali to West Papua in E. Indonesia, Philippines and S.W. Japan.

TORPEDO WRASSE *Pseudocoris heteroptera*
SIZE: to 20 cm (8 in.) Wrasses – Labridae
ID: TP – Blue-green head and forebody with **alternating yellow and blackish bars** behind; dark patch behind eye. Form groups well above bottom. Mixed sand, rubble and coral areas with periodic strong currents in 15-24 m.
East Indo-Asian Pacific: Chagos to Sabah, Malaysia, Indonesia, Philippines, S. Japan, Micronesia and Solomon Is.

Torpedo Wrasse – IP
SIZE: to 16 cm (6 1/4 in.)
ID: Dark greenish brown; **red anal fin,** dark spot on rear gill cover above pectoral fin, row of pale spots in dorsal fin, dark tail borders.

Wrasses

REDSPOT WRASSE   *Pseudocoris yamashiroi*
SIZE: to 15 cm (6 in.)   Wrasses – Labridae
ID: TP – Shades of green with white belly; black spots and blotches on green areas, black borders on tail. Form aggregations above bottom. Edge of reef slopes in 5-30 m.
East Indo-West Pacific: Maldives to Indonesia, Philippines, Micronesia, Papua New Guinea and Solomon Is. – S. Japan to Australia and Fiji.

Redspot Wrasse – IP
SIZE: to 15 cm (6 in.)
ID: Lavender-gray with wavy lines; pair of silver-white stripes on head above and below eye, **reddish spot on pectoral fin base.** Form large plankton-feeding aggregations with IP greatly outnumbering TP.

FLORAL WRASSE   *Cheilinus chlorourus*
SIZE: to 36 cm (14 1/4 in.)   Wrasses – Labridae
ID: Highly variable from orange-brown to green-brown; usually prominent horizontal rows of pink or white dots, often white to whitish blotches on back and tail. Solitary. Mixed sand, rubble and coral areas of lagoons and coastal reefs in 2-30 m.
Indo-Pacific: E. Africa to Indonesia, Philippines, Micronesia and Solomon Is. to French Polynesia. – S.W. Japan to Australia.

REDBREASTED WRASSE   *Cheilinus fasciatus*
SIZE: to 36 cm (14 1/4 in.)   Wrasses – Labridae
ID: Red-orange rear head and forebody; alternating white and blackish bars on body and tail. Solitary. Mixed sand, rubble and coral areas of lagoons and outer reefs in 3-40 m.
Indo-West Pacific: Red Sea and E. Africa to Indonesia, Philippines, Micronesia and Solomon Is.– S.W. Japan to Australia, New Caledonia and Fiji.

TRIPLETAIL WRASSE   *Cheilinus trilobatus*
SIZE: to 40 cm (16 in.)   Wrasses – Labridae
ID: TP – Shades of green; ornate pattern of pink lines and spots cover head, two white to whitish bars on tail base, rounded tail fin with protruding upper and lower lobes. Solitary and wary. Lagoons, passes and outer reefs to 30 m.
Indo-Pacific: E. Africa to Indonesia, Philippines, Micronesia and French Polynesia. – S.W. Japan to Australia and New Caledonia.

Tripletail Wrasse – IP
SIZE: to 40 cm (16 in.)
ID: Whitish undercolor with greenish brown head and 4 wide brownish bars on body; lack elongate lobes of TP.

HUMPHEAD WRASSE *Cheilinus undulatus*

SIZE: to 170 cm (5 1/2 ft.) Wrasses – Labridae

ID: TP – Blue head with maze-like markings; green body with dark vertical streaks, pronounced hump above eyes. Solitary or occasionally in pairs; wary. Lagoon and outer reefs to 60 m.

Indo-Pacific: Red Sea and E. Africa to Indonesia, Philippines, Micronesia, Papua New Guinea, Solomon Is. and French Polynesia. – S.W. Japan to Australia.

Humphead Wrasse – IP

SIZE: to 60 cm (2 ft.)

ID: Olive to bluish or greenish gray with dark vertical streaks on body; dark diagonal streak extends from front of lower eye, 2 dark lines extend from rear eye. An endangered species in many areas due to overfishing. Also commonly known as Napoleon Wrasse.

Humphead Wrasse – JP

SIZE: to 20 cm (8 in.)

ID: Pale brown to yellowish green with longitudinal rows of vertically elongate brown spots; 4-5 narrow white bars, diagonal line markings through eye. Unlike adults, JP are shy and inconspicuous. Shelter among dense stands of branching corals, gorgonians and seagrass.

SNOOTY WRASSE *Cheilinus oxycephalus*

SIZE: to 17 cm (6 3/4 in.) Wrasses – Labridae

ID: Variable from greenish brown to red; **black spot on front of dorsal fin,** often a trio of small black spots on tail base, may have some red or white spots and/or whitish bars. Solitary or in pairs; secretive. Corals of lagoons and outer reefs to 40 m.

Indo-Pacific: E. Africa to Indonesia, Philippines, Micronesia, Solomon Is. to French Polynesia. – S.W. Japan to Australia.

GRAPHIC TUSKFISH *Choerodon graphicus*

SIZE: to 46 cm (18 in.) Wrasses – Labridae

ID: Tan to yellow undercolor; dark irregular spoke-like markings radiating from eye, dark interconnected bars on sides. Solitary. Sand and rubble patches of coastal, lagoon and seaward reefs in 2-30 m.

Localized: Great Barrier Reef and New Caledonia.

BLUE TUSKFISH *Choerodon cyanodus*

SIZE: to 70 cm (2 1/4 ft.) Wrasses – Labridae

ID: Shades of gray often with yellowish or greenish tints; white chin and **white spot on rear back,** frequently display 4-5 faint bars on back. Solitary. Sand and rubble areas near coastal reefs in 2-35 m.

Localized: E. to W. Australia.

Wrasses

HARLEQUIN TUSKFISH *Choerodon fasciatus*
SIZE: to 30 cm (12 in.) Wrasses – Labridae
ID: White undercolor, rear body becomes dark with age; 6-9 blue edged orange bars extend from head to tail base, tail white with orangish margin. Solitary, territorial and often in caves. Coastal outer reefs to 15 m.
Localized: Taiwan and S.W. Japan to Milne Bay in Papua New Guinea, Great Barrier Reef and S.E. Australia.

ZAMBOANGA TUSKFISH *Choerodon zamboangae*
SIZE: to 38 cm (15¼ in.) Wrasses – Labridae
ID: Similar to Blackwedge Trunkfish [next], but has **blue markings on chin and beneath eye;** dark wedge-shaped marking on side has reddish-brown hue. Solitary or small groups. Sand rubble areas near coastal and lagoon reefs in 25-60 m.
Asian Pacific: Indonesia, Philippines and Papua New Guinea.

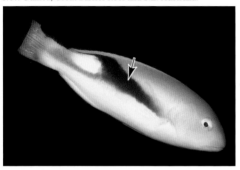

BLACKWEDGE TUSKFISH *Choerodon jordani*
SIZE: to 17 cm (6¾ in.) Wrasses – Labridae
ID: Gray; **blackish wedge-shaped mark** surrounds large white spot below rear dorsal fin. Solitary or form small groups. Sand and rubble areas near reefs in passes and on outer slopes in 20-40 m.
West Pacific: E. Australia to New Caledonia and Fiji.

BLACKSADDLE TUSKFISH *Choerodon zosterophorus*
SIZE: to 25 cm (10 in.) Wrasses – Labridae
ID: Gray upper body, yellowish midbody, white below; **dark-edged white band from pectoral to rear dorsal** with black blotch behind. Solitary or form small groups. Sand and rubble areas near coastal and lagoon reefs in 10-40 m.
Asian Pacific: Brunei, Sabah in Malaysia, Indonesia, Philippines and Papua New Guinea.

ANCHOR TUSKFISH *Choerodon anchorago*
SIZE: to 50 cm (20 in.) Wrasses – Labridae
ID: Gray head, lower body white to darkish gray; **rectangular black marking on back** surrounded by pearly white to yellow area. Solitary. Mixed seagrass, sand, rubble and coral areas of coastal reefs, shallow flats and lagoons to 25 m.
Asian Pacific: Indonesia, Philippines, Micronesia and Solomon Is. – S.W. Japan to Great Barrier Reef and New Caledonia.

Anchor Tuskfish – JP
SIZE: to 10 cm (4 in.)
ID: Greenish brown with white bars and blotches on belly; white rear of ventral fin.

DARKSTRIPE TUSKFISH *Choerodon vitta*
SIZE: to 20 cm (8 in.)   Wrasses – Labridae
ID: Light gray to yellowish with yellow head; **dark midlateral stripe, dark spot on tail base.** Solitary or form small groups. Flat sand and rubble bottoms in vicinity of coastal reefs in 10-40 m.
Localized: Aru Is. in Indonesia and N. Australia.

SINGAPORE TUSKFISH *Choerodon oligacanthus*
SIZE: to 28 cm (11 in.)   Wrasses – Labridae
ID: Tan; pale yellow stripes on lower body, **elongate white patch on middle of back,** small black spot on tail base. Solitary or in pairs. Sand and rubble bottoms next to coastal reefs in 2-15 m.
Asian Pacific: Malaysia, Indonesia and Philippines.

BLACKSPOT TUSKFISH *Choerodon schoenleinii*
SIZE: to 90 cm (3 ft.)   Wrasses – Labridae
ID: Greenish gray with vertical blue streaks on body; **small dark spot on base of mid-dorsal fin.** Solitary. Sand, rubble and weedy areas on flat bottoms of lagoons and seaward reefs in 10-60 m.
Asian Pacific: E. Malaysia and Indonesia to Papua New Guinea. – S.W. Japan to N. Australia.

DARKSPOT TUSKFISH *Choerodon monostigma*
SIZE: to 25 cm (10 in.)   Wrasses – Labridae
ID: Light orange brown with vague broad dusky bars; **narrow blue stripe from front of spinous dorsal fin to large dark spot on rear** occasionally ringed with white, blue line markings on head. Sand and weedy rubble areas in 10-40 m.
Localized: Fakfak Peninsula, West Papua in Indonesia to N. Australia.

PENNANT FAIRY WRASSE *Cirrhilabrus joanallenae*
SIZE: to 6 cm (2 ¼ in.)   Wrasses – Labridae
ID: **TP –** Red upper head and body darkening toward rear, white below, blue tail; elongate 1st dorsal spine, **large black ventral fins.** Solitary or form small groups. Rubble bottoms in 12-25 m.
Localized: Sri Lanka to Andaman Sea and N. Sumatra in Indonesia.

BACKSTRIPED FAIRY WRASSE *Cirrhilabrus katherinae*
SIZE: to 9 cm (3 ½ in.)   Wrasses – Labridae
ID: **TP –** Reddish to greenish brown with greenish stripe centered on back, pale belly; black dorsal and anal fins with red borders. Solitary or form aggregations. Patch reefs on outer reef slopes and in clear lagoons in 10-40 m.
Asian Pacific: Micronesia from Palau to Pohnpei and Ryukyu and Izu Is. south of Japan.

Wrasses

BLUESIDED FAIRY WRASSE *Cirrhilabrus cyanopleura*

SIZE: to 11 cm (4 1/4 in.) Wrasses – Labridae

ID: **TP** – Blue to blue-green upper head and forebody, orange to brown to green rear body, pale belly; **blue band across pectoral fin base.** Solitary or form small groups. Mixed coral and rubble areas of lagoons and coastal reefs 5-30 m.

East Indo-Asian Pacific: Andaman Sea to Java, Kalimantan and Komodo and Raja Ampat in Indonesia.

Bluesided Fairy Wrasse – TP Variation

ID: Blue-green nape and forebody, orange back and purple elongate midbody patch; all variations have blue band across pectoral fin base.

YELLOWFLANKED FAIRY WRASSE *Cirrhilabrus ryukyuensis*

SIZE: to 11 cm (4 1/4 in.) Wrasses – Labridae

ID: Variable. **TP** – Often triangular blue to purple patch behind head; **yellow to orange patch behind pectoral fin,** or broad orange stripe on lower half of body. Form aggregations. Rubble in lagoons and seaward reef slopes in 2-25 m.

Asian-Pacific: Sabah in Malaysia, Indonesia, Philippines, Taiwan, and S.W. Japan.

Yellowflanked Fairy Wrasse – IP

SIZE: to 8 cm (3 1/4 in.)

ID: Orange to reddish purple upper body and white below; **small black spot on upper tail base.**

SOLOR FAIRY WRASSE *Cirrhilabrus solorensis*

SIZE: to 12 cm (4 3/4 in.) Wrasses – Labridae

ID: **TP** – Upper body orange in highly variable amounts, head and lower body violet to lavender. Solitary or with several females. Rubble areas near coral reefs in 5-35 m.

Localized: Flores and Alor in S. Indonesia.

Solor Fairy Wrasse – IP

SIZE: to 8 cm (3 1/4 in.)

ID: Red to maroon head and body with pale chin and belly; **1-2 black spots on tail base.**

MELANESIAN FAIRY WRASSE   *Cirrhilabrus beauperryi*
SIZE: to 12 cm (4 3/4 in.)   Wrasses – Labridae
ID: TP – Purplish with numerous small blue spots, shading to blue on lower body; yellowish brown on back, occasionally yellowish area on forehead, **purple stripe on base dorsal fin**; long ventral fins. Rubble areas next to reefs in lagoons and on seaward slopes in 5-32 m.
Localized: N. Papua New Guinea to Solomon Is. and Vanuatu.

Melanesian Fairy Wrasse – IP
SIZE: to 8 cm (3 1/4 in.)
ID: Dull purplish red with **small black spot on upper tail base;** pair of thin purple bars across pectoral fin base.

JAVAN FAIRY WRASSE   *Cirrhilabrus filamentosus*
SIZE: to 8 cm (3 1/4 in.)   Wrasses – Labridae
ID: TP – Red upper body, white below; **yellow dorsal, anal and ventral fins,** rounded blue tail, long filament extends from mid-dorsal fin. **IP –** Red with small black spot at upper tail base. Flat rubble bottoms next to coral patches in 10-35 m.
Localized: Java, Bali and Lesser Sunda Is. in Indonesia.

IRIAN JAYA FAIRY WRASSE   *Cirrhilabrus tonozukai*
SIZE: to 7.5 cm (3 in.)   Wrasses – Labridae
ID: TP – Red upper body, white below; long filament extends from mid-dorsal fin, **magenta anal fin.** Solitary or form groups. Rubble areas near seaward reefs in 15-40 m.
Localized: Gulf of Tomini and Lembeh Straits, Sulawesi to West Papua in Indonesia and Palau in Micronesia.

HUMANN'S FAIRY WRASSE   *Cirrhilabrus humanni*
SIZE: to 7 cm (2 3/4 in.)   Wrasses – Labridae
ID: TP – Golden brown with short **backward-arching foredorsal fin** (not always held erect); large red ventral fins, red rear dorsal and anal fins outlined with blue. Rubble area inside lagoon in 7-45 m.
Localized: Alor and W. Timor in S. Indonesia.

Humann's Fairy Wrasse – IP
SIZE: to 5 cm (2 in.)
ID: Red to orange-brown with **rows of blue spots forming blue lines on side.**

Wrasses

CORAL SEA FAIRY WRASSE *Cirrhilabrus bathyphilus*
SIZE: to 11 cm (4 1/4 in.) Wrasses – Labridae
ID: **TP –** Red upper body, yellowish to white below; purplish dorsal fin with black border and black blotch on front. Form aggregations over rubble bottoms. Outer reefs in 5-30 m.
Localized: Coral Sea, off N.E. Australia.

SAMOA FAIRY WRASSE *Cirrhilabrus walshi*
SIZE: to 10 cm (4 in.) Wrasses – Labridae
ID: **TP –** Blotchy orange-red with yellowish to whitish underside; yellowish dorsal fin and red tail with white spotted borders. Form aggregations over rubble bottoms. Seaward reefs in 5-30 m.
Localized: Known only from Samoa.

REDFIN FAIRY WRASSE *Cirrhilabrus rubripinnis*
SIZE: to 9 cm (3 1/2 in.) Wrasses – Labridae
ID: **TP –** Brownish red with pale red to white belly, brilliant scarlet dorsal, anal and ventral fins. **IP –** Similar, but less vivid; black saddle or spot on tail base. Form groups. Rubble and coral areas at base of coastal reef slopes in 16-40 m.
Asian Pacific: Kalimantan (Borneo) and N. Sulawesi in Indonesia and Philippines.

MAGENTA-STREAKED FAIRY WRASSE *Cirrhilabrus laboutei*
SIZE: to 11 cm (4 1/4 in.) Wrasses – Labridae
ID: **TP –** Pattern of curved red and yellow-orange lines on upper body, white below; diagonal magenta band at pectoral fin base. Solitary or form small groups. Rubble bottoms near coral reefs in 8-55 m.
Localized: Coral Sea, New Caledonia including Loyalty Is., Vanuatu and Fiji.

REDBACK FAIRY WRASSE *Cirrhilabrus condei*
SIZE: to 8 cm (3 1/4 in.) Wrasses – Labridae
ID: **TP –** Bright red upper body, white below; **broad black outer margin on dorsal fin,** older TP develop long filamentous ventral fins. Solitary or form small groups. Mixed coral and rubble areas of seaward slopes in 5-70 m.
Localized: West Papua in Indonesia, N.E. Papua New Guinea to N. Great Barrier Reef.

Redback Fairy Wrasse – IP
SIZE: to 6 cm (2 1/4 in.)
ID: Reddish upper body, pale below; dark margin on red dorsal fin, red tail with blue spots; red anal fin with row of blue bands extending from base and spots on mid fin.

EXQUISITE FAIRY WRASSE   *Cirrhilabrus exquisitus*
SIZE: to 11 cm (4 1/2 in.)   Wrasses – Labridae
ID: TP – Highly variable, commonly green to red, yellow and blue; **diagonal white band below pectoral fin another above,** bluish stripe from snout to upper pectoral fin, a 2nd from midbody to tail. Solitary. Seaward reefs in current in 5-30 m.
Indo-Pacific: E. Africa to Indonesia, Philippines, Micronesia and French Polynesia. – S. Japan to Great Barrier Reef.

Exquisite Fairy Wrasse – Variation
ID: Photo taken in Fiji.

Exquisite Fairy Wrasse – IP
SIZE: to 7 cm (2 3/4 in.)
ID: Shades of brownish red; spot on upper tail base, **white spot on tip of snout.**

DOTTED FAIRY WRASSE   *Cirrhilabrus punctatus*
SIZE: to 13 cm (5 in.)   Wrasses – Labridae
ID: TP – Gray to blue-green with numerous light blue body spots; dark bar on base of pectoral fin; **dark stripe along base of dorsal and anal fins,** very long ventral fins. Solitary or from small groups. Rubble areas next to reefs in lagoons and on seaward slopes in 2-32 m.
West Pacific: Papua New Guinea and E. Australia to Fiji.

Dotted Fairy Wrasse – TP Fiji Variation
ID: Reddish brown to purplish brown head and body; dark stripe along base of dorsal and anal fins; light blue spots on body, bluish tail.

Dotted Fairy Wrasse – IP/JP
SIZE: to 9 cm (3 1/2 in.)
ID: Shades of red to bluish pink with pale to blue speckling; yellow to yellowish tail, **dark bar on base of pectoral fin,** often dark spot on upper tail base of varying size.

Wrasses

REDTAILED FAIRY WRASSE   *Cirrhilabrus scottorum*
SIZE: to 12 cm (4³/₄ in.)   Wrasses – Labridae
ID: TP – Shades of green with pale yellow to reddish underside; red to reddish tail, red to yellow dorsal fin and dark blue anal fin, black speckles on upper forebody. Solitary or form groups. Coral or rubble of outer slopes in 3-40 m.
South Pacific: Coral Sea to Society and Tuamotu Is. in French Polynesia and east to Pitcairn Is.

YELLOWBAND FAIRY WRASSE   *Cirrhilabrus luteovittatus*
SIZE: to 12 cm (4³/₄ in.)   Wrasses – Labridae
ID: TP – Variable with maroon to red head and upper body, red to violet belly; broad pale red to yellow stripe from pectoral fin to tail base, black wedge-shaped mark below pectoral fin base. Form aggregations. Rubble around lagoon patch reefs in 7-30 m.
Localized: E. Caroline and Marshall Is. in Micronesia (rare at Johnston Atoll).

RED-MARGINED FAIRY WRASSE   *Cirrhilabrus rubrimarginatus*
SIZE: to 15 cm (6 in.)   Wrasses – Labridae
ID: TP – Blue to lavender to pink; fine yellow lines on head, often rows of red spots on body, **broad red margin on tail** and rear half of dorsal fin. **IP –** Similar, but lack red tail margin. Form small groups. Sand and rubble with patch reefs in 25-52 m.
West Pacific: Indonesia, Philippines, Micronesia and Papua New Guinea. – S.W. Japan to Fiji.

PEACOCK FAIRY WRASSE   *Cirrhilabrus temminckii*
SIZE: to 10 cm (4 in.)   Wrasses – Labridae
ID: TP – Brown to red back, whitish belly; **two blue to green body stripes, the 1st extends from below foredorsal fin, the 2nd below from mid-body to tail**, red dorsal and anal fins with blackish bases. Form groups. Rubble of seaward reefs in 6-20 m.
Asian Pacific: Central Indonesia and Philippines. – S. Japan and Taiwan to W. Australia.

PURPLE-LINED FAIRY WRASSE   *Cirrhilabrus lineatus*
SIZE: to 12 cm (4³/₄ in.)   Wrasses – Labridae
ID: TP – Pale purple to magenta body; dorsal, anal and tail fins yellow to orange with blue spots, bright blue lines and spots on head and forebody. **IP –** Similar, but less vivid. Solitary or form groups. Rubble bottoms near reefs in 8-55 m.
Localized: Great Barrier Reef, Coral Sea, New Caledonia and Loyalty Is.

CENDERAWASIH FAIRY WRASSE   *Cirrhilabrus cenderawasih*
SIZE: to 8.5 cm (3¹/₄ in.)   Wrasses – Labridae
ID: TP – Pinkish with whitish belly, broad yellow stripe at midbody, black blotches along back. Form groups on rubble bottoms of steep slopes along coastal reefs from 22-60 m. Closely related to Walindi Fairy Wrasse [next].
Localized: Cenderawasih Bay and northern Raja Ampat in West Papua, Indonesia.

WALINDI FAIRY WRASSE *Cirrhilabrus walindi*
SIZE: to 8.5 cm (3 1/4 in.) Wrasses – Labridae
ID: **TP** – Pink with yellow wash on back; dorsal, anal and ventral fins yellowish with blue or magenta borders, **two black blotches on dorsal fin.** Solitary or form small groups. Rubble next to coral patches on seaward slopes in 10-65 m. Closely related to Cenderawasih Fairy Wrasse [previous].
Localized: N. E. Papua New Guinea.

Walindi Fairy Wrasse – IP
SIZE: to 6 cm (2 1/4 in.)
ID: Salmon to pinkish; small black spot on upper tail base. Form groups, often accompanied by a single male.

YELLOWBACK FAIRY WRASSE *Cirrhilabrus lubbocki*
SIZE: to 8 cm (3 1/4 in.) Wrasses – Labridae
ID: **TP – Orange to yellow upper head,** back and dorsal fin, purplish white below; row of maroon to purple spots on back and another on middle of tail base with white stripe between. Solitary or form groups. Rubble bottoms next to seaward coral reefs in 4-45 m.
Asian Pacific: Sabah in Malaysia to Indonesia and Philippines.

Yellowback Fairy Wrasse – TP Variation
ID: Bright yellow to orange upper head, back and dorsal fin, red to purple below; **both variations have yellow pectoral fin base. IP –** Both variations have red area with small black spot on upper tail base.

MARJORIE'S FAIRY WRASSE *Cirrhilabrus marjorie*
SIZE: to 7 cm (2 3/4 in.) Wrasses – Labridae
ID: **TP** – Bright red upper body and whitish gray below; yellowish tail base, bold black margin on dorsal and tail fins. **IP** – Red with black dot on tail base. Solitary or form small groups. Outer reefs in 15-40 m.
Localized: Fiji.

YELLOWFIN FAIRY WRASSE *Cirrhilabrus flavidorsalis*
SIZE: to 6.5 cm (2 1/2 in.) Wrasses – Labridae
ID: **TP** – White to pinkish with red upper head; **2 wide red bars on forebody,** broad yellow margin on dorsal fin. **IP** – Red; small black spot on upper tail base. Rubble bottoms among coral patches in 12-30 m.
Asian Pacific: Indonesia and Philippines.

Wrasses

FILAMENTED FLASHER *Paracheilinus filamentosus*
SIZE: to 8 cm (3 1/4 in.) Wrasses – Labridae
ID: TP – Orange-red to purple with 5-6 violet stripes on side; **5-6 filaments extend from soft dorsal fin rays,** deep lunate tail with filaments. **Nuptial Male –** Bright yellow spinous dorsal and pale yellow dorsal fin filaments. Form aggregations in 5-35 m.
Asian Pacific: Indonesia and Philippines to Micronesia, Solomon Is. and N. Great Barrier Reef.

RED FILAMENTED FLASHER *Paracheilinus paineorum*
SIZE: to 8 cm (3 1/4 in.) Wrasses – Labridae
ID: TP – Orange-red with yellowish hue on forebody and pink stripes; **4-7 red filamentous dorsal rays,** yellow on foredorsal fin, deep lunate tail with filamentous lobes. Form small to large groups. Rubble, fringing reefs and outer slopes in 15-50 m.
Localized: Seribu Is. (Java) to Flores and N. Sulawesi in Indonesia.

NURSALIM FLASHER *Paracheilinus nursalim*
SIZE: to 8 cm (3 1/4 in.) Wrasses – Labridae
ID: TP – Golden brown body and head with thin, blue stipes; **dark area just above rear anal fin and below foredorsal fin,** four elongate, white filamentous rays; deep lunate tail. Complex bottoms of fringing reefs and outer slopes in 5-50 m.
Localized: Raja Ampat south to Triton Bay, West Papua in Indonesia.

Nursalim Flasher – Young TP
ID: Pale red with thin dark red lines on side of body; **darkish back beneath dorsal fin.**

WALTON'S FLASHER *Paracheilinus walton*
SIZE: 6 cm (2 1/4 in.) Wrasses – Labridae
ID: TP – Courtship pattern consists of **white dorsal fin and backward projecting rounded filamentous rays;** bright red to black along rear back. Smaller than other flashers. **IP –** Plain red with faint horizontal lines. Form aggregations. Rubble bottoms near coral reefs in 18-45 m.
Localized: Cenderawasih Bay, N.W. West Papua in Indonesia.

Walton's Flasher – Young TP
ID: Reddish with dark thin red lines and white belly.

SOUTH CHINA SEA FLASHER *Paracheilinus xanthocirritus*
SIZE: to 7 cm (2 3/4 in.)   Wrasses – Labridae
ID: **TP –** Orange upperhead and forebody shading to red to maroon rear body and tail; 3-4 thin bluish body stripes, 4-6 orange to red dorsal fin filamentous rays, deep lunate tail. Form aggregations. Outer reef slopes in 15-30 m.
Localized: Borneo and Vietnam, South China Sea.

BLUE FLASHER *Paracheilinus cyaneus*
SIZE: to 8 cm (3 1/4 in.)   Wrasses – Labridae
ID: **TP –** Red, turn bluish green on back and forebody during courtship display; **small dots between lines on head,** red dorsal fin with 5 long filamentous rays turns white during display. Complex bottoms of fringing reefs and outer slopes in 6-20 m.
Asian Pacific: N.E. Kalimantan (Borneo) east to Raja Ampat in Indonesia.

SCARLET-FIN FLASHER *Paracheilinus lineopunctatus*
SIZE: 7 cm (2 3/4 in.)   Wrasses – Labridae
ID: **TP – Deep red body and head with blue patches** and occasional spotting; 6 to 7 elongate, white dorsal filaments, round red tail and anal fins edged in blue. Complex bottoms of fringing reefs and outer slopes in 15-40 m.
Asian Pacific: N. Indonesia and Philippines.

CARPENTER'S FLASHER *Paracheilinus carpenteri*
SIZE: to 7 cm (2 3/4 in.)   Wrasses – Labridae
ID: **TP –** Yellow-orange body and head with thin bluish stripes; 4-6 long filamentous rays extend from dorsal fin, **dark patch on base of rear dorsal often more prominent than in photo,** rounded tail. Form aggregations above complex bottoms at base of steep outer reef slopes in 12-40 m.
Localized: Brunei and Philippines to S. Japan.

YELLOWFIN FLASHER *Paracheilinus flavianalis*
SIZE: to 7 cm (2 3/4 in.)   Wrasses – Labridae
ID: **TP –** Orange-red or yellowish with violet to blue stripes, **anal fin yellow;** single long red filamentous ray extends from mid-dorsal fin, round tail. Form aggregations above complex bottoms and base of seaward slopes in 6-35 m.
Localized: N. Sulawesi to Bali and the Scott and Hibernia Reefs in the Timor Sea in Indonesia.

MCCOSKER'S FLASHER *Paracheilinus mccoskeri*
SIZE: to 7 cm (2 3/4 in.)   Wrasses – Labridae
ID: **TP –** Orange-red or yellowish with violet to blue stripes, outer half of anal fin red; single long filamentous ray extends from dorsal fin, round tail. Form aggregations above complex bottoms in 6-50 m.
Indian Ocean: E. Africa to Comoro, Maldives and Andaman Sea to N.W. Sumatra in Indonesia.

Wrasses

RED-TAILED FLASHER   *Paracheilinus rubricaudalis*
SIZE: to 8 cm (3¼ in.)   Wrasses – Labridae
ID: TP – Single yellow filamentous ray at middle of dorsal fin; distinguished from McCosker's Flasher [previous] by location, bright red tail with bluish marginal bands. Solitary or form small groups on outer reefs in 15-45 m.
West Pacific: E. Papua New Guinea to Solomon Is., Great Barrier Reef, Vanuatu and Fiji.

Red-tailed Flasher – TP Variation
ID: With yellow anal fin from Solomon Is.

ROUNDFIN FLASHER   *Paracheilinus togeanensis*
SIZE: to 7 cm (2¾ in.)   Wrasses – Labridae
ID: TP – Unlike most members of genus have a **rounded, unfilamented dorsal fin**; yellow anal fin and lunate tail. Solitary or join mixed groups with other flasher species in 16-40 m.
Localized: Togean Is. and Lembeh Straits, N. Sulawesi in Indonesia.

ALFIAN'S FLASHER   *Paracheilinus alfiani*
SIZE: to 6 cm (2¼ in.)   Wrasses – Labridae
ID: TP – Orange body with faint bluish stripes; bluish lower body, red dorsal and anal fins with narrow blue border and orange on rear of each, orange front of dorsal fin. Form aggregations. Rubble slopes in 15-21 m.
Localized: Currently known only from Alor in Indonesia.

RENNY'S FLASHER   *Paracheilinus rennyae*
SIZE: to 6 cm (2¼ in.)   Wrasses – Labridae
ID: TP – Red to orange body with faint narrow stripes; red dorsal and anal fins with narrow blue border, lack extensions of dorsal fin rays, rounded tail with wide yellow margin. Form aggregations. Coral reefs and adjacent areas in 15-21 m.
Localized: Currently known only from S.W. Flores and Komodo Is. in Indonesia.

ANGULAR FLASHER   *Paracheilinus angulatus*
SIZE: to 7 cm (2¾ in.)   Wrasses – Labridae
ID: TP – Red with magenta stripes; pale belly; **elongate rear dorsal and anal fins with blue spots and edging give an angular profile.** Form aggregations over complex bottoms of seaward reefs and lagoons in 10-40 m.
Asian Pacific: Sabah in Malaysia, Brunei, N.E. Kalimantan in Indonesia and Philippines.

TANAKA'S WRASSE *Wetmorella tanakai*
SIZE: to 5.5 cm (2 1/4 in.) Wrasses – Labridae
ID: TP – Orange red forebody and purplish red rear body with
4 narrow bluish white bars on head and body; large white
ringed, black spots on dorsal anal and ventral fins. Cryptic.
Coral reefs in 10-30 m.
Asian Pacific: Flores, S. Sulawesi and W. Papua in Indonesia
and Philippines.

WHITEBANDED PYGMY WRASSE *Wetmorella albofasciata*
SIZE: to 6 cm (2 1/4 in.) Wrasses – Labridae
ID: TP – Brown; pair of inward slanting white bands on
midside, pair of white bands radiate from eye, thin white bar
on tail base, black spots on dorsal, anal and ventral fins. Solitary.
Crevices and recesses of lagoon and seaward reefs in 8-42 m.
Indo-Pacific: E. Africa to Indonesia, Philippines, Micronesia,
Hawaii and French Polynesia. – S.W. Japan to Great Barrier Reef.

SHARPNOSE WRASSE *Wetmorella nigropinnata*
SIZE: to 8 cm (3 1/4 in.) Wrasses – Labridae
ID: TP – Red to brownish; yellow bar behind eye and another
across tail base, white-edged black spot on ventral, rear dorsal
and anal fins. Solitary. Recesses, caves and crevices of lagoon
and seaward reefs to 30 m.
Indo-Pacific: Red Sea and E. Africa to Indonesia, Philippines,
Micronesia to French Polynesia. – S.W. Japan to Australia.

LEOPARD WRASSE *Macropharyngodon meleagris*
SIZE: to 15 cm (6 in.) Wrasses – Labridae
ID: TP – Dull orange-red to purple or green undercolor with
black or blue-edged green spot on each body scale; blue-edged
green bands on head, small black spot above pectoral. Solitary
or form small groups. Coral and rubble bottoms in 2-30 m.
East Indo-Pacific: Cocos-Keeling Is. to Indonesia, Micronesia
and French Polynesia. – S. Japan to S.E. Australia.

Leopard Wrasse – IP
SIZE: to 10 cm (4 in.)
ID: White with leopard-like pattern of close-set brown to
black spots; irregular red bands on front of head. Lagoon and
seaward reefs in areas of mixed sand, rubble and coral.

Leopard Wrasse – JP
SIZE: to 2.5 cm (1 in.)
ID: Pale gray undercolor with network of brownish lines
covering head and body; **blue to black ocellated spot on rear
of dorsal fin and another on rear of anal fin,** 1st three spines
of dorsal fin are long and about the same length.

Wrasses

BLACK LEOPARD WRASSE *Macropharyngodon negrosensis*
SIZE: to 12 cm (4³/₄ in.)   Wrasses – Labridae
ID: TP – Greenish with black scale margins; **medial fins black with translucent tail that may have dark borders,** several pale band markings on head. Solitary or form small groups. Mixed sand, rubble and coral areas in 8-32 m.
West Pacific: Andaman Sea to Indonesia, Philippines, Micronesia, Solomon Is. – S.W. Japan to Great Barrier Reef and Fiji.

Black Leopard Wrasse – IP
SIZE: to 8 cm (3¹/₄ in.)
ID: Black with numerous small white spots, pale dorsal fin, translucent tail. **JP –** Similar with about 4 white saddles below dorsal fin.

ORNATE LEOPARD WRASSE *Macropharyngodon ornatus*
SIZE: to 11 cm (4¹/₄ in.)   Wrasses – Labridae
ID: TP – Black to orange or reddish body with **blue to green spot on each scale;** dark borders on tail. **IP –** Similar, but lack dark tail borders. Solitary or form small groups. Sand and rubble bottoms of coral reefs in 3-30 m.
East Indo-Asian Pacific: Maldives and Andaman Sea to West Papua in Indonesia. – S.W. Japan to W. Australia.

KUITER'S WRASSE *Macropharyngodon kuiteri*
SIZE: to 10 cm (4 in.)   Wrasses – Labridae
ID: TP – Whitish with orange scale margins; pale-edged black spot on rear gill cover, green bands on head. **IP –** Similar, but lack head bands. Solitary or form small groups. Base of dropoffs in deeper water to 55 m.
Southwestern Pacific: S. Great Barrier Reef and New Caledonia to New South Wales in S.E. Australia.

CHOAT'S WRASSE *Macropharyngodon choati*
SIZE: to 10 cm (4 in.)   Wrasses – Labridae
ID: TP – White with red blotches and red stripes on lower head; large dark spot surrounded by yellow on gill cover. Solitary or small groups. Lagoons, passes and seaward reefs in 6-28 m.
Localized: S. Great Barrier Reef to N. New South Wales in Australia.

LINED WRASSE *Anampses lineatus*
SIZE: to 12 cm (4³/₄ in.)   Wrasses – Labridae
ID: TP – Orange brown with pale stripes on body and **bluish spots on head;** black spot on rear edge of gill cover, yellow blotch behind pectoral fin. Solitary or form small groups. Lagoons and outer reefs in 20-45 m.
Indo-Asian Pacific: Red Sea and E. Africa to Andaman Sea, Sumatra, Java and Bali in Indonesia.

Lined Wrasse – IP
SIZE: to 12 cm (4 ³/₄ in.)
ID: Blackish with white stripes on body and white spots on head; white bar on tale base and black tail.

WHITESPOTTED WRASSE *Anampses melanurus*
SIZE: to 14.5 cm (5 ³/₄ in.) Wrasses – Labridae
ID: TP – Black with rows of bluish white spots, yellow foretail; **broad yellow irregular stripe extends from pectoral fin to tail.** Solitary or in pairs; bury in sand at night. Outer reefs and lagoons, most common in shallow but occasionally to 30 m.
Pacific: Indonesia, Philippines, S. Micronesia and Solomon Is. to French Polynesia. – S.W. Japan to Australia and Vanuatu.

Whitespotted Wrasse – IP
SIZE: to 10 cm (4 in.)
ID: Black with white spots or horizontal lines on sides; yellow foretail, black spot behind upper gill cover (also present on TP).

Whitespotted Wrasse – JP
SIZE: to 2.5 cm (1 in.)
ID: Dark brown to black with numerous white to yellow intricate spot and line markings; white wavey narrow bar on forebody behind pectoral fin; white lower edge on dorsal and tail fins.

YELLOWTAIL WRASSE *Anampses meleagrides*
SIZE: to 21 cm (8 ¹/₄ in.) Wrasses – Labridae
ID: TP – Reddish to greenish brown with vertical blue streak on each scale; **bluish crescent followed by white margin on tail.** Usually solitary or form small groups with females. Coral, rubble and sand of seaward reefs in 4-60 m.
Indo-Pacific: Red Sea and E. Africa to Indonesia, Philippines, Micronesia, Solomon Is., French Polynesia. – S. Japan to Australia.

Yellowtail Wrasse – IP
SIZE: to 10 cm (4 in.)
ID: Black with horizontal rows of white spots, yellow tail; young have white ringed spots on rear dorsal and anal fins.

Wrasses

FEMININE WRASSE *Anampses femininus*
SIZE: to 24 cm (9 ½ in.) Wrasses – Labridae
ID: TP – Dusky yellow to greenish brown body with blue line markings on scales; dark tail with blue markings, dark head with blue stripes. Form small groups. Coral and rocky reefs in 10-30 m.
Southern Pacific: S.E. Australia to Easter I. in southeastern Pacific.

Feminine Wrasse – IP
SIZE: to 24 cm (9 ½ in.)
ID: Yellow-orange becoming blue on tail base and tail; brilliant blue stripes on head and body. Named *femininus* because female more beautiful than male, an unusual characteristic in wrasses.

BLUE-SPOTTED WRASSE *Anampses caeruleopunctatus*
SIZE: to 42 cm (17 in.) Wrasses – Labridae
ID: TP – Green to brownish green with blue to blue-green vertical streak on each scale; yellow to lime-green bar on forebody. Solitary or with one or more females. Bury in sand at night. Shallow coral or rocky reefs to 30 m.
Indo-Pacific: Red Sea and E. Africa to Indonesia, Micronesia and French Polynesia. – S. Japan to Great Barrier Reef and W. Australia.

Blue-spotted Wrasse – IP
SIZE: to 42 cm (17 in.)
ID: Green to brownish green with horizontal rows of blue spots on side; blue bands on head, blue margins on fins. **JP –** Similar, but spots are white and more pronounced; wide white bar on tail base.

NEW GUINEA WRASSE *Anampses neoguinaicus*
SIZE: to 17 cm (6 ³/₄ in.) Wrasses – Labridae
ID: TP – Yellowish white with pale blue vertical lines on scales; **upper head dark with blue line markings,** salmon, blue and black spot behind edge of upper gill cover. Solitary. Coral-rich seaward slopes in 4-30 m.
West Pacific: E. Indonesia, Philippines and Micronesia. – S.W. Japan to Great Barrier Reef, New Caledonia and Fiji.

New Guinea Wrasse – IP
SIZE: to 13 cm (5 in.)
ID: Yellowish white with pale blue vertical lines or spots on scales and black back, dorsal and anal fins; blue ringed black spot behind upper edge of gill cover and on rear dorsal and anal fins. **JP –** Similar, but the row of blue spots along the base of the dorsal fin are larger than those of IP.

YELLOW-BREASTED WRASSE   *Anampses twistii*

SIZE: to 18 cm (7 in.)   Wrasses – Labridae

ID: TP – Purplish brown with horizontal rows of small white spots, **yellow lower head and forebody;** ragged yellow bar on midside. Solitary or in pairs. Lagoons and outer reefs in 3-30 m.

Indo-Pacific: Red Sea and E. Africa to Micronesia and French Polynesia. – S. Japan to Great Barrier Reef and W. Australia.

Yellow-breasted Wrasse – IP

SIZE: to 14 cm (5 1/2 in.)

ID: Purplish brown with horizontal rows of small white spots, yellow lower head and forebody; twin spots on rear of dorsal and anal fins; blue ringed black spot on rear dorsal and anal fins.

GEOGRAPHIC WRASSE   *Anampses geographicus*

SIZE: to 24 cm (9 1/2 in.)   Wrasses – Labridae

ID: TP – Brown to olive with thin dark bluish outlines on scales; **labyrinth of line markings on head,** narrow bluish borders on dorsal, anal and tail fins. Solitary. Mix with algae and soft corals on reef tops and slopes to 25 m.

West Pacific: Borneoin Indonesia to Micronesia. – S. Japan to Great Barrier Reef, W. Australia and Fiji.

Geographic Wrasse – IP

SIZE: to 15 cm (6 in.)

ID: Brown to olive with darkish outlines on scales; pale ringed black spot on rear dorsal and anal fins.

SHOULDERSPOT WRASSE   *Leptojulis cyanopleura*

SIZE: to 13 cm (5 in.)   Wrasses – Labridae

ID: TP – Green to blue-green; **diffuse orangish blotch above and behind pectoral fin,** blue-edged orange bands on head and tail. Form aggregations; feed on plankton 1-2 m above substrate. Sand and rubble areas and reefs in 6-75 m.

East Indo-Asian Pacific: Persian Gulf to Indonesia, Philippines, Solomon Is. and Great Barrier Reef.

Shoulderspot Wrasse – IP

SIZE: to 10 cm (4 in.)

ID: White with pair of yellowish brown stripes. Form aggregations mixed with a few TP.

Wrasses

SLINGJAW WRASSE   *Epibulus insidiator*
SIZE: to 35 cm (14 in.)   Wrasses – Labridae
ID: TP – Deep body; white head and dark body with orange back from head to midbody; diffuse yellow midbody bar, black outline around scales, black streak through eye. Solitary. Coral-rich areas of lagoons and outer or seaward reefs to 42 m.
Indo-Pacific: Red Sea and E. Africa to Indonesia, Micronesia, Solomon Is., Hawaii and French Polynesia. – S. Japan to Australia.

Slingjaw Wrasse – IP Yellow Variation
SIZE: to 26 cm (10 in.)
ID: Tan to dark brown or yellow. When feeding this species has the ability to extend its jaw structure forward a third the length of its body to form a suction tube.

LATENT SLINGJAW WRASSE   *Epibulus brevis*
SIZE: to 22 cm (8 3/4 in.)   Wrasses – Labridae
ID: TP – Tan to charcoal gray; lobes of tail often yellow or orange; **often yellow blotch below front of dorsal fin,** smaller lighter phases show dark blotch on pectoral fin which is absent in adults. Protected inshore waters to 18 m.
West Pacific: Indonesia, Philippines, Micronesia, Papua New Guinea, Solomon Is. and Fiji.

Latent Slingjaw Wrasse – IP
SIZE: 10 cm (4 in.)
ID: Yellow to white body with yellow head and fins also can be light to dark brown; dark margin on large body scales, smaller than male, **black pectoral fins;** lighter color phase show large dark blotch on pectoral fins, yellow blotch below front of dorsal fin.

PACIFIC BIRD WRASSE   *Gomphosus varius*
SIZE: to 32 cm (13 in.)   Wrasses – Labridae
ID: TP – Slender body with greatly elongate snout; blue-green head, green body, dark streak on pectoral fin. Solitary. Coral-rich areas of lagoons and seaward reefs to 35 m.
Pacific: Brunei, Indonesia, Philippines, Micronesia, Papua New Guinea, Solomon Is. to Hawaii and French Polynesia. – S. Japan to Australia and Fiji.

Pacific Bird Wrasse – IP
SIZE: to 15 cm (6 in.)
ID: Slender body with greatly elongate orangish snout; whitish head and breast, body gray gradating to a nearly black tail base. Similar appearing Indian Ocean Bird Wrasse, *G. caeruleus*, in Indian Ocean, including Andaman and Java Seas. JP on following page.

Pacific Bird Wrasse – JP
SIZE: to 15 cm (6 in.)
ID: Green upper body; wide white stripe from mouth to tail with wide black borders, wide black outer edge on dorsal and anal fins; white underside of head and belly.

BLACKEYE THICKLIP   *Hemigymnus melapterus*
SIZE: to 90 cm (3 ft.)   Wrasses – Labridae
ID: **TP –** White to greenish or goldish brown head and forebody, remainder of body dark; large dark spot behind eye, pale streak on most scales, extremely thick fleshy lips. Solitary. Mixed sand, rubble and coral areas to 30 m.
Indo-Pacific: Red Sea and E. Africa to Indonesia, Micronesia, Solomon Is. to French Polynesia. – S.W. Japan to Australia and Fiji.

Blackeye Thicklip – IP
SIZE: to 15 cm (6 in.)
ID: Pale gray head and dark gray to black remainder of body; yellow tail, often pale streak on most scales.

Blackeye Thicklip – JP
SIZE: to 5 cm (2 in.)
ID: **TP –** Elongate green body with numerous narrow violet bars and bands on head; often yellow bar on body behind pectoral fin. Solitary. Offshore reef slopes in 8-40m.

BARRED THICKLIP   *Hemigymnus fasciatus*
SIZE: to 40 cm (16 in.)   Wrasses – Labridae
ID: **TP/IP –** Black body with 5 narrow white bars; green head with pink bands, thick lips. Solitary or form small groups. Mixed sand, rubble and coral areas of lagoons, passes and outer slopes to 25 m, more common on sheltered reefs.
Indo-Pacific: Red Sea and E. Africa to Indonesia, Micronesia, Solomon Is., French Polynesia. – S. Japan to Great Barrier Reef and Fiji.

Barred Thicklip – JP
SIZE: to 6 cm (2 1/4 in.)
ID: Light green head and brownish to dark green body; 5-6 narrow pale bars. Solitary. Among branching corals of lagoons and outer reefs.

Wrasses

CIGAR WRASSE *Cheilio inermis*
SIZE: to 50 cm (20 in.) Wrasses – Labridae
ID: **TP –** Elongate; shades of green with several white spots along back. Solitary or with several females; often shadowed by other fish predators. Usually in weedy areas or seagrass beds of lagoons, reef flats and coastal reefs to 30 m.
Indo-Pacific: Red Sea to Indonesia, Philippines, Micronesia, Solomon Is., Hawaii and French Polynesia. – S.W. Japan to Australia.

Cigar Wrasse – IP Female
ID: Elongate. Often unmarked green to yellow or brown; occasionally with dark midlateral stripe.

Cigar Wrasse – JP
SIZE: to 10 cm (4 in.)
ID: Wide brown to green on top of body and at midbody separated by broad white stripe.

RING WRASSE *Hologymnosus annulatus*
SIZE: to 40 cm (16 in.) Wrasses – Labridae
ID: **TP –** Elongate bright green body with numerous narrow violet bars; violet to dark green bands and spots on head and below pectoral fin; occasionally yellow bar or vertical area on midbody. Solitary. Offshore reef slopes in 8-40 m.
Indo-Pacific: Red Sea and E. Africa to Indonesia, Philippines, Micronesia, French Polynesia. – S. Japan to E. & W. Australia.

Ring Wrasse – IP
SIZE: to 20 cm (8 in.)
ID: Elongate dark green body with numerous narrow dark purple bars; medium green head.

Ring Wrasse – JP
SIZE: to 12 cm (4³/₄ in.)
ID: Elongate; white to pale yellowish on back with narrow black stripe along dorsal-fin base, broad dark stripe covering most of side. Solitary. Sand and rubble areas.

PASTEL RING WRASSE *Hologymnosus doliatus*
SIZE: to 50 cm (20 in.) Wrasses – Labridae
ID: **TP** – Elongate; light green, blue and green markings on head, numerous blue body bars, wide pale bar with blue borders on forebody, bicolor spot on edge of gill cover. Solitary. Mixed sand, rubble and coral areas to 30 m.
Indo-West Pacific: E. Africa to Indonesia, Micronesia, Solomon Is. and Line Is. in eastern Central Pacific. – S. Japan to Australia and Fiji.

Pastel Ring Wrasse – IP
SIZE: to 25 cm (10 in.)
ID: Elongate; pale green, blue and green markings on head, numerous blue body bars (lack pale bar with blue borders of TP), bicolor spot on upper edge of gill cover.

Pastel Ring Wrasse – Intermediate JP/IP
SIZE: to 10 cm (4 in.)
ID: Elongate; whitish to pale green, 3 reddish stripes or rows of spots and thin bars on side. Solitary or form small groups. Sand and rubble bottoms.

Pastel Ring Wrasse – JP
SIZE: to 7 cm (2 3/4 in.)
ID: Elongate; yellowish with **3 red stripes.** Frequently form small groups. Sand and rubble bottoms.

HILOMEN'S WRASSE *Halichoeres hilomeni*
SIZE: to 12 cm (4 3/4 in.) Wrasses – Labridae
ID: **TP** – Dark greenish brown upper body often with 5-6 narrow pale to whitish bars; whitish lower body with two rows of squarish red spots, often darkish stripe from snout through eye to tail. Solitary or form small groups. Sheltered fringing reefs in 1-3 m.
Asian Pacific: Sabah in Malaysia to Philippines.

KNER'S WRASSE *Halichoeres kneri*
SIZE: to 10 cm (4 in.) Wrasses – Labridae
ID: **TP** – Gray back, white lower body; wide dusky midbody stripe from snout to tail, prominent ocellated spot on mid-dorsal fin, small black spot on pectoral fin base. Solitary or form small groups. Coastal reefs, often in turbid areas, in 2-12 m.
Asian Pacific: Brunei and W. Indonesia to Philippines and Japan.

Wrasses

CANARY WRASSE *Halichoeres chrysus*
SIZE: to 12 cm (4 3/4 in.) Wrasses – Labridae

ID: TP – Bright golden yellow; black spot on front of dorsal fin, occasionally dark spot behind eye, faint orange bands on head and breast. Form small groups. Sand and rubble edge of reefs in 2-70 m.

Asian Pacific: Brunei, Indonesia and Philippines to Micronesia. – S. Japan to N.W. and Great Barrier Reef.

CANARYTOP WRASSE *Halichoeres leucoxanthus*
SIZE: to 11 cm (4 1/4 in.) Wrasses – Labridae

ID: TP – Bright golden-yellow upper body, **white below. IP –** Similar, but with 3 black spots on dorsal fin. Form small groups. Sand and rubble fringes of coral reefs in 20-40 m.

East Indian Ocean: Maldives and Andaman Sea to Java and Bali in Indonesia.

PALE WRASSE *Halichoeres pallidus*
SIZE: to 8 cm (3 1/4 in.) Wrasses – Labridae

ID: TP – Pink with **yellow tail base;** large black spot on front of dorsal fin. **IP/JP –** Similar, but with 3 pale-rimmed black spots on dorsal fin. Form small groups. Steep outer reef slopes in 30-70 m.

West Pacific: E. Indonesia to Micronesia, Line Is. in eastern Central Pacific.

PASTEL-GREEN WRASSE *Halichoeres chloropterus*
SIZE: to 19 cm (7 1/2 in.) Wrasses – Labridae

ID: TP – Pastel green with lavender scale spots; lavender markings on head, may display small yellow patch behind eye. Solitary. Protected, silty reefs of lagoons, sheltered coasts and adjacent sand and rubble bottoms to 10 m.

Asian Pacific: Andaman Sea to Micronesia and Solomon Is. – Philippines to Great Barrier Reef.

Pastel-green Wrasse – IP
SIZE: to 12 cm (4 3/4 in.)

ID: Pale greenish to yellowish brown with silvery white belly; small black spot centered on most scales, faint, rear pointing "V" line markings on belly, yellow to green stripe from corner of mouth to rear gill cover, often dark oval spot on side. Solitary or form small groups.

Pastel-green Wrasse – JP
SIZE: to 6 cm (2 1/4 in.)

ID: White to pale green with 3-4 dark green or brown stripes; often horizontal rows of small black spots are evident, which persist into the subadult and IP stages. Form aggregations. Sand and rubble fringes of reefs.

PINSTRIPED WRASSE   *Halichoeres melanurus*
SIZE: to 11 cm (4 ¹/₄ in.)   Wrasses – Labridae
ID: TP – Alternating green to blue-green and orange stripes; 3-6 narrow blue-green bars on upper side, large yellow spot on pectoral fin base, **black tail tip.** Solitary or form small groups. Sheltered reefs to 15 m.
West Pacific: Brunei, Indonesia, Philippines and Micronesia. – S.W. Japan to Great Barrier Reef and Fiji.

Pinstriped Wrasse – IP
SIZE: to 11 cm (4 ¹/₄ in.)
ID: Alternating yellow to orange and blue stripes; large blue-edged black spot on mid-dorsal fin and another on upper tail base, **small black spot on front of dorsal fin,** similar IP Chain-lined Wrasse and Richmond's Wrasse [below] lack this spot.

CHAIN-LINED WRASSE   *Halichoeres leucurus*
SIZE: to 11 cm (4 ¹/₄ in.)   Wrasses – Labridae
ID: TP – Greenish brown with orange spots on scales forming stripes on side; blue submarginal band on tail, blue and green banding over orange head, may display blue blotch behind pectoral fin. Solitary or in pairs. Silty coastal reefs to 15 m.
Asian Pacific: Brunei to Indonesia, Philippines, Micronesia, Papua New Guinea and Solomon Is.

Chain-lined Wrasse – IP
SIZE: to 11 cm (4 ¹/₄ in.)
ID: Alternating green to blue and orange stripes from head to tail; pale-rimmed black spot on mid-dorsal fin, similar but smaller spot on upper tail base. IP Pinstripe Wrasse [above] and Richmond's Wrasse [below] may be best distinguished by presence of nearby males of same species.

RICHMOND'S WRASSE   *Halichoeres richmondi*
SIZE: to 19 cm (7 ¹/₂ in.)   Wrasses – Labridae
ID: TP – Pale yellowish green; blue stripes on head, blue-green stripes on body, **pectoral fin base yellow,** margin of tail blue. Solitary or form small groups. Sheltered reefs of lagoons, channels and shorelines to 15 m.
West Pacific: Brunei, Indonesia, Philippines, Micronesia, Papua New Guinea, Solomon Is., Vanuatu and Fiji.

Richmond's Wrasse – IP
SIZE: to 13 cm (5 in.)
ID: Alternating orange to brown with blue stripes; somewhat elongate and pointed snout, large blue-edged black spot on mid-dorsal fin, similar but slightly smaller marking on upper tail base. IP Chain-lined Wrasse and Pinstriped Wrasse are nearly identical and are best distinguished by presence of nearby males.

Wrasses

YELLOWFACE WRASSE   *Halichoeres solorensis*
SIZE: to 14 cm (5 1/2 in.)   Wrasses – Labridae
ID: TP – Mauve to gray to nearly black body, yellowish head with pink bands; black spot on pectoral fin base, **yellow-ringed black spot on front of dorsal fin.** Coral-rich areas of shoreline, lagoon and seaward reefs in 10-40 m.
Asian Pacific: Indonesia and Philippines.

Yellowface Wrasse – IP
SIZE: to 10 cm (4 in.)
ID: Slate gray to purplish with yellow head; dull yellow fins, ocellated spot on front of dorsal fin and another near middle, ocellated spot on center of tail base.

GREEN-HEADED WRASSE   *Halichoeres chlorocephalus*
SIZE: to 12 cm (4 3/4 in.)   Wrasses – Labridae
ID: IP – Purple with narrow orange stripes. **TP –** Dark purple with orange stripes on lime-green head. Form small groups. Silty lagoon and coastal reefs in 15-30 m.
Localized: N. Papua New Guinea.

BLACK-EARED WRASSE   *Halichoeres melasmapomus*
SIZE: to 14 cm (5 1/2 in.)   Wrasses – Labridae
ID: TP – Reddish to greenish body and red tail, green head with blue-edged orange bands; **large blue-edged black spot behind eye. IP –** Similar, but has gray body and spot on upper tail base. Solitary or form small groups. Steep outer slopes in 10-56 m.
Indo-Pacific: Cocos-Keeling to Indonesia, Philippines, Micronesia and French Polynesia. – Great Barrier Reef and W. Australia.

GOLDSTRIPE WRASSE   *Halichoeres hartzfeldii*
SIZE: to 20 cm (8 in.)   Wrasses – Labridae
ID: TP – Green; **bright yellow patch above pectoral fin base,** orange midbody stripe with 3 or 4 black spots above on rear body; purple stripe on base of dorsal fin. Form small groups. Sand and rubble bottoms near reefs in 10-40 m.
West Pacific: Brunei, Indonesia, Philippines, Micronesia, Great Barrier Reef and Fiji.

Goldstripe Wrasse – IP
SIZE: to 10 cm (4 in.)
ID: Pinkish upper body, whitish below; broad yellow to orange stripe from eye to tail base with black spot centered on tail base. Solitary or form small groups. Similar Goldstripe Wrasse, *H. zeylonicus*, restricted to Indian Ocean and W. Indonesia.

ZIGZAG WRASSE *Halichoeres scapularis*
SIZE: to 20 cm (8 in.) Wrasses – Labridae
ID: **TP** – Pale green with blue to lavender scale margins; blue to lavender bands on head, **large black diffuse stripe on forebody.** Solitary or form groups. Sand and rubble bottoms and seagrass beds near reefs to 20 m.
Indo-Asian Pacific: Red Sea and E. Africa to Micronesia and Solomon Is. – S. Japan to Great Barrier Reef and Vanuatu.

Zigzag Wrasse – IP/JP
SIZE: to 15 cm (6 in.)
ID: Whitish or pale green with continuous or interrupted zipper-like black stripe from head to tail base. Form groups. Sand and rubble bottoms or in weedy areas.

THREESPOT WRASSE *Halichoeres trimaculatus*
SIZE: to 27 cm (11 in.) Wrasses – Labridae
ID: **TP** – Pale yellowish green with lavender vertical streak on most scales; lavender bands on head, black spot on upper tail base, **dusky blotch on forebody with dark spot above.** Solitary or form groups. Sand and rubble with isolated coral heads to 18 m.
Pacific: Indonesia, Micronesia, Papua New Guinea, Solomon Is. to French Polynesia. – S. Japan to Great Barrier Reef.

Threespot Wrasse – IP
SIZE: to 20 cm (8 in.)
ID: White to pale green to pinkish scales with bluish markings, indistinct to green or lavender band markings on head, dark spot on upper tail base. Form aggregations. Sand and rubble areas.

REDHEAD WRASSE *Halichoeres rubricephalus*
SIZE: to 10 cm (4 in.) Wrasses – Labridae
ID: **TP** – Red head, often with green nape and behind gill cover, and bluish black body; narrow bright blue outer edge on all fins. Solitary or form small groups. Rich coral areas of lagoons, patch reefs and outer slopes in 2-40 m.
Localized: Togean Is. and Flores I. to West Papua in Indonesia.

Redhead Wrasse – IP
SIZE: to 8.5 cm (3 1/4 in.)
ID: Dull green to blue body with numerous orange stripes from head to tail; pair of black spots on dorsal fin, narrow bright blue outer edge on anal fin and bands on tail.

Wrasses

AXILSPOT WRASSE *Halichoeres podostigma*
SIZE: to 19 cm (7 1/2 in.)   Wrasses – Labridae
ID: IP – Pale brown upper head, white below and white tail base; black scales on body with contrasting yellow-brown margins, black spot on pectoral fin base and ventral fins. **TP –** Similar, but with green and blue wash on head. Solitary. Coastal and lagoon reefs in 2-12 m.
Asian Pacific: Sabah in Malaysia to Indonesia and Philippines.

Axilspot Wrasse – JP
SIZE: to 5 cm (2 in.)
ID: Green with black-edged white stripes on head and forebody; black spot on pectoral fin base and ventral fins. Solitary. Sand and rubble fringe of coral reefs in 2-12 m.

TWOTONE WRASSE *Halichoeres prosopeion*
SIZE: to 15 cm (6 in.)   Wrasses – Labridae
ID: TP – Purplish head and forebody gradating to yellow rear body; black spot on front of dorsal fin. **IP –** Similar in appearance. Solitary or form groups. Coral-rich areas of lagoons, patch reefs and outer slopes in 2-40 m.
West Pacific: Brunei, Indonesia, Philippines, Micronesia and Solomon Is. – S.W. Japan to Great Barrier Reef and Fiji.

Twotone Wrasse – JP
SIZE: to 5 cm (2 in.)
ID: White to yellow with 4 black stripes often have thin brown stripes between; large black spot on front of dorsal fin. Sand and rubble fringe of coral reefs.

ARGUS WRASSE *Halichoeres argus*
SIZE: to 11 cm (4 1/4 in.)   Wrasses – Labridae
ID: TP – Green with red margin around each scale; intricate pattern of curved pink, red or orange bands on head, often dark to nearly black margin on tail. Form aggregations. Seagrass and weed-covered reefs near shore to 5 m.
West Pacific: Andaman Sea, Indonesia, Philippines and Micronesia. – Taiwan to Great Barrier Reef and Fiji.

CIRCLE-CHEEK WRASSE *Halichoeres miniatus*
SIZE: to 14 cm (5 1/2 in.)   Wrasses – Labridae
ID: TP – Variegated pattern similar to Nebulous [next] and Weedy Surge Wrasse [following], but has **red to orange circular rather than diagonal orange marking below eye.** Occurs in groups. Sheltered rubble and coral reefs to 10 m.
Asian Pacific: Philippines to Solomon Is. – S. Japan to Great Barrier Reef.

NEBULOUS WRASSE   *Halichoeres nebulosus*
SIZE: to 12 cm (4³/₄ in.)   Wrasses – Labridae
ID: TP – Shades of green to mauve; ocellated spot on mid-dorsal fin, wide reddish blotch on belly. Boomerang-shaped orange marking below eye. Form groups. Shallow weedy areas near reefs 3-40 m.
Indo-Asian Pacific: Red Sea and E. Africa to Indonesia, Philippines, Papua New Guinea and Solomon Is. – S. Japan to Australia.

Nebulous Wrasse – IP
SIZE: to 10 cm (4 in.)
ID: Lack band marking below eye and has pale-edged dark spot on middorsal fin; two squarish red blotches on white belly.

WEEDY SURGE WRASSE   *Halichoeres margaritaceus*
SIZE: to 12 cm (4³/₄ in.)   Wrasses – Labridae
ID: TP – Shades of green with mauve scale spots forming a series of blotches along sides; **orange diagonal band on cheek. IP/JP –** Similar in appearance. Form groups. Reef flats, shallow tops of patch reefs in areas exposed to surge to 3 m.
Pacific: Brunei to Indonesia, Philippines, Micronesia, Solomon Is. and French Polynesia. – S. Japan to N.W. & S.E. Australia.

DUSKY WRASSE   *Halichoeres marginatus*
SIZE: to 12 cm (4³/₄ in.)   Wrasses – Labridae
ID: TP – Green to brownish, tail green with ornate central bar; narrow blue bands on head, stripes of joined dark blue spots on body. Solitary or form small groups. Coral-rich areas of lagoons and outer reefs to 30 m.
Indo-Pacific: Red Sea and E. Africa to Indonesia, Philippines, Micronesia and French Polynesia. – S. Japan to Great Barrier Reef.

BUBBLEFIN WRASSE   *Halichoeres nigrescens*
SIZE: to 14 cm (5¹/₂ in.)   Wrasses – Labridae
ID: TP – Green head and body with brownish purple bars on sides; lavender bands on head, **small dark spot between 6th-7th dorsal fin spines and small dark spot above pectoral fin base.** Form aggregations. Shallow weedy reefs in 3-10 m.
Indo-Asian Pacific: E. Africa to Micronesia. – Taiwan to Australia. Absent Papua New Guinea and Solomon Is.

Bubblefin Wrasse – IP
SIZE: to 10 cm (4 in.)
ID: Dark brown to nearly black upper body and whitish lower body; often row of 5-6 white spots on midbody, **small black spot on pectoral fin base.**

Wrasses

TIMOR WRASSE   *Halichoeres timorensis*
SIZE: to 12 cm (4 ³/₄ in.)   Wrasses – Labridae
ID: TP – Green with orange stripes and bands on head; several broken orange stripes on body, about 4 vague darkish bars on side. Solitary or form small groups. Coral and rocky reefs dominated by algal and soft coral growth in 5-15 m.
East Indo-Asian Pacific: Maldives Is. to Andaman Sea and West Papua in Indonesia.

Timor Wrasse – JP
SIZE: to 5 cm (2 in.)
ID: Bluish gray with orange stripes and bands on head; several broken orange stripes on body, ocellated spot on middle of dorsal fin and another on upper tail base.

CHECKERBOARD WRASSE   *Halichoeres hortulanus*
SIZE: to 27 cm (11 in.)   Wrasses – Labridae
ID: TP – Green with blue bar on each scale; mauve or orange bands on head, pale green area behind head, yellow spot below front of dorsal fin. Solitary. Sand patches of lagoons and seaward reefs to 35 m.
Indo-Pacific: Red Sea and E. Africa to Indonesia, Philippines, Micronesia, Solomon Is., French Polynesia. – S. Japan to Australia.

Checkerboard Wrasse – IP
SIZE: to 20 cm (8 in.)
ID: Bluish white body with blue bar on each scale, yellow tail; green and mauve, pink or orange bars on head, 2-3 yellow saddle spots on back, black patch under front of dorsal fin. Solitary. Sand patches of lagoons and seaward reefs to 35 m.

Checkerboard Wrasse – JP
SIZE: to 5 cm (2 in.)
ID: Starting with white snout, wide alternating black and white bars encircle body; yellowish tail, gold ringed black spot on mid-dorsal fin. Solitary. Sand patches of lagoons and seaward reefs to 35 m.

ORANGEFIN WRASSE   *Halichoeres melanochir*
SIZE: to 11 cm (4 ¼ in.)   Wrasses – Labridae
ID: TP – Purple with black spots and black scale margins on side; prominent black spot covering pectoral fin base. **JP –** Blue stripes and orange ventral fins. Solitary or in pairs or small groups. Sand and rubble fringe of coral reefs in 5-25 m.
Asian Pacific: Indonesia and Philippines. – S. Japan and Taiwan to W. Australia.

WISATA WRASSE   *Halichoeres binotopsis*
SIZE: to 12 cm (4³/₄ in.)   Wrasses – Labridae
ID: **TP** – Light green with 4-5 squarish dark bars on back; **red stripes on lower side,** bands on head and tail. Solitary or form small groups. **IP** – Body bars extend to underside; yellow tail. Coral-rich areas of coastal reefs and lagoons in 2-8 m.
Asian Pacific: Bali in Indonesia to Philippines and Papua New Guinea.

WEED WRASSE   *Halichoeres papilionaceus*
SIZE: to 10 cm (4 in.)   Wrasses – Labridae
ID: **TP** – Green with orange stripes on head and forebody; 4-5 broad dark bars on side and wide black margin on tail. Solitary or form small groups. Seagrass beds and weedy areas to 4 m.
Asian Pacific: Indonesia, Philippines, Micronesia, Papua New Guinea and Solomon Is.

DOUBLESPOT WRASSE   *Halichoeres biocellatus*
SIZE: to 12 cm (4³/₄ in.)   Wrasses – Labridae
ID: **TP** – Alternating red and green stripes on head fade onto forebody; darkish toward rear with 4 wide dusky bars. Form small groups. Seaward reefs in 6-35 m.
West Pacific: Indonesia, Philippines and Micronesia. – S.W. Japan to N.W. Australia and S. Great Barrier Reef to Fiji.

Doublespot Wrasse – IP
SIZE: to 5 cm (2 in.)
ID: Reddish with whitish to yellowish stripe from snout and above eye to forebody and another from snout and below eye to forebody; large ocellated spots on mid and rear dorsal fin. **JP** – Similar to IP, but white stripe extending from mouth, under eye to tail.

CLAUDIA'S WRASSE   *Halichoeres claudia*
SIZE: to 15 cm (6 in.)   Wrasses – Labridae
ID: **IP** – Reddish brown undercolor with greenish to white stripes extending from head to tail base; **two dark spots on dorsal fin, often ocellated;** small dark green spot behind eye. **TP** – Similar, but pale stripes are composed of rows of spots; has only one dark spot on dorsal fin.
East Indo-Pacific: Indonesia to Solomon Is. and French Polynesia.

ARENATUS WRASSE   *Oxycheilinus arenatus*
SIZE: to 21 cm (8¹/₄ in.)   Wrasses – Labridae
ID: Salmon to brown; **dark stripe from eye to tail,** orange base of pectoral fin, may change to white below dark stripe. Solitary and cryptic. Hide inside caves, recesses and protected areas of outer reefs and dropoffs in 25-46 m.
Indo-West Pacific: Red Sea to Indonesia, Philippines and Micronesia to Fiji and French Polynesia.

Wrasses

CHEEK-LINED WRASSE *Oxycheilinus digrammus*
SIZE: to 30 cm (12 in.)
Wrasses – Labridae
ID: Variable color and markings; most commonly shades of brown, green and red with wide bar markings on back and dark bar markings on each scale, **diagonal lines on lower gill cover**. Solitary. Coral-rich lagoons and seaward reefs in 3-60 m.
Indo-West Pacific: Red Sea and E. Africa to Micronesia. – S.W. Japan to Great Barrier Reef, New Caledonia and Fiji.

Cheek-lined Wrasse – IP
ID: Usaually shades of gray to brown and reddish brown with dark edges on scales, pale stripe from above eye and another from below eye to tail; white ventral fin with large red spot; **diagonal lines on lower gill cover.**

Cheek-lined Wrasse – JP
SIZE: to 5 cm (2 in.)
ID: Broad orange to brown midbody stripe from snout to tail fin base; two black spots toward rear center of body.

RINGTAIL WRASSE *Oxycheilinus unifasciatus*
SIZE: to 46 cm (18 in.)
Wrasses – Labridae
ID: Shades of green to brown with white belly (capable of rapid and intense color change); **brown band with red borders from eye to gill cover,** white bar across tail base. Solitary. Lagoon and outer reefs to 160 m.
East Indo-Pacific: Cocos-Keeling to Indonesia, Philippines, Micronesia, Hawaii and French Polynesia. – S.W. Japan to Australia.

Ringtail Wrasse – IP
SIZE: to 28 cm (11 in.)
ID: Purplish brown; wide band with red borders from eye to over gill cover, pink or reddish bands on head, white bar across tail base, dark patch on ventral fins.

Ringtail Wrasse – JP
SIZE: to 5 cm (2 in.)
ID: Red-brown with dark red scale outlines; pale band with red borders from eye to gill cover; often dark spot on tail base and narrow white bar at end.

CELEBES WRASSE *Oxycheilinus celebicus*
SIZE: to 20 cm (8 in.) Wrasses – Labridae
ID: **IP** – Variegated shades of brown with **several dark blotches on rear body;** pink to orange lines radiating from eye, elongate snout. Solitary. Coral-rich areas on slopes of lagoon and seaward reefs in 3-30 m.
Asian Pacific: Andaman Sea, Indonesia, Philippines, Papua New Guinea, Solomon Is. – S. Japan to W. Australia and New Caledonia.

ORIENTAL WRASSE *Oxycheilinus orientalis*
SIZE: to 18 cm (7 in.) Wrasses – Labridae
ID: Whitish with broad diffuse **reddish brown central stripe formed by squarish blotches;** small dark spot on rear body and another on tail base. Solitary or form small groups. Sand and rubble areas near reefs in 10-30 m.
Indo-Asian Pacific: Red Sea and E. Africa to Indonesia, Philippines, Micronesia and Solomon Is. – Japan to Australia.

BLACKMARGINED WRASSE *Oxycheilinus nigromarginatus*
SIZE: to 14 cm (5 ½ in.) Wrasses – Labridae
ID: Slender red-brown body; 5 thin white bars on upper body, white spotting on belly and **black tail margin.** Solitary. Rubble bottoms, frequently with soft corals in 8-16 m.
West Pacific: E. Australia to New Caledonia and Fiji.

TWOSPOT WRASSE *Oxycheilinus bimaculatus*
SIZE: to 15 cm (6 in.) Wrasses – Labridae
ID: **TP** – Variable from shades of red to brown to green; dark spot bordered in red on front of dorsal fin, black midbody spot, **yellow base of pectoral,** pointed tail. Form small groups. Rubble and weedy areas around rocky outcrops in 2-100 m.
Indo-Pacific: Red Sea and E. Africa to Indonesia, Philippines, Micronesia, Hawaii, French Polynesia. – S. Japan to Great Barrier Reef.

BLACKBAR WRASSE *Thalassoma nigrofasciatum*
SIZE: to 20 cm (8 in.) Wrasses – Labridae
ID: **TP** – Black head with white lower head and belly; short yellow band above pectoral fin, **2 white body bands with yellow edges,** yellow tail base. **IP** – Similar, with no yellow on bands. Solitary or form groups. Coastal and outer reefs to 10 m.
West Pacific: E. Papua New Guinea to Great Barrier Reef, New Caledonia and Fiji.

JANSEN'S WRASSE *Thalassoma jansenii*
SIZE: to 20 cm (8 in.) Wrasses – Labridae
ID: **TP** – Black upper head and body with white lower head and yellow belly; **5 narrow yellow bars on older adults** 2 wide bars on young (occasionally 1st bar white), narrow bar above pectoral fin; yellow front anal fin. Form groups in 1-10 m.
East Indo-Asian Pacific: Maldives to Indonesia, Philippines and Papua New Guinea. – S. Japan to W. Australia.

Wrasses

LADDER WRASSE   *Thalassoma trilobatum*
SIZE: to 30 cm (12 in.)   Wrasses – Labridae
ID: **TP –** Salmon-pink to orange; **two stripes formed by vertically elongate green to blue-green rectangles.** Solitary or in pursuit of females. Shallow reefs and rocky shores exposed to wave action to 10 m.
Indo-Pacific: E. Africa to Indonesia, Philippines, Micronesia, Solomon Is., French Polynesia. – S.W. Japan to Great Barrier Reef.

Ladder Wrasse – IP
SIZE: to 20 cm (8 in.)
ID: Greenish gray to pale green to green with 5-6 distinct or diffuse dark saddles on back; 2 distinct to diffuse dark stripes on side, dark vertical line markings on scales, line and spot markings on head, but lack the V-shaped mark on snout of similar Surge Wrasse IP [next page]. Form small fast-swimming groups.

SUNSET WRASSE   *Thalassoma lutescens*
SIZE: to 25 cm (10 in.)   Wrasses – Labridae
ID: Pink head with green bands, bluish green forebody, green rear and yellow-green tail; **yellow pectoral fin with blue outer edge.** Form groups. Sand, rubble and coral patches of lagoon and seaward reefs to 30 m.
Pacific: S. Japan, Indonesia, Papua New Guinea, Micronesia, E. Australia to French Polynesia. Absent Philippines.

Sunset Wrasse – IP
SIZE: to 20 cm (8 in.)
ID: Yellow head with red bands and light green body. **JP/Small IP –** Yellow. Both TP and IP have faint vertical lines on scales.

FIVESTRIPE WRASSE   *Thalassoma quinquevittatum*
SIZE: to 17 cm (6 ³/₄ in.)   Wrasses – Labridae
ID: **TP – Purple head with green bands,** green upper body, yellow below; pair of purple wavy-edged stripes on upper body, purple tail borders. Form aggregations. Lagoon and seaward reefs to 18 m, often in surge channels less than 5 m.
Indo-Pacific: E. Africa to Indonesia, Philippines, Micronesia, Solomon Is., Hawaii and French Polynesia. – S. Japan to Australia.

Fivestripe Wrasse – IP
SIZE: to 13 cm (5 in.)
ID: Green with purple and green bands on head, upper body and belly; pair of violet stripes on upper body, 3-4 faint white diagonal bars; red curving band extends from lower eye to gill cover edge. Form groups, frequently accompanied by a male.

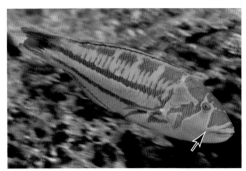

SURGE WRASSE *Thalassoma purpureum*

SIZE: to 43 cm (17 in.) Wrasses – Labridae

ID: TP – Green to blue; pink to **lavender bands on head form V-shaped marking on snout,** irregular purple stripe on back and pair of reddish stripes on side. Form groups. Surge zone of reef flats and rocky coasts to 5 m.

Indo-Pacific: Red Sea and E. Africa to Indonesia, Philippines, Micronesia, Solomon Is., Hawaii and Easter I. – S. Japan to Australia.

SIXBAR WRASSE *Thalassoma hardwicke*

SIZE: to 18 cm (7 in.) Wrasses – Labridae

ID: TP/IP – Pale green to whitish with **5-6 black saddles that gradually decrease in size towards tail;** pink bands on head, purple to black band on rear edge of gill cover. Form groups. Coastal, lagoon and outer reefs to 15 m.

Indo-Pacific: E. Africa to Indonesia, Philippines, Micronesia, Papua New Guinea, Solomon Is., French Polynesia. – S. Japan to Australia.

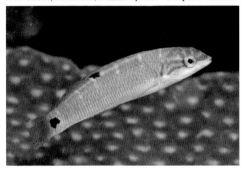

CRESCENT WRASSE *Thalassoma lunare*

SIZE: to 25 cm (10 in.) Wrasses – Labridae

ID: TP – Blue to blue-green; lavender to green bands on head, lavender pectoral fins with blue margin, **deep lunate tail with yellow center. IP –** Similar, but more green. Solitary or form groups. Coastal, lagoon and outer reefs to 20 m.

Indo-Pacific: Red Sea and E. Africa to Indonesia, Philippines, Micronesia to Line Is. – S. Japan to Australia and Fiji.

Crescent Wrasse – JP

SIZE: to 8 cm (3 in.)

ID: Medium orangish-brown and bluish underside; large black spot on base of tail and another on mid-dorsal fin; about 5 narrow pale bars on upper body (lost with age); green bands on head.

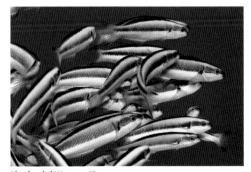

BLUNTHEADED WRASSE *Thalassoma amblycephalum*

SIZE: to 16 cm (6 1/4 in.) Wrasses – Labridae

ID: TP – Blue to green head, **wide yellow "collar,"** bluish to red body with vertical green lines on scales; two thin lines under eye, blue edge on outer pectoral fin. Form groups. Upper edge of lagoon and seaward reefs to 15 m.

Indo-Pacific: E. Africa to Andaman Sea, Indonesia, Philippines, Micronesia, Solomon Is., French Polynesia. – S. Japan to Australia.

Bluntheaded Wrasse – IP

SIZE: to 10 cm (4 in.)

ID: Pale green back with broad dark brown midlateral stripe, white belly. Form harems consisting of a single TP and numerous IP. Commonly spawn in large groups in the late afternoon.

Wrasses

REDSHOULDER WRASSE   *Stethojulis bandanensis*
SIZE: to 12 cm (4 3/4 in.)   Wrasses – Labridae
ID: TP – Green to brownish gray upper body, pale belly; bright red-orange patch above pectoral fin, **blue to green stripe markings curve to touch top of eye.** Form groups. Reef flats and coastal shallows to 20 m, but usually less than 3 m.
East Indo-Pacific: Cocos-Keeling Is. and W. Australia to Indonesia, Micronesia and French Polynesia. – S. Japan to Great Barrier Reef.

Redshoulder Wrasse – IP
SIZE: to 9 cm (3 1/2 in.)
ID: Dark gray with small white dots on upper body; markings on scales of lower body form diamond-like netted pattern, often pair of whitish stripes extend from head onto body, small bright red-orange patch above pectoral fin base. Form groups.

THREELINE WRASSE   *Stethojulis strigiventer*
SIZE: to 11 cm (4 1/4 in.)   Wrasses – Labridae
ID: TP – Green back, pale belly with **dark green to yellow midbody stripe bordered with thin blue stripes** and another on back; black spot on rear gill cover, another on tail base. Form groups. Sand mixed with seagrass or algae or reef flats to 3 m.
Indo-Pacific: E. Africa to Indonesia, Philippines, Micronesia and French Polynesia. – S. Japan to Australia.

Threeline Wrasse – IP
SIZE: to 10 cm (4 in.)
ID: Greenish with several thin white stripes on lower body; pale stripe under eye from snout tip to edge of gill cover, black spot on rear dorsal fin. **JP –** White stripe from above eye to upper tail base.

FOURLINE WRASSE   *Stethojulis trilineata*
SIZE: to 14 cm (5 1/2 in.)   Wrasses – Labridae
ID: TP – Shades of green with red dorsal fin; yellow-orange around pectoral fin and occasionally on back, **4 blue stripes, 3 from head to tail and one from snout to above pectoral.** Solitary. Shallow reefs to 8 m.
East Indo-West Pacific: Maldives to Indonesia, Philippines, Micronesia and Solomon Is. – S. Japan to Australia and Fiji.

Fourline Wrasse – IP
SIZE: to 10 cm (4 in.)
ID: Dark gray with small white dots on upper body, **pale gray lower body;** white stripe below eye merges into a salmon stripe with a middle row of black spots from pectoral fin to lower tail base.

CUTRIBBON WRASSE   *Stethojulis interrupta*
SIZE: to 12 cm (4 3/4 in.)   Wrasses – Labridae
ID: TP – Orangish rust upper body, green below; pair of blue to green stripes, one on back, other extends from snout to tail usually interrupted on forebody. Form groups. Reef flats and coastal shallows to 18 m, but usually less than 3 m.
Indo-Asian Pacific: Red Sea and E. Africa to Indonesia, Papua New Guinea and Solomon Is. – Philippines to Australia.

Cutribbon Wrasse – IP
SIZE: to 12 cm (4 3/4 in.)
ID: Green to yellow upper body with pale spots, white below with darkish spots; thin white, yellow to orange stripe from snout passes below eye and onto forebody.

Cutribbon Wrasse – JP
SIZE: to 5 cm (2 in.)
ID: Brown back white below with dark brown to black midlateral stripe.

SOUTHERN WRASSE   *Stethojulis notialis*
SIZE: to 10 cm (4 in.)   Wrasses – Labridae
ID: IP – Green back gradating to pale underside with fine dark reticulations; broad yellow stripe from upper head to upper tail base. **TP –** Similar but with 4 dark bars on midside and orange marking above pectoral fin. Small groups. Mixed bottoms of sand, rubble, weeds, and coral, usually less than 6 m.
West Pacific: New Caledonia and Norfolk I. to Fiji

TAILSPOT WRASSE   *Pseudocheilinus ocellatus*
SIZE: to 8.5 cm (3 1/4 in.)   Wrasses – Labridae
ID: TP – Magenta to red body; often display thin vertical white lines on side, white-ringed black ocellated spot on tail base, curved magenta spot surrounded by yellow markings below eye. Solitary. Outer reef areas in 20-60 m.
Pacific: S.W. Japan, Micronesia and Coral Sea to Pitcairn Is. east of French Polynesia.

DISAPPEARING WRASSE   *Pseudocheilinus evanidus*
SIZE: to 9 cm (3 1/2 in.)   Wrasses – Labridae
ID: Red to orange with thin white lines on side; white band under eye, bluish vertical streak on cheek, occasionally with 5-6 diffuse pale bars. Solitary and cryptic. Rubble and coral patches on seaward slopes in 6-61 m.
Indo-Pacific: Red Sea and E. Africa to Indonesia, Philippines, Hawaii and French Polynesia. – S. Japan to Australia.

Wrasses

SIXSTRIPE WRASSE   *Pseudocheilinus hexataenia*
SIZE: to 7.5 cm (3 in.)   Wrasses – Labridae
ID: Violet with 6 orange stripes on side; **small black spot on upper tail base.** Solitary or form small groups; cryptic, hide among coral branches. Lagoons and seaward reefs in 2-35 m.
Indo-Pacific: Red Sea and E. Africa to Indonesia, Philippines, Solomon Is., Hawaii and French Polynesia. – S. Japan to Australia.

FOURSTRIPE WRASSE   *Pseudocheilinus tetrataenia*
SIZE: to 7.5 cm (3 in.)   Wrasses – Labridae
ID: Four orange to green stripes alternate with 3 blue stripes on back; bluish to greenish to brownish belly. Solitary; cryptic, hide in small recesses of coral or rubble. Seaward reefs in 6-44 m.
Pacific: Bonin Is. in S. Japan to Micronesia, Hawaii and French Polynesia.

EIGHTSTRIPE WRASSE   *Pseudocheilinus octotaenia*
SIZE: to 11.5 cm (4 1/2 in.)   Wrasses – Labridae
ID: Reddish brown to yellowish with eight thin dark stripes extending from behind gill cover to base of tail; numerous small yellow spots on cheeks. Solitary and cryptic. Coral or rubble patches of coastal and offshore reefs in 5-50 m.
Indo-Pacific: Madagascar to Indonesia, Philippines, Micronesia, Solomon Is., Hawaii and French Polynesia. – S. Japan to Australia.

MIDDLESPOT WRASSE   *Pseudojuloides mesostigma*
SIZE: to 9 cm (3 1/2 in.)   Wrasses – Labridae
ID: TP – Greenish brown with **large black patch on side that extends onto pale yellow dorsal fin;** black tail. **Nuptial Male –** Displays neon blue scales on upper body and lines on head. Solitary or pairs. Mixed rubble and coral areas in 25-45 m.
Localized: Bali, Triton Bay, West Papua in Indonesia and Luzon in the Philippines.

ROYAL WRASSE   *Pseudojuloides severnsi*
SIZE: to 12 cm (4 3/4 in.)   Wrasses – Labridae
ID: TP – Large dark patch from eyes to midbody, blue, green and yellow behind; violet snout joins violet stripe below eye to beyond pectoral base, blue margins on tail. **IP –** Pink. Sand and rubble bottoms in 20-40 m.
East Indo-Asian Pacific: Sri Lanka to Indonesia and Philippines, north to S.W. Japan.

ATAVIA WRASSE   *Pseudojuloides atavai*
SIZE: to 13 cm (5 in.)   Wrasses – Labridae
ID: IP – Upper head and back reddish brown, white below separated by bicolor black and pale blue stripe. **TP –** Pink forebody with orange spots or vertical streaks and bands on head, gray rear body and tail with white triangular rear margin. Solitary or groups. Outer reefs in 12-31 m.
South Central Pacific: Micronesia to French Polynesia.

SPLENDID PENCIL WRASSE   *Pseudojuloides splendens*
SIZE: to 12 cm (4³/₄ in.)   Wrasses – Labridae
ID: **TP –** Green upper body, blue below; **blue and yellow midbody stripes,** wide black tail margin. Form small groups. Rubble, weed and coral areas of lagoon and seaward reefs to 61 m, usually over 20 m.
West Pacific: Indonesia, Philippines and Micronesia to Fiji. – S. Japan to Australia and Samoa.

Splendid Pencil Wrasse – IP
SIZE: to 10 cm (4 in.)
ID: Red to pink without markings.

KALEIDOS WRASSE   *Pseudojuloides kaleidos*
SIZE: to 10 cm (4 in.)   Wrasses – Labridae
ID: **TP –** Green to bluish; **broad blackish stripe on upper side with narrower blue stripe immediately below,** bright blue snout followed by salmon stripe down back. Solitary or with groups of IP. Rubble bottoms in 15-40 m.
East Indo-Asian Pacific: Maldives and Andaman Sea to Indonesia.

Kaleidos Wrasse – IP
SIZE: to 10 cm (4 in.)
ID: Red upper body, yellow to whitish below; dark reddish upper head. Form groups on rubble bottoms, often accompanied by at least one TP.

MIDGET WRASSE   *Pseudocheilinops ataenia*
SIZE: to 5 cm (2 in.)   Wrasses – Labridae
ID: **TP –** Red with 7-8 rose to orange stripes; **blue on ventral fins.** Form small groups; cryptic, stay close to shelter. Rubble and coral patches on protected reefs in 5-15 m.
Asian Pacific: Sulawesi and Flores in Indonesia, Philippines, Micronesia, Papua New Guinea and Solomon Is.

Midget Wrasse – IP
SIZE: to 4 cm (1¹/₂ in.)
ID: Rose to pink with whitish lower head and belly; 7-8 lavender stripes, darkish spot on ventral fins and bright yellow iris.

CRYPTIC WRASSE *Pteragogus cryptus*
SIZE: to 9.5 cm (3 3/4 in.) Wrasses – Labridae
ID: **TP/IP** – Mottled shades of red-brown with scattered dark spots; **white stripe from upper eye to above pectoral,** pale ocellated spot on gill cover. Solitary. Cryptic; branching stony corals, soft corals and weeds to 67 m.
Indo-West Pacific: E. Africa to Indonesia, Philippines, Micronesia and Solomon Is. – Taiwan to Australia and Fiji.

COCKEREL WRASSE *Pteragogus enneacanthus*
SIZE: to 15 cm (6 in.) Wrasses – Labridae
ID: **TP/IP** – Mottled shades of red-brown; **dark spots on lateral line (may join to form thin stripe),** ocellated brown spot on gill cover. **TP** – Filaments on 1st two dorsal rays. Solitary and cryptic. Among stony and soft corals and weeds in 3-25 m.
Asian Pacific: Indonesia, Philippines, Papua New Guinea and Solomon Is. – Taiwan to Great Barrier Reef.

FLAGFIN WRASSE *Pteragogus flagellifer*
SIZE: to 20 cm (8 in.) Wrasses – Labridae
ID: **TP/IP** – Shades of green to yellow-orange with grayish scale spots; yellow-green lips and purplish markings on cheek, deep body. **TP** – 1st 2 dorsal spines with long filaments. Solitary and cryptic. Among branches of soft coral or weedy areas in 2-30 m.
Indo-Asian Pacific: Red Sea and E. Africa to Papua New Guinea. – S. Japan to Great Barrier Reef and Vanuatu.

NEILL'S HOGFISH *Bodianus neilli*
SIZE: to 20 cm (8 in.) Wrasses – Labridae
ID: Reddish brown head and upper forebody, white lower and rear body; large blackish and red blotch on middle of dorsal and anal fins. Solitary. Coastal reefs in 2-15 m.
Indian Ocean: Maldives to Andaman Sea.

AXILSPOT HOGFISH *Bodianus axillaris*
SIZE: to 22 cm (8 3/4 in.) Wrasses – Labridae
ID: Purplish brown head and forebody, white rear body; **large black spot on base of pectoral fin and on rear dorsal and anal fins.** Solitary, rarely form small groups. Clear water lagoons and outer reefs in 2-40 m.
Indo-Pacific: E. Africa to Andaman Sea, Indonesia, Philippines, Micronesia, Solomon Is. to French Polynesia. – S. Japan to Australia.

Axilspot Hogfish – JP
SIZE: to 6 cm (2 1/4 in.)
ID: Dark brown to black; double row of large white spots and white snout. Solitary; occasionally act as cleaners. Shelter inside caves and crevices.

BLACKBELT HOGFISH   *Bodianus mesothorax*
SIZE: to 19 cm (7 ¹/₂ in.)   Wrasses – Labridae
ID: Purplish brown head with **wide black band on forebody,** white to yellowish rear body; black spot on pectoral fin base. Lack black spots like similar Axilspot Hogfish [previous]. Solitary or small groups. Outer reef slopes and passes in 4-40 m.
West Pacific: Andaman Sea to Indonesia, Micronesia, Papua New Guinea and Solomon Is. – S. Japan to Great Barrier Reef and Fiji.

Blackbelt Hogfish – JP
SIZE: to 6 cm (2 ¹/₄ in.)
ID: Purple to nearly black; double row of large black-edged yellow spots. Solitary. Shelter in caves and under ledges.

SADDLEBACK HOGFISH   *Bodianus bilunulatus*
SIZE: to 55 cm (22 in.)   Wrasses – Labridae
ID: Pale reddish with red stripes on head and behind eye; white and black patches below eye, **large black spot below rear dorsal fin,** yellowish tail. Solitary; feed on benthic invertebrates. Lagoons and outer reef slopes in 8-108 m.
Indo-Pacific: E. Africa to Indonesia, Philippines and Micronesia. – S.W. Japan to Australia.

Saddleback Hogfish – JP
SIZE: to 10 cm (4 in.)
ID: Yellow upper head and forebody; white below with thin red stripes, black rear body and white tail base. Solitary; occasional act as cleaners.

BLACKFIN HOGFISH   *Bodianus loxozonus*
SIZE: to 40 cm (16 in.)   Wrasses – Labridae
ID: Red to yellow upper head and back, pale below; numerous pale thin stripes run from head to tail, **large diagonal black area across tail base,** ventral fins and border of anal fin black. Solitary. Lagoon and seaward reef slopes in 3-40 m.
Pacific: Vietnam to Indonesia, Philippines, Micronesia, Solomon Is. and French Polynesia. – S.W. Japan to E. Australia.

Blackfin Hogfish – JP
SIZE: to 10 cm (4 in.)
ID: Front half of body yellow, dark rear body; whitish tail base, body marked with small white spots arranged into numerous thin stripes.

Wrasses

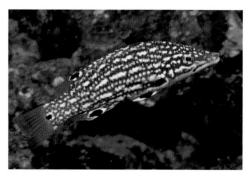

REDFIN HOGFISH *Bodianus dictynna*
SIZE: to 25 cm (10 in.) Wrasses – Labridae
ID: Purple to red head and back, yellow to whitish below; 4-5 white spots on back, large black spots on ventral and anal fins. Formerly identified as *B. diana*, an Indian Ocean species. Solitary or in pairs. Outer reef slopes in 9-35 m.
West Pacific: Indonesia, Philippines, Micronesia and Papua New Guinea. – Japan to Great Barrier Reef and Fiji.

Redfin Hogfish – JP
SIZE: to 7.5 cm (3 in.)
ID: Maroon undercolor with indistinct stripes formed by numerous white spots and blotches; black spots on ventral and anal fins. Solitary. Commonly around black coral or gorgonian fans. A closely related and nearly identical species, *B. diana*, occurs in Indian Ocean and Red Sea.

IZU HOGFISH *Bodianus izuensis*
SIZE: to 11 cm (4 1/4 in.) Wrasses – Labridae
ID: Red upper head and back, white lower half of head; large black spot on upper rear gill cover, pair of irregular black stripes on back and midbody, red to yellow stripe from pectoral fin to yellowish tail. Solitary. Sand and rubble adjacent outer reefs in 30-70 m.
Asian Pacific: Indonesia, S. Japan, E. Australia and New Caledonia.

LYRETAIL HOGFISH *Bodianus anthioides*
SIZE: to 21 cm (8 1/4 in.) Wrasses – Labridae
ID: Brown head and forebody, white rear body; scattered brown spots on white areas, deeply forked tail. Solitary; feed on benthic invertebrates. Steep outer slopes, often adjacent to dropoffs in 6-25 m.
Indo-Pacific: Red Sea and E. Africa to Indonesia, Philippines, Micronesia, Solomon Is., French Polynesia. – S. Japan to Australia.

TWOSPOT SLENDER HOGFISH *Bodianus bimaculatus*
SIZE: to 8 cm (3 1/4 in.) Wrasses – Labridae
ID: Small and slender; variable from yellowish upper body and pink below to red upper body and white below; may have thin red stripes, **black spot on rear gill cover** and often another on tail base. Form small groups. Steep outer reef slopes in 30-70 m.
Indo-Asian Pacific: Mauritius to Philippines, Micronesia, Papua New Guinea. – S. Japan to Great Barrier Reef and Fiji.

GOLDSPOT HOGFISH *Bodianus perditio*
SIZE: to 15 cm (6 in.) Wrasses – Labridae
ID: JP – Yellow; large black patch on rear back preceded by white bar. **TP –** Reddish; same markings as JP except bar is yellowish, to 80 cm (2 3/4 ft.). Solitary. Sand and rubble patches of seaward reefs in 13-40 m.
West Pacific: Australia to New Caledonia and French Polynesia.

REDSTRIPE HOGFISH *Bodianus opercularis*
SIZE: to 18 cm (7 in.) Wrasses – Labridae
ID: Slender body; bold red and white candy-stripe pattern, black spot on rear edge of gill cover. Solitary or form small groups. Rubble bottoms on or adjacent to steep outer reef slopes in 35-70 m.
Indian Ocean: Red Sea, Kenya, Madagascar, Mauritius and Christmas I. to Micronesia.

CRESCENT-TAIL HOGFISH *Bodianus sepiacaudus*
SIZE: to 10 cm (4 in.) Wrasses – Labridae
ID: Red with white belly; 2 narrow white stripes from snout to tail, large black spot on tail base extends onto tail and is bordered by crescents of yellow, red and black, black spot on upper rear gill cover. Deep reefs in 25-75 m.
Pacific: Indonesia to Fiji and Line Is.

ROCKMOVER WRASSE *Novaculichthys taeniourus*
SIZE: to 30 cm (12 in.) Wrasses – Labridae
ID: TP – Pale gray head, dark brown to black body with pale spot on each scale; white bar on tail base, may display lines radiating from eye. Solitary; turn over rocks in search of prey. Rubble bottoms next to reefs to 20 m.
Indo-Pacific: Red Sea and E. Africa to Micronesia, Solomon Is., Hawaii and French Polynesia. – S.W. Japan to Australia.

Rockmover Wrasse – JP
SIZE: to 8 cm (3 ¼ in.)
ID: "Antlers" at front of dorsal fin; green to brown to maroon body with white irregular spots and 3 black bars; white banded head. Effectively mimic bits of drifting weed with "to and fro" swimming motion.

SEAGRASS WRASSE *Novaculoides macrolepidotus*
SIZE: to 16 cm (6 ¼ in.) Wrasses – Labridae
ID: TP – Bright green with dark band and lines radiating from eye; several black spots usually just before tail base. **IP –** Similar, with dark midlateral stripe. Form aggregations in seagrass and weed beds. Lagoons and coastal areas to 10 m.
Indo-West Pacific: Red Sea and E. Africa to Micronesia. – S.W. Japan to Great Barrier Reef and Fiji.

KNIFE RAZORFISH *Cymolutes praetextatus*
SIZE: to 20 cm (8 in.) Wrasses – Labridae
ID: TP – Blunt snout; white, often with green to brown tint; faint irregular brown bars with white line running down back, may display broad yellow stripe. Solitary; dive into sand when alarmed and to sleep at night. Extensive sand or weedy areas near reefs or lagoons in 2-10 m.
Indo-Pacific: E. Africa to Micronesia and French Polynesia.

Wrasses

COLLARED RAZORFISH   *Cymolutes torquatus*
SIZE: to 20 cm (8 in.)   Wrasses – Labridae
ID: **TP –** Blunt snout; pale green to brown with numerous thin body bars, slanting dark bar just above pectoral fin "collar". Solitary; dive into sand when alarmed and to sleep at night. Extensive sand bottoms near reefs or lagoons in 2-15 m.
Indo-Asian Pacific: E. Africa to Indonesia, Philippines and Papua New Guinea. – S. Japan to Australia and New Caledonia.

Collared Razorfish – Variation
ID: Shades of brown with numerous narrow dusky brown bars; dark "collar" bar blends in with other bars.

WHITEPATCH RAZORFISH   *Iniistius aneitensis*
SIZE: to 24 cm (9 1/2 in.)   Wrasses – Labridae
ID: **TP –** Steep blunt snout; pale gray; **large white patch on lower forebody,** may display 3-4 dusky bars. Dive into sand when alarmed and to sleep at night. Open sand areas near reefs in 12-92 m.
East Indo-West Pacific: Maldives to Indonesia, Micronesia and Hawaii. – S.W. Japan and to Great Barrier Reef and Fiji.

Whitepatch Razorfish – JP
SIZE: to 6 cm (2 1/4 in.)
ID: Pale gray; 3 dark body bars and another behind eye, dark foretail, single or double spots on rear dorsal fin above two rear bars, some elongation of spinous dorsal fin forms "flag" when raised. Dive into sand when alarmed and to sleep at night.

Whitepatch Razorfish – JP Variation
SIZE: to 6 cm (2 1/4 in.)
ID: Green or pale yellow or dark brown to black without distinctive markings.

FINSPOT RAZORFISH   *Iniistius melanopus*
SIZE: to 25 cm (10 in.)   Wrasses – Labridae
ID: **IP –** Light gray with large **yellow-edged white patch** on lower side just behind pectoral fin. **TP –** Similar, but with elongate black spot on rear anal fin. Solitary or form loose groups. Open sand areas near reefs in 6-15 m.
Asian Pacific: Bali, Sulawesi and West Papua in Indonesia to Philippines, north to Japan.

FIVEFINGER RAZORFISH *Iniistius pentadactylus*
SIZE: to 25 cm (10 in.) Wrasses – Labridae
ID: TP – Greenish gray; **4-5 overlapping red spots** (appear dark underwater) above pectoral fin, often whitish area followed by dark blotch on forebody, steep blunt head. Solitary or form loose groups. Sand to mud slopes in 4-30 m.
Indo-Asian Pacific: Red Sea and E. Africa to Indonesia, Micronesia and Papua New Guinea. – S. Japan to Great Barrier Reef.

Fivefinger Razorfish – IP
SIZE: to 20 cm (8 in.)
ID: Steep blunt head; green to gray; **large white patch above belly with yellow to red-edged scales.** Similar TP Whitepatch Razorfish [previous page] lack these markings. Solitary or form loose groups. Both TP/IP dive into sand or mud bottom when alarmed and to sleep at night.

PEACOCK RAZORFISH *Iniistius pavo*
SIZE: to 35 cm (14 in.) Wrasses – Labridae
ID: TP – Whitish steep blunt head; light gray; **white patch on lower forebody with one or two black spots above,** may display dusky bars. Solitary; dive into sand when alarmed and to sleep at night. Open sand areas near reefs in 2-100 m.
Indo-Pacific: Red Sea and E. Africa to Indonesia, Philippines, Papua New Guinea, Solomon Is., French Polynesia. – S. Japan to Australia.

Peacock Razorfish – Subadult
SIZE: to 13 cm (5 in.)
ID: Long 1st dorsal spines; steep blunt head, whitish, white patch on lower forebody with dark spot above, faint bars, dark spots on mid and rear dorsal fin. Solitary or form loose groups. Open sand areas near reefs in 2-20 m.

Peacock Razorfish – JP Variation
SIZE: to 8 cm (3 1/4 in.)
ID: Long 1st dorsal fin spines; steep blunt head, brown with 3-4 dark bars on head, occasionally scattered black spots and 2-3 vertical rows of white ocellated spots on body. Solitary or form loose groups. Young mimic drifting plant debris. Open sand near reefs in 2-20 m.

Peacock Razorfish – Young JP
SIZE: to 2.5 cm (1 in.)
ID: Long 1st dorsal fin spines; translucent white with a few gold to brown line markings on head and body. Solitary or form loose groups. Open sand areas near reefs in 2-20 m.

Wrasses

CELEBES RAZORFISH   *Iniistius celebicus*
SIZE: to 16.5 cm (6 ½ in.)   Wrasses – Labridae
ID: TP – Pale gray back and white below; large black patch on center of forebody, horizontally **elongate black streak on tail base.** Solitary or form loose groups. Open sand areas near reefs in 7-15 m.
West Pacific: Indonesia, Philippines, Taiwan, Micronesia, Fiji and Hawaii.

Celebes Razorfish – JP
SIZE: to 8 cm (3 ¼ in.)
ID: Large dark brown patch behind head; elongate dark spot on tail base, often two dark bars toward rear of body.

CHISELTOOTH WRASSE   *Pseudodax moluccanus*
SIZE: to 25 cm (10 in.)   Wrasses – Labridae
ID: TP – Reddish brown to blue-green undercolor with dark spots on scales; orange to red to rust wash on forebody and back, yellow upper lip, black tail, occasionally pale bar on tail base. Solitary. Seaward reefs and outer slopes in 3-40 m.
Indo-Pacific: Red Sea and E. Africa to Indonesia, Philippines, Micronesia, Solomon Is., French Polynesia. – S. Japan to Australia.

Chiseltooth Wrasse – Subadult
SIZE: to 8 cm (3 ¼ in.)
ID: Bluish green undercolor with dark scale row stripes; blue to white stripe from snout runs on back to near black tail, a 2nd stripe on lower head and belly, pale bar across tail base. Act as cleaners.

YELLOWTAIL TUBELIP   *Diproctacanthus xanthurus*
SIZE: to 10 cm (4 in.)   Wrasses – Labridae
ID: TP/IP – White with yellow tail; dark brown stripe on back, black stripe from snout through eye to tail. **JP –** Similar, but black tail; act as cleaners. Solitary. Sheltered reefs in 3-25 m.
Asian Pacific: Indonesia, Philippines, Micronesia to Solomon Is., Great Barrier Reef and New Caledonia.

TUBELIP WRASSE   *Labrichthys unilineatus*
SIZE: to 17.5 cm (7 in.)   Wrasses – Labridae
ID: TP – Shades of green from dark olive to brownish green; wide yellowish to white bar on forebody behind pectoral fin, numerous thin blue stripes on body and narrow blue borders on fins. Solitary or form small groups. Sheltered coral-rich reefs to 20 m.
Indo-West Pacific: E. Africa to Indonesia, Philippines, Micronesia and Solomon Is. – S.W. Japan to Australia to Fiji.

Tubelip Wrasse – IP
SIZE: to 14 cm (5 1/2 in.)
ID: Shades of blue to green or brown, yellow lips; numerous thin blue stripes on body, narrow blue border on tail. Forage in coral-rich area.

Tubelip Wrasse – JP
SIZE: to 8 cm (3 1/4 in.)
ID: Dark brown to nearly black with thin white stripe from snout to tail; small juveniles have an additional stripe along lower edge of body. Forage around branching coral patches.

ALLEN'S TUBELIP *Labropsis alleni*
SIZE: to 10 cm (4 in.) Wrasses – Labridae
ID: TP – Slender; brown head, yellow green midbody, white rear and tail, large ocellated spot at pectoral base, small black spot on front of dorsal fin and rear belly. Solitary or in pairs. Steep slopes of lagoon and seaward reefs in 4-52 m.
West Pacific: Indonesia, Philippines, Micronesia, Solomon Is. to New Caledonia, Vanuatu and Fiji.

Allen's Tubelip – IP
SIZE: to 6 cm (2 1/4 in.)
ID: Slender; brown head, yellow green midbody, white rear body and tail, large pale-edged black spot at pectoral base and on front of dorsal fin and rear belly, pair of yellow stripes from snout to tail base.

WEDGE-TAILED WRASSE *Labropsis xanthonota*
SIZE: to 13 cm (5 in.) Wrasses – Labridae
ID: TP – Bluish gray to brown with yellow spot on each scale; blue markings on head, yellow edge on gill cover, **white triangular marking centered on tail.** Solitary. Coral-rich areas of clear lagoons to seaward reefs in 7-55 m.
Indo-West Pacific: E. Africa to Indonesia, Philippines, Micronesia, Solomon Is. – S.W. Japan to Great Barrier Reef and Fiji.

Wedge-tailed Wrasse – IP
SIZE: to 4 cm (1 1/2 in.)
ID: Dark blue to black with numerous thin bluish white stripes and yellow dorsal fin; tail rounded with wide black margin.
JP – Similar with white stripes on dorsal fin.

Wrasses

MICRONESIAN TUBELIP   *Labropsis micronesica*
SIZE: to 13 cm (5 in.)   Wrasses – Labridae
ID: TP – Orange-brown with dark scale margins; gray to blue-gray head with white lips, black tail with light margin. Solitary; occasionally act a cleaners. Clear lagoon or seaward reefs in 7-33 m.
Localized: Micronesia.

Micronesian Tubelip – JP
SIZE: to 4 cm (1 ¹/₂ in.)
ID: White with three dark brown to black stripes from snout to black tail. **IP –** Retain stripes, but head and body become gray to brownish. JP and IP best distinguished from similar Southern and Northern Tubelips [following] by location. Juveniles act as cleaners.

SOUTHERN TUBELIP   *Labropsis australis*
SIZE: to 10.5 cm (4 in.)   Wrasses – Labridae
ID: TP – Dark gray head, golden-brown to orange body; pale fleshy lips, black-edged orange spot on pectoral fin base. Solitary or occasionally in pairs. Coral-rich areas of lagoons, outer reefs and passes in 2-55 m.
West Pacific: Papua New Guinea, Solomon Is. to Great Barrier Reef, New Caledonia and Fiji.

Southern Tubelip – JP/IP
SIZE: to 5 cm (2 in.)
ID: White with three dark brown to black stripes from snout to black tail. **IP –** Retain stripes but head and body gray to brownish. Juveniles act as cleaners.

NORTHERN TUBELIP   *Labropsis manabei*
SIZE: to 13 cm (5 in.)   Wrasses – Labridae
ID: TP – Dark gray head, light brown body; blue lips, **large yellow patch at base of blue tail,** black spot on pectoral fin base. Coral-rich areas to 15-30 m.
Asian Pacific: Indonesia, Philippines to S.W. Japan.

Northern Tubelip – JP/IP
SIZE: to 5 cm (2 in.)
ID: White with three dark brown to black stripes from snout to black tail. **IP –** Retain stripes but head and body gray to brownish. Juveniles act as cleaners.

BLUESTREAK CLEANER WRASSE *Labroides dimidiatus*

SIZE: to 11.5 cm (4 1/2 in.) Wrasses – Labridae

ID: TP – White to yellowish head and forebody becomes bluish toward tail; stripe from snout becomes progressively wider toward tail. Solitary or in pairs; establish cleaning stations, swim with jerky motion to attract clients. Coral reefs in 2-40 m.

Indo-Pacific: Red Sea and E. Africa to Indonesia, Philippines, Micronesia and French Polynesia. – S. Japan to Australia.

Bluestreak Cleaner Wrasse – TP Variation

ID: Yellowish head and forebody; black stripe from snout to tail, blue areas on upper and lower body extend tail.

Bluestreak Cleaner Wrasse – JP

SIZE: to 5 cm (2 in.)

ID: Neon blue stripe from snout to upper border of tail, also neon blue border on lower tail.

BLACKSPOT CLEANER WRASSE *Labroides pectoralis*

SIZE: to 8 cm (3 1/4 in.) Wrasses – Labridae

ID: TP – Yellow head and back, white belly; dark stripe from snout becomes progressively wider toward tail, black spot below pectoral fin. Solitary cleaner; swim with jerky motion to attract clients. Coral reefs in 2-28 m.

East Indo-West Pacific: Cocos-Keeling to Indonesia, Micronesia and Solomon Is. – Philippines to Australia and New Caledonia.

BICOLOR CLEANER WRASSE *Labroides bicolor*

SIZE: to 14 cm (5 1/2 in.) Wrasses – Labridae

ID: TP – Slender; blue lips gradating to black forebody, pale yellow to white rear body and tail, blue crescent on tail. Solitary or in pairs; cleaner, swim with jerky motion to attract clients. Coral reefs in 2-25 m.

Indo-Pacific: E. Africa to Indonesia, Philippines, Micronesia, Solomon Is. and French Polynesia – S. Japan to Great Barrier Reef.

Bicolor Cleaner Wrasse – IP

SIZE: to 8 cm (3 1/4 in.)

ID: Gray forebody becoming white to yellow from rear body to tail; wide black stripe from snout to midbody. **JP –** Black with yellow stripe from snout extends down back to tail.

Reddish/Big Eyes
Soldierfishes & Squirrelfishes – Bigeyes

This ID Group consists of moderate-sized, predominantly reddish fishes with large eyes.

FAMILY: Soldierfishes & Squirrelfishes – Holocentridae
4 Genera – 34 Species Included

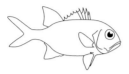

Soldierfishes – Typical Shape

Squirrelfishes – Typical Shape

As their reddish color and large eyes indicate, family members are nocturnal feeders that typically lurk near or just inside reef pockets or branching corals during the day. Large squirrel-like eyes and a tall upright second dorsal fin reminiscent of a squirrel's tail provide squirrelfishes with their common name. Soldierfishes, represented by two genera in this book, can be distinguished from squirrelfishes by blunter snouts and the lack of a prominent pre-gill cover spine that are venomous in a few species. Although spine wounds can be quite painful they are not believed to be life threatening.

Soldierfishes feed in the water column after dark on large zooplankton, including crab larvae, while squirrelfishes forage the sea floor primarily in search of crabs, shrimps and small fishes. Soldierfishes in the genus *Myripristis* are major sound producers. However, virtually all their vocalizations, including pops, grunts and clicks, believed to function as intraspecies communications, are of such a low frequency that they are inaudible to divers.

FAMILY: Bigeyes – Priacanthidae
2 Genera – 5 Species Included

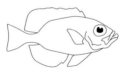

Typical Shape

Bigeyes have deep, compressed bodies, large upturned mouths with projecting lower jaws and large ventral fins connected to the belly by a membrane. Like soldierfishes, bigeyes are also nocturnal zooplankton feeders that move out and away from the reef after sunset to dine in open water on small squids and octopuses and a variety of crabs, shrimps, fishes and polychaete worm larvae. Primarily red by day as they shelter near the reef's base, they change to a pale or blotched pattern at night. Occasionally, during daylight hours, they acquire these alternate patterns, which can be switched on and off quite rapidly.

ROUGHSCALE SOLDIERFISH *Plectrypops lima*
SIZE: to 16 cm (6 1/4 in.) Squirrelfishes – Holocentridae
ID: Bright red with clearish to reddish fins; small scales compared to other soldierfishes, **no dark margin on rear gill cover or white margins on fins.** Solitary; hide in deep recesses. Outer reef slopes in 5-40 m.
Indo-Pacific: E. Africa to Indonesia, Papua New Guinea, Solomon Is., Hawaii and Easter I. – S. Japan to E. Australia.

ROBUST SOLDIERFISH *Myripristis robusta*
SIZE: to 22 cm (8 3/4 in.) Squirrelfishes – Holocentridae
ID: Reddish orange with darker scale margins; white edges on fins (except pectorals), blackish tips on dorsal and anal fins, broad blackish margin on gill cover. Sheltered reefs around rock and coral outcroppings or sea fans in 30-45 m.
Asian Pacific: Philippines to West Papua, Indonesia and Papua New Guinea.

WHITETIP SOLDIERFISH *Myripristis vittata*
SIZE: to 20 cm (8 in.) Squirrelfishes – Holocentridae
ID: Orange-red; **white tips on dorsal fin spines,** narrow white margins on fins, slightly darkened margin on gill cover. Commonly form groups. Steep outer reef slopes in 15-80 m.
Indo-Pacific: Seychelles to Indonesia, Philippines, Micronesia, Papua New Guinea, Solomon Is., Hawaii and French Polynesia. – S.W. Japan to Australia and New Caledonia.

DOUBLETOOTH SOLDIERFISH *Myripristis hexagona*
SIZE: to 20 cm (8 in.) Squirrelfishes – Holocentridae
ID: Red with **broad pink scale margins;** pale reddish fins have no white margins, broad dark red band on rear margin of gill cover. Coastal reefs and seaward slopes, often in turbid areas in 3-40 m.
Indo-West Pacific: E. Africa to Indonesia, Philippines, Micronesia and Solomon Is. – S.W. Japan to E. Australia, New Caledonia and Fiji.

SCARLET SOLDIERFISH *Myripristis pralinia*
SIZE: to 20 cm (8 in.) Squirrelfishes – Holocentridae
ID: Red back shading to silvery on side with red scale margins and silver breast; red fins with narrow white margins, **brown margin confined to upper gill cover.** Coastal, lagoon and outer reefs in 2-40 m.
Indo-Pacific: E. Africa to Indonesia, Philippines, Micronesia, Solomon Is. and French Polynesia. – S.W. Japan to Australia.

YELLOWFIN SOLDIERFISH *Myripristis chryseres*
SIZE: to 25 cm (10 in.) Squirrelfishes – Holocentridae
ID: Red with darkish scale margins; **all fins (except pectorals) bright yellow,** dark brown rear margin on gill cover. Solitary or form groups. Seaward reef slopes in 30-200 m, rarely to 12 m.
Indo-West Pacific: E. Africa to Indonesia, Philippines, Micronesia, Papua New Guinea, Solomon Is. and French Polynesia. – S. W. Japan to Great Barrier Reef and Samoa.

Squirrelfishes

EPAULETTE SOLDIERFISH *Myripristis kuntee*
SIZE: to 19 cm (7 1/2 in.)   Squirrelfishes – Holocentridae
ID: Orange-red with pearly scale centers; **scales small compared with other soldierfishes,** red fins with narrow white margins, dusky brown band along rear gill cover. Form loose groups. Coastal, lagoon and outer reefs in 2-35 m.
Indo-Pacific: E. Africa to Indonesia, Philippines, Micronesia, Solomon Is., Hawaii and French Polynesia. – S.W. Japan to Australia.

SPLENDID SOLDIERFISH *Myripristis botche*
SIZE: to 30 cm (12 in.)   Squirrelfishes – Holocentridae
ID: White with red scale margins; **red dorsal, anal and tail fins with dark lobe tips** and narrow white margins, dark brown edge on gill cover. Bottom of slopes, dead reef areas and isolated coral formations in 20-65 m.
Indo-Pacific: E. Africa to Indonesia, Philippines, Micronesia, Solomon Is., Hawaii and French Polynesia. – S.W. Japan to Australia.

BLOTCHEYE SOLDIERFISH *Myripristis murdjan*
SIZE: to 27 cm (10 3/4 in.)   Squirrelfishes – Holocentridae
ID: Pink to silvery white with red scale margins; red spiny dorsal fin, narrow white margins on all fins (except pectorals), **large black spot around base of pectoral fin.** Solitary; hide in caves during day. Coastal, lagoon and outer reefs in 2-40 m.
Indo-West Pacific: Red Sea and E. Africa to Indonesia, Micronesia, Solomon Is. – S.W. Japan to Australia and Fiji.

VIOLET SOLDIERFISH *Myripristis violacea*
SIZE: to 22 cm (8 3/4 in.)   Squirrelfishes – Holocentridae
ID: Silvery with violet sheen and prominent dark scale margins; fins red with narrow white margins (except pectorals), **light red band along margin of gill cover.** Coastal, lagoon and outer reefs in 3-30 m.
Indo-Pacific: E. Africa to Indonesia, Philippines, Micronesia, Solomon Is. and French Polynesia. – S.W. Japan to Australia.

BRICK SOLDIERFISH *Myripristis amaena*
SIZE: to 32 cm (13 in.)   Squirrelfishes – Holocentridae
ID: Red with dark scale margins; **red dorsal, anal and tail fins without white margins,** dark margin on rear gill cover. During day often form large aggregations under ledges or inside caves. Lagoons and outer slopes in 2-52 m.
Pacific: Alor in Indonesia, Micronesia to Hawaii and French Polynesia. – S.W. Japan to New Caledonia.

SHADOWFIN SOLDIERFISH *Myripristis adusta*
SIZE: to 32 cm (13 in.)   Squirrelfishes – Holocentridae
ID: Pale salmon-pink with dark scale margins; **black margin on rear dorsal and tail fins,** black spot on rear gill cover. Solitary or form small groups. Coastal, lagoon and outer reefs to 25 m.
Indo-Pacific: E. Africa to Indonesia, Philippines, Micronesia, Solomon Is. to French Polynesia. – S.W. Japan to N. Great Barrier Reef and New Caledonia.

BIGSCALE SOLDIERFISH *Myripristis berndti*
SIZE: to 30 cm (12 in.)   Squirrelfishes – Holocentridae
ID: White with reddish tints and red scale margins; yellow spiny dorsal fin, narrow white margins on all fins (except pectorals), dark margin on rear gill cover. Lagoon and seaward reefs in 3-50 m.
Indo-Pacific: E. Africa to Indonesia, Philippines, Micronesia, Solomon Is. and French Polynesia. – S.W. Japan to E. Australia.

EARLE'S SOLDIERFISH *Myripristis earlei*
SIZE: to 30 cm (12 in.)   Squirrelfishes – Holocentridae
ID: Yellow dorsal fin; red margins on whitish body scales, dark margin on rear gill cover. Usually in groups that shelter in caves and under ledges during day. Seaward reefs slopes in 10-20 m.
Central Pacific: Phoenix Is. in Central Pacific and Marquesas Is. in French Polynesia.

WHITESPOT SOLDIERFISH *Myripristis woodsi*
SIZE: to 21 cm (8 1/4 in.)   Squirrelfishes – Holocentridae
ID: Reddish with red scale margins; **white spot at base of pectoral fin,** red fins with white margins (except pectoral), red-brown band on rear margin of gill cover to rear pectoral. Reef flats, lagoons and seaward reefs to 30 m.
Northwest to South Central Pacific: Japan's Bonin and Marcus Is. to Micronesia, Line Is. and French Polynesia.

BLACKFIN SQUIRRELFISH *Neoniphon opercularis*
SIZE: to 35 cm (14 in.)   Squirrelfishes – Holocentridae
ID: Silvery with dark red to blackish scale spots; **black spiny dorsal fin with white tips** and white marks along base. Solitary or form small groups. Often within branches of staghorn corals in 3-25 m.
Indo-Pacific: E. Africa to Indonesia, Philippines, Micronesia, Solomon Is., French Polynesia. – S.W. Japan to Australia and New Caledonia.

SPOTFIN SQUIRRELFISH *Neoniphon sammara*
SIZE: to 30 cm (12 in.)   Squirrelfishes – Holocentridae
ID: Silvery with thin dark red to blackish stripes; **reddish spiny dorsal fin with large black spot at front,** white spine tips and white spots along base. Solitary. Most common among staghorn coral patches in 3-40 m.
Indo-Pacific: Red Sea and E. Africa to Indonesia, Micronesia, Solomon Is., Hawaii and French Polynesia – S.W. Japan to Australia.

CLEARFIN SQUIRRELFISH *Neoniphon argenteus*
SIZE: to 19 cm (7 1/2 in.)   Squirrelfishes – Holocentridae
ID: Silvery with thin dark red to blackish stripes; **translucent spiny dorsal fin.** Solitary; commonly among branches of large staghorn corals. Coral-rich areas of reef flats, lagoons and sheltered seaward reefs to 20 m.
Indo-Pacific: E. Africa to Indonesia, Philippines, Papua New Guinea and French Polynesia. – S.W. Japan to Australia.

Squirrelfishes – Bigeyes

GOLDLINED SQUIRRELFISH *Neoniphon aurolineatus*
SIZE: to 22 cm (8 ³/₄ in.)   Squirrelfishes – Holocentridae
ID: Silvery white to pinkish with yellow stripes between scale rows; yellow pectoral fin base, **red bar across nape and gill cover.** Solitary or form small groups. Steep outer reef slopes in 30-160 m.
Indo-Asian Pacific: Mauritius to Indonesia, E. Papua New Guinea and Hawaii. – S.W. Japan to Great Barrier Reef.

BLACKSPOT SQUIRRELFISH *Sargocentron melanospilos*
SIZE: to 25 cm (10 in.)   Squirrelfishes – Holocentridae
ID: Pale golden to silvery white with orange scale margins; **black spot at base of rear dorsal,** rear anal, pectoral base and on tail fins, pair of thin white bars behind eye. Seaward slopes in 10-35 m.
Indo-West Pacific: E. Africa to Indonesia, Philippines, Micronesia, Solomon Is. – S.W. Japan to Great Barrier Reef and Fiji.

THREESPOT SQUIRRELFISH *Sargocentron cornutum*
SIZE: to 16 cm (6 ¹/₄ in.)   Squirrelfishes – Holocentridae
ID: Alternating red and white stripes; **darkish spot on middle tail base,** 1st spine of anal fin white. Solitary or form small groups. Dropoffs and outer slopes in 6-40 m.
Asian Pacific: Indonesia, Philippines, Papua New Guinea, Solomon Is. to Great Barrier Reef. – Philippines to Great Barrier Reef.

CROWN SQUIRRELFISH *Sargocentron diadema*
SIZE: to 17 cm (6 ³/₄ in.)   Squirrelfishes – Holocentridae
ID: Alternating red and white stripes; **dark red to black spiny dorsal fin** with white spine tips and white streak through middle. Solitary or form small groups. Tidal flats and deeper areas of lagoon and seaward reefs to 40 m.
Indo-Pacific: Red Sea and E. Africa to Indonesia, Micronesia, Solomon Is., Hawaii, French Polynesia. – S.W. Japan to Australia.

SAMURAI SQUIRRELFISH *Sargocentron ittodai*
SIZE: to 20 cm (8 in.)   Squirrelfishes – Holocentridae
ID: Alternating red and white stripes; **red spiny dorsal fin with row of white spots through middle.** Solitary or form groups. Outer reef slopes in 5-70 m.
Indo-Pacific: Red Sea and E. Africa to Indonesia, Philippines, Micronesia, Papua New Guinea, Solomon Is. and French Polynesia. – S.W. Japan to Great Barrier Reef.

SMALLMOUTH SQUIRRELFISH *Sargocentron microstoma*
SIZE: to 19 cm (7 ¹/₂ in.)   Squirrelfishes – Holocentridae
ID: Alternating red and white stripes; **black markings on front of 1st dorsal fin** and very long white margin on anal fin spine. Solitary or form loose groups. Reef flats and deeper coastal reefs, lagoons and outer slopes in 2-183 m.
East Indo-Pacific: Maldives to Indonesia, Micronesia, Solomon Is. and French Polynesia. – Philippines to N. Great Barrier Reef.

PEPPERED SQUIRRELFISH *Sargocentron punctatissimum*
SIZE: to 13 cm (5 in.)   Squirrelfishes – Holocentridae
ID: Pink shading to silvery belly with **fine spotting and indistinct stripes;** spiny dorsal fin with white tips, red border and row of white spots below. Solitary or form groups. Tide pools, reef flats and subtidal reefs to 30 m, usually less than 3 m.
Indo-Pacific: Red Sea and E. Africa to Indonesia, Philippines, Micronesia, Solomon Is., Hawaii. – S.W. Japan to Great Barrier Reef.

YELLOW-STRIPED SQUIRRELFISH *Sargocentron ensifer*
SIZE: to 25 cm (10 in.)   Squirrelfishes – Holocentridae
ID: Red with narrow yellow stripes dorsally and white stripes ventrally, yellow spiny dorsal fin with red margin. Solitary or form groups. Usually on seaward coral reefs or rocky bottoms in 18-50 m.
West & Central Pacific: Micronesia to Hawaii and French Polynesia. – S.W. Japan to New Caledonia and Fiji.

PINK SQUIRRELFISH *Sargocentron tiereoides*
SIZE: to 19.5 cm (7 3/4 in.)   Squirrelfishes – Holocentridae
ID: Alternating silvery pink and red stripes; **bright red spiny dorsal fin** with white spine tips, other fins mainly transparent to pinkish. Solitary or form small groups. Seaward and lagoon reefs in 6-45 m, usually below 15 m.
Indo-Pacific: E. Africa to Indonesia, Philippines, Micronesia, Solomon Is., French Polynesia. – S.W. Japan to E. Australia.

DWARF SQUIRRELFISH *Sargocentron iota*
SIZE: to 9.5 cm (3 3/4 in.)   Squirrelfishes – Holocentridae
ID: Small; red without distinctive markings, tail lobes rounded, scales with serrated edges. Solitary and cryptic; lurk in caves and recesses. Steep outer reef slopes to 34 m.
West Pacific: Indonesia, Micronesia, Papua New Guinea, New Caledonia, Coral Sea, Fiji and Hawaii.

TAHITIAN SQUIRRELFISH *Sargocentron tiere*
SIZE: to 33 cm (13 in.)   Squirrelfishes – Holocentridae
ID: Red with **iridescent blue stripes** (more evident on lower side); red spiny dorsal fin with white tips and white streaks on center of dorsal fin. Solitary; hide inside caves and crevices during day. Lagoon and seaward reefs to 183 m.
Indo-Pacific: Madagascar to Indonesia, Micronesia, Solomon Is., Hawaii and Pitcairn Is. – S.W. Japan to Australia.

TAILSPOT SQUIRRELFISH *Sargocentron caudimaculatum*
SIZE: to 25 cm (10 in.)   Squirrelfishes – Holocentridae
ID: Red with variable amount of silvery white on rear body and tail base; silver-white streak above upper edge of gill cover. Solitary or form loose groups. Coral-rich areas of outer reefs, frequently on steep dropoffs in 6-40 m.
Indo-Pacific: Red Sea and E. Africa to Indonesia, Micronesia, Solomon Is. and French Polynesia. – S. Japan to Great Barrier Reef.

Squirrelfishes – Bigeyes

VIOLET SQUIRRELFISH *Sargocentron violaceum*
SIZE: to 25 cm (10 in.) Squirrelfishes – Holocentridae
ID: Red head, body purplish to brownish red with vertical bluish streak on each scale; rear margin of gill cover blackish. Coral-rich clear water lagoons and seaward reefs to 25 m.
Indo-Pacific: E. Africa to Andaman Sea, Indonesia, Philippines, Micronesia, Papua New Guinea, Solomon Is. and Line Is. – S.W. Japan, Great Barrier Reef to Fiji.

SABRE SQUIRRELFISH *Sargocentron spiniferum*
SIZE: to 45 cm (18 in.) Squirrelfishes – Holocentridae
ID: Large; red with vertical silvery streak on each scale, anal and ventral fins often yellowish; **prominent cheek spine.** Solitary or in pairs; often inside caves and ledges. Lagoon and seaward reefs to 122 m.
Indo-Pacific: Red Sea and E. Africa to Indonesia, Micronesia, Solomon Is., Hawaii and French Polynesia. – S. Japan to Australia.

REDCOAT SQUIRRELFISH *Sargocentron rubrum*
SIZE: to 27 cm (10 3/4 in.) Squirrelfishes – Holocentridae
ID: Alternating reddish brown and white stripes; often dark streak on tail base and on bases of rear dorsal and anal fins. Solitary or form small groups. Frequently on silty reefs and wrecks in lagoons, bays and harbors to 84 m.
Indo-Asian Pacific: Red Sea and E. Africa to Indonesia, Micronesia, Solomon Is. – S.W. Japan to Australia and Vanuatu.

DARKSTRIPED SQUIRRELFISH *Sargocentron praslin*
SIZE: to 21 cm (8 1/4 in.) Squirrelfishes – Holocentridae
ID: Alternating dark reddish brown and silvery white stripes; top two stripes merge to form elongate dark spot under rear dorsal fin and dark spot above anal fin, yellowish fins; white bar on gill cover. Seaward reefs to 20 m.
Indo-Pacific: Red Sea and E. Africa to Indonesia, Micronesia, Solomon Is., French Polynesia. – S. Japan to Australia, Vanuatu.

GLASSEYE *Heteropriacanthus cruentatus*
SIZE: to 32 cm (13 in.) Bigeyes – Priacanthidae
ID: Variable red to silvery with distinct to obscure bars or blotches; **fins lightly spotted or mottled,** tail slightly rounded. Solitary or form small groups; drift next to coral heads during day. Lagoon and seaward reefs in 3-20 m.
Circumtropical.

Glasseye – Phase
ID: Dark red with silvery tints and narrow bars. Have ability to rapidly intensify or diminish color and markings. Tend to be around islands rather than continental coastlines.

BLOCH'S BIGEYE   *Priacanthus blochii*
SIZE: to 35 cm (14 in.)   Bigeyes – Priacanthidae
ID: Variable red to silvery with red blotches; **fins generally plain without spots or mottling,** slightly rounded tail. Similar Arrow Bigeye [next] distinguished by longer ventral fins and taller anal and dorsal fins.
Indo-West Pacific: S. Red Sea to Indonesia, Philippines, Micronesia and Solomon Is. – S. W. Japan to Great Barrier Reef and Fiji.

Bloch's Bigeye – Blotched Phase
ID: Silvery blotched pattern. Have ability to rapidly intensify or fade color and markings, most commonly display some body blotches, but occasionally may be red to silver without blotches. Solitary or form small groups. Under ledges or hover next to coral heads during day. Lagoon and seaward reefs in 15-30 m.

ARROW BIGEYE   *Priacanthus sagittarius*
SIZE: to 28.5 cm (11 1/2 in.)   Bigeyes – Priacanthidae
ID: Red to silvery often with 2-5 bars or blotches; **black spot on ventral fin base,** somewhat longer ventral fins and taller dorsal and anal fins than similar species. Solitary, often hover next to coral outcroppings. Outer slopes in 15-80 m.
Indo-West Pacific: Red Sea to Indonesia and Philippines. – S. Japan to Australia to Fiji.

SPOTTEDFIN BIGEYE   *Priacanthus macracanthus*
SIZE: to 33 cm (13 in.)   Bigeyes – Priacanthidae
ID: Reddish silver with red irregular blotches of varying size; dark basal spot on ventral fin, **rusty brown to red spots on dorsal, ventral and anal fins.** Hover under ledges or near recesses in seaward reefs in 12-400 m.
Indo-Asian Pacific: E. Africa to Indonesia and Philippines.– S. Japan to Great Barrier Reef.

CRESCENT-TAIL BIGEYE   *Priacanthus hamrur*
SIZE: to 40 cm (16 in.)   Bigeyes – Priacanthidae
ID: Variable red to silvery occasionally with about 6 red bars or large spots; fins without spots or mottling, **crescent tail.** Solitary. Hover under ledges or next to coral heads during day. Lagoon and seaward reefs in 15-250 m.
Indo-Pacific: Red Sea and E. Africa to Indonesia, Philippines, Micronesia, Solomon Is., French Polynesia. – S. Japan to Australia.

Cresent-tail Bigeye – Barred Phase
ID: Pinkish sliver with red bar under eye. Have ability to rapidly intensify or diminish color and markings.

Cardinalfishes
and Glassfishes

This ID Group consists of cardinalfishes and a small family known as glassfishes.

FAMILY: Cardinalfishes – Apogonidae
24 Genera – 125 Species Included

Typical Shape   Typical Shape   Typical Shape

Cardinalfishes are relatively small (from 5 to 15 cm [2 to 6 in.]), compared to most families of reef fishes, have large eyes, short snouts, moderately large oblique mouths, two short, widely separated dorsal fins, and double-edged preopercula. The family received its common name from the reddish color common to many species, however, shades of black, white, brown, silver and yellow are well represented. By day cardinalfishes shelter in the protection of corals, undercuts and crevices; a few associate with urchins, sea anemones, and gorgonian branches. Although typically solitary, in pairs or loose clusters, species in genus *Rhabdamia* occur in dense aggregations over isolated coral bommies. At dusk cardinalfishes leave their daytime refuges to feed throughout the night on zooplankton and small, bottom-dwelling crustaceans.

Male cardinalfishes are known for the unusual behavior of incubating egg masses inside their mouths. During the few days before hatching, brooding males can be recognized by their expanded jaws. Often their mouths are so engorged that the egg masses are clearly visible between open lips. At intervals brooding males shift their mouthfuls allowing the masses to aerate more evenly.

Glassfishes are a small family of small fishes that resemble semitransparent cardinalfishes, except for their larger more obvious scales. They often form large, closely packed aggregations above structures and among mangroves in the shallow water of estuaries.

CAPRICORN CARDINALFISH   *Ostorhinchus capricornis*
SIZE: to 10 cm (4 in.)   Cardinalfishes – Apogonidae
ID: Yellowish undercolor; pink to blue **scale margins form wavy vertical lines**, pair of blue stripes extend from snout through eye, black spot on tail base. In pairs or small groups. Coral and rocky reefs in 2-20 m.
Southwestern Pacific: S. Great Barrier Reef and Coral Sea to S. New South Wales, Australia.

SPOTNAPE CARDINALFISH   *Ostorhinchus jenkinsi*
SIZE: to 9 cm (3 ½ in.)   Cardinalfishes – Apogonidae
ID: Purplish brown; black stripe from snout through eye and another on base of dorsal fin, **black spot on side of nape** and larger black spot on tail. Form aggregations. Coastal, lagoon and seaward reefs in 8-45 m.
Asian Pacific: Indonesia, Philippines, Micronesia to Solomon Is., Great Barrier Reef and New Caledonia.

MOLUCCAN CARDINALFISH *Ostorhinchus moluccensis*
SIZE: to 8.5 cm (3 1/4 in.) Cardinalfishes – Apogonidae
ID: Reddish brown; **white spot below rear dorsal fin,** pair of pearl-white stripes through eye and white stripes on upper head, often brown stripe from eye to tail and bars on lower side. Solitary. Sheltered reefs to 25 m.
Asian Pacific: Andaman Sea to Indonesia, Philippines, Papua New Guinea and Solomon Is.

REDSTRIPED CARDINALFISH *Ostorhinchus margaritophorus*
SIZE: to 5.5 cm (2 1/4 in.) Cardinalfishes – Apogonidae
ID: Tan to red with white belly; broad white **midlateral stripe with 2 narrower stripes above and fragmented stripe below.** Form aggregations. Around rocks and crevices of weedy areas in coastal reefs and lagoons to 5 m.
Asian Pacific: Indonesia, Philippines, Papua New Guinea and Solomon Is.

TALBOT'S CARDINALFISH *Apogon talboti*
SIZE: to 9 cm (3 1/2 in.) Cardinalfishes – Apogonidae
ID: Bright red with **dusky brownish scale margins;** no distinctive markings. Solitary and cryptic. Inside deep caves and crevices of seaward reef slopes and dropoffs in 8-35 m.
Indo-Asian Pacific: E. Africa to Brunei, Sabah in Malaysia, Indonesia, Philippines, N. Papua New Guinea and Solomon Is.

SLENDERLINE CARDINALFISH *Ostorhinchus leptofasciatus*
SIZE: to 7 cm (2 3/4 in.) Cardinalfishes – Apogonidae
ID: Reddish brown with violet tint and white lower head; dark stripe extends from snout onto gill cover, **dark oval bar on tail base,** black margin on 1st dorsal fin, black stripe just above base of 2nd dorsal fin. Form aggregations. Coastal reefs in 10-20 m.
Localized: Sulawesi to West Papua, Indonesia and Great Barrier Reef.

MANYLINED CARDINALFISH *Ostorhinchus multilineatus*
SIZE: to 10 cm (4 in.) Cardinalfishes – Apogonidae
ID: Brown head and white body with numerous narrow dark brown stripes; may have faint dark spot on tail base. Solitary or form small groups. Around large coral heads of coastal reefs and lagoons in 2-25 m.
Asian Pacific: Sumatra in Indonesia to Philippines and Solomon Is.

Manylined Cardinalfish – Juvenile
SIZE: to 5 cm (2 in.)
ID: White with alternating wide and narrow dark stripes; black spot on middle of tail base. Solitary or form small groups. Shelter at night beneath ledges or inside crevices; feed over adjacent sand patches.

Cardinalfishes

SPLITBAND CARDINALFISH   *Ostorhinchus compressus*
SIZE: to 10 cm (4 in.)   Cardinalfishes – Apogonidae
ID: Pinkish with reddish brown stripes; blue eye, **short white stripe from upper eye to above pectoral fin.** Form aggregations. Shelter within branching corals or inside crevices of protected lagoon and seaward reefs to 10 m.
Asian Pacific: Andaman Sea to Indonesia, Philippines, Micronesia, Papua New Guinea, Solomon Is. – S.W. Japan to N. Great Barrier Reef.

Splitband Cardinalfish – Juvenile
SIZE: to 5 cm (2 in.)
ID: White with 4 black stripes; yellow tail base with centered black spot. Form groups.

BLACKSTRIPE CARDINALFISH   *Ostorhinchus nigrofasciatus*
SIZE: to 8 cm (3 1/4 in.)   Cardinalfishes – Apogonidae
ID: Alternating black and pale yellow stripes; dark midlateral stripe does not extend onto tail. Solitary or in pairs. Under ledges or inside caves of lagoon and seaward reefs and along steep dropoffs in 3-50 m.
Indo-Pacific: Red Sea and E. Africa to Indonesia, Philippines, Micronesia and French Polynesia. – S.W. Japan to E. Australia.

Blackstripe Cardinalfish – Variation
ID: Alternating black and white stripes. Generally as common as yellow-striped variation, but the two variations usually not together.

SIXSTRIPE CARDINALFISH   *Ostorhinchus endekataenia*
SIZE: to 14 cm (5 1/2 in.)   Cardinalfishes – Apogonidae
ID: Whitish with 6 red-brown stripes with **incomplete stripes or rows of spots** between; large black spot on tail base. Form small to large aggregations. Sheltered inshore reefs, often in silty areas, to 15 m.
Asian Pacific: Singapore, Brunei, Sabah in Malaysia and Indonesia.

STRIPED CARDINALFISH   *Ostorhinchus angustatus*
SIZE: to 9 cm (3 1/2 in.)   Cardinalfishes – Apogonidae
ID: White with 5 dark brown stripes; **dark midlateral stripe enlarges into black spot on tail base.** Solitary or form groups. Inside caves and crevices of clear water seaward reefs in 5-65 m.
Indo-Pacific: Red Sea to Indonesia, Philippines, Micronesia, Papua New Guinea, Solomon Is., French Polynesia. – Taiwan to Australia.

SHORTSTRIPE CARDINALFISH   *Ostorhinchus cookii*
SIZE: to 10 cm (4 in.)   Cardinalfishes – Apogonidae
ID: Alternating dark and white stripes; dark midlateral stripe enlarges to form black spot on tail base, **short stripe from upper eye to middle of body.** Solitary or form small groups. Shallow inshore reefs and protected lagoons to 10 m.
Indo-Asian Pacific: Red Sea and E. Africa to Indonesia, Philippines. – S.W. Japan to Great Barrier Reef and New Caledonia.

REEF-FLAT CARDINALFISH   *Ostorhinchus taeniophorus*
SIZE: to 11.5 cm (4 1/2 in.)   Cardinalfishes – Apogonidae
ID: Alternating dark and light stripes; dark midlateral stripe does not extend onto tail. Habitat useful in distinguishing from Blackstripe Cardinalfish [previous page]. Shallow reef flats, usually in surge areas to 3 m.
Indo-Pacific: E. Africa to Indonesia, Philippines, Micronesia and French Polynesia. – S.W. Japan to Great Barrier Reef.

COPPERSTRIPED CARDINALFISH   *Ostorhinchus holotaenia*
SIZE: to 8 cm (3 1/4 in.)   Cardinalfishes – Apogonidae
ID: Pinkish; 6 copper-brown stripes, midlateral stripe extends onto tail, short stripe from upper eye to middle of body, **row of pearly spots on lower head and belly.** Form aggregations. Coastal reefs and silty areas in 15-35 m.
East Indo-Asian Pacific: Persian Gulf to Andaman Sea, Indonesia, Philippines and S.W. Japan.

RIFLE CARDINALFISH   *Ostorhinchus kiensis*
SIZE: to 8 cm (3 1/4 in.)   Cardinalfishes – Apogonidae
ID: Silvery white; **pair of narrow dark stripes and adjacent white stripes on upper body,** the midlateral stripe extends onto tail. Solitary or form small groups. Sandy slopes of coastal reefs and lagoons in 8-39 m.
Asian Pacific: Indonesia and Philippines, north to S. Japan.

SEVENSTRIPE CARDINALFISH   *Ostorhinchus novemfasciatus*
SIZE: to 9 cm (3 1/2 in.)   Cardinalfishes – Apogonidae
ID: Alternating dark and light stripes; dark midlateral stripe extends onto tail and **dark stripes immediately above and below extends onto tail diagonally.** Form small groups. Reef flats and lagoons to 4 m.
East Indo-Pacific: Cocos-Keeling Is. to Indonesia, Philippines, Solomon Is. and Line Is. – S.W. Japan to Great Barrier Reef, Samoa.

FOURLINED CARDINALFISH   *Ostorhinchus fasciatus*
SIZE: to 7 cm (2 3/4 in.)   Cardinalfishes – Apogonidae
ID: Pink to whitish; brown midlateral stripe darkens and extends onto tail and another on back, both bordered by thin white lines, **white spots and lines on lower body.** Sand and mud bottoms, sometimes far from reefs in 8-80 m.
Indo-Asian Pacific: Red Sea and E. Africa to Indonesia, Philippines and north to Japan.

Cardinalfishes

YELLOWLINED CARDINALFISH *Ostorhinchus chrysotaenia*
SIZE: to 10 cm (4 in.)   Cardinalfishes – Apogonidae
ID: Pinkish yellow with indistinct bronze stripes; pair of silvery streaks through eye, **blue stripe across lower cheek,** very tall 2nd dorsal fin. Solitary. Reef flats, lagoons and outer slopes to 10 m.
Asian Pacific: Brunei, Indonesia, Philippines, Papua New Guinea and Solomon Is. – S.W. Japan to N.W. Australia.

Yellowlined Cardinalfish – Juvenile
SIZE: to 5 cm (2 in.)
ID: Well defined brown stripes of variable width; pair of pearl-white stripes from snout through eye, **blue stripe on lower cheek,** and darkish spot on middle of tail base.

WASSINKI CARDINALFISH *Ostorhinchus wassinki*
SIZE: to 6 cm (2 1/4 in.)   Cardinalfishes – Apogonidae
ID: Dusky yellow orange; 5 silver-gray stripes, **lower stripe extends from head onto lower body.** Solitary, in pairs or small groups. Coastal reefs in 2-15 m, often in silty conditions.
Asian Pacific: N.W. Australia, Sabah in Malaysia, Indonesia and Philippines.

SILTY CARDINALFISH *Ostorhinchus properuptus*
SIZE: to 6 cm (2 1/4 in.)   Cardinalfishes – Apogonidae
ID: Yellow orange; 5 narrow pale gray stripes, **lower stripe only on head.** In pairs. Sheltered rocky areas and coral reefs to 14 m, often in silty conditions.
West Pacific: E. Papua New Guinea, Australia, Coral Sea and Fiji.

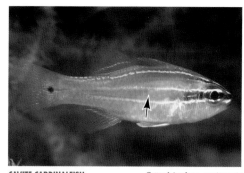

YELLOWSTRIPED CARDINALFISH *Ostorhinchus cyanosoma*
SIZE: to 8 cm (3 1/4 in.)   Cardinalfishes – Apogonidae
ID: Bluish silver to pinkish white; **6 orange-yellow stripes, stripe behind upper eye short ending on midbody.** Solitary, in pairs or aggregations. Around coral heads of sheltered coastal and lagoon reefs and seagrass beds to 40 m.
Indo-Asian Pacific: Red Sea and E. Africa to Indonesia, Micronesia, Solomon Is. – S.W. Japan to Great Barrier Reef and New Caledonia.

CAVITE CARDINALFISH *Ostorhinchus cavitensis*
SIZE: to 7.5 cm (3 in.)   Cardinalfishes – Apogonidae
ID: Pinkish; **bronze midlateral stripe with silvery white margins** and narrow yellow stripe on back, small dark spot on front middle of tail. Solitary or form small groups. Silty coastal reefs to 20 m.
Asian Pacific: Singapore, Brunei and Indonesia. – Philippines to Australia.

RINGTAILED CARDINALFISH *Ostorhinchus aureus*
SIZE: to 14.5 cm (5 3/4 in.) Cardinalfishes – Apogonidae
ID: Yellow-orange shading to pale upper back and tail base; pair of blue stripes through eye, series of dark spots along lateral line, **black bar on tail base narrows in middle.** Form aggregations. Sheltered coastal, lagoon and outer reefs to 40 m.
Indo-West Pacific: E. Africa to Indonesia, Philippines, Papua New Guinea and Solomon Is. – S.W. Japan to E. Australia and Fiji.

FLOWER CARDINALFISH *Ostorhinchus fleurieu*
SIZE: to 13 cm (5 in.) Cardinalfishes – Apogonidae
ID: Mauve upper back shading to yellowish orange side; pair of blue stripes through eye, series of dark spots along lateral line, **black bar (more rounded in middle) or round spot on tail base.** Form aggregations. Sheltered lagoon and outer reefs to 30 m.
Indo-Asian Pacific: Red Sea and E. Africa to Indonesia, Philippines, Papua New Guinea, Solomon Is., north to Taiwan.

KOMODO CARDINALFISH *Ostorhinchus komodoensis*
SIZE: to 8 cm (3 1/4 in.) Cardinalfishes – Apogonidae
ID: Golden yellow shading to coppery brown on back and rear body; pair of bluish white stripes extend from snout through eye, **red streak on upper margin of gill cover.** Form groups. Sheltered reefs with rich coral in 5-12 m.
Asian Pacific: Philippines to Komodo Is. in S.E. Indonesia.

GOLDBELLY CARDINALFISH *Ostorhinchus apogonides*
SIZE: to 10 cm (4 in.) Cardinalfishes – Apogonidae
ID: Mauve back shading to yellow-orange below; **scattered blue spots on side.** Form aggregations. Shelter among branching corals of coastal reefs and outer slopes in 3-40 m.
Indo-West Pacific: E. Africa to Indonesia, Philippines, Micronesia, Papua New Guinea, Solomon Is. to French Polynesia. – S. Japan to Great Barrier Reef, Coral Sea and Fiji.

YELLOW CARDINALFISH *Ostorhinchus luteus*
SIZE: to 5 cm (2 in.) Cardinalfishes – Apogonidae
ID: Yellow-orange with **1-2 somewhat indistinct dusky stripes on back;** pair of silver-blue stripes through eye. Form aggregations. Hide under ledges, inside holes or among sea urchin spines in sheltered lagoon and seaward reefs to 50 m.
Asian Pacific: Palau to Marshall Is. in Micronesia, north to Bonin Is. in S. Japan.

FROSTFIN CARDINALFISH *Ostorhinchus hoevenii*
SIZE: to 5 cm (2 in.) Cardinalfishes – Apogonidae
ID: Pale yellowish brown head and mauve body; **white-edged 1st dorsal fin,** small white spot below base of 1st dorsal ray. Form groups. Shelter near corals, weeds and sea urchins of sheltered reef slopes to 30 m.
Asian Pacific: E. Indonesia, Philippines, Papua New Guinea and Solomon Is. – Taiwan to Great Barrier Reef.

253

Cardinalfishes

SPOTGILL CARDINALFISH *Ostorhinchus chrysopomus*
SIZE: to 9 cm (3 1/2 in.)   Cardinalfishes – Apogonidae
ID: Pale yellow; pair of brown stripes on upper side, small dark spot on middle of tail base, **orange spots on lower gill cover.** Form aggregations. Often shelter among branching corals of protected reefs in 2-25 m.
Asian Pacific: Indonesia, Philippines and Solomon Is.

BARGILL CARDINALFISH *Ostorhinchus sealei*
SIZE: to 9 cm (3 1/2 in.)   Cardinalfishes – Apogonidae
ID: Pale yellow; pair of thin brown stripes on upper side, small dark spot on middle of tail base, **2 short orange bars in bluish white patch on gill cover.** Form aggregations; shelter among branches of *Acropora* corals. Protected reefs in 2-25 m.
Asian Pacific: Indonesia, Philippines and Micronesia to N. Australia.

ORANGESPOT CARDINALFISH *Ostorhinchus rubrimacula*
SIZE: to 5.5 cm (2 1/4 in.)   Cardinalfishes – Apogonidae
ID: Bluish silver to mauve to pinkish white; 6 orange-yellow stripes, stripe behind upper eye short, **orange to red spot at end of midlateral stripe** more intense at night. Form groups. Clearwater lagoon and seaward reefs to 33 m.
West Pacific: Indonesia, Philippines, Papua New Guinea to Solomon Is. – S.W. Japan to Great Barrier Reef and Fiji.

HOOKFIN CARDINALFISH *Ostorhinchus griffini*
SIZE: to 14 cm (5 1/2 in.)   Cardinalfishes – Apogonidae
ID: Golden brown head, pinkish to purplish midbody with dark scale margins and pale tail base, **orange to yellowish pectoral and anal fins;** first rays of 2nd dorsal fin elongate. Around rock and weed areas with scattered coral to 10 m.
Localized: Sabah in Malaysia, Brunei and Philippines.

LINESPOT CARDINALFISH *Ostorhinchus lineomaculatus*
SIZE: to 6.5 cm (2 1/2 in.)   Cardinalfishes – Apogonidae
ID: Pinkish sliver; thin dark stripe from eye to **black spot on tail base with pearly white leading edge,** black tip on 1st dorsal fin. Solitary or form small groups inside crevices; feed over open sand at night. Seaward slopes in 8-32 m.
Localized: Bali to Komodo in S.E. Indonesia.

GORGONIAN CARDINALFISH *Ostorhinchus cladophilos*
SIZE: to 6 cm (2 1/2 in.)   Cardinalfishes – Apogonidae
ID: Whitish to translucent; tapering dark brown, **often fragmented midlateral stripe,** large black spot on tail base. Form groups. Shelter among urchin spines, black coral and gorgonian branches of protected reefs in 20-35 m.
Asian Pacific: Flores I. in Indonesia and Great Barrier Reef to New Caledonia.

WHITESPOT CARDINALFISH *Ostorhinchus dispar*
SIZE: to 5 cm (2 in.)   Cardinalfishes – Apogonidae
ID: Pinkish silver, translucent; thin reddish brown midlateral stripe, brown spot on middle of tail base with white spot above. Form aggregations. Shelter inside black coral thickets on steep outer reef slopes in 15-50 m.
East Indo-West Pacific: Cocos-Keeling Is. to Indonesia, S.W. Japan, Papua New Guinea, Solomon Is. and Fiji.

BLACKBLOTCH CARDINALFISH *Ostorhinchus melanoproctus*
SIZE: to 5 cm (2 in.)   Cardinalfishes – Apogonidae
ID: Pinkish translucent with reddish brown stripe from snout to tail base; red spot on middle of tail base with white spot above, **black blotch around anus.** Form aggregations within black coral and gorgonians. Steep outer reef slopes in 15-50 m.
Asian Pacific: Indonesia, Philippines, Papua New Guinea to Solomon Is.

FAINTBANDED CARDINALFISH *Ostorhinchus franssedai*
SIZE: to 7 cm (2 3/4 in.)   Cardinalfishes – Apogonidae
ID: Pinkish translucent; **3 dark stripes on head,** occasional stripe through eye extends to tail, black spot on middle of tail base, may display an additional thin stripe on upper back. Form groups. Inside crevices and caves of seaward reefs in 12-40 m.
Asian Pacific: Indonesia, Philippines and Micronesia.

TAILSPOT CARDINALFISH *Ostorhinchus ocellicaudus*
SIZE: to 6 cm (2 1/2 in.)   Cardinalfishes – Apogonidae
ID: Pinkish translucent; dark stripe on snout through eye and short distance behind, usually another from corner of mouth to under eye; **white-edged black spot on middle of tail base.** In pairs or small groups. Beneath ledges or around coral outcroppings of sheltered coastal reefs in 15-30 m.
West Pacific: Indonesia, Papua New Guinea, Solomon Is. to Fiji.

SILVERLINED CARDINALFISH *Ostorhinchus hartzfeldii*
SIZE: to 12 cm (4 3/4 in.)   Cardinalfishes – Apogonidae
ID: Purplish brown; pair of silver-white stripes through eye and thin white stripes on upper head and back, black spot on middle of tail base. Form small groups. Shelter among sea urchin spines on coastal reefs and lagoons to 12 m.
Asian Pacific: Indonesia, Philippines, Micronesia and Solomon Is.

Silverlined Cardinalfish – Juvenile
SIZE: to 5 cm (2 in.)
ID: Brown; pair of white eye stripes continue onto body and white stripe on upper head and back, black spot on middle of tail base. Shelter among sea urchin spines or under rocky ledges of shallow reef flats in areas of mixed sand, weed and coral.

Cardinalfishes

LARVAL CARDINALFISH *Ostorhinchus neotes*
SIZE: to 3.3 cm (1¼ in.) Cardinalfishes – Apogonidae
ID: Translucent silvery mauve; black stripe intermittently bordered with yellow or silver runs from above eye to black spot on middle of tail base. Form aggregations. Shelter in soft corals and sea fans of coastal, lagoon and seaward reefs in 15-25 m.
Asian Pacific: Indonesia, Philippines, Micronesia, Papua New Guinea and Solomon Is.

BANDSPOT CARDINALFISH *Ostorhinchus selas*
SIZE: to 5.5 cm (2¼ in.) Cardinalfishes – Apogonidae
ID: Mauve back; brown midlateral stripe with yellowish borders and **dusky to reddish belly with speckles**, large black spot on tail base. Form small groups. Inside or near caves and crevices of sheltered coastal reefs in 3-45 m.
Asian Pacific: Indonesia, Philippines, Micronesia, N. Papua New Guinea and Solomon Is. to New Caledonia and Vanuatu.

REDSPOT CARDINALFISH *Ostorhinchus parvulus*
SIZE: to 4 cm (1½ in.) Cardinalfishes – Apogonidae
ID: Silvery translucent; thin dark midlateral stripe often bordered with silvery or golden streaks, **bright red spot on middle of tail base**. Commonly form large midwater aggregations. Coastal and lagoon reefs in 2-12 m.
Asian Pacific: Brunei, Sabah in Malaysia, Indonesia, Philippines, north to S. Japan.

TINY CARDINALFISH *Ostorhinchus nanus*
SIZE: to 3.5 cm (1½ in.) Cardinalfishes – Apogonidae
ID: Silvery mauve and translucent to golden brown; midlateral stripe extends onto tail with pearly white stripe below. Form midwater aggregations. Sheltered areas of silty coastal reefs and lagoons in 5-20 m.
Asian Pacific: Sabah in Malaysia, Indonesia, Philippines, Papua New Guinea and Solomon Is.

SHARPLINE CARDINALFISH *Ostorhinchus oxygrammus*
SIZE: to 5 cm (2 in.) Cardinalfishes – Apogonidae
ID: Whitish translucent with broad dark brown to black tapering stripe from eye to middle of tail; white stripe or row of spots below dark stripe from eye to near tail base. Solitary or form small groups. Rubble and areas of *Halimeda* algae near small formations in 35-50 m.
Localized: Raja Ampat, Cenderawasih Bay in West Papua, Indonesia.

BRYX CARDINALFISH *Ostorhinchus bryx*
SIZE: to 7.5 cm (3 in.) Cardinalfishes – Apogonidae
ID: Silvery white to light gray; black stripe with narrow white borders and **thin black stripe above eye to below 2nd dorsal fin**. Sand and silt in 15-155 m.
Indo-Asian Pacific: E. Africa to Andaman Sea, Indonesia and Philippines.

GOBBLEGUTS *Ostorhinchus rueppellii*
SIZE: to 12 cm (4 ³/₄ in.)   Cardinalfishes – Apogonidae
ID: Light silvery brown to mauve; **row of dark spots along lateral line** and another row near base of second dorsal fin, narrow brown band below eye. Form aggregations. In weedy areas of coastal reefs and estuaries to 10 m.
Localized: W. Australia to S. Papua New Guinea.

NORFOLK CARDINALFISH *Ostorhinchus norfolcensis*
SIZE: to 10 cm (4 in.)   Cardinalfishes – Apogonidae
ID: Brown with darker scale margins; silvery stripe through eye, **faint bar on side**, black spot on middle of tail base, tall 2nd dorsal fin. Solitary or form groups. Lagoon and outer reefs in 3-25 m.
Localized: New Caledonia.

YELLOWEYE CARDINALFISH *Ostorhinchus monospilus*
SIZE: to 9.5 cm (3 ³/₄ in.)   Cardinalfishes – Apogonidae
ID: Reddish brown; **yellow iris**, pair of pearl-white stripes through eye and bluish stripe below eye. In pairs or groups. Weedy areas of sheltered coastal reefs to 30 m.
Asian Pacific: Indonesia to Philippines, N. Papua New Guinea to Great Barrier Reef and W. Australia.

AMBON CARDINALFISH *Fibramia amboinensis*
SIZE: to 10 cm (4 in.)   Cardinalfishes – Apogonidae
ID: Light brown; dark midlateral stripe and **thin dark stripe follows lateral line**, dark leading edge on first dorsal fin, black spot on tail base. Rarely on reefs, usually in fresh or brackish water of river mouths to 3 m.
Asian Pacific: Singapore to Papua New Guinea.

SANGI CARDINALFISH *Fibramia thermalis*
SIZE: to 8 cm (3 ¹/₄ in.)   Cardinalfishes – Apogonidae
ID: Brown to shades of gray except for pale lower head and belly; **broad black stripe through eye**, black leading edge on first dorsal fin, 3 tiny black spots along dorsal-fin base. Form aggregations. Sheltered, silty coastal reefs to 12 m.
Indo-Asian Pacific: E. Africa to Indonesia and Philippines. - S.W. Japan to Australia and Vanuatu.

Sangi Cardinalfish – Juvenile
SIZE: to 3.5 cm (1 ¹/₂ in.)
ID: Brownish translucent; markings similar to those of adults, large black spot on middle of tail base. Form small to large aggregations. Hover close above coral heads adjacent to mangroves or in seagrass habitat.

Cardinalfishes

MANGROVE CARDINALFISH *Fibramia ceramensis*
SIZE: to 8 cm (3 1/4 in.)   Cardinalfishes – Apogonidae
ID: Translucent with silver to gold sheen; small black spot on middle of tail base, **black leading edge on 1st dorsal fin.** Form aggregations in mangroves to 3 m. Distinguished from similar Coastal Cardinalfish [next] by habitat.
Asian Pacific: Indonesia, Philippines and Papua New Guinea.

COASTAL CARDINALFISH *Fibramia lateralis*
SIZE: to 10 cm (4 in.)   Cardinalfishes – Apogonidae
ID: Translucent with silvery reflection; narrow dark midlateral stripe, small spot on middle of tail base. Form aggregations. Coastal reefs to 15 m. Distinguished from similar Mangrove Cardinalfish [previous] by reef habitat. Form aggregations. Coastal reefs to 15 m.
Indo-West Pacific: E. Africa to Micronesia. – Taiwan to Samoa.

NARROWSTRIPE CARDINALFISH *Pristiapogon exostigma*
SIZE: to 12 cm (4 3/4 in.)   Cardinalfishes – Apogonidae
ID: Pinkish to silvery gray; tapering solid midlateral stripe, **dark spot centered above middle of tail base.** Solitary or groups. Near coral heads and ledges of lagoon and seaward reefs in 3-20 m.
Indo-Pacific: Red Sea to Line Is. and French Polynesia. – Australia to S.W. Japan.

IRIDESCENT CARDINALFISH *Pristiapogon kallopterus*
SIZE: to 15 cm (6 in.)   Cardinalfishes – Apogonidae
ID: Pale pinkish brown; dark midlateral stripe of uniform width, black spot centered above middle of tail base, **yellow leading edge on first dorsal fin.** Solitary or form small groups. Lagoon and seaward reefs in 3-45 m.
Indo-Pacific: Red Sea and E. Africa to Indonesia, Micronesia, Solomon Is. to Hawaii and Pitcairn Is. – S. Japan to Australia.

LATERALSTRIPE CARDINALFISH *Pristiapogon abrogramma*
SIZE: to 9.5 cm (3 3/4 in.)   Cardinalfishes – Apogonidae
ID: Pinkish to silvery gray; tapering solid midlateral stripe, lack spot on tail base like similar Narrowstripe Cardinalfish [above] Solitary or loose groups. Caves and beneath ledges of outer reef slopes in 15-40 m.
Indo-Asian Pacific: Madagascar to Indonesia, Philippines, Papua New Guinea and Solomon Is. – S.W. Japan to Australia.

SPURCHEEK CARDINALFISH *Pristiapogon fraenatus*
SIZE: to 11 cm (4 1/4 in.)   Cardinalfishes – Apogonidae
ID: Pinkish silver to pale gray; **yellow snout,** tapering solid midlateral stripe, dark spot centered on tail base. Solitary or form groups. Near base of coral heads or under ledges of lagoon and seaward reefs in 3-25 m.
Indo-WestPacific: Red Sea and E. Africa to Indonesia, Philippines, Micronesia and French Polynesia. – S.W. Japan to Australia and Fiji.

REARBAR CARDINALFISH   *Apogon posterofasciatus*
SIZE: to 6 cm (2¹/₄ in.)   Cardinalfishes – Apogonidae
ID: Translucent red with **diffuse darkish bar from below 2nd dorsal fin to anal fin base;** broad diffuse darkish bar on tail base. Solitary or form small groups. Caves and beneath ledges in 18-37 m.
West Pacific: Indonesia and Philippines to Solomon Is. and Fiji.

OBLIQUEBANDED CARDINALFISH   *Apogon semiornatus*
SIZE: to 5 cm (2 in.)   Cardinalfishes – Apogonidae
ID: Pinkish translucent; broad reddish to dark brown and **whitish bicolor band from eye to anus,** broad brownish stripe on rear body. Solitary and cryptic. Inside caves and recesses of coastal reefs and seaward slopes in 3-30 m.
Indo-West Pacific: Red Sea and E. Africa to Indonesia, Philippines, Micronesia and Phoenix Is. – S. Japan to N. Australia and Fiji.

DARKTAIL CARDINALFISH   *Apogon seminigracaudus*
SIZE: to 5 cm (2 in.)   Cardinalfishes – Apogonidae
ID: Reddish translucent; **wide darkish band from below nape to tail base extends onto lower lobe of tail.** Solitary and cryptic. Near shore patch and fringing reefs, often in silty conditions to 30 m.
West Pacific: Indonesia and Philippines to S. Japan and Fiji.

RUBY CARDINALFISH   *Apogon crassiceps*
SIZE: to 5 cm (2 in.)   Cardinalfishes – Apogonidae
ID: Reddish translucent; may display diffuse dusky midlateral stripe. Solitary and cryptic; rarely in open except at night. Coastal and offshore reefs to 20 m.
Pacific: Red Sea to Andaman Sea, Indonesia, Philippines, Micronesia, Papua New Guinea, Solomon Is. and French Polynesia. – S. Japan to Australia.

THREESPOT CARDINALFISH   *Pristicon trimaculatus*
SIZE: to 19.5 cm (8 in.)   Cardinalfishes – Apogonidae
ID: Pale red to silver with darkish scale centers; **darkish spot on gill cover** and 2 smaller spots on tail base, ragged band below 1st and 2nd dorsal fins with saddle between. Solitary or in pairs. Lagoons and outer slopes in 3-15 m.
West Pacific: Andaman Sea to Indonesia, Philippines, Micronesia and Solomon Is. – S.W. Japan to Australia and Fiji.

Threespot Cardinalfish – Juvenile
SIZE: to 5 cm (2 in.)
ID: Whitish to pale gray; markings similar to those of adult, but much darker and bars extend farther down side. Solitary. Both adults and juveniles feed at night over sand or rubble near reefs; lurk inside reef recesses during day.

Cardinalfishes

FALSE THREESPOT CARDINALFISH *Pristicon rhodopterus*
SIZE: to 6 cm (2 1/4 in.)   Cardinalfishes – Apogonidae
ID: Pale with darkish scale edges; spot on rear tail base (**no spot on gill cover**), ragged band below 1st and 2nd dorsal fins often with saddle between. Solitary. Crevices of lagoon and seaward reefs in 8-25 m.
Asian Pacific: Brunei, Indonesia, Philippines, Micronesia, Papua New Guinea and Solomon Is.

False Threespot Cardinalfish – Juvenile
SIZE: to 5 cm (2 in.)
ID: Pale brown; markings similar to adults, but are much darker and bars extend farther down side. Solitary. Both adults and juveniles feed at night over sand or rubble near reefs; lurk inside reef recesses during day.

RUFUS CARDINALFISH *Pristicon rufus*
SIZE: to 11 cm (4 1/4 in.)   Cardinalfishes – Apogonidae
ID: Reddish brown with pair of dusky bars on side, one below each dorsal fin; dark brown spot on tail base, thin white edge on rear of 1st dorsal fin, **narrow darkish band extends from lower eye.** Solitary. Outer reef slopes in 15-80 m.
West Pacific: Indonesia, Philippines, S.W. Japan, Micronesia, Papua New Guinea to Fiji.

ODDSCALE CARDINALFISH *Zapogon evermanni*
SIZE: to 12 cm (4 3/4 in.)   Cardinalfishes – Apogonidae
ID: Red with hint of 3-4 broad pale bars on side; dark stripe from snout to gill cover edge, **dark-edged white spot at base of last dorsal fin rays.** Solitary or in pairs; cryptic. Shelter inside caves of outer slopes in 5-70 m.
Indo-Pacific: E. Africa to Indonesia, Philippines, Micronesia, Papua New Guinea, Solomon Is., Hawaii and French Polynesia.

OCELLATED CARDINALFISH *Apogonichthys ocellatus*
SIZE: to 6 cm (2 1/4 in.)   Cardinalfishes – Apogonidae
ID: Dark brown; 3 dark brown bands radiate from rear eye, **large ocellated black spot on first dorsal fin.** Solitary and cryptic. Exposed reef flats and sheltered lagoons to 3 m.
Indo-Pacific: E. Africa to Indonesia, Philippines, Micronesia, Solomon Is. and French Polynesia. – S. Japan to Australia.

CAMOUFLAGE CARDINALFISH *Apogonichthys perdix*
SIZE: to 6 cm (2 1/4 in.)   Cardinalfishes – Apogonidae
ID: Mottled reddish brown to olive with brownish blotches and white flecks; darkish bands radiate from eye, whitish to transparent edge on tail. Solitary and cryptic. Sheltered lagoons and bays with abundant algae and seagrass to 4 m.
Indo-Pacific: Red Sea and E. Africa to Micronesia, Hawaii and French Polynesia. – S. Japan to Australia.

GILBERT'S CARDINALFISH   *Zoramia gilberti*
SIZE: to 5.5 cm (2 1/4 in.)   Cardinalfishes – Apogonidae
ID: Pale brown to mauve with bluish belly; **blue spots on head and front of body,** often black marks on tips of tail lobes. Form large aggregations. Shelter among branching corals of protected reefs in 2-10 m.
Asian Pacific: Sabah in Malaysia, Indonesia, Philippines, Micronesia, Papua New Guinea and Solomon Is.

THREADFIN CARDINALFISH   *Zoramia leptacantha*
SIZE: to 6 cm (2 1/4 in.)   Cardinalfishes – Apogonidae
ID: Whitish translucent with frosty iridescence on back; blue iris, **orange-edged blue bands and bars on rear head and front of body,** long 1st dorsal-fin tip. Form groups. Shelter among branching corals of protected reefs in 2-12 m.
Indo-West Pacific: Red Sea and E. Africa to Andaman Sea, Indonesia and Micronesia and Solomon Is. – S.W. Japan to Fiji.

PEARLY CARDINALFISH   *Zoramia perlita*
SIZE: to 5.3 cm (2 1/4 in.)   Cardinalfishes – Apogonidae
ID: Light brown with blue and yellow belly; **blackish stripe along anal fin base,** dusky area and small spot within dusky area on tail base. Form aggregations. Shelter among branching corals of protected reefs in 3-12 m.
Asian Pacific: Andaman Sea to Indonesia, Philippines, Micronesia and Papua New Guinea. – S.W. Japan to Australia.

Pearly Cardinalfish – Variation
ID: Pale grayish translucent; small black spot on tail base, white spot under the rear of the 2nd dorsal fin is often present on all variations.
Localized: Andaman Sea to N.W. Sumatra in Indonesia.

FRAGILE CARDINALFISH   *Zoramia viridiventer*
SIZE: to 5.7 cm (2 1/4 in.)   Cardinalfishes – Apogonidae
ID: Translucent with silvery reflections; scattered blue spots and markings on head and front of body, small black spot on middle of tail base, often black spot on tips of tail lobes. Solitary or in pairs. Sheltered reefs to 15 m.
Indo-West Pacific: Indonesia and Philippines to Micronesia. – S.W. Japan to Great Barrier Reef, Vanuatu and Fiji.

BELTED CARDINALFISH   *Taeniamia leai*
SIZE: to 8 cm (3 1/4 in.)   Cardinalfishes – Apogonidae
ID: Mauve to brown to olive often with heavy yellowish tan speckling; 3 to 5 bars, the widest and most distinct below 2nd dorsal fin, pair of white stripes from snout through eye. Solitary. Shallow coastal and lagoon reefs to 15 m.
Localized: Great Barrier Reef and Coral Sea to New Caledonia.

Cardinalfishes

TWINSPOT CARDINALFISH　　　*Taeniamia biguttata*
SIZE: to 8.6 cm (3¼ in.)　　Cardinalfishes – Apogonidae
ID: Mauve to pinkish brown with thin orange bars; **black spot above upper edge of gill cover and on middle of tail base,** black bar beneath eye. Form aggregations. Shelter inside caves and crevices of protected coastal reefs and lagoons in 3-18 m.
West Pacific: Indonesia, Philippines, Micronesia, Papua New Guinea, Solomon Is. to Vanuatu and Fiji.

BLACKSPOT CARDINALFISH　　*Taeniamia melasma*
SIZE: to 8.6 cm (3¼ in.)　　Cardinalfishes – Apogonidae
ID: Mauve to pinkish with thin orange bars; **black spot above upper edge of gill cover with diffuse bar below,** dusky bar under eye. Form small aggregations. Shelter among branching corals of protected coastal reefs in 2-12 m, often in silty areas.
Asian Pacific: Indonesia, Philippines and Papua New Guinea.

ORANGELINED CARDINALFISH　　*Taeniamia fucata*
SIZE: to 9.5 cm (3¾ in.)　　Cardinalfishes – Apogonidae
ID: Mauve undercolor with orange head; numerous narrow orange bars, pair of blue stripes through eye, **black spot on tail base.** Form aggregations. Inside caves or above branching corals of coastal and lagoon reefs in 2-60 m.
Indo-West Pacific: Red Sea and E. Africa to Indonesia, Micronesia and Solomon Is. – S.W. Japan to Australia and Fiji.

DUSKYTAIL CARDINALFISH　　*Taeniamia macroptera*
SIZE: to 9 cm (3½ in.)　　Cardinalfishes – Apogonidae
ID: Pale gray with numerous close-set thin orange bars; **blackish tail base.** Form dense aggregations. Hover above beds of *Porites* or other branching corals of sheltered coastal reefs and lagoons in 2-20 m.
East Indo-West Pacific: Sri Lanka to Indonesia, Philippines and Micronesia. – S.W. Japan to Australia and Samoa.

GIRDLED CARDINALFISH　　*Taeniamia zosterophora*
SIZE: to 6.5 cm (2½ in.)　　Cardinalfishes – Apogonidae
ID: Pale gray; **pair of narrow reddish bars on gill cover,** broad dark brown bar on middle of side, small black spot on middle of tail base. Form aggregations. Shelter among branching corals inside protected bays and lagoons in 2-15 m.
Asian Pacific: Andaman Sea, Indonesia, Philippines, Micronesia, Papua New Guinea, Solomon Is. – S.W. Japan to Australia, Vanuatu.

Girdled Cardinalfish – Variation
ID: Pale gray; pair of narrow reddish bars on gill cover and hint of broad body bar in middle of body formed by pepper-like dark spots, small black spot on middle of tail base.

BRACELET CARDINALFISH   *Nectamia viria*
SIZE: to 9.5 cm (3 ³/₄ in.)   Cardinalfishes – Apogonidae
ID: Bronze to brownish; either **uniform or with few faint bars on side,** wedge shaped dark mark beneath eye, whitish tail base with broad black bar. Solitary or form small groups. Shelter in branching corals to 8 m.
West Pacific: Sabah in Malaysia, Indonesia, Philippines, Micronesia to Great Barrier Reef, New Caledonia and Fiji.

BANDA CARDINALFISH   *Nectamia bandanensis*
SIZE: to 10 cm (4 in.)   Cardinalfishes – Apogonidae
ID: Brown; wedge-shaped dark mark below eye, whitish tail base with dark bar, frequently broad brown bar below each dorsal fin. Solitary or form small groups. Shelter in branching corals on protected reefs to 12 m.
East Indo-West Pacific: Cocos-Keeling Is. to Indonesia, Philippines, Micronesia and Solomon Is. – S.W. Japan to Australia and Fiji.

SIMILAR CARDINALFISH   *Nectamia similis*
SIZE: to 10 cm (4 in.)   Cardinalfishes – Apogonidae
ID: Brown with 7-8 thin pale bars; wedge-shaped dark mark below eye, **white margin on second dorsal, anal and tail fins.** Form aggregations among branching corals on protected coastal reefs and lagoons to 15 m.
Asian Pacific: Sabah in Malaysia, Indonesia, Philippines to Micronesia and Papua New Guinea.

MULTI-BARRED CARDINALFISH   *Nectamia luxuria*
SIZE: to 10 cm (4 in.)   Cardinalfishes – Apogonidae
ID: Brown with 8-9 thin pale bars; wedge-shaped dark mark below eye, **yellow margin on second dorsal, anal and tail fins.** Form loose aggregations among branching corals during day. Sheltered coastal reefs and lagoons to 10 m.
East Indo-Pacific: Maldives to Indonesia, Philippines, Micronesia, Solomon Is. and French Polynesia. – S.W. Japan to Australia.

SAMOAN CARDINALFISH   *Nectamia savayensis*
SIZE: to 11.3 cm (4 ¹/₂ in.)   Cardinalfishes – Apogonidae
ID: Brown, darker on back, usually with coppery or silvery reflections; wedge-shaped bar below eye, **dark saddle on upper tail base.** Form aggregations. Often shelter within staghorn corals on protected seaward slopes to 22 m.
Indo-Pacific: E. Africa to Andaman Sea, Indonesia, Philippines and French Polynesia. – S.W. Japan to Australia.

GHOST CARDINALFISH   *Nectamia fusca*
SIZE: to 11 cm (4 ¹/₄ in.)   Cardinalfishes – Apogonidae
ID: Pale to brown; narrow dark band below eye, may display dark bar on tail base. Solitary or in pairs; cryptic, ventures in the open at night. Reef flats and shallow lagoons to 3 m.
East Indo-Pacific: Maldives to Indonesia, Philippines, Micronesia, Papua New Guinea, Solomon Is. to French Polynesia – S.W. Japan to Australia.

Cardinalfishes

YELLOWMOUTH CARDINALFISH *Archamia bleekeri*
SIZE: to 8.5 cm (3¼ in.)   Cardinalfishes – Apogonidae
ID: Whitish translucent with silvery reflections; **yellowish wash on snout,** black spot on middle of tail base. Form aggregations. Hover above rocky outcroppings of coastal reefs in 10-30 m, typically in silty areas.
Asian Pacific: Brunei, Indonesia, Philippines, north to Singapore and Taiwan.

ONESPOT CARDINALFISH *Apogonichthyoides uninotatus*
SIZE: to 9 cm (3½ in.)   Cardinalfishes – Apogonidae
ID: Pale pinkish brown; faint spoke-like marks radiate from rear half of eye, **large dark brown spot on middle of body.** Solitary. Dead reefs and mangrove fringed estuaries and brackish lakes to 10 m.
Localized: Philippines.

TIMOR CARDINALFISH *Apogonichthyoides timorensis*
SIZE: to 7 cm (2¾ in.)   Cardinalfishes – Apogonidae
ID: Shades of brown; fins yellowish, **short pale band behind eye with dark borders,** usually dark brown bar extends below 1st dorsal fin and another below 2nd dorsal fin. Solitary. Shallow reef flats and seagrass beds to 15 m.
Indo-Asian Pacific: Red Sea and E. Africa to Indonesia, Philippines, Micronesia and Solomon Is. – S.W. Japan to Australia.

BLACK CARDINALFISH *Apogonichthyoides melas*
SIZE: to 12.5 cm (5 in.)   Cardinalfishes – Apogonidae
ID: Dark brown; tall rounded dorsal and anal fins and broad rounded tail lobes, **black ocellated spot on base of second dorsal fin.** Solitary. Sheltered coastal reefs and bays to 15 m.
East Indo-Asian Pacific: Cocos-Keeling Is. to Andaman Sea, Indonesia, Philippines, Micronesia, Papua New Guinea and Solomon Is. – S.W. Japan to Australia.

TWINBAR CARDINALFISH *Apogonichthyoides sialis*
SIZE: to 14 cm (5½ in.)   Cardinalfishes – Apogonidae
ID: Pale reddish brown to pinkish gray; **dark leading edge on 1st and 2nd dorsal fins connect to dark body bars,** black spot on tail base. Solitary or form small groups around rocks and ledges in 8-15 m.
Asian Pacific: Brunei and Philippines to Japan.

CRYPTIC CARDINALFISH *Apogonichthyoides umbratilis*
SIZE: to 5.2 cm (2 in.)   Cardinalfishes – Apogonidae
ID: Pale gray with 5 irregular bars on side; **dark oblique band below eye and 3-4 lines extend from back of eye,** blackish blotch on 1st dorsal fin. Solitary or pairs. Rubble, crevices and outcropping in 20-60 m.
Asian Pacific: Brunei, West Papua in Indonesia, Micronesia and offshore reefs of N.W. Australia.

HUMPBACKED CARDINALFISH   *Yarica hyalosoma*
SIZE: to 20 cm (8 in.)   Cardinalfishes – Apogonidae
ID: Whitish to light pinkish brown; black leading edge on first dorsal fin, black spot on tail base. Form aggregations. Mangrove shores, estuaries and stream mouths to 3 m.
Asian Pacific: Sumatra in Indonesia to E. Australia and Solomon Is., north to S. Japan.

SOLITARY CARDINALFISH   *Cercamia eremia*
SIZE: to 5 cm (2 in.)   Cardinalfishes – Apogonidae
ID: Slender; translucent with silvery reflections, **pinkish brown gill cover.** Solitary. During day shelter inside deep caves and crevices of clear water seaward reefs in 2-40 m.
Indo-West Pacific: Red Sea to Indonesia, Philippines, Micronesia, Papua New Guinea and Solomon Is. – S.W. Japan to Australia and Fiji.

CIRCULAR CARDINALFISH   *Neamia articycla*
SIZE: to 4.5 cm (1 3/4 in.)   Cardinalfishes – Apogonidae
ID: Reddish orange head, body and fins; **large yellowish to brown spot with incomplete pale yellowish outer ring on gill cover.** Solitary or form small groups. Rubble bottoms with *Halimeda* algae and small branching corals in 10-40 m.
West Pacific: Indonesia and Philippines. – S.W. Japan to Great Barrier Reef and Fiji.

GILLSPOT CARDINALFISH   *Neamia notula*
SIZE: to 4.5 cm (1 3/4 in.)   Cardinalfishes – Apogonidae
ID: Reddish orange with **pale-edged black spot on gill cover;** faint speckling on fins. Rubble bottoms in 17-70 m.
Localized: Known only from Mauritius and Bali in Indonesia.

EIGHTSPINE CARDINALFISH   *Neamia octospina*
SIZE: to 5 cm (2 in.)   Cardinalfishes – Apogonidae
ID: Whitish to pinkish or beige; translucent fins, **3 short brown bands radiating from rear eye.** Solitary and cryptic. Hide deep inside crevices of sheltered reefs in 3-5 m.
Indo-West Pacific: Red Sea and E. Africa to Andaman Sea, Indonesia, Micronesia, Papua New Guinea and Solomon Is. – S.W. Japan to Australia and Fiji.

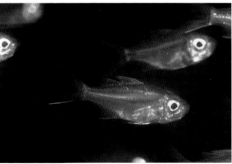

LARGESPINED GLASSFISH   *Ambassis macracanthus*
SIZE: to 12 cm (4 3/4 in.)   Glassfishes – Ambassidae
ID: Translucent with narrow dark scale margins and silvery reflections on head and belly region. Form aggregations. Near stream mouths of coastal waters to 4 m.
Asian Pacific: Sumatra in Indonesia to N. Papua New Guinea.

Cardinalfishes

TIGER CARDINALFISH   *Cheilodipterus macrodon*
SIZE: to 25 cm (10 in.)   Cardinalfishes – Apogonidae
ID: Large; pale with 8 red-brown stripes, whitish tail base with dark to dusky bar, **large prominent pointed teeth.** Solitary; male shown mouth brooding eggs. Hover inside caves and under ledges of coastal, lagoon and outer reefs to 40 m.
Indo-Pacific: Red Sea and E. Africa to Andaman Sea, Micronesia and French Polynesia. – S.W. Japan to Great Barrier Reef and Fiji.

Tiger Cardinalfish – Juvenile
SIZE: to 4 cm (1 1/2 in.)
ID: Yellow head shading to whitish translucent rear body; 4 black stripes and large black spot covering entire tail base.

ALLEN'S CARDINALFISH   *Cheilodipterus alleni*
SIZE: to 13 cm (5 in.)   Cardinalfishes – Apogonidae
ID: Silvery gray with 8 red to brown stripes; **black upper half of first dorsal fin.** Solitary. Hover inside entrances of caves and under ledges of seaward reef slopes in 10-40 m.
West Pacific: Brunei, Indonesia, Philippines, Papua New Guinea, Solomon Is. and Fiji.

INTERMEDIATE CARDINALFISH   *Cheilodipterus intermedius*
SIZE: to 11 cm (4 1/4 in.)   Cardinalfishes – Apogonidae
ID: Slender; pale undercolor with 8 stripes, **white tail base with no hint of a bar.** Solitary or form small groups. Sheltered coastal, lagoon and outer reefs in 2-15 m.
Asian Pacific: Indonesia, Micronesia, Papua New Guinea and Solomon Is. – S.W. Japan to Great Barrier Reef.

WOLF CARDINALFISH   *Cheilodipterus artus*
SIZE: to 12 cm (4 3/4 in.)   Cardinalfishes – Apogonidae
ID: Pale undercolor with about 8 reddish brown stripes; **black tail base.** Form loose aggregations. Hover inside caves or just above coral branches of sheltered bays and lagoons in 3-25 m.
Indo-Pacific: E. Africa to Andaman Sea, Indonesia, Micronesia and French Polynesia. – S.W. Japan to Australia.

Wolf Cardinalfish – Phase
ID: Whitish to bluish undercolor with about 8 reddish brown stripes; black spot on yellow tail base. Adults capable of quickly changing black tail base phase to yellow. Young similar to yellow tail base variation, but have larger black spot.

FIVELINED CARDINALFISH *Cheilodipterus quinquelineatus*
SIZE: to 12.5 cm (5 in.) Cardinalfishes – Apogonidae
ID: Light gray to whitish with 5 black stripes; **yellow tail base with black spot in line with midlateral stripe**, lack canine teeth at front of lower jaw. Form aggregations. Coastal, lagoon and outer reefs to 40 m.
Indo-Pacific: Red Sea and E. Africa to Indonesia, Micronesia, Solomon Is. and French Polynesia. – S. Japan to E. Australia.

TOOTHY CARDINALFISH *Cheilodipterus isostigmus*
SIZE: to 10 cm (4 in.) Cardinalfishes – Apogonidae
ID: Light gray to whitish with 5 black stripes; yellow tail base with **black spot slightly above line of midlateral stripe**, canine teeth at front of lower jaw. Form groups. Hover among branching corals of sheltered reefs in 4-40 m.
West Pacific: Sabah in Malaysia, Indonesia, Philippines, Micronesia to Vanuatu and Fiji.

BLACKSTRIPED CARDINALFISH *Cheilodipterus nigrotaeniatus*
SIZE: to 8 cm (3 1/4 in.) Cardinalfishes – Apogonidae
ID: Yellow head and white body with 4 black stripes; **often broken line of black spots on tail base.** Solitary or in pairs; swim with quick darting motion mimicking Striped Fangblenny. Sheltered reefs in 2-25 m.
Localized: Indonesia and Philippines.

SINGAPORE CARDINALFISH *Cheilodipterus singapurensis*
SIZE: to 22 cm (8 3/4 in.) Cardinalfishes – Apogonidae
ID: Pale undercolor with 5 dark stripes with diffuse edges; small black spot at middle of tail base, **small white spot below last dorsal fin ray.** Solitary; nocturnal feeder. Sheltered coastal and lagoon reefs in 2-10 m.
Asian Pacific: Sumatra in Indonesia, Philippines, Micronesia, Papua New Guinea and Solomon Is. – S.W. Japan to Australia.

MIMIC CARDINALFISH *Cheilodipterus parazonatus*
SIZE: to 7 cm (2 3/4 in.) Cardinalfishes – Apogonidae
ID: Light gray to whitish with **white-edged black stripe from snout to tail base.** Solitary or in pairs; swim with quick darting motion mimicking Blackstripe Fangblenny. Coastal reefs and sheltered lagoons in 2-25 m.
Asian Pacific: Indonesia, Papua New Guinea, Solomon Is. to N. Great Barrier Reef.

YELLOWBELLY CARDINALFISH *Cheilodipterus zonatus*
SIZE: to 7 cm (2 3/4 in.) Cardinalfishes – Apogonidae
ID: Gray back and yellow belly; white stripe above black midlateral stripe snout to tail. Solitary or in pairs; swim with quick darting motion mimicking Twin Fangblenny. Coastal reefs and sheltered lagoons in 2-25 m.
Asian Pacific: Sabah in Malaysia to Philippines.

Cardinalfishes

VARIEGATED CARDINALFISH *Fowleria variegata*
SIZE: to 7 cm (2 3/4 in.) Cardinalfishes – Apogonidae
ID: Variegated shades of brown; wide pale bands radiate from rear eye, may display faint dark bars on side, **dark ocellated spot on gill cover.** Solitary or form small groups. Seagrass beds, mangroves and coastal reefs to 27 m.
Indo-West Pacific: Red Sea and E. Africa to Indonesia, Micronesia. – S.W. Japan to Great Barrier Reef and Samoa.

SPOTLESS CARDINALFISH *Fowleria vaiulae*
SIZE: to 5 cm (2 in.) Cardinalfishes – Apogonidae
ID: Variegated shades of brown; often display alternating dark and light bars, alternating light and dark bands radiate from rear eye, **no spot gill cover.** Solitary. Sheltered coastal reefs and lagoons in 3-15 m.
Indo-Pacific: Red Sea and E. Africa to Indonesia, Micronesia and French Polynesia. – S.W. Japan to Great Barrier Reef.

BARRED CARDINALFISH *Fowleria marmorata*
SIZE: to 9 cm (3 1/2 in.) Cardinalfishes – Apogonidae
ID: Pale reddish brown with about 5-8 dusky reddish brown bars on side; **thin dark bands radiate from rear eye,** large ocellated black spot on gill cover. Solitary and cryptic. Hide in crevices and corals of coastal, lagoon and outer reefs to 20 m.
Indo-Pacific: Red Sea and E. Africa to Andaman Sea, Indonesia, Micronesia and French Polynesia. – S.W. Japan to Australia.

FLAME CARDINALFISH *Fowleria flammea*
SIZE: to 5 cm (2 in.) Cardinalfishes – Apogonidae
ID: Red head and body with yellowish to reddish iris. Solitary and secretive; nocturnal. Costal, lagoon and outer reefs in 10-40 m.
Asian Pacific: Indonesia, Philippines and Papua New Guinea.

HYALINE CARDINALFISH *Foa hyalina*
SIZE: to 5 cm (2 in.) Cardinalfishes – Apogonidae
ID: Pinkish to mauve; **distinct concave nape,** red-brown bands radiate from eye, several red-brown bars of irregular width on front of body. Solitary. Shelter among *Sinularia* leather corals of coastal and lagoon reefs in 2-15 m.
Asian Pacific: N. Sulawesi to Alor in Indonesia, Philippines, Micronesia and Papua New Guinea.

WEEDY CARDINALFISH *Foa fo*
SIZE: to 6.8 cm (2 3/4 in.) Cardinalfishes – Apogonidae
ID: Brown with numerous white blotches; **whitish irregular area on leading edge of dorsal and ventral fins.** Solitary or form small groups. Seagrass beds, weed, sand and rubble to 15 m.
Indo-Pacific: Red Sea and E. Africa to Indonesia, Philippines, Micronesia, Solomon Is., French Polynesia. – S. Japan to Australia.

HAYASHI'S CARDINALFISH *Pseudamia hayashii*
SIZE: to 7.5 (3 in.) Cardinalfishes – Apogonidae
ID: Slender body, long tail base and large rounded tail; reddish brown with silvery reflections on cheek. Solitary and nocturnal. Coastal, lagoon and outer reefs in 2-64 m. Most common member of genus encountered on coral reefs.
West Pacific: Indonesia, Philippines, Micronesia and Solomon Is. – S. Japan to N. Great Barrier Reef and Samoa.

GELATINOUS CARDINALFISH *Pseudamia gelatinosa*
SIZE: to 10 cm (4 in.) Cardinalfishes – Apogonidae
ID: Very slender body, long tail base and large rounded tail; shades of mauve to pink with stripes formed by rows of small dark spots, dusky spot on tail base. Solitary and nocturnal. Lagoons and sheltered coastal reefs to 40 m.
Indo-Pacific: Red Sea and E. Africa to Indonesia, Philippines, Micronesia and French Polynesia. – S. Japan to Australia.

PADDLEFIN CARDINALFISH *Pseudamia zonata*
SIZE: to 11.5 cm (4 1/2 in.) Cardinalfishes – Apogonidae
ID: Slender body, long tail base, large fan-shaped tail; **wide dark bars extend between dark 1st dorsal and ventral fin and 2nd dorsal and anal fin,** narrower bar on tail base. Solitary. Hover inside caves and crevices of seaward reef slopes in 10-35 m.
West Pacific: Bali, Sulawesi, West Papua in Indonesia to Philippines, Micronesia, Papua New Guinea, Solomon Is., Fiji.

BLACK-NOSED CARDINALFISH *Verulux cypselurus*
SIZE: to 6 cm (2 1/4 in.) Cardinalfishes – Apogonidae
ID: Translucent body with yellowish sheen on head and belly; **black spot or short stripe on snout,** darkish margins on tail. Form aggregations. Hover above coral patches and inside caves of sheltered coastal reefs and lagoons in 2-15 m.
Indo-West Pacific: Red Sea and E. Africa to Micronesia. – S.W. Japan to Australia, Vanuatu, New Caledonia and Fiji.

SLENDER CARDINALFISH *Rhabdamia gracilis*
SIZE: to 6 cm (2 1/4 in.) Cardinalfishes – Apogonidae
ID: Translucent body with silver sheen on head and belly; usually display **small black spot on lower tail base** and tips of tail lobes. Form aggregations. Sheltered coastal, lagoon and outer reefs in 2-15 m.
Indo-West Pacific: E. Africa to Indonesia, Philippines, Micronesia and Papua New Guinea. – S. Japan to Australia and Fiji.

GLASSY CARDINALFISH *Rhabdamia spilota*
SIZE: to 6 cm (2 1/4 in.) Cardinalfishes – Apogonidae
ID: Translucent body with silver sheen on head and belly; **small dark spot above pectoral fin base,** 1-2 faint brownish stripes on side of front of body. Form small groups. Seaward reefs in 17-45 m.
Localized: Known from Bali, Flores, Sulawesi and West Papua in Indonesia.

Cardinalfishes

BANGGAI CARDINALFISH *Pterapogon kauderni*
SIZE: to 8 cm (3 1/4 in.)   Cardinalfishes – Apogonidae
ID: White to cream with white spotting; 3 white-edged black bars; black tail lobes bordered with white spots. Form aggregations; males brood eggs and incubate young inside mouths. Shelter among urchin spines or anemones in protected rubble areas to 16 m.
Localized: Banggai Is., Lembeh Strait and N.W. Bali in Indonesia.

ORBICULAR CARDINALFISH *Sphaeramia orbicularis*
SIZE: to 12 cm (4 3/4 in.)   Cardinalfishes – Apogonidae
ID: Pale gray with silvery reflections; narrow dark midbody bar and variable-sized dark spots on rear body. From aggregations. Mangroves, near debris and beneath docks in sheltered bays and shoreline reefs to 3 m.
Indo-West Pacific: E. Africa to Indonesia, Philippines, Micronesia and Papua New Guinea. – S.W. Japan to New Caledonia and Fiji.

PAJAMA CARDINALFISH *Sphaeramia nematoptera*
SIZE: to 9 cm (3 1/2 in.)   Cardinalfishes – Apogonidae
ID: White with yellow wash on head; **red iris**, dark midbody bar and spots on rear body, elongate 2nd dorsal-fin tip. Form aggregations. Shelter among branching coral thickets of protected bays and lagoons to 14 m.
West Pacific: Sabah to Indonesia, Philippines, Micronesia and Solomon Is. – S.W. Japan to Great Barrier Reef and Fiji.

Pajama Cardinalfish – Night Phase
ID: At night the spots on rear body tend to disappear.

JEBB'S SIPHONFISH *Siphamia jebbi*
SIZE: to 3 cm (1 1/4 in.)   Cardinalfishes – Apogonidae
ID: Silvery translucent head and body peppered with small dark dots and overlaid with numerous orange dots; large eye and mouth. Form small groups. Sheltered bays and lagoons and reefs; shelter in small branching corals in 14-29 m.
West Pacific: Indonesia, Philippines, Papua New Guinea to Australia and Fiji.

CORAL SIPHONFISH *Siphamia corallicola*
SIZE: to 3.8 cm (1 1/2 in.)   Cardinalfishes – Apogonidae
ID: Silvery undercolor with large red-brown blotches; hint of large darkish spot on tail base. Form loose aggregations. Commonly shelter among thin branched coral colonies, *Seriatopora hystrix*. Lagoons and coastal inlets in 6-22 m.
Asian Pacific: Sabah in Malaysia to Indonesia, Philippines and Papua New Guinea.

TUBED SIPHONFISH *Siphamia tubifer*
SIZE: to 5 cm (2 in.) Cardinalfishes – Apogonidae
ID: Alternating silver-white and blackish stripes; two upper black stripes about equal width of white, **lower black stripe narrow.** Form groups. Shelter among *Diadema* and *Asthenosoma* urchin spines on sheltered reefs to 20 m.
Asian Pacific: Andaman Sea, Indonesia, Micronesia and Papua New Guinea. – S.W. Japan to Great Barrier Reef.

Tubed Siphonfish – Dark Phase
ID: Can rapidly change from striped phase to dark brown to black with transparent fins; **small white dot at rear of each dorsal fin.**

DUSKYLINED SIPHONFISH *Siphamia fuscolineata*
SIZE: to 3.5 cm (1¹/₂ in.) Cardinalfishes – Apogonidae
ID: Alternating dark and silver-white **stripes of equal width;** transparent fins, can rapidly change to black. Form groups. Shelter among *Diadema* urchin spines on sheltered reefs to 18 m.
Asian Pacific: Indonesia and Philippines to Micronesia, north to S. Japan.

ORANGE-SPOTTED SIPHONFISH *Siphamia cyanophthalma*
SIZE: to 3 cm (1¹/₄ in.) Cardinalfishes – Apogonidae
ID: Semi-transparent pinkish with silvery reflections; pair of narrow blue lines across eye iris, orange spots of variable size scattered on head and body, and orange stripe along upper edge of abdominal light organ. Solitary or small groups. Sheltered coral reefs in 15-30 m.
Localized N. Sulawesi, Flores and West Papua in Indonesia.

ELONGATE SIPHONFISH *Siphamia elongata*
SIZE: to 4.4 cm (1³/₄ in.) Cardinalfishes – Apogonidae
ID: Silvery undercolor with large red-brown blotches; hint of large darkish spot on tail base. Form aggregations. Sheltered lagoon and coastal reefs in 5-25 m.
Asian Pacific: Brunei, N.E. Kalimantan in Indonesia to Philippines.

NARROWLINED SIPHONFISH *Siphamia stenotes*
SIZE: to 3 cm (1¹/₄ in.) Cardinalfishes – Apogonidae
ID: Silvery translucent with narrow dark stripe from eye to tail base; darkish stripe from above eye to middle of back and another from below eye to tail base. Form aggregations among crinoids. Sheltered, silty coastal reefs in 10-15 m.
Localized: Known only from Triton Bay, West Papua, Indonesia.

Cryptic Crevice Dwellers
Basslets – Dottybacks – Devilfishes
Brotulas – Cusk-eels

This ID Group consists of secretive fishes that live primarily within the confines of the reef's structure.

SUBFAMILY: Basslets – Serranidae/Liopropomatinae
Single Genus – 4 Species Included

Typical Shape

This small family of slender, diminutive seldom-sighted sea basses live secretive lives within recesses of the reef. Reddish coloration and large eyes indicate the shadowed existence of these small, invertebrate-eating carnivores. The shallow-water species included have distinctly separate two-part dorsal fins and pinstriped bodies. Several additional members of the subfamily have been collected from depths below 100 meters. Unlike most sea basses, basslets are believed to remain the same sex throughout life, although their gonad structure indicates a hermaphroditic heritage.

FAMILY: Dottybacks – Pseudochromidae
10 Genera – 30 Species Included

Typical Shape

These, small, elongate, and often brightly colored, crevice-dwellers only venture into the open momentarily from their cryptic haunts inside the reef's structure. The color of dottybacks is often variable within a species, which is frequently believed to relate to an individual's sex. Their diet consists of small crustaceans including zooplankton and polychaete worms. Sex reversal appears to be a common family trait. Females lay adhesive egg masses, which are guarded and regularly picked up in the mouths of the males to mix and aerate.

FAMILY: Devilfishes (Longfins) – Plesiopidae
5 Genera – 7 Species Included

Typical Shape   Comet

Devilfishes, also commonly known as longfins, are typically an unfamiliar family because of a limited number of family representatives and their secretive nature. Family members are characterized by a single, long dorsal fin, often with deeply indented membrane between the spines, and elongate ventral fins. Those in genus *Assessor* commonly swim upside down orienting their bellies to the ceiling of caves and overhangs, and the males incubate egg bundles inside their mouths (a behavior possibly common to other members of the family). Probably the most recognizable and sought-after member of the family is the elegant Comet, which is occasionally glimpsed gliding through crevices on walls and dropoffs. Comets are known for their ability to mimic the white-spotted head of the Whitemouth Moray, *Gymnothorax meleagris,* by inserting their heads inside crevices when frightened. This ploy leaves their elongate rear bodies and fins, which bear a prominent false eye-spot, exposed. The Comet's adhesive eggs are deposited on the ceiling of a crevice where they are attended by the male until hatching.

FAMILY: Viviparous Brotulas – Bythitidae
2 Genera – 3 Species Included

Typical Shape

Brotulas are one of the few families of bony fishes known to be viviparous (bearing live offspring). The 85 species of these curious little fishes inhabit a variety of habitats worldwide, including freshwater caves, estuaries, reefs and the deep sea. Brotulas swim by undulating fins that encircle the length of their bodies. The male's copulatory organs, located behind the anus and surrounded by two pairs of pseudoclaspers, are apparently derived from the anal fin. Embryos, closely packed like cordwood, develop inside the female's ovaries. It is not known whether the offspring disperse in the currents or remain near their birthplace.

FAMILY: Cusk-eels – Ophidiidae
Single Genus – Single Species Included

Typical Shape

These cryptic fishes have eel-like bodies with long dorsal and anal fins that join at the tail to encircle the body. Ventral fins are absent or consist of one or two filamentous rays. Some species have barbels extending from around their mouths. Unlike the similar-appearing members of family Bythitidae, cusk-eels lay eggs. At night they occasionally leave the confines of the reef's nooks and crannies to hunt for crabs and fishes.

REDSTRIPED BASSLET   *Liopropoma tonstrinum*
SIZE: to 8 cm (3 1/4 in.)   Basslets – Liopropomatinae
ID: Pale reddish to yellowish head; two wide red stripes from head to tail with narrower white stripes between and on back. Solitary and cryptic. Caves and recesses of steep outer reef slopes in 11-50 m.
East Indo-West Pacific: Christmas I. to Indonesia, Philippines, Micronesia, Fiji and French Polynesia.

BLACKSTRIPED BASSLET   *Liopropoma latifasciatum*
SIZE: to 16 cm (6 1/4 in.)   Basslets – Liopropomatinae
ID: Yellow back and tail, pale below; black stripe from snout to tail. Solitary and cryptic deepwater dweller. Caves and recesses of outer slopes and dropoffs below 30 m.
North Asian Pacific: S. Japan and S. Korea to Micronesia.

MANYLINE BASSLET   *Liopropoma multilineatum*
SIZE: to 7.7 cm (3 in.)   Basslets – Liopropomatinae
ID: Pink head, yellowish body with longitudinal lines on scale rows; red rear body and tail base with central white stripe. Solitary and cryptic. Caves and recesses of steep outer reef slopes in 20-46 m.
West Pacific: Rowley Shoals, 260 km west of N.W. Australia to Philippines, Solomon Is., Coral Sea and Fiji.

STRIPED BASSLET   *Liopropoma susumi*
SIZE: to 9 cm (3 1/2 in.)   Basslets – Liopropomatinae
ID: Brownish gray to pale reddish with eight yellowish brown stripes. Solitary and cryptic. Caves and recesses in lagoon and outer reefs in 2-34 m.
Indo-West Pacific: Red Sea and E. Africa, Brunei, Sabah in Malaysia, Indonesia, Philippines, Micronesia, Solomon Is. and Line Is. – S.W. Japan to E. Australia and to Fiji.

OBLIQUE-LINED DOTTYBACK   *Cypho purpurascens*
SIZE: to 8 cm (3 1/4 in.)   Dottybacks – Pseudochromidae
ID: Male – Red to orange; fine blue scale margins form diagonal lines on body, may have one or two ocellated spots on dorsal fin. Solitary or in pairs. Coral reefs in 5-35 m.
West Pacific: Papua New Guinea, Solomon Is. to Great Barrier Reef, Coral Sea, Vanuatu and Fiji.

Oblique-lined Dottyback – Female
ID: Bluish gray on head and front of body shading to pale orange-yellow behind; fine blue scale margins form diagonal lines on body, **yellow patch on gill cover.**

CHECKERED DOTTYBACK   *Cypho zaps*
SIZE: to 6.3 cm (2 1/2 in.)   Dottybacks – Pseudochromidae
ID: Male – Orangish red forebody gradating to crimson or purple toward rear; fine, blue scale margins, **blue ring around back of eye** and blue bar on cheek. Solitary or pairs. Coral reefs in 5-35 m.
West Pacific: Central and E. Indonesia to Philippines and S.W. Japan.

Checkered Dottyback – Female
ID: Orange to purplish gray head graduating to reddish brown toward rear body; **blue ring around back of eye** with yellow to orange crescent behind.

FIRETAIL DOTTYBACK   *Labracinus cyclophthalmus*
SIZE: to 23.5 cm (9 1/2 in.)   Dottybacks – Pseudochromidae
ID: Highly variable; the variation shown has bright red body and fins, dark blotch on middle of 1st dorsal fin, **all variations have blue-gray diagonal lines on head and front of body.** Solitary or in loose pairs. Coastal, lagoon and outer reefs to 15 m.
Asian Pacific: Indonesia and Philippines to N.E. Papua New Guinea. – S.W. Japan to N.W. Australia.

Firetail Dottyback – Female
ID: Red with 8-18 thin blackish stripes on side; wavy blue oblique lines on head, occasionally a female has orangish brown head and pinkish body.

Firetail Dottyback – Variation
ID: Brick red with bright red fins; about 10 narrow black stripes; all variations, blue-gray diagonal lines on head and front of body, most variations display a set of whitish bars on forebody.
Localized: Indonesia, Sabah in Malaysia and Philippines.

Firetail Dottyback – Variation
ID: Dark brown to nearly black with large reddish patch on belly and midside; 1-2 pale bars under front of dorsal fin, blue-gray diagonal lines on head and front of body. Lurk near crevices and recesses in reefs.
Asian Pacific: S.W. Japan, Indonesia and Philippines to N.W. Australia.

Dottybacks

QUEENSLAND DOTTYBACK *Ogilbyina queenslandiae*
SIZE: to 15 cm (6 in.) Dottybacks – Pseudochromidae
ID: Male – Reddish on head and front of body and purplish behind; bluish dorsal, anal and tail fins. Solitary or form small groups. Lagoon and seaward reefs in 10-20 m.
Localized: Great Barrier Reef.

Queensland Dottyback – Female
SIZE: to 15 cm (6 in.)
ID: Brownish to grayish head, yellow-orange midbody shading to pinkish red behind; 5-6 brown bars on upper front of body.

SAILFIN DOTTYBACK *Oxycercichthys veliferus*
SIZE: to 12 cm (4 3/4 in.) Dottybacks – Pseudochromidae
ID: Male – Pale gray to yellowish or reddish with bluish upper head and pale yellow anal and tail fins; **blue blotch on front of dorsal fin,** large and long pointed tail. Solitary or form small groups. Lurk near and inside crevices and recesses in lagoon and seaward reefs in 12-35 m.
Localized: Great Barrier Reef.

Sailfin Dottyback – Young/Female
ID: Light mauve to pinkish with yellowish top of head extends to front of dorsal fin; bluish rear dorsal, anal and tail fins.

BLACKSTRIPE DOTTYBACK *Pseudochromis perspicillatus*
SIZE: to 10.4 cm (4 in.) Dottybacks – Pseudochromidae
IID: White to slightly yellowish; **dark gray to black stripe from snout to end of dorsal fin base.** Solitary. Coral outcroppings in sand and rubble areas of coastal reefs in 3-18 m.
Asian Pacific: Sabah in Malaysia to Indonesia, Philippines and Papua New Guinea.

Blackstripe Dottyback – Variation
ID: Pale yellowish with dusky gray to brown upper head and back; black stripe from snout tip to below beginning of dorsal fin often concealed on the dark upper head and back, **line formed by yellow spot on each lateral line scale becoming progressively smaller toward rear.**

PURPLETOP DOTTYBACK *Pictichromis diadema*
SIZE: to 6.2 cm (2 1/2 in.)   Dottybacks – Pseudochromidae
ID: Bright yellow to orange, usually display magenta band from snout to end of dorsal fin, but can be shorter or absent. Solitary or small groups. Base of cliffs or inside crevices and caves on steep slopes and coastal and seaward reefs in 10-30 m.
North Asian Pacific: Brunei, Sabah in Malaysia and Philippines.

ROYAL DOTTYBACK *Pictichromis paccagnellae*
SIZE: to 5.7 cm (2 1/4 in.)   Dottybacks – Pseudochromidae
ID: Magenta head and front half of body and yellow to orange rear body. Solitary or form small, scattered groups. Base of steep dropoffs or in caves and recesses on steep slopes and coastal and outer reefs in 5 to 40 m, usually below 15 m.
Asian Pacific: N.W. Australia, E. Indonesia, Papua New Guinea and Solomon Is.

FLAME DOTTYBACK *Pictichromis caitlinae*
SIZE: to 5.5 cm (2 1/4 in.)   Dottybacks – Pseudochromidae
ID: Bright yellow head and nape to mid-dorsal fin; magenta over remainder of body. Solitary or small groups. Silty or turbid reefs in 10-55 m.
Localized: Known only from Cenderawasih Bay, West Papua in Indonesia.

MAGENTA DOTTYBACK *Pictichromis porphyrea*
SIZE: to 6.3 cm (2 1/2 in.)   Dottybacks – Pseudochromidae
ID: Magenta. Solitary or form small, loosely scattered groups. Base of steep dropoffs or in caves and crevices in steep slopes and coastal and outer reefs in 5 to 40 m, usually below 15 m.
West Pacific: S.W. Japan, Indonesia, Philippines, Micronesia, N. Papua New Guinea to Fiji.

RAJA DOTTYBACK *Pseudochromis ammeri*
SIZE: to 9 cm (3 1/2 in.)   Dottybacks – Pseudochromidae
ID: Male – Whitish to light gray with bluish gray upper head and yellowish brown back; faint blue spot on most scales and dark mark behind eye (also on female). Solitary. Coral outcroppings on fringing reefs in 10-45 m.
Localized: Raja Ampat and Fakfak Peninsula in Indonesia.

Raja Dottyback – Female
ID: Bluish gray upper body, white below; dark gray to yellowish brown stripe from gill cover to tail base, yellow stripe on belly, dark mark behind eye.

Dottybacks

SURGE DOTTYBACK   *Pseudochromis cyanotaenia*
SIZE: to 6.2 cm (2 1/2 in.)   Dottybacks – Pseudochromidae
ID: Male – Blue with yellow area on lower head to pectoral fin; yellow stripe on front lateral line; may display light blue bars on side. **Female –** Reddish gray, orange tail with yellow margin. Solitary or pairs. Shallow surge of outer reef flats and slopes to 10 m.
West Pacific: Indonesia, Philippines, Micronesia, Papua New Guinea and Solomon Is. – S.W. Japan to E. Australia to Fiji.

HORSESHOE-TAILED DOTTYBACK   *Pseudochromis tapeinosoma*
SIZE: to 7.4 cm (3 in.)   Dottybacks – Pseudochromidae
ID: Male – Bright yellow to tan; broad greenish blue stripe from top of head along back to tail, scattered blue flecks on side; **horseshoe-shaped greenish blue border on tail.** Solitary or pairs, often in crevices. Coral reefs to 20 m.
Asian Pacific: Indonesia and Philippines to Solomon Is. and north to S. Japan.

DOUBLESTRIPED DOTTYBACK   *Pseudochromis bitaeniatus*
SIZE: to 7.2 cm (2 3/4 in.)   Dottybacks – Pseudochromidae
ID: Bluish gray to brown with yellowish brown head; **broad pale to white stripe from gill cover to end of tail.** Solitary. Crevices and caves in steep slopes and coastal and seaward reefs to 20 m.
Asian Pacific: Indonesia, Philippines, Micronesia to Papua New Guinea and Solomon Is. to N.E. Australia.

PLAIN DOTTYBACK   *Pseudochromis litus*
SIZE: to 6.3 cm (2 1/2 in.)   Dottybacks – Pseudochromidae
ID: Yellowish brown to bluish gray with pale to whitish lower body; bluish gray tail base, yellow spot on each scale on upper back. Usually hide beneath rocks and inside recesses. Coral outcrops with encrusting sponges and sea fans on current swept slopes in 15-50 m.
Localized: Indonesia.

DUSKY DOTTYBACK   *Pseudochromis fuscus*
SIZE: to 8.8 cm (3 1/2 in.)   Dottybacks – Pseudochromidae
ID: Dark Variation – Dark brown with white to translucent tail; blue spots on scales form stripes. Solitary. Lurk near crevices and recesses in coastal, lagoon and seaward reefs to 30 m.
East Indo-Asian Pacific: India to Andaman Sea, Brunei, Indonesia, Philippines, Micronesia, Papua New Guinea and Solomon Is. – S.W. Japan to Australia, New Caledonia and Vanuatu.

Dusky Dottyback – Yellow Variation
ID: Pale tan to brilliant yellow; blue spots on scales form stripes.

YELLOW DOTTYBACK *Pseudochromis moorei*

SIZE: to 10.5 cm (4¹/₄ in.) Dottybacks – Pseudochromidae
ID: Male – Yellow-orange with dark spot on central body scales; translucent fins, dusky band at front of gill cover and **dark spot on upper rear gill cover.** Solitary. Sand and rubble and low rocky outcroppings on coastal reefs in 12-30 m.
Localized: Philippines.

Yellow Dottyback – Female
ID: Dark gray body; yellowish tint on fins.

ERDMANN'S DOTTYBACK *Pseudochromis erdmanni*
SIZE: to 10.5 cm (4¹/₄ in.) Dottybacks – Pseudochromidae
ID: Male – Orange head and forebody gradating to dark brown to gray rear body; **narrow white border around rear eye, dark spot on upper gill cover,** dark spot on body scales. Solitary or pairs around coral outcroppings. Coastal reef slopes in 20-40 m.
Localized: N. Sulawesi to Halmahera, Ambon and Raja Ampat in Indonesia.

Erdmann's Dottyback – Female
ID: Dark brown to gray; narrow white border on rear eye, pale patch on gill cover with two darkish bars, darkish borders on tail.

STEENE'S DOTTYBACK *Pseudochromis steenei*
SIZE: to 10 cm (4 in.) Dottybacks – Pseudochromidae
ID: Male – Pinkish orange head and forebody with brown scale spots shading to dark brown rear body; **narrow white bar behind eye.** Lack dark gill spot like Erdmann's Dottyback [above]. Solitary or pair with female. Recesses in coral outcroppings and coastal reef slopes in 15-100 m.
Localized: E. Indonesia.

Steene's Dottyback – Female
SIZE: to 12 cm (4³/₄ in.)
ID: Dark brown to nearly black with yellow tail; narrow white bar behind eye, faint whitish patch on gill cover. Commonly near males.

Dottybacks

SOLAR DOTTYBACK *Pseudochromis matahari*
SIZE: to 5.6 cm (2 1/4 in.)   Dottybacks – Pseudochromidae
ID: Yellowish head and upper forebody, yellow lower head, pinkish red from midbody to tail base, purplish gray tail; **blue semicircular ring around lower half of eye.** Solitary. Hide around base of large barrel sponges in 32-45 m.
Localized: Halmahera and Raja Ampat in Indonesia.

ELONGATE DOTTYBACK *Pseudochromis elongatus*
SIZE: to 5.2 cm (2 in.)   Dottybacks – Pseudochromidae
ID: Orange to yellow head; purplish gray body with pinkish belly; **orange edged black tail.** Solitary. Coastal and seaward reefs in caves and around rocky outcrops in 5-60 m.
Localized: E. Indonesia.

ZIPPERED DOTTYBACK *Pseudochromis jace*
SIZE: to 7.8 cm (3 1/4 in.)   Dottybacks – Pseudochromidae
ID: White head and body; yellow dorsal fin, dark stripe from snout extends down back to tail base with **short vertical extensions below giving a "zipper" appearance.** Solitary or pairs. Around isolated rock and coral outcroppings in 35-52 m.
Localized: Halmahera, Balanta I. and Triton Bay, West Papua in Indonesia.

ORANGESPOTTED DOTTYBACK *Pseudochromis marshallensis*
SIZE: to 7 cm (2 3/4 in.)   Dottybacks – Pseudochromidae
ID: Brown to brownish gray or bluish gray with yellowish tail; **pale orange spots on scales form body stripes.** Solitary and cryptic. Caves, recesses and beneath ledges of coastal, lagoon and seaward reefs to 15 m.
Asian Pacific: Indonesia, Philippines, Micronesia, Papua New Guinea, Solomon Is. – S. Japan to Australia and Vanuatu.

MANY-SCALED DOTTYBACK *Lubbockichthys multisquamatus*
SIZE: to 7 cm (2 3/4 in.)   Dottybacks – Pseudochromidae
ID: Red to reddish brown, purple or pink body, usually yellowish upper head and pink lower head extending to belly; fins translucent to reddish or yellow. Outer reef slopes in caves and recesses 12-60 m, usually below 20 m.
East Indo-Asian Pacific: Cocos-Keeling to Brunei, Indonesia, Philippines, Solomon Is. and N. Great Barrier Reef.

MIDNIGHT DOTTYBACK *Manonichthys paranox*
SIZE: to 8.6 cm (3 1/4 in.)   Dottybacks – Pseudochromidae
ID: Black with no distinctive markings. Flare dorsal and anal fins to mimic Pygmy Angelfish, (the two species occur in same area). Solitary. Lurk near recesses and crevices on coastal slopes in 5-30 m.
Asian Pacific: Papua New Guinea, Solomon Is. and Great Barrier Reef.

LONGFIN DOTTYBACK *Manonichthys polynemus*
SIZE: to 10 cm (4 in.) Dottybacks – Pseudochromidae
ID: Brown to gray-brown with pale spots on scales forming scale row stripes; triangular orange mark below eye, **orange blotch at base of ventral fins,** black margins on dorsal and anal fins. Solitary. Outer slopes in 5-25 m.
Asian Pacific: Halmahera and Sulawesi in Indonesia, Philippines and Micronesia.

SPLENDID DOTTYBACK *Manonichthys splendens*
SIZE: to 9 cm (3 1/2 in.) Dottybacks – Pseudochromidae
ID: Gray undercolor with orange to yellow spots on scales forming scale row stripes, orange tail, yellow snout; black eye bar. Solitary. Lurk near recesses and tube sponges on steep reef slopes in 5-40 m.
Localized: E. Indonesia from Sulawesi to West Papua.

COLLARED DOTTYBACK *Pseudoplesiops collare*
SIZE: to 3.8 cm (1 1/2 in.) Dottybacks – Pseudochromidae
ID: Pale grayish pink to pale yellow or green body; pale head with red band from eye to nape and red bar from front of dorsal fin to pectoral fin, long white ventral fins. Around rock outcroppings in sandy fringe around coral reefs in 30-70 m.
Localized: Bali to Flores and West Papua in Indonesia.

BEARDED DOTTYBACK *Pseudoplesiops immaculatus*
SIZE: to 4 cm (1 1/2 in.) Dottybacks – Pseudochromidae
ID: Greenish yellow to bright yellow and pink; **skin flap extends from chin.** Solitary and cryptic. Holes and recesses of seaward reefs to 15 m.
East Indo-Asian Pacific: Maldives to Great Barrier Reef and Coral Sea.

RINGEYED DOTTYBACK *Pseudoplesiops typus*
SIZE: to 3.7 cm (1 1/2 in.) Dottybacks – Pseudochromidae
ID: Red Variation – Red body and median fins with thin blue borders on dorsal and anal fins; white ventral fins, **dark ring around eye.** Cryptic inhabitant of reef crevices in 6-20 m.
Asian Pacific: N. Kalimantan and Komodo in Indonesia to S. Philippines, Micronesia, Papua New Guinea and Solomon Is. to Great Barrier Reef and Coral Sea.

Ringeyed Dottyback – Yellow-gray Variation
ID: Light gray with yellowish median fins and yellow hue on lower head and pectoral fin base.

Devilfishes – Viviparous Brotulas

RED-TIPPED LONGFIN   *Plesiops coeruleolineatus*
SIZE: to 8.5 cm (3 1/4 in.)   Devilfishes – Plesiopidae
ID: Brown; blue-edged black stripe on dorsal fin, red to yellowish dorsal fin spine tips, dark line below eye and pair of dark spots behind eye. Solitary and cryptic. Crevices and recesses in outer reefs to 23 m.
Indo-West Pacific: Red Sea and E. Africa to Indonesia, Philippines and Micronesia. – S.W. Japan to Australia and Samoa.

SPOTGILL LONGFIN   *Plesiops corallicola*
SIZE: to 18 cm (7 in.)   Devilfishes – Plesiopidae
ID: Dark brown with small blue spots on head, body and fins; **thin blue submarginal band on dorsal fin,** dark edged blue spot on gill cover. Cryptic. Crevices on outer reefs to 23 m.
East Indo-Pacific: Cocos-Keeling Is. toIndonesia, Philippines, and Micronesia. – S.W. Japan to Fiji and Cook Is.

BLUE DEVILFISH   *Assessor macneilli*
SIZE: to 6 cm (2 1/4 in.)   Devilfishes – Plesiopidae
ID: Dark blue; forked tail. Form small cryptic groups; often swim upside down orienting to the ceiling of caves; brood eggs in mouth. Caves, recesses and under ledges of coastal reefs in 5-20 m.
West Pacific: Great Barrier Reef, Coral Sea and New Caledonia.

YELLOW DEVILFISH   *Assessor flavissimus*
SIZE: to 6 cm (2 1/4 in.)   Devilfishes – Plesiopidae
ID: Bright yellow to brownish yellow; forked tail. Form small cryptic groups; often swim upside down orienting belly to the ceiling of caves; brood eggs in mouth. Caves, recesses and under ledges of coastal reefs and outer slopes in 5-20 m.
Localized: N. Great Barrier Reef and S.E. Papua New Guinea.

BANDED LONGFIN   *Belonepterygion fasciolatum*
SIZE: to 5 cm (2 in.)   Devilfishes – Plesiopidae
ID: Shades of brown with pale lower head; thin dark bars on lower body, broad white stripe on top of head from lip to dorsal fin, pale-edged dark spot on gill cover. Solitary and cryptic. Caves and crevices in coastal reefs and lagoons to 10 m.
Asian Pacific: E. Indonesia and Philippines. – S.W. Japan to E. Australia and New Caledonia.

CHRISTMAS LONGFIN   *Steeneichthys nativitatus*
SIZE: to 3.2 cm (1 1/2 in.)   Devilfishes – Plesiopidae
ID: Brown body with vague stripes formed by dark centered scale rows, all fins except pectorals dark brown; **white margin on all fins except ventrals.** Reef crevices in 15-40 m.
East Indo-Asian Pacific: Christmas I. to Indonesia.

COMET   *Calloplesiops altivelis*

SIZE: to 16 cm (6 ¼ in.)   Devilfishes – Plesiopidae

ID: Dark brown to nearly black; head, body and fins covered with small white spots, white-edged black spot on dorsal fin. Solitary and cryptic. Crevices on outer reefs in 3-45 m.

Indo-Pacific: Red Sea and E. Africa to Indonesia, Philippines, Micronesia, Papua New Guinea, Solomon Is. and French Polynesia. – S.W. Japan to Australia.

Comet – Variation

ID: Variation with striped dorsal, ventral and anal fins from Philippines. Appearance of rear body with prominent eye spot is believed to mimic head of Whitespotted Moray when only the Comet's tail extends from a hiding hole.

DUSKY VIVIPAROUS BROTULA   *Diancistrus fuscus*

SIZE: to 7.5 cm (3 in.)   Viviparous Brotulas – Bythitidae

ID: Brown elongate body with long dorsal and anal fins and narrow tail; ventral fins are long, thin, barbel-like strands used for sensing food. Solitary; common but cryptic and seldom seen by divers. Caves and crevies of coast reef to 20 m.

North Asian Pacific: S.W. Japan, Taiwan, and Philippines.

ALLEN'S VIVIPAROUS BROTULA   *Diancistrus alleni*

SIZE: to 8 cm (3 ¼ in.)   Viviparous Brotulas – Bythitidae

ID: Uniform yellow elongate body with long dorsal and anal fins and narrow tail; ventral fins long, thin barbel-like strands used for sensing food. Solitary and cryptic. Coral reef crevices in 5-17 m.

East Indo-West Pacific: Maldives to Indonesia, Great Barrier Reef, Vanuatu and Samoa.

SLIMY VIVIPAROUS BROTULA   *Brosmophyciops pautzkei*

SIZE: to 7.5 cm (3 in.)   Viviparous Brotulas – Bythitidae

ID: Pale grayish with translucent fins; darkish stripe under dorsal fin. Body covered with thick clear mucous. Coastal and offshore reefs to 55 m.

Indo-Pacific: Madagascar to Indonesia, Philippines, Micronesia, Papua New Guinea, Solomon Is., Hawaii and Pitcairn Is. – S.W. Japan to E. Australia.

BEARDED BROTULA   *Brotula multibarbata*

SIZE: to 60 cm (2 ft.)   Cusk-eels – Ophidiidae

ID: Shades of gray to brownish gray; eel-like body with barbels "whiskers" around mouth, thread-like ventral fins. Solitary and cryptic; occasionally in open at night. Deep recesses and caves in coastal and outer reefs to 220 m.

Indo-Pacific: Red Sea to Indonesia, Micronesia, Solomon Is., Hawaii and French Polynesia. – S.W. Japan to E. Australia.

Elongate Sand & Burrow Dwellers
Dartfishes – Tilefishes – Dragonets – Sandperches
Lizardfishes – Others

This ID Group consists of fishes that typically inhabit sandy bottoms and often live in burrows. Shrimp gobies and sand gobies, which also live on the sand and in burrows, are presented in the next ID Group.

FAMILY: Dartfishes – Ptereleotridae
5 Genera – 21 Species Included

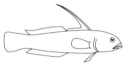

Typical Shape

Dartfishes are elegant little, elongate fishes with two-part dorsal fins, long anal fins and small upturned mouths. They generally hover alone, in pairs or small groups just above the bottom where they feed on water-borne zooplankton. When approached, they dart into burrows or under rocks.

FAMILY: Tilefishes – Malacanthidae
2 Genera – 11 Species Included

Typical Shape

Tilefishes, also known as sand tilefishes, are moderate-sized fishes with long, continuous, unnotched dorsal fins. They commonly occur in pairs on outer sand or rubble reef slopes. Those in genus *Hoplolatilus* pick plankton from the current a few feet above their bottom shelters where they dive for protection when threatened. The Flagtail Blanquillo, *Malacanthus brevirostris*, lives in rock mounds of their own construction.

FAMILY: Dragonets – Callionymidae
5 Genera – 17 Species Included

Typical Shape

Dragonets are small, charismatic bottom-dwelling fishes with two-part dorsal fins, somewhat flattened heads and bodies, and protrusible mouths used for snapping up tiny benthic invertebrates. Although generally cryptically patterned, several species are quite colorful. Males often display a slightly different pattern from the smaller females. The males also have taller, more elaborate first dorsal fins that are erected intermittently as they dart over the bottom and during evening courtship that culminates just after sunset in a brief spawning rise.

FAMILY: Sandperches – Pinguipedidae
Single Genus – 14 Species Included

Typical Shape

These elongate sand and rubble dwellers live in small harems with single dominant, territorial males. A few species have proven to be hermaphroditic, changing from females to males with age. The appealing fishes are easily approached and often perch near divers watching their activities.

FAMILY: Lizardfishes – Synodontidae
3 Genera – 8 Species Included

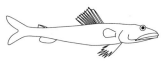

Typical Shape

Lizardfishes are voracious lie-in-wait predators with large, tooth-filled mouths. They have pointed snouts and long, cylindrical bodies bearing small, high first dorsal fins followed by tiny rear dorsal fins toward the tail. Experts at camouflage, pairs and solitary individuals rest motionless on the bottom, blending with their surroundings. Some species bury in the sand with only their heads protruding as they wait for unsuspecting prey. At times they attack quite large prey and are capable of taking fishes several meters above the bottom with lightning-fast strikes. Because they are similar in color and markings, especially in their pale phases, lizardfishes can be difficult to identify to species.

FAMILY: Others

Wormfishes – Microdesmidae

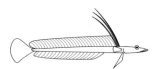

Sand Divers – Trichonotidae

Bandfishes – Cepolidae

Convict Blennies – Pholidichthyidae

Jawfishes – Opistognathidae

Pearlfishes – Carapidae

285

Dartfishes

DECORATED DARTFISH *Nemateleotris decora*
SIZE: to 8.5 cm (3¼ in.) Dartfishes – Ptereleotridae
ID: Pale head and forebody shading to purplish rear body; violet snout and stripe to dorsal fin, violet and red marked fins, elongate 1st dorsal fin. Solitary or in pairs; shelter in sandy burrows. Sand and rubble patches at base of reefs in 28-68 m.
Indo-West Pacific: Red Sea and Mauritius to Micronesia. – S.W. Japan to N. Great Barrier Reef, New Caledonia and Fiji.

HELFRICH'S DARTFISH *Nemateleotris helfrichi*
SIZE: to 6.3 cm (2½ in.) Dartfishes – Ptereleotridae
ID: Yellow head and lavender body, yellowish rear dorsal, anal and tail fins; violet stripe on top of head, black ventral fin tips, long whitish and red 1st dorsal fin. Solitary or in pairs. Steep outer reef slopes in 25-70 m.
Pacific: S.W. Japan, Micronesia to Fiji and French Polynesia.

FIRE DARTFISH *Nemateleotris magnifica*
SIZE: to 8 cm (3¼ in.) Dartfishes – Ptereleotridae
ID: Yellow head, white forebody shading to reddish brown rear body; dark brown tail, very long 1st dorsal fin. Solitary or in pairs. Hover above burrows on patches of sand and rubble on outer reef slopes in 6-60 m.
Indo-Pacific: E. Africa to Indonesia, Micronesia, Hawaii and Pitcairn Is. – S. Japan to Great Barrier Reef, New Caledonia.

LINED DARTFISH *Ptereleotris grammica*
SIZE: to 10 cm (4 in.) Dartfishes – Ptereleotridae
ID: Pale blue-gray with greenish highlights, orange and blue stripes on body and fins; first dorsal fin relatively high and fan-shaped. Solitary or in pairs that share burrows. Sand and rubble slopes in 36-60 m.
Asian Pacific: E. Indonesia, Philippines and Papua New Guinea. – S. Japan to N. Great Barrier Reef.

TWOTONE DARTFISH *Ptereleotris evides*
SIZE: to 13.8 cm (5½ in.) Dartfishes – Ptereleotridae
ID: Pale bluish gray head and forebody abruptly shading to black rear body; pale forked tail with dark borders, iridescent blue markings on gill cover. Pairs share sandy burrows. Exposed lagoon and outer reef slopes in 2-25 m.
Indo-Pacific: Red Sea and E. Africa to Indonesia, Philippines, Micronesia and Line Is. – S.W. Japan to Great Barrier Reef.

Twotone Dartfish – Juvenile/Young Adult
SIZE: 3-6 cm (1¼-2¼ in.)
ID: Silvery gray with yellow-green wash; dark border on 2nd dorsal and anal fins and margins of tail, black spot on tail base. Pairs share sandy burrows.

THREADFIN DARTFISH *Ptereleotris hanae*
SIZE: to 12 cm (4 ³/₄ in.) Dartfishes – Ptereleotridae
ID: Pale bluish gray to bluish green upper body and darker blue below; tail of larger adults trail 1-6 filaments. Solitary or in pairs. Shelter under rocks or inside burrows shared with shrimp gobies. Sand and rubble bottoms near reefs in 3-50 m.
Pacific: Indonesia, Philippines, Micronesia, Solomon Is. and Line Is. – S. Japan to Australia and Fiji.

BANDTAIL DARTFISH *Ptereleotris uroditaenia*
SIZE: to 10 cm (4 in.) Dartfishes – Ptereleotridae
ID: Pale blue; iridescent blue markings on head; stripe along base of dorsal fin, **pair of black bands with yellow between on tail.** Solitary or form small groups. Sand and rubble areas in 18-40 m.
Asian Pacific: Brunei, Indonesia, Solomon Is. and Great Barrier Reef.

SPOTTAIL DARTFISH *Ptereleotris heteroptera*
SIZE: to 12 cm (4 ³/₄ in.) Dartfishes – Ptereleotridae
ID: Bright to pale blue to bluish gray with iridescent blue marks on head; yellow to bluish **tail with elongate black area on center.** Solitary in pairs or colonies. Shelter in burrows on sand and rubble near reefs in 7- 46 m.
Indo-Pacific: Red Sea and E. Africa to Indonesia, Micronesia, Hawaii and French Polynesia. – S.W. Japan to Great Barrier Reef.

REDSPOT DARTFISH *Ptereleotris rubristigama*
SIZE: to 10.5 cm (4 ¹/₄ in.) Dartfishes – Ptereleotridae
ID: Pale blue, usually with wide dark blue stripe extending from behind pectoral fin onto tail; **large red spot on upper pectoral fin base** (not always obvious); tail rounded without filaments. Solitary or small groups. Sometimes shelter in burrow with shrimp gobies. Sand bottoms in 15-60 m.
Localized: E. Indonesia from Bali to W. Papua.

MONOFIN DARTFISH *Ptereleotris monoptera*
SIZE: to 15.5 cm (6 in.) Dartfishes – Ptereleotridae
ID: Pale blue to green to yellow with wash of blue over belly; **broad blackish bar angles from mouth to eye,** tail of adults trail filaments. Form loose colonies. Shelter in burrows on sand and rubble or hard bottoms in 6-15 m.
Indo-Pacific: Seychelles to Indonesia, Philippines, Micronesia, Solomon Is. and French Polynesia. – S. Japan to Great Barrier Reef.

PEARLY DARTFISH *Ptereleotris microlepis*
SIZE: to 12 cm (4 ³/₄ in.) Dartfishes – Ptereleotridae
ID: Pale bluish to pinkish gray; usually several iridescent bluish stripes on head, **narrow dark bar on base of pectoral fin,** numerous indistinct body bars. Form small to large colonies. Shelter in burrows on sand and rubble near reefs to 22 m.
Indo-Pacific: Red Sea and E. Africa to Indonesia, Philippines, Micronesia and French Polynesia. – S.W. Japan to S.E. Australia.

ZEBRA DARTFISH   *Ptereleotris zebra*
SIZE: to 11.4 cm (4¹/₂ in.)   Dartfishes – Ptereleotridae
ID: Green to greenish gray with about 20 narrow orange to pink bars; darkish bar below eye and across pectoral fin base. Blue chin flap extened during courtship. Form aggregations along rock walls affected by surge in 2-10 m.
Indo-Pacific: Red Sea and Seychelles to Indonesia, Philippines and French Polynesia. – S.W. Japan to Great Barrier Reef.

RAINFORD'S DARTFISH   *Parioglossus rainfordi*
SIZE: to 4.5 cm (1³/₄ in.)   Dartfishes – Ptereleotridae
ID: Grayish, white lower head and belly; thin yellow stripe from eye to tail with wide gray to black areas below, **black bar on upper tail base.** Form aggregation among mangrove roots and adjacent coral reefs to 7 m.
Asian Pacific: West Papua in Indonesia, Philippines, Micronesia, Papua New Guinea, N. Great Barrier Reef and New Caledonia.

YELLOW-STRIPED DARTFISH   *Parioglossus formosus*
SIZE: to 4 cm (1¹/₂ in.)   Dartfishes – Ptereleotridae
ID: Broad yellowish **stripe from snout to tail wedged between pair of dark brown to blackish stripes that converge to form point on tail.** Form dense aggregations near surface. Sheltered bays, mangroves and marine lakes.
West Pacific: Indonesia, Philippines, Micronesia, Papua New Guinea. – S.W. Japan to Great Barrier Reef, Vanuatu and Fiji.

INTERRUPTED DARTFISH   *Parioglossus interruptus*
SIZE: to 3 cm (1¹/₄ in.)   Dartfishes – Ptereleotridae
ID: Brownish with **white belly;** yellow midlateral stripe, thin blackish stripe extends along lower body to black marking on tail. Forms aggregations near the surface. Marine lakes, tidal streams and mangrove swamps.
Asian Pacific: Indonesia, Philippines and Papua New Guinea, north to S. Japan.

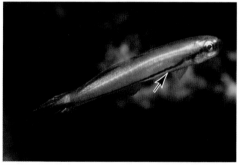

NAKED DARTFISH   *Parioglossus nudus*
SIZE: to 2.3 cm (1 in.)   Dartfishes – Ptereleotridae
ID: Translucent with light orange-brown midlateral stripe and white belly; slightly concave tail. Form aggregations. Sheltered coastal reefs in 5-35 m.
West Pacific: Indonesia, Philippines, Micronesia to Papua New Guinea, Solomon Is. and Fiji.

RAO'S DARTFISH   *Parioglossus raoi*
SIZE: to 3 cm (1¹/₄ in.)   Dartfishes – Ptereleotridae
ID: Light brown; yellow midlateral stripe, **dark brown to blackish stripe extends diagonally from eye to light brown belly and along lower body to tail base.** Form aggregations near surface. Sheltered rocky shorelines and mangroves.
East Indo-West Pacific: Persian Gulf to Andaman Sea, Indonesia, Philippines and Micronesia – S.W. Japan and Fiji.

NEW GUINEA MINIDARTFISH *Aioliops novaeguineae*
SIZE: to 2.3 cm (1 in.) Dartfishes – Ptereleotridae
ID: Translucent pale gray; yellowish snout and top of head, thin black stripe extends from top of head beneath dorsal fin to tail base, diffuse black stripe on lower side, **large black spot on translucent tail edged on top with yellow.** Aggregations hover above coral thickets. Inshore reefs to 10 m.
Localized: West Papua in Indonesia and Papua New Guinea.

BIGSPOT MINIDARTFISH *Aioliops megastigma*
SIZE: to 3 cm (1¼ in.) Dartfishes – Ptereleotridae
ID: Yellowish head, mauve back and brown lower side, bluish belly; black stripe along back and **large black spot on tail bordered with white.** Form groups that hover over coral thickets. Sheltered shoreline and lagoon reefs in 2-10 m.
Asian Pacific: Indonesia and Philippines.

ROBUST RIBBONGOBY *Oxymetopon compressus*
SIZE: to 20 cm (8 in.) Dartfishes – Ptereleotridae
ID: Laterally compressed body; light blue-gray, dusky lower body and **blackish lower tail,** faint blue diagonal body bands, widely scattered blue markings on head. Form small groups. Shelter in burrows on mud bottoms in 15-40 m.
Asian Pacific: S. China, Indonesia, Philippines and Papua New Guinea. – S.W. Japan to Great Barrier Reef.

BLUE-BARRED RIBBONGOBY *Oxymetopon cyanoctenosum*
SIZE: to 20 cm (8 in.) Dartfishes – Ptereleotridae
ID: Light blue-gray with **wavy body bands;** pale yellowish gill cover and front of belly, blue markings on head. Solitary or form small groups. Shelter in burrows on silt or mud slopes in 20-45 m.
Asian Pacific: Indonesia and Philippines.

YELLOW-SPOTTED BANDFISH
Acanthocepola abbreviata
SIZE: to 20 cm (8 in.)
Bandfishes – Cepolidae
ID: Eel-like body with short ventral fins; red with silvery white lower head; yellow spots or short bars on back and another series along lower side. Form colonies in separate burrows on sand or mud bottoms to 40 m.
Indo-Asian Pacific: Persian Gulf to Indonesia, Philippines, Papua New Guinea and Australia.

CONVICT FISH *Pholidichthys leucotaenia*
SIZE: to 10 cm (4 in.) Convict Fish – Pholidichthyidae
ID: Juvenile – Blackish with whitish stripe from eye to rear body. **Adult –** Black saddles and white bars, to 34 cm (13 in.) remain inside sand burrows. Form large to huge aggregations over slopes; mimic Striped Catfish. Coastal reefs in 3-30 m.
Asian Pacific: Brunei, E. Indonesia, Philippines, Papua New Guinea, Solomon Is. and New Caledonia.

THREADFIN SAND DIVER *Trichonotus elegans*
SIZE: to 17 cm (6 ³/₄ in.)   Sand Divers – Trichonotidae
ID: Male – Pale brown to gray with horizontal rows of spots; large whitish, yellow or black ventral fins displayed during courtship; 3-4 long rays extend from dark foredorsal fin. **Female –** Lack long dorsal fin rays. Loose aggregations form just above sand; bury quickly when alarmed. Sand slopes to 40 m.

East Indo-West Pacific: Maldives to Indonesia, Micronesia and Fiji.

GOLDBAR SAND DIVER *Trichonotus halstead*
SIZE: to 14 cm (5 ¹/₂ in.)   Sand Divers – Trichonotidae
ID: Male – Tan with numerous white to gold spots on body and fins; 8-9 brown bars across back, black shaded ventral fins, **1-2 large ocellated spots on dorsal fin,** 5-6 long dorsal fin rays. Sand slopes in 12-35 m.

Localized: Sulawesi in Indonesia to Papua New Guinea.

SPOTTED SAND DIVER *Trichonotus setiger*
SIZE: to 19 cm (7 ¹/₂ in.)   Sand Divers – Trichonotidae
ID: Male – Tan to golden with numerous white spots; **10-11 brown body bars;** dorsal fin translucent with white spots on rays; long foredorsal fin rays. Form loose aggregations just above sand; quickly bury when alarmed. Sandy bottoms to 20 m.

Asian Pacific: Indonesia and Philippines to Solomon Is., Coral Sea, Great Barrier Reef and New Caledonia.

PEARLY SIGNALFISH *Pteropsaron springeri*
SIZE: to 3 cm (1¹/₄ in.)   Sand Divers – Trichonotidae
ID: Male – Silvery to pinkish brown; row of white spots on back and row of blue dashes below, very tall 1st dorsal fin. **Female –** Similar, but with much shorter black 1st dorsal fin. Both sexes rest on long ventral fins. Sand slopes in 30-50 m.

Asian Pacific: Indonesia and Philippines.

ONESTRIPE WORMFISH *Gunnellichthys pleurotaenia*
SIZE: to 9 cm (3¹/₂ in.)   Wormfishes – Microdesmidae
ID: Elongate worm-like body; pale gray to yellow-white with **narrow orange-brown to black stripe from snout to tail base.** Solitary or in pairs. Hover or swim with undulating motion just above sand bottoms of lagoons and sheltered reefs in 3-15 m.

Indo-West Pacific: E. Africa to Indonesia, Philippines, Micronesia and Solomon Is. – S.W. Japan to Great Barrier Reef and Fiji.

ORANGESTRIPE WORMFISH *Gunnellichthys viridescens*
SIZE: to 7.2 cm (3 in.)   Wormfishes – Microdesmidae
ID: Elongate worm-like body; **orange-brown to salmon with pearly midlateral stripe,** orange lips and orange stripe through eye, bright orange stripe on midtail. Solitary or pairs. Hover or swim with undulating motion above sand slopes in 3-50 m.

East Indo-West Pacific: Persian Gulf to Indonesia, Philippines, Micronesia and Solomon Is. to Great Barrier Reef and Samoa.

CURIOUS WORMFISH *Gunnellichthys curiosus*
SIZE: to 12 cm (4 3/4 in.) Wormfishes – Microdesmidae
ID: Elongate worm-like body; bluish white with **broad orange stripe from snout to tail base,** black spot on tail. Solitary or in pairs. Hover or swim with undulating motion just above sand and rubble at base of outer slopes in 9-60 m.
Indo-Pacific: Madagascar to Indonesia, Micronesia, Solomon Is., Hawaii and French Polynesia.– S.W. Japan to Great Barrier Reef.

ONESPOT WORMFISH *Gunnellichthys monostigma*
SIZE: to 11 cm (4 1/4 in.) Wormfishes – Microdesmidae
ID: Elongate worm-like body; light blue to tan, **small dark spot on rear gill cover,** blue line from snout to top of nape. Solitary or in pairs. Hover or swim with undulating motion just above sand and rubble bottoms of lagoons in 6-20 m.
Indo-Pacific: East Africa to Indonesia, Philippines, Solomon Is., Micronesia to French Polynesia. – S.W. Japan to N. Australia.

CHLUPATY'S TILEFISH *Hoplolatilus chlupatyi*
SIZE: to 15 cm (6 in.) Tilefishes – Malacanthidae
ID: Sky blue with pair of darker blue stripes on cheek; **yellow bar on iris,** occasionally broad yellow area on back or with narrow yellow area on base of dorsal fin. Solitary. Deep rubble slopes in 30-55 m.
Asian Pacific: Indonesia and Philippines.

STOCKY TILEFISH *Hoplolatilus randalli*
SIZE: to 19 cm (7 1/2 in.) Tilefishes – Malacanthidae
ID: Pale blue to green with **long blue patch from end of rear dorsal fin to tail;** stout/deep body compared to other tilefishes. Groups hover above huge constructed mounds of sand and rubble on deep sand and rubble slopes in 35-85 m.
Asian Pacific: E. Indonesia, Philippines, Micronesia to Solomon Is.

PALE TILEFISH *Hoplolatilus cuniculus*
SIZE: to 16 cm (6 1/4 in.) Tilefishes – Malacanthidae
ID: Pale gray to yellowish with **fine blue margin on dorsal and anal fins.** Solitary, in pairs or loose colonies; shy, quickly retreat into burrows when threatened. Rubble slopes on outer reefs in 25-115 m.
Indo-Pacific: E. Africa to Indonesia, Philippines, Micronesia, Solomon Is., French Polynesia. – S.W. Japan to Great Barrier Reef.

YELLOW TILEFISH *Hoplolatilus luteus*
SIZE: to 13 cm (5 in.) Tilefishes – Malacanthidae
ID: Bright yellow, undersides often whitish; **blue line above eye extends to black spot on upper gill cover.** Solitary or form small groups hovering above mud, silt and fine sandy bottoms well away from reefs in 30-40 m.
Asian Pacific: Andaman Sea to Indonesia.

291

BLUEHEAD TILEFISH *Hoplolatilus starcki*
SIZE: to 15 cm (6 in.) Tilefishes – Malacanthidae
ID: Yellowish tan with bright blue head and bright yellow tail.
Juvenile – Sky blue without marking. Similar to Purple Queen
Anthias; often mix with anthias. Usually in pairs. Hover above
burrows in sand and rubble on outer reef slopes in 20-105 m.
Pacific: Indonesia, Papua New Guinea, Micronesia to French
Polynesia. – Philippines to Great Barrier Reef and Fiji.

PURPLE TILEFISH *Hoplolatilus purpureus*
SIZE: to 12 cm (4 ³/₄ in.) Tilefishes – Malacanthidae
ID: Lavender with **broad dark red to black upper and lower
borders on tail.** Solitary or form small groups. Shy; retreat to
sandy burrow when threatened. Sand and rubble slopes of
seaward reefs in 18-80 m.
Asian Pacific: Halmahera and West Papua in Indonesia,
Philippines and Solomon Is.

REDLINED TILEFISH *Hoplolatilus marcosi*
SIZE: to 11 cm (4 ¹/₄ in.) Tilefishes – Malacanthidae
ID: White with **vivid arching red stripe** that narrows on rear
body before widening on tail. Solitary or form small groups.
Shy; retreat to sandy burrow when threatened. Sand and
rubble slopes of seaward reefs in 18-60 m.
Asian Pacific: Indonesia, Philippines, Papua New Guinea
and Solomon Is.

ERDMANN'S TILEFISH *Hoplolatilus erdmanni*
SIZE: to 16 cm (6 ¹/₄ in.) Tilefishes – Malacanthidae
ID: Bluish gray with bright blue tail base; upper back
reddish to greenish brown with 15-17 bars extending to
lower side, **bright red patch on central tail.** Solitary, pairs
or trios. Rubble bottoms in 42-60 m.
Localized: Known only from Raja Ampat, Triton Bay and
Cenderawasih Bay, West Papua in Indonesia.

FOURMANOIR'S TILEFISH *Hoplolatilus fourmanoiri*
SIZE: to 14 cm (5 ¹/₂ in.) Tilefishes – Malacanthidae
ID: Lilac body with yellow upper head extending to back; white
lower head and forebody extends past pectoral fin base, **black
tail base and center of tail,** black spot on upper corner of gill
cover; yellowish patch behind pectoral fin. Solitary or pairs. Sand
and rubble in 18-55 m.
Asian Pacific: Vietnam to Brunei, Philippines and Solomon Is.

FLAGTAIL BLANQUILLO *Malacanthus brevirostris*
SIZE: to 30 cm (12 in.) Tilefishes – Malacanthidae
ID: Pale yellow head and pale blue body; **pair of black stripes
on tail,** vague body bars. Usually in pairs; hover with
undulating motion near sandy burrows typically constructed
under rocks. Sand and rubble of seaward reefs in 14-45 m.
Indo-Pacific: Red Sea and E. Africa to Indonesia, Micronesia,
Hawaii and French Polynesia. – S. Japan to E. Australia.

BLUE BLANQUILLO *Malacanthus latovittatus*

SIZE: to 44 cm (17 1/4 in.) Tilefishes – Malacanthidae

ID: Blue head with blue to blue-green back and pale underside; broad black midlateral stripe extends onto tail. Solitary or in pairs; swim away when threatened rather than retreating to burrows. Sand and rubble areas in 5-30 m.

Indo-Pacific: Red Sea and E. Africa to Indonesia, Micronesia, Solomon Is. and Line Is. – S. Japan to New Caledonia and Fiji.

Blue Blanquillo – Juvenile

SIZE: to 12 cm (4 3/4 in.)

ID: Primarily black with white stripe on back that extends onto upper edge of tail; with increased growth white area expands and the black area is reduced to a midlateral stripe.

FINGERED DRAGONET *Dactylopus dactylopus*

SIZE: to 15 cm (6 in.) Dragonets – Callionymidae

ID: Male – Mottled in shades of brown with dark blotches on side; separated 1st ray of ventral fin used for "walking"; 1st dorsal fin with long filamentous rays and **blue spot between 1st and 2nd rays.** Solitary or in pairs on sandy bottoms in 3-55 m.

Asian Pacific: Andaman Sea to Indonesia and Philippines. – S.W. Japan to W. Australia and Great Barrier Reef.

Fingered Dragonet – Female

ID: Mottled in shades of brown with dark blotches on side; alternating dark and light bands on tail, **black ocellated spot on lower rear of long 1st dorsal fin,** dark horizontal streaks on 2nd dorsal fin.

ORANGE & BLACK DRAGONET *Dactylopus kuiteri*

SIZE: to 15 cm (6 in.) Dragonets – Callionymidae

ID: Mottled in shades of brown with dark blotches on side; long 1st dorsal fin with large ocellated spot near rear base, (Similar Finger Dragonet female [previous] has shorter 1st dorsal.); blue spots on dark anal fin. Sand bottoms to 40 m.

Asian Pacific: Sabah in Malaysia, Indonesia and Philippines.

Orange & Black Dragonet – Juvenile

ID: White with black and orange marked dorsal fins. **Female –** Body spotted in shades of brown; most of tail yellow with dark margin.

Dragonets

MANDARINFISH *Synchiropus splendidus*
SIZE: to 6.7 cm (2 1/2 in.) Dragonets – Callionymidae
ID: Orange with ornate pattern of dark-edged green and blue bands and spots and a few yellow line markings on lower head. Small groups shelter among coral rubble. Come out of hiding at dusk to spawn. Sheltered coastal reefs and lagoons to 18 m.
Asian Pacific: Indonesia, Philippines, Micronesia and Solomon Is.– S.W. Japan to Great Barrier Reef and New Caledonia.

PICTURESQUE DRAGONET *Synchiropus picturatus*
SIZE: to 6 cm (2 1/4 in.) Dragonets – Callionymidae
ID: Light brown to greenish covered with large dark spots ringed with orange and green, blue-green band markings around eye. Form small loose groups. Rubble patches near living corals of coastal reefs in 2-10 m.
Asian Pacific: Sabah in Malaysia, Indonesia and Philippines.

MOYER'S DRAGONET *Synchiropus moyeri*
SIZE: to 8.3 cm (3 1/4 in.) Dragonets – Callionymidae
ID: Male – Whitish undercolor with reddish brown blotches; large fan-shaped dorsal fin marked with concentric bands and pair of dark "eye" spots. Solitary or form small groups dominated by a single large male. Algae covered rocks in 3-30 m.
Asian Pacific: Indonesia, Philippines, Micronesia, Papua New Guinea and Solomon Is – S. Japan to Great Barrier Reef.

Moyer's Dragonet – Female
ID: Reddish mottling with white spots of varying size and shape; **1st dorsal fin mostly black with wide whitish outer margin.** Solitary or form small groups dominated by a single large male.

STARRY DRAGONET *Synchiropus stellatus*
SIZE: to 6 cm (2 1/4 in.) Dragonets – Callionymidae
ID: Male – Red with white and black blotches and saddles, **yellowish snout tip;** blue spots and lines on head, tall dorsal fin ornately banded with pale-edged spot in middle. Solitary or form small groups. Sandy areas in 5-20 m.
Indian Ocean: E. Africa to Andaman Sea and N.W. Sumatra in Indonesia.

Starry Dragonet – Juvenile
SIZE: to 3 cm (1 1/4 in.)
ID: White with red blotches; **red bands on 2nd dorsal and anal fin,** red bands around eye, short black first dorsal fin. **Female –** Similar, but red blotches more extensive.

BARTELS' DRAGONET *Synchiropus bartelsi*
SIZE: to 4.5 cm (1 3/4 in.) Dragonets – Callionymidae
ID: Male – Shades of brown to red; **row of white midlateral spots with black semicircular edge on upper portion,** tall 1st dorsal fin. Solitary or in pairs. Lagoon and seaward reefs in 6-35 m.
Asian Pacific: Indonesia, Philippines, and Papua New Guinea.

Bartels' Dragonet – Female/Red Phase
ID: Adults in all phases have tiny blue ocellated spots on back and the midlateral semicircular markings. Note low 1st dorsal fin characteristic of female.

MORRISON'S DRAGONET *Synchiropus morrisoni*
SIZE: to 8 cm (3 1/4 in.) Dragonets – Callionymidae
ID: Male – Red to pink with random dark blotches; **large black blotch at pectoral fin base,** tall 1st dorsal fin with vertical dark brown bands. Solitary, in pairs or small groups. Algal-turf rocks of seaward reefs in 10-33 m.
West Pacific: Indonesia, Philippines and Micronesia. – S. Japan to Great Barrier Reef and Fiji.

Morrison's Dragonet – Female
ID: Red to pink with random dark blotches and pale spots; large dark blotch at pectoral base, dusky fan-shaped first dorsal fin with vertical streaks.

OCELLATED DRAGONET *Synchiropus ocellatus*
SIZE: to 8.9 cm (3 1/2 in.) Dragonets – Callionymidae
ID: Male – Large fan-shaped dorsal fin with concentric bands and 2-4 dark ocellated spots. Solitary or small groups. Sand and rocks to 30 m.
Pacific: Indonesia, Philippines, Micronesia and Papua New Guinea to French Polynesia. – S.W. Japan to Great Barrier Reef.

Ocellated Dragnet – Female
ID: Dark greenish brown to golden brown with irregular whitish saddles and blotches; small blue dots on head and body; **1st dorsal fin dark with white outer margin.**

GORAM DRAGONET   *Diplogrammus goramensis*
SIZE: to 9 cm (3 1/2 in.)   Dragonets – Callionymidae
ID: Male – Shades of brown with flower-like spots; blue-edged brown bars on cheek, irregular blue marking on gill cover, **distinct horizontal ridge on lower side.** Solitary or small groups. Sand and rubble in lagoon and seaward reefs in 5-40 m.
Asian Pacific: Indonesia to Papua New Guinea, Micronesia and Cook Is. – Philippines to Great Barrier Reef.

LEAFY DRAGONET   *Anaora tentaculata*
SIZE: to 5.9 cm (2 1/4 in.)   Dragonets – Callionymidae
ID: Mottled and spotted in shades of brown and gray; **numerous small skin flaps on side of body.** Solitary, experts at camouflage. Sand, rubble and weedy bottoms near coastal, lagoon and outer reefs in 5-40 m.
Asian Pacific: Indonesia, Philippines, Micronesia and Papua New Guinea. – S.W. Japan to Great Barrier Reef.

SPOTTED DRAGONET   *Callionymus pleurostictus*
SIZE: to 5.2 cm (2 in.)   Dragonets – Callionymidae
ID: Male – Mottled yellowish tan with white belly; row of irregular blotches along back, row of 8-10 diffuse brown blotches on side, **1st dorsal fin spine extremely long,** 2-3 rows of small blue spots along side. Sand bottoms to 22 m.
Asian Pacific: Vietnam, Brunei, Indonesia, N. Australia and New Caledonia.

SUPERB DRAGONET   *Callionymus superbus*
SIZE: to 30 cm (12 in.)   Dragonets – Callionymidae
ID: Male – Blotchy brown with two irregular rows small pale spots; extremely long tail, **1st dorsal fin triangular with 3 long rays extending past fin,** yellow on lower head and dorsal fin. **Female –** Similar, but with black dorsal fin. Solitary or in pairs. Sand and rubble in 2-15 m.
Localized: Indonesia and Philippines.

LONG FILAMENT DRAGONET   *Callionymus keeleyi*
SIZE: to 6 cm (2 1/4 in.)   Dragonets – Callionymidae
ID: Mottled shades of brown; wavy dark bands on dorsal fins, **blue spots and lines on head,** 1st two dorsal fin spines long and filamentous. Solitary or form small groups. Sand and rubble near reefs of coastal, lagoon and outer reefs in 5-60 m.
Asian Pacific: E. Indonesia, Philippines and Papua New Guinea.

MANGROVE DRAGONET   *Callionymus enneactis*
SIZE: to 7.5 cm (3 in.)   Dragonets – Callionymidae
ID: Male – Mottled tan and white; row of vague white blotches along side, **dark "moustache"' below eyes,** 3rd and 4th spines of tall ornate dorsal fin with orange rear edge. Pairs or small groups. Sand near mangroves and silty reefs to 12 m.
Asian Pacific: Singapore to Micronesia, Solomon Is. and New Caledonia. – S.W. Japan to Great Barrier Reef.

WONGAT DRAGONET *Callionymus zythros*
SIZE: to 9 cm (3 1/2 in.) Dragonets – Callionymidae
ID: **Male –** Light brown with various-sized blotches and spots on head and body; long thin tail, **1st dorsal fin yellowish with black bands separated by white.** Solitary or small groups. Silty sand in 15-28 m.
Localized: Raja Ampat in Indonesia to Madang in Papua New Guinea.

NEPTUNE DRAGONET *Callionymus neptunius*
SIZE: to 23 cm (9 in.) Dragonets – Callionymidae
ID: **Male –** Grayish brown with numerous fine white spots; **yellow 1st dorsal fin with irregular dark vertical lines,** long brown and white tail taller in front when raised. Sand bottoms in 5-37 m.
East Indo-Asian Pacific: Sri Lanka to Indonesia, Philippines, Papua New Guinea and Solomon Is.

LATTICED SANDPERCH *Parapercis clathrata*
SIZE: to 17.5 cm (7 in.) Sandperches – Pinguipedidae
ID: **Male –** Grayish tan back with darkish blotches, white below with row of black centered orangish brown blotches on lower body; **dark ocellated spot above gill cover.** Solitary or small groups. Sand and rubble of lagoon and seaward reefs in 3-50 m.
Indo-West Pacific: Andaman Sea to Micronesia, Phoenix Is. – S.W. Japan to Australia and Fiji.

Latticed Sandperch – Female
ID: Similar to male, but lack ocellated spot; both have dark spots on top of snout and cheek and **white or yellow streak on center of tail.**

SPECKLED SANDPERCH *Parapercis hexophtalma*
SIZE: to 23 cm (9 in.) Sandperches – Pinguipedidae
ID: White with grayish back; numerous dark lines, spots and speckles, **large black spot on tail. Male –** [top] Wavy lines on cheek. Solitary or form small loose groups. Sand or rubble bottoms of coastal, lagoon and outer reefs in 8-25 m.
West Pacific: Red Sea and E. Africa to Micronesia and Solomon Is. – S.W. Japan to Great Barrier Reef and Fiji.

YELLOWLINE SANDPERCH *Parapercis flavolineata*
SIZE: to 14 cm (5 1/2 in.) Sandperches – Pinguipedidae
ID: White to pinkish with 10 brown bars, every other bar narrow and irregular; irregular pale to yellowish midlateral stripe from rear gill cover to tail base. Solitary. Sand bottoms of coastal slopes in 25-70 m.
Localized: Known only from Bali and N. Sulawesi, Indonesia.

Sandperches

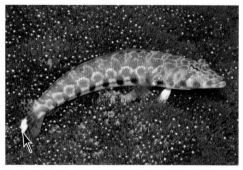

DIAGONAL SANDPERCH *Parapercis diagonalis*

SIZE: to 15 cm (6 in.)   Sandperches – Pinguipedidae

ID: Pale whitish to gray; 6 triangular-shaped brown to reddish **saddles on back connecting to six wide bars on lower body;** diagonal brown band below eye. Solitary. Sandy sheltered bottoms to 25 m.

Localized: Known only from Bali and Solor in Indonesia.

SPOTTED SANDPERCH *Parapercis millepunctata*

SIZE: to 18 cm (7 in.)   Sandperches – Pinguipedidae

ID: Pale undercolor with intermingled dark and light brown blotches on back; double row of large brown to blackish spots on middle and lower side, **white blotch on tail.** Solitary or small groups. Sand and rubble near reefs in 4-30 m.

East Indo-Pacific: Maldives to Indonesia, Philippines, Micronesia and Pitcairn Is. – S.W. Japan to Great Barrier Reef.

RETICULATED SANDPERCH *Parapercis tetracantha*

SIZE: to 26 cm (10 in.)   Sandperches – Pinguipedidae

ID: Brown with double row of 8 nearly white square blotches along middle and lower side and corresponding light brown saddles above on back; **pair of dark brown bars on head.** Sand and rubble bottoms near reefs in 12-25 m.

Asian Pacific: Andaman Sea to S.W. Japan, Indonesia, Philippines, Micronesia and Solomon Is.

TWOSPOT SANDPERCH *Parapercis bimacula*

SIZE: to 12.5 cm (5 in.)   Sandperches – Pinguipedidae

ID: Pale gray to whitish with brown to reddish mottling; 10 oval bars on side, **reddish bar below eye encloses two small black spots.** Solitary. Sand and scattered rock bottoms in 2-20 m.

Asian Pacific: Andaman Sea to Sumatra, Java, Bali, and Komodo I. in Indonesia.

DOUBLESPOT SANDPERCH *Parapercis diplospilus*

SIZE: to 10 cm (4 in.)   Sandperches – Pinguipedidae

ID: Pale tan on back shading to white on lower body; about 9 brown saddles on back and row of 8-9 large brown spots on lower side, **pair of large blackish spots on base of tail.** On mud and sand bottoms near reefs in 20-50 m.

Asian Pacific: Indonesia, Philippines and Papua New Guinea.

REDSPOTTED SANDPERCH *Parapercis schauinslandii*

SIZE: to 13 cm (5 in.)   Sandperches – Pinguipedidae

ID: White with 8-9 red to brownish bars or paired spots; **red 1st dorsal fin,** pair of small dark spots on tail base. Solitary or form groups, occasionally hover. Sand and rubble bottoms of seaward reef slopes in 10-50 m.

Indo-Pacific: E. Africa to Indonesia, Philippines, Micronesia, Hawaii and Pitcairn Is. – S. Japan to Great Barrier Reef.

REDBARRED SANDPERCH   *Parapercis multiplicata*

SIZE: to 12 cm (4 3/4 in.)   Sandperches – Pinguipedidae

ID: Whitish with 8 narrow red bars below back each containing a pair of small dark spots; **red or black spot on belly above ventral fin base.** Solitary or form small groups. Sand and rubble bottoms of clear water reefs in 25-40 m.

Pacific: Indonesia, Papua New Guinea to Pitcairn Is.– S.W. Japan to Great Barrier Reef and New Caledonia.

Redbarred Sandperch – Variation

ID: Without bars on side, but small dark spots combine with yellow dashes to form broken line on lower side and red or black spot above ventral fin base.

NOSESTRIPE SANDPERCH   *Parapercis lineopunctata*

SIZE: to 10 cm (4 in.)   Sandperches – Pinguipedidae

ID: White with darkish saddles on back and narrow bars aligned between saddles on sides; **black stripe from snout through eye.** Solitary or form small groups. Usually on clean sand near reefs in 5-35 m.

Asian Pacific: Indonesia, Philippines, Papua New Guinea and Solomon Is. to Great Barrier Reef.

BLACKFIN SANDPERCH   *Parapercis snyderi*

SIZE: to 10.5 cm (4 in.)   Sandperches – Pinguipedidae

ID: Five brown to reddish saddles on back; **spotted tail fin with red margin,** 8-9 darkish bars on side, black 1st dorsal fin. Sand and rubble near reefs in 10-40 m.

East Indo-Asian Pacific: Indonesia, Philippines, Micronesia and Papua New Guinea. – S.W. Japan to Australia.

SHARPNOSE SANDPERCH   *Parapercis cylindrica*

SIZE: to 12 cm (4 3/4 in.)   Sandperches – Pinguipedidae

ID: Whitish undercolor; about 8 dark squarish bars on back and about 8 dark bars on lower body aligned between bars on back, **narrow dark bar below eye,** tail yellow or translucent. Solitary or form small groups. Sand, rubble and weeds to 20 m.

Asian Pacific: Gulf of Thailand to Indonesia, Philippines, Micronesia and Papua New Guinea. – S. Japan to Australia.

WHITESTRIPE SANDPERCH   *Parapercis xanthozona*

SIZE: to 23 cm (9 in.)   Sandperches – Pinguipedidae

ID: **Male –** Thin white bars on cheek; about 6 U-shaped saddles on back; 10 bars on lower side, wavy white stripe on side extends onto tail. **Female –** Similar, but lack bars on cheek. Silty bottoms of coastal reefs and lagoons in 10-25 m.

West Pacific: Indonesia to Micronesia and Solomon Is. – S. Japan to Australia, New Caledonia and Fiji.

CLOUDED LIZARDFISH *Saurida nebulosa*
SIZE: to 28 cm (11 in.) Lizardfishes – Synodontidae
ID: Very similar to Slender Lizardfish [next], but **distinguished by shorter pectoral fins, rear edge does not extend to above ventral fin base.** Solitary. Sand and silt bottoms in 2-60 m.
Indo-Pacific: Mauritius to Indonesia, Philippines, Micronesia, Hawaii and French Polynesia. – S.W. Japan to Great Barrier Reef.

SLENDER LIZARDFISH *Saurida gracilis*
SIZE: to 28 cm (11 in.) Lizardfishes – Synodontidae
ID: Mottled gray to brown; **3 dark or diffuse bars on rear body,** line pattern on lips; teeth visible when mouth closed. Solitary. Sand or silty bottoms near protected reefs to 12 m.
Indo-Pacific: Red Sea and E. Africa to Indonesia, Philippines, Micronesia, Solomon Is., Hawaii and French Polynesia. – S.W. Japan to Great Barrier Reef.

REDMARBLED LIZARDFISH *Synodus rubromarmoratus*
SIZE: to 8.4 cm (3 ¼ in.) Lizardfishes – Synodontidae
ID: Relatively small compared to other lizardfishes; reddish brown with red or pink head markings, **series of irregular hourglass-shaped red body saddles.** Solitary or form small groups. Sand and rubble, mainly near seaward reefs in 5-50 m.
West Pacific: Indonesia, Philippines and Papua New Guinea. – S.W. Japan to Great Barrier Reef, New Caledonia and Fiji.

CLEARFIN LIZARDFISH *Synodus dermatogenys*
SIZE: to 23 cm (9 in.) Lizardfishes – Synodontidae
ID: Mottled gray to brown with 6 saddles across back; **8-9 dark diamond-shaped spots frequently with pale centers along middle of side,** cluster of 6 dark spots on snout tip. Solitary, in pairs or small groups. Sand or rubble bottoms to 70 m.
Indo-Pacific: Red Sea and E. Africa to Indonesia, Micronesia, Hawaii and French Polynesia. – S.W. Japan to Australia.

TWOSPOT LIZARDFISH *Synodus binotatus*
SIZE: to 18 cm (7 in.) Lizardfishes – Synodontidae
ID: Whitish to pale brown with 6-7 irregular dark red to brown bars; series of large dark spots with small spot between on lower side; **pair of small dark spots on snout tip.** Solitary or in pairs. Often on hard bottoms of seaward reefs in 3-20 m.
Indo-Pacific: Red Sea and E. Africa to Indonesia, Micronesia, Hawaii and French Polynesia. – S.W. Japan to Australia.

BLACKBLOTCH LIZARDFISH *Synodus jaculum*
SIZE: to 20 cm (8 in.) Lizardfishes – Synodontidae
ID: Mottled gray to brown with bars across back and diamond-shaped spots on sides; **black band on tail base.** Solitary, in pairs or small groups; occasionally swim above bottom. Sand and rubble bottoms in 2-88 m.
Indo-Pacific: Red Sea and E. Africa to Indonesia, Solomon Is. and French Polynesia. – S.W. Japan to Australia.

REEF LIZARDFISH *Synodus variegatus*
SIZE: to 28 cm (11 in.) Lizardfishes - Synodontidae
ID: Gray to whitish with about 6 irregular dark bars on upper side with smaller blotches between; whitish stripe on lower side interrupted by dark red to brown bars. Most frequently seen lizardfish on coral reefs in 5-60 m.
Indo-Pacific: Red Sea and E. Africa to Indonesia, Micronesia, Solomon Is., Hawaii and French Polynesia. – S.W. Japan to Australia.

SNAKEFISH *Trachinocephalus myops*
SIZE: to 32 cm (13 in.) Lizardfishes – Synodontidae
ID: Brown to brownish yellow stripes alternate with pale blue stripes; dark vague bars, **black spot on upper rear gill opening,** short rounded upturned snout. Solitary; often bury leaving eyes exposed. Sand bottoms in 3-400 m.
Circumtropical: Absent in E. Pacific.

VARIABLE JAWFISH
Opistognathus variabilis
SIZE: to 10 cm (4 in.)
Jawfishes – Opistoganthus
ID: Whitish; during courtship males turn blue and hover above burrows [pictured], females orange, eight evenly spaced double blotches on lower dorsal fin, 7-8 dark blotches on side. Solitary. Inhabit rock-lined burrows on sand and rubble bottoms.
Asian Pacific: Indonesia, Philippines and Micronesia.

Variable Jawfish – Variation
ID: Highly variable. The pictured form is relatively common and has a mottled yellowish brown pattern sometimes with a hint of blue on the posterior body and fins. Another form is mottled pale brown with about 7 squarish dark brown blotches on middle of side.

YELLOWBARRED JAWFISH *Opistognathus randalli*
SIZE: to 11 cm (4¼ in.) Jawfishes – Opistognathidae
ID: Dark brown upper head and white body; 8-10 pale yellow to orange body bars, **bright yellow mark on front of upper iris,** yellowish dorsal, anal and tail fins with black spot on front of dorsal fin. Inhabit rock-lined burrows on sand and rubble bottoms near reefs in 5-30 m.
Asian Pacific: Sabah in Malaysia, Indonesia and Philippines.

DENDRITIC JAWFISH *Opistognathus dendriticus*
SIZE: to 28 cm (11 in.) Jawfishes – Opistognathidae
ID: Large; yellowish brown with dark brown spots and branching blotches, **brown "mask" across eye,** narrow white margin on dorsal fin and black blotch between 3rd and 5th spines. Solitary. Inhabit rock-lined burrows on sand and rubble bottoms near reefs in 2-40 m.
Asian Pacific: Sabah in Malaysia, Indonesia and Philippines.

Jawfishes – Pearlfishes

DARWIN JAWFISH
*Opistognathus
darwiniensis*
SIZE: 45 cm (18 in.)
Jawfishes –
Opistognathidae
ID: Large; **tan with brown spots and blotches,** yellowish fins with prominent bands or spots, pale-rimmed black spot on front of dorsal fin. Solitary. Inhabit rock-lined burrows on tidal flats and coastal reefs to 10 m.
Localized:
N.W. Australia to Gulf of Carpentaria, Queensland in Australia.

ANDAMAN JAWFISH   *Opistognathus cyanospilotus*
SIZE: to 13 cm (5 ¹/₂ in.)   Jawfishes – Opistognathidae
ID: Dark brown head with blue spots, yellow-brown front body shading to blue rear body and yellow tail. Solitary. Inhabit rock-lined burrows on sand and rubble bottoms near reefs in 5-25 m.
Localized: E. Andaman Sea, Malaysia, Sumatra and Bali in Indonesia.

DOTTED JAWFISH   *Opistognathus papuensis*
SIZE: to 45 cm (18 in.)   Jawfishes – Opistognathidae
ID: Large; pale brown to light gray with **black rounded elongate spots.** Solitary. Inhabit rock-lined burrows on sand and rubble bottoms near reefs in 5-60 m.
Localized: N. Australia.

CHINSTRAP JAWFISH   *Opistognathus sp.*
SIZE: to 7.9 cm (3 ¹/₄ in.)   Jawfishes – Opistognathidae
ID: Red brown; white blotches on head and **black bar beneath chin,** prominent white-edged black spot on front of dorsal fin. Solitary. Inhabit rock-lined burrows on sand and rubble bottoms in 10-45 m.
West Pacific: Sabah in Malaysia, Indonesia, Philippines to Solomon Is. – S.W. Japan to Great Barrier Reef and Fiji.

SOLOR JAWFISH   *Opistognathus solorensis*
SIZE: to 9 cm (3 ¹/₂ in.)   Jawfishes – Opistognathidae
ID: Mottled tan to yellow-brown to red-brown; **1-2 dark oblong spots on front of dorsal fin,** 4-5 white spots below dorsal fin, 2 faint bars on upper lip extend into upper mouth. Solitary. Inhabit rock-lined burrows in 10-45 m.
Asian Pacific: Brunei, Philippines, Indonesia, Micronesia and Papua New Guinea.

Solor Jawfish – Yellow-brown Variation
ID: Variation with a single dark spot on foredorsal fin.

CASTELNAU'S JAWFISH   *Opistognathus castelnaui*
SIZE: to 28 cm (11 in.)   Jawfishes – Opistognathidae
ID: Dull white with network of brown interconnected wavy lines and small blotches; row of brown rectangular spots on base of dorsal fin. Solitary or loose groups. Seldom seen by divers. Inhabit sand-rubble bottoms between 15-100 m.
Asian Pacific: Malaysia, Indonesia and Philippines. – S.W. Japan to Australia.

MIMIC JAWFISH   *Stalix* sp.
SIZE: to 6 cm (2 1/4 in.)   Jawfishes – Opistognathidae
ID: Dark brown with yellowish blotches on upper back and tail base; **scrawled yellowish-white markings on rear gill cover,** irregular white band on rear dorsal, anal and tail fins. Build burrows not lined with rocks on sand slopes from 5-20 m.
Asian Pacific: Known only from Lembeh Strait in Indonesia and Dumaguete in Philippines.

VERSLUYS' JAWFISH   *Stalix versluysi*
SIZE: to 4 cm (1 1/4 in.)   Jawfishes – Opistognathidae
ID: Brown marbling on head, body and dorsal fin; black spot followed by small white spot on rear spiny dorsal fin, **pale-edged brown patch on rear jaw.** Silt and sand in 15-45 m.
Asian Pacific: Sulawesi, Molucca Is. and West Papua in Indonesia and N. Palawan in Philippines.

YELLOWFIN JAWFISH   *Stalix* sp.
SIZE: to 6 cm (2 1/4 in.)   Jawfishes – Opistognathidae
ID: Gray back with white scrawl pattern; pink and white mottled head; white blotch on lower midside, yellowish soft tissue on dorsal, anal and tail fins. Build burrows not lined with rocks on sand slopes in 7 to 20 m.
Localized: Known only from Lembeh Strait in Indonesia.

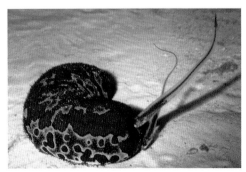

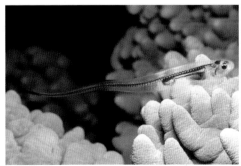

SILVER PEARLFISH   *Encheliophis homei*
SIZE: to 19 cm (7 1/2 in.)   Pearlfishes – Carapidae
ID: Translucent with silvery head, dark body organs and backbone visible; elongate tapering body. Live in body cavity of large sea cucumbers, entering through anus, leave host to feed at night. Sand and rubble to 30 m.
Indo-Pacific: Red Sea and E. Africa to Hawaii and Society Is. in French Polynesia. – S.W. Japan to Australia.

Silver Pearlfish
ID: Occasionally sighted feeding in the open at night.

Small, Elongate Bottom Dwellers – Gobies

This ID Group consists of small, elongate, typically bottom-dwelling species in the family Gobiidae.

FAMILY: Gobies – Gobiidae
54 Genera – 250 Species Included

| Typical Shape | Typical Shape | Genus *Gobiodon* |

Because of similar elongate bodies and small size, species in the goby and blenny families, Blenniidae (following ID Group), are often confused with one another. Blennies can be easily distinguished from gobies by noting their single, long continuous dorsal fins (except for a small group known as triplefins), ventral fins situated clearly in front of their pectoral fins, and their habit of resting on the bottom with curved bodies. Gobies have distinct, two-part dorsal fins and ordinarily perch with straight bodies.

Gobies are the largest family of marine fishes with more than 1,600 species in 220 genera currently classified. Approximately 500 species in 212 genera inhabit the Indo-Pacific region; however, numerous species still remain undescribed.

Most of the small carnivores live on bottom habitats varying from coral reefs to sand flats. Gobies feed on tiny crustaceans, including shrimps, copepods, worms, sponges and mollusks. A few species dine on drifting plankton just above the bottom. Gobies have distinct two-part dorsal fins and ventral fins that join, or nearly join, beneath the body forming cup-shaped discs.

A large group of gobies live on the open sand. Because of the danger from roving predators, a few species excavate subterranean dwellings; however, a large number of species, known as shrimp gobies, survive by sharing burrows with snapping shrimps from genus *Alpheus*. The nearly-blind shrimps need the sharp-eyed gobies to warn of danger, while the gobies need a ready-made place to hide. Throughout the day the shrimps industriously maintain extensive winding tunnels connecting two or three enlarged chambers by continually hauling dirt up from below. Passages are typically one inch in diameter, two to four feet long and run a few inches beneath the surface. The instability of shifting sands in the upper sections requires constant maintenance and new burrow openings must be dug frequently. While the shrimps toil, the gobies perch near the burrow openings acting as sentinels, except, now and again, when they dart a few inches away to grab a mouthful of sand that is filtered for food. When the shrimps appear above ground, they nearly always keep at least one antenna in contact with the gobies' bodies, usually near their tails. Warning signals range from a slight tail twitch, indicating caution, to a thrash for alarm. Once the warning is given, the time it takes for the duo to disappear can be measured in tenths of a second.

Because of their vigilant nature, the burrowmates are difficult, but not impossible, to approach closely. Once a goby is sighted, remain still for a minute or two before moving slowly in its direction. To get a good view of the shrimps bulldozing their entranceways is worth the effort. They battle the shifting sands like power robots – heaving and hoisting, poking and plowing. Using their single enlarged snapping claws, the mini-titans lift shell fragments twice their weight, often to have the tunnel collapse for their efforts.

RANDALL'S SHRIMPGOBY   *Amblyeleotris randalli*
SIZE: to 11 cm (4 1/4 in.)   Gobies – Gobiidae
ID: White with 6-7 orange bars on head and body; tall rounded 1st dorsal fin with pale-edged black spot. Share burrow with alpheid shrimp beneath overhangs and shallow caves on steep outer reef slopes and dropoffs in 15-48 m.
West Pacific: Sabah in Malaysia, Indonesia, Philippines, Papua New Guinea and Solomon Is. – S.W. Japan to Great Barrier Reef and Fiji.

FLAGTAIL SHRIMPGOBY
Amblyeleotris yanoi
SIZE: to 10 cm (4 in.)
Gobies – Gobiidae
ID: White with 5 light brown diffuse bars; yellow and orange tail with blue line markings, blue stripes on 2nd dorsal and anal fins, dark bar markings on white iris. Share burrow with alpheid shrimp. Lagoon and seaward reef slopes in 3-35 m.
Asian Pacific: Indonesia, Micronesia, Solomon Is. – S.W. Japan to Fiji.

SPOTTED SHRIMPGOBY   *Amblyeleotris guttata*
SIZE: to 10 cm (4 in.)   Gobies – Gobiidae
ID: White; bright orange spots on head and body, dark band below back of head and another on belly, whitish iris with black markings. Share burrow with alpheid shrimp. Sandy bottoms of lagoons and seaward reef slopes in 4-35 m.
West Pacific: Brunei and Sabah in Malaysia, Indonesia, Philippines and Micronesia. – S.W. Japan to Australia and Fiji.

OGASAWARA SHRIMPGOBY   *Amblyeleotris ogasawarensis*
SIZE: to 14 cm (5 1/2 in.)   Gobies – Gobiidae
ID: Five well-defined reddish brown bars over pearl-white body with vague blue spots; small, dark mark at top edge of gill cover. Seaward reef slopes in 20-30 m.
Pacific: Philippines, Papua New Guinea to Line Is. and Great Barrier Reef.

AURORA SHRIMPGOBY   *Amblyeleotris aurora*
SIZE: to 9 cm (3 1/2 in.)   Gobies – Gobiidae
ID: White; 5 brown bands with diffuse margins, yellow tail with red markings, **oblique reddish brown band at rear corner of mouth.** Share burrow with alpheid shrimp. Sandy slopes of lagoons and seaward reefs in 5-40 m.
Indian Ocean: E. Africa and Aldabra to Maldives and Andaman Sea.

WIDE-BARRED SHRIMPGOBY   *Amblyeleotris latifasciata*
SIZE: to 10 cm (4 in.)   Gobies – Gobiidae
ID: Brown; may have 3-4 broad dark bars; bluish spots on head, orange spots on body, **pale-edged orange spots on 1st dorsal fin and on tips of 2nd.** Share burrow with alpheid shrimp. Sand bottoms of lagoons and seaward reefs in 5-40 m.
Asian Pacific: Andaman Sea to Sabah in Malaysia, Indonesia and Philippines.

STEINITZ' SHRIMPGOBY   *Amblyeleotris steinitzi*
SIZE: to 8 cm (3 1/4 in.)   Gobies – Gobiidae
ID: White with 5 wide brown bars on head and body; yellow spots on both dorsal fins, **frequently thin orange bars between dark body bars.** Share burrow with alpheid shrimp. Lagoon and outer reef slopes in 6-35 m.
Indo-West Pacific: Red Sea to Indonesia, Philippines and Micronesia. – S.W. Japan to Great Barrier Reef and Fiji.

MAIUSI'S SHRIMPGOBY   *Amblyeleotris masuii*
SIZE: to 10.2 cm (4 in.)   Gobies – Gobiidae
ID: Pale gray with whitish underside; 4 wide brown bars on body with **brown reticulations and/or thin bars between,** narrow brown band below eye and wider band on rear gill cover. Share burrow with alpheid shrimp. Sand bottoms 12-25 m.
Asian Pacific: Indonesia and Philippines to S. Japan.

SLANTBAR SHRIMPGOBY   *Amblyeleotris diagonalis*
SIZE: to 9 cm (3 1/2 in.)   Gobies – Gobiidae
ID: White with 5 brown bands on side and **2 narrow bands on head;** yellow anal fin with blue and red margin, blue and red streaks on ventral fin. Share burrow with alpheid shrimp. Sandy slope of lagoon and seaward reef slopes in 6-30 m.
Indo-Asian Pacific: Red Sea and E. Africa to Indonesia, Philippines and Solomon Is. – S.W. Japan to Great Barrier Reef.

Slantbar Shrimpgoby – Variation
ID: Width and darkness of 5 body bands variable; occasionally dark band behind eye has blue borders, band in front of eye not always distinct.

NAKEDHEAD SHRIMPGOBY   *Amblyeleotris gymnocephala*
SIZE: to 10 cm (4 in.)   Gobies – Gobiidae
ID: Tan with 5 brown diffuse bars on side; **cluster of dark markings on back between bars,** thin dark brown stripe behind eye to above pectoral fin. Share burrow with alpheid shrimp. Lagoons and seaward slopes in 5-35 m.
Asian Pacific: Brunei, Indonesia, Philippines, Micronesia, Papua New Guinea and Solomon Is. – S.W. Japan to N.E. Australia.

ARCFIN SHRIMPGOBY   *Amblyeleotris arcupinna*
SIZE: to 10 cm (4 in.)   Gobies – Gobiidae
ID: Tan with 5 brown bars with diffuse edges; **brown and reddish arc-shaped marking on first dorsal fin,** dark wedge-shaped mark below eye. Share burrow with alpheid shrimp. Sandy lagoons and sheltered reefs in 12-25 m.
West Pacific: Bali in Indonesia to New Britain and Milne Bay Province in E. Papua New Guinea and Fiji.

Small, Elongate Bottom Dwellers – Gobies

BROAD-BANDED SHRIMPGOBY *Amblyeleotris periophthalma*
SIZE: to 9 cm (3 1/2 in.)   Gobies – Gobiidae
ID: Tan with six dark brown bars on head and side; **irregular brown blotches on tan area between bars**, dark-ringed orange spots on head. Share burrow with alpheid shrimp. Sand bottoms of lagoons and seaward slopes in 3-35 m.
Indo-West Pacific: Red Sea and E. Africa to Indoneisa, Philippines and Micronesia. – S.W. Japan to Great Barrier Reef and Fiji.

Broad-banded Shrimpgoby – Variation
ID: No markings between 6 red-brown bars; **pale-edged reddish spots on dorsal fins;** bluish tinted ventral and anal fins.

GIANT SHRIMPGOBY *Amblyeleotris fontanesii*
SIZE: to 20 cm (8 in.)   Gobies – Gobiidae
ID: Large, about twice the length of similar species; white with 5 brown bars, **small orange spots on head.** Share burrow with alpheid shrimp. Sand bottoms of lagoons and sheltered reefs in 5-30 m.
West Pacific: Andaman Sea, Indonesia and Micronesia. – S.W. Japan to N.E. Australia, New Caledonia and Fiji.

Giant Shrimpgoby – Juvenile
ID: Tan with 5 darker brown bars compared to those of adult, **elongate blackish spot at base of first dorsal fin.** Usually seen in same locations as adults in about 5-30 m.

RED-MARGIN SHRIMPGOBY *Amblyeleotris rubrimarginata*
SIZE: to 11 cm (4 1/4 in.)   Gobies – Gobiidae
ID: Pale brown with 5-6 light brown bars and dark speckles on back; upper iris and head behind eyes dark brown, **red margin on dorsal fins and frerquently upper tail.** Share burrow with alpheid shrimp. Coastal bays and estuaries in 2-20 m.
Asian Pacific: Indonesia, Philippines, Papua New Guinea and Solomon Is. to New Caledonia and Vanuatu.

GORGEOUS SHRIMPGOBY *Amblyeleotris wheeleri*
SIZE: to 8 cm (3 1/4 in.)   Gobies – Gobiidae
ID: Six dark red bars with yellowish spaces between; scattered red spots on head and blue spots on body, blue-edged red stripe on anal fin. Share burrow with alpheid shrimp. Lagoon and outer reef slopes in 2-28 m.
Indo-West Pacific: Red Sea and E. Africa to Indonesia and Micronesia. – S.W. Japan to Great Barrier Reef and Fiji.

Gobies

RED-BANDED SHRIMPGOBY *Amblyeleotris fasciata*
SIZE: to 8 cm (3 1/4 in.)   Gobies – Gobiidae
ID: Whitish with 7 relatively narrow reddish bars including one through eye and another on tail base; yellow spots on head and body. Share burrow with alpheid shrimp. Patches of coarse sand in lagoons and on outer slopes in 5-25 m.
Pacific: Micronesia to Great Barrier Reef, Fiji and French Polynesia.

VOLCANO SHRIMPGOBY *Ambyeleotris rhyax*
SIZE: to 10 cm (4 in.)   Gobies – Gobiidae
ID: White with 3 or more reddish brown body bars; **wide rear slanting reddish brown band below eye,** yellow spots on head and body. Share burrow with alpheid shrimp. Rubble and sand inside caves and canyons of seaward reefs and slopes in 30-45 m.
Asian Pacific: Indonesia, Philippines, Micronesia and Papua New Guinea to Great Barrier Reef.

BARRED SHRIMPGOBY *Cryptocentrus fasciatus*
SIZE: to 9.5 cm (3 3/4 in.)   Gobies – Gobiidae
ID: Pale with 4 brown occasionally irregular bars; white or bluish streaks and spots on head, **thin blue stripes on dark anal fin and blue dots on ventral fin.** Share burrow with alpheid shrimp. Sheltered sand and mud bottoms to 15 m.
Indo-Asian Pacific: Red Sea and E. Africa to Indonesia, Philippines and Solomon Is.– S.W. Japan to Great Barrier Reef.

Barred Shrimpgoby – Variation
ID: Dark brown with several whitish saddles on top of head and along back; small white or blue spots or streaks on head and ventral fin. All variations can be distinguished by the thin blue stripes on dark anal fin and blue dots on ventral fin.

BLUE-SPECKLED SHRIMPGOBY *Cryptocentrus caeruleomaculatus*
SIZE: to 8 cm (3 1/4 in.)   Gobies – Gobiidae
ID: About 8 irregular greenish brown bars with narrow pale spaces between; **tiny blue spots and pink to reddish markings on head and body.** Share burrow with alpheid shrimp. Lagoon and sheltered reefs to 6 m.
Asian Pacific: Andaman Sea to Indonesia, Philippines, Micronesia and Solomon Is. – S.W. Japan to Great Barrier Reef.

Blue-speckled Shrimpgoby – Variation
ID: Greenish dorsal fins with reddish markings; body bars diffuse and indistinct. Both variations have tiny blue spots and pink to red markings on head and body and darkish bars on anal fin.

Banded Shrimpgoby – Variation

ID: Dark brown; white spotting on head, pectoral, ventral and dorsal fins. All variations usually with one of two shrimp species: *Alpheus djeddensis* or *A. ochrostriatus;* shrimp commonly in pairs.

BANDED SHRIMPGOBY *Cryptocentrus cinctus*
SIZE: to 7.5 cm (3 in.) Gobies – Gobiidae
ID: Yellow to gold with **white to blue spotting on head, upper body and dorsal fins;** may display faint dark bars. Share burrow with alpheid shrimp. Sheltered sand and mud bottoms in 2-15 m.
Asian Pacific: Andaman Sea to Indonesia, Philippines and Micronesia. – S.W. Japan to Great Barrier Reef.

Banded Shrimpgoby – Variation
ID: Pale with 4-5 dark bars, occasionally with pale narrow bars between; white to blue spotting on head, dorsal, pectoral and ventral fins. Commonly in pairs of the same variation, but occasionally pair with other variations.

LAGOON SHRIMPGOBY *Cryptocentrus multicinctus*
SIZE: to 12 cm (4¾ in.) Gobies – Gobiidae
ID: Gray with **14-15 narrow white bars;** white line and spot markings on head, yellow margin on 1st dorsal fin, blue streaks on 2nd dorsal, anal and tail fins. Share burrow with alpheid shrimp. Sheltered reefs to 10 m.
Asian Pacific: Indonesia, Micronesia and Papua New Guinea.

VENTRAL-BARRED SHRIMPGOBY *Cryptocentrus sericus*
SIZE: to 10 cm (4 in.) Gobies – Gobiidae
ID: Pale with dark brown bars (may run from back to belly or only on back); **pair of brown spots behind mouth and on gill cover,** has yellow variation. Share burrow with alpheid shrimp. Lagoon and shoreline reefs to 20 m.
Asian Pacific: Andaman Sea to Indonesia, Philippines, Micronesia and Solomon Is. – S.W. Japan to Great Barrier Reef.

Ventral-barred Shrimpgoby – Variation
ID: Yellow to yellowish brown often with darker bar markings on rear body; blue spotting on head; first dorsal fin with bright yellow spots surrounded by darker markings.

Gobies

INEXPLICABLE SHRIMPGOBY *Cryptocentrus inexplicatus*
SIZE: to 9.5 cm (3 ³/₄ in.) Gobies – Gobiidae
ID: Grayish with several dark bars on side; 3-4 black spots on rear edge of gill cover, irregular dark markings on top of head (sometimes cheek) and upper body. Share burrow with alpheid shrimp. Sheltered reefs to 15 m.
Asian Pacific: Indonesia, Philippines, Micronesia and Solomon Is., north to S.W. Japan.

BLUESPOT SHRIMPGOBY *Cryptocentrus cyanospilotus*
SIZE: to 7.5 cm (3 ¹/₄ in.) Gobies – Gobiidae
ID: About 8 irregular greenish brown bars with narrow pale spaces between; **tiny blue spots and pink to reddish markings on head and body.** Share burrow with alpheid shrimp. Lagoon and sheltered reefs to 6 m.
Asian Pacific: Andaman Sea to Indonesia, Philippines, Micronesia and Solomon Is. – S.W. Japan to Great Barrier Reef.

PINK-SPOTTED SHRIMPGOBY *Cryptocentrus leptocephalus*
SIZE: to 10 cm (4 in.) Gobies – Gobiidae
ID: Pale brown to gray; small pale and red spots on head and body, large red pale ringed spots on dorsal fins, 6-7 brown bars, **the 3rd from front extends from 2nd dorsal to front of barred anal fin.** Sheltered sand and mud to 10 m, often near mangroves.
West Pacific: Indonesia, Philippines to Micronesia, Solomon Is. – S.W. Japan to Australia, New Caledonia and Fiji.

SINGAPORE SHRIMPGOBY *Cryptocentrus melanopus*
SIZE: to 10 cm (4 in.) Gobies – Gobiidae
ID: Pale brown to gray with small pale and red spots on head and body, large red pale ringed spots on dorsal fins; 6-7 brown bars, **the 3rd from front extends from 2nd dorsal to belly in front of bluish anal fin.** Sheltered sand bottoms to about 10 m.
Asian Pacific: Brunei, Sabah in Malaysia, Indonesia and Philippines to N. Australia.

BLUE-TAILED SHRIMPGOBY *Cryptocentrus cebuanus*
SIZE: to 10 cm (4 in.) Gobies – Gobiidae
ID: Light to dark brown with faint narrow yellowish body bars, yellow anal fin; white to blue spots on head, blue streaks on tail, **black spot near tip of 1st dorsal fin.** Share burrow with alpheid shrimp. Sand near inshore reefs in 2-10 m.
Asian Pacific: Indonesia, Philippines and N. Australia.

SADDLED SHRIMPGOBY *Cryptocentrus leucostictus*
SIZE: to 11 cm (4 ¹/₄ in.) Gobies – Gobiidae
ID: Dark brown with several white saddles on back; white lips, snout and top of head. Share burrow with a large undescribed alpheid shrimp. Sand bottoms of lagoons and seaward reefs in 2-25 m.
West Pacific: Andaman Sea to Indonesia, Philippines, Micronesia and Solomon Is. – S.W. Japan to Australia and Fiji.

TARGET SHRIMPGOBY   *Cryptocentrus strigilliceps*
SIZE: to 12 cm (4 ³/₄ in.)   Gobies – Gobiidae
ID: Pale brown with broad irregular brown bands; several pale-edged black midlateral spots, 1st bordered with white spots. Share burrow with alpheid shrimp. Lagoon and shallow shoreline reefs to 10 m.
Indo-West Pacific: E. Africa to Indonesia, Philippines, Micronesia and Solomon Is. – S.W. Japan to Australia, New Caledonia and Fiji.

BLACKSPOT SHRIMPGOBY   *Cryptocentrus nigrocellatus*
SIZE: to 13 cm (5 in.)   Gobies – Gobiidae
ID: Dark brown to blackish with 5 white saddles; white snout, white speckling on lower side, **prominent white ringed black spot in front of pectoral fin base.** Share burrow with alpheid shrimp. Sand and rubble littered bottoms in 3-10 m.
Asian Pacific: Brunei, Sabah in Malaysia, Indonesia and Philippines to S. Japan.

MAUDE'S SHRIMPGOBY   *Cryptocentrus maudae*
SIZE: to 10 cm (4 in.)   Gobies – Gobiidae
ID: Dark brown covered with numerous white spots and short line markings; 5 or more whitish saddles across back, bright white snout. Share burrow with alpheid shrimp. Sheltered sand bottoms to about 10 m.
East Indo-Asian Pacific: Sri Lanka to Indonesia, Philippines, Micronesia and Papua New Guinea. – S.W. Japan to Australia.

FLAGFIN SHRIMPGOBY   *Mahidolia mystacina*
SIZE: to 7 cm (2 ³/₄ in.)   Gobies – Gobiidae
ID: Male – Yellow to gray with 6 dark slightly diagonal body bands; **tall pointed 1st dorsal fin** with several dark spots on margins. Share burrow with alpheid shrimp. Fine silt bottoms of lagoon and seaward reef slopes in 5-25 m.
Indo-Pacific: E. Africa to Indonesia, Philippines, Micronesia and Solomon Is. – S.W. Japan to Great Barrier Reef and New Caledonia.

Flagfin Shrimpgoby – Female Variation
ID: Yellow-gold with dark body bands; large sail-like 1st dorsal fin with several band markings.

Flagfin Shrimpgoby – Female Variation
ID: Gray to brown with dark body bands; large sail-like 1st dorsal fin with band markings.

Gobies

GOLD-STREAKED SHRIMPGOBY　　*Ctenogobiops aurocingulus*
SIZE: to 8.8 cm (3¹/₂ in.)　　　　　　Gobies – Gobiidae
ID: Pale gray with diffuse dusky blotches and numerous orange lines and spots; white spot on upper pectoral fin base and another on fin. Share burrow with alpheid shrimp. Lagoon and seaward reef slopes in 2-20 m.
West Pacific: Indonesia, Philippine, Papua New Guinea and Solomon Is. – Great Barrier Reef to Fiji.

SILVERSTREAK SHRIMPGOBY　　*Ctenogobiops maculosus*
SIZE: to 7 cm (2³/₄ in.)　　　　　　Gobies – Gobiidae
ID: Whitish with 3 longitudinal rows of large oblong brown spots; 3 diagonal rows of orange to brown lines from below eye to rear gill cover, white streak on pectoral fin. Share burrow with alpheid shrimp. Sand and rubble littered bottoms in 1-15 m.
Indo-West Pacific: Red Sea to Indonesia, Philippines, Micronesia and Solomon Is. – S.W. Japan to Great Barrier Reef and Fiji.

SAND SHRIMPGOBY　　　　　　*Ctenogobiops feroculus*
SIZE: to 5.5 cm (2¹/₄ in.)　　　　　　Gobies – Gobiidae
ID: Whitish with dusky elliptical spots encircled by tiny blue spots; 1st dorsal ray elongate in adults, white spot on pectoral fin, lack orange spots on head as similar Silverspot Shrimpgoby [previous]. Lagoon and coastal reefs to 10 m.
Asian Pacific: Red Sea to Indonesia, Philippines and Micronesia. – S.W. Japan to Great Barrier Reef, New Caledonia and Fiji.

GOLD-SPECKLED SHRIMPGOBY　　*Ctenogobiops pomastictus*
SIZE: to 7 cm (2³/₄ in.)　　　　　　Gobies – Gobiidae
ID: Whitish with brown elliptical spots encircled by tiny blue spots; white spot on pectoral fin, **smaller orange spot between elliptical spots on side.** Share burrow with alpheid shrimp. Sheltered reefs in 2-10 m.
Asian Pacific: Andaman Sea to Indonesia, Philippines, Micronesia and Solomon Is. – S.W. Japan to Great Barrier Reef and Fiji.

THREAD SHRIMPGOBY　　　　　*Ctenogobiops mitodes*
SIZE: to 7 cm (2³/₄ in.)　　　　　　Gobies – Gobiidae
ID: Whitish with 4 rows of brown spots **(midlateral row form dashes)**; white spot on pectoral fin base; very tall 2nd dorsal fin. Share burrow with alpheid shrimp. Lagoon and outer reef slopes in 1-15 m.
West Pacific: Indonesia to Micronesia, Papua New Guinea and Solomon Is. – Great Barrier Reef to Fiji.

TANGAROA SHRIMPGOBY　　　*Ctenogobiops tangaroai*
SIZE: to 6.5 cm (2¹/₂ in.)　　　　　　Gobies – Gobiidae
ID: Whitish with orange spots interspersed with tiny white or bluish spots; white spot on pectoral fin base, very tall dark edge on 1st dorsal fin. Share burrow with alpheid shrimp. Lagoon and outer reef slopes in 2-40 m.
West Pacific: Indonesia, Philippines, Papua New Guinea and Solomon Is. – S.W. Japan to Great Barrier Reef and Fiji.

BLACK-RAYED SHRIMPGOBY *Stonogobiops nematodes*
SIZE: to 6 cm (2 1/4 in.)   Gobies – Gobiidae
ID: White to pale brown with yellow snout and 4 pale edged black bands; tall black-edged 1st dorsal spine. Share burrow with alpheid shrimp. Lagoon and outer reef slopes in 15-25 m.
Indo-Asian Pacific: Seychelles and Andaman Sea to Indonesia, Philippines and Micronesia. – S.W. Japan to Fiji.

YELLOWNOSE SHRIMPGOBY *Stonogobiops xanthorhinica*
SIZE: to 6.4 cm (2 1/2 in.)   Gobies – Gobiidae
ID: White with yellow snout and 4 black bands; **sail-like 1st dorsal fin,** often with black rear border extending from 2nd band. Share burrow with alpheid shrimp. Sand slopes of seaward reefs in 3-45 m.
West Pacific: Indonesia, Philippines, Micronesia, Papua New Guinea, Solomon Is. – S.W. Japan to Great Barrier Reef and Fiji.

MARQUESAS SHRIMPGOBY *Stonogobiops medon*
SIZE: to 6 cm (2 1/4 in.)   Gobies – Gobiidae
ID: White with yellow snout and nape; arching dark brown to blackish marking from gill cover to above pectoral fin base, row of dark spots along back. Share burrow with alpheid shrimp. Seaward sand slopes in 20-40 m.
Localized: Marquesas Is. in French Polynesia.

RED-STRIPED SHRIMPGOBY *Stonogobiops yasha*
SIZE: to 6 cm (2 1/4 in.)   Gobies – Gobiidae
ID: White with red stripes; tall 1st dorsal fin spine with triangular dark spot on middle of fin, remaining fins yellowish or with yellow spotting. Share burrow with alpheid shrimp. Seaward sand slopes in 15-40 m.
West Pacific: Indonesia and Micronesia. – S. Japan to New Caledonia and Fiji.

BLACK SPEAR SHRIMPGOBY *Myersina lachneri*
SIZE: to 5 cm (2 in.)   Gobies – Gobiidae
ID: Grayish brown; very tall dark 1st dorsal spine with back front edge, pale stripe from eye to upper tail base, blue edge on 2nd dorsal and upper tail fins. Share burrow with alpheid shrimp. Sand and silt bottoms of sheltered reefs in 2-10 m.
Asian Pacific: Known only from Bali in Indonesia and New Britain in Papua New Guinea.

BLACKLINE SHRIMPGOBY *Myersina nigrivirgata*
SIZE: to 10 cm (4 in.)   Gobies – Gobiidae
ID: Whitish to pale gray to pale brown or bright yellow; darkish brown to blackish stripe from eye to tail base, orange to bluish spots on head. Share burrow with alpheid shrimp. Sheltered bays and lagoons in 2-20 m.
Asian Pacific: Indonesia and Philippines. – S.W. Japan to N. Australia.

313

Gobies

WHITECAP SHRIMPGOBY *Lotilia klausewitzi*
SIZE: to 4.5 cm (1 ³/₄ in.) Gobies – Gobiidae
ID: Dark brown with white band from snout to dorsal fin; dark ocellated spot on 1st dorsal fin, clear pectoral and tail fins with large brown spots. Share burrow with alpheid shrimp. Sand bottoms of lagoon and coastal reefs in 5-40 m.
West Pacific: Indonesia to Papau New Guinea. –S.W. Japan to Great Barrier Reef and Fiji.

MAGNIFICENT SHRIMPGOBY *Tomiyamichthys* sp.
SIZE: to 6 cm (2 ¹/₄ in.) Gobies – Gobiidae
ID: Dark body, often black, with pale spots and bars; large orangish or gray fan-shaped 1st dorsal fin with dark mosaic markings, large orange fan-shaped tail with blue lines radiating from tail base. Share burrow with alpheid shrimp. Sand bottoms in 15-40 m.
Asian Pacific: Bali in Indonesia, Philippines to S. Japan.

LANCEOLATE SHRIMPGOBY *Tomiyamichthys lanceolatus*
SIZE: to 5 cm (2 in.) Gobies – Gobiidae
ID: Pale gray to whitish with row of several large brown spots on middle of side; 2 dark spots on 1st dorsal fin, brown mottling on back, **long tapering pointed tail**. Share burrow with alpheid shrimp. Sand in lagoons and sheltered bays in 10-25 m.
Asian Pacific: Indonesia, Philippines, Micronesia and Papua New Guinea. –S. Japan to Great Barrier Reef.

BLACKTIP SHRIMPGOBY *Tomiyamichthys russus*
SIZE: to 12 cm (4 ³/₄ in.) Gobies – Gobiidae
ID: Pale gray with about 5 dark midlateral blotches; dark brown bar below eye, brown spotting on back, **single pale-edged ocellated spot on rear of 1st dorsal fin.** Share burrow with shrimp. Silty coastal reefs, often stream mouths to 5 m.
Asian Pacific: Indonesia and Philippines to Papua New Guinea, including New Britain.

LONGSPOT SHRIMPGOBY *Tomiyamichthys tanyspilus*
SIZE: to 5 cm (2 in.) Gobies – Gobiidae
ID: Pale gray to whitish; large brown midlateral blotches, light brown spotting on back, **tall 1st first dorsal fin with pair of pale-edged black spots.** Share burrow with alpheid shrimp. Lagoons and sheltered bays in 3-10 m.
Localized: Maumere Bay, Flores and Cenderawasih Bay, West Papua in Indonesia.

HIGHFIN SHRIMPGOBY *Tomiyamichthys alleni*
SIZE: to 5 cm (2 in.) Gobies – Gobiidae
ID: Pale gray with 4-5 irregular brown saddles on upper body; white below with black blotches and brown bars, dark-edged orange spots on head and body and 1st and 2nd dorsal fins, short filaments and **2-3 black spots on 1st dorsal fin.** Share burrow with alpheid shrimp. Sheltered sand slopes in 15-40 m.
Asian Pacific: Known only from Indonesia, S. Japan and Fiji.

MONSTER SHRIMPGOBY *Tomiyamichthys oni*
SIZE: to 11 cm (4 1/4 in.) Gobies – Gobiidae
ID: **Male –** White with 4-5 large blackish blotches and small brown spots; dark band below eye, **tall rounded 1st dorsal fin with pale-edged black spot.** Share burrow with alpheid shrimp. Sheltered sand slopes in 10-30 m.
Asian Pacific: Indonesia, Philippines, Micronesia and Papua New Guinea. –S.W. Japan to N. Australia and New Caledonia.

Monster Shrimpgoby – Female
ID: White with scattering of large brown blotches and small spots; dark band extends from lower eye; **tall triangular 1st dorsal fin with small black and white spot.**

YELLOWFIN SHRIMPGOBY *Tomiyamichthys latruncularius*
SIZE: to 5.5 cm (2 1/4 in.) Gobies – Gobiidae
ID: **Male –** Whitish undercolor with 3-4 wide brown body bars; brown-edged orange spots on head, **tall filamentous 1st dorsal fin,** yellow ventral fins. Share burrow with alpheid shrimp. Sand and rubble bottoms in 2-45 m.
Indo-Asian Pacific: Red Sea and Maldives to Lembeh Strait, Sulawesi, Indonesia.

Yellowfin Shrimpgoby – Female
ID: Similar to male, but tall 1st dorsal fin not filamented.

RAYED SHRIMPGOBY *Tomiyamichthys nudus*
SIZE: to 5 cm (2 in.) Gobies – Gobiidae
ID: Mottled dark brown to nearly black front body and pale rear with dark blotches or bars; **fan-shaped 1st dorsal fin with protruding filaments and blue spot.** Share burrow with alpheid shrimp. Sheltered sand and rubble bottoms in 3-20 m.
Localized: Brunei, Sabah in Malaysia to Lembeh Strait and West Papua in Indonesia.

REDEYED SHRIMPGOBY *Tomiyamichthys smithi*
SIZE: to 15 cm (6 in.) Gobies – Gobiidae
ID: Pale with 12 brown diffuse bars on side with narrow white spaces between; dense dark blotches on head and 1st dorsal fin, red pupil, elongate 1st dorsal spine and **elongate nostril tubes.** Share burrow with alpheid shrimp. Sand/mud of sheltered bays in 10-23 m.
Asian Pacific: Sabah in Malaysia and Papua New Guinea.

Gobies

BROWNBAND SHRIMPGOBY   *Tomiyamichthys zonatus*
SIZE: to 4 cm (1 1/2 in.)   Gobies – Gobiidae
ID: Female – Dark bands below eye and on snout, dark chin marking, large rectangular blackish spots on side, black spot on rear first dorsal fin. **Male –** Similar but with golden brown bars on ventral side below each rectangular spot.
Localized: Milne Bay, PNG, but probably widespread in Asia Pacific region.

SCALYCHEEK SHRIMPGOBY   *Vanderhorstia lepidobucca*
SIZE: to 5.5 cm (2 1/4 in.)   Gobies – Gobiidae
ID: Neon-blue stripe on top of head, orange bands on snout and cheek, dark spot on gill cover, and numerous, short brown bars on side. Lives in burrow with alpheid shrimp. Silty or muddy bottoms of sheltered bays with periodic strong currents in 25-40 m.
Localized: Lembeh Strait, Indonesia, but probaly more widespread in Asia Pacific region.

MAJESTIC SHRIMPGOBY   *Vanderhorstia nobilis*
SIZE: to 7 cm (2 1/2 in.)   Gobies – Gobiidae
ID: Male – Much the same as smaller adults, but colors and markings more brilliant and distinct; yellow upper body, **bright blue margins on dorsal, anal and tail fins,** white ventral fin; large tail with rear margin tapering to central point.
Asian Pacific: Indonesia and Philippines.

Majestic Shrimpgoby – Female
ID: Much the same as smaller adult, except dark marking on base of 1st dorsal fin has become yellow; other colors and markings more brilliant and distinct, note the bright blue margins on dorsal, anal and tail fins, white ventral fin, long tapering tail.

SPANGLED SHRIMPGOBY   *Vanderhorstia dorsomacula*
SIZE: to 6 cm (2 1/4 in.)   Gobies – Gobiidae
ID: Male – Pale bluish **covered with yellow and blue spots;** neon-blue stripe from snout to large 1st dorsal fin. Share burrow with alpheid shrimp. Estuaries and sheltered reefs in 5-35 m.
Asian Pacific: Indonesia, Philippines and Papua New Guinea to S. Japan.

Spangled Shrimpgoby – Female
ID: Pale bluish covered with yellow and blue spots; tall large dorsal fin marked with yellow streaks and dark blotch.

WAYAG SHRIMPGOBY *Vanderhorstia wayag*
SIZE: to 4.5 cm (1 ³/₄ in.) Gobies – Gobiidae
ID: Pale gray to whitish with **irregular row of large orangish spots ringed with blue on middle of side** and smaller row of smaller spots above on lower back; diagonal blue and yellow lines on head. Sandy slopes in 8-15 m.
Localized: Known only from Raja Ampat and West Papua in Indonesia.

YELLOWFOOT SHRIMPGOBY *Vanderhorstia phaeosticta*
SIZE: to 5 cm (2 in.) Gobies – Gobiidae
ID: Color variable, often pale gray to yellowish with row of large diffuse dark blotches on lower side; blue-edged yellow spots often prominent on lower side between dark blotches, yellowish ventral fin. Share burrow with alpheid shrimp. Sandy shores, often in seagrass beds in 8-20 m.
Asian Pacific: E. Indonesia and S. W. Japan to New Caledonia.

AMBANORO SHRIMPGOBY *Vanderhorstia ambanoro*
SIZE: to 7.3 cm (2 ³/₄ in.) Gobies – Gobiidae
ID: Pale gray or whitish; midateral row of large black spots, smaller black spots and blotches on back, **blue bordered dark stripe on 2nd dorsal fin extents onto tail.** Share burrow with alpheid shrimp. Lagoon and seaward slopes in 4-25 m.
Indo-Asian Pacific: Red Sea and E. Africa to Indonesia, Micronesia. – S.W. Japan to N. Australia and New Caledonia.

BELLA SHRIMPGOBY *Vanderhorstia belloides*
SIZE: to 6 cm (2 ¹/₄ in.) Gobies – Gobiidae
ID: Pale gray with white lower head and belly; yellow to orange spots on head and back that are larger below 1st dorsal fin, **broad dusky stripe on center of anal fin.** Share burrow with alpheid shrimp. Sand and rubble bottoms in 12-25 m.
Localized: N. Papua New Guinea.

YELLOW-SPOTTED SHRIMPGOBY *Vanderhorstia papilio*
SIZE: to 6 cm (2 ¹/₄ in.) Gobies – Gobiidae
ID: **Male –** Whitish or pale gray with 4 broad tapering brown saddles; numerous dark edged yellow spots on head and body, blue stripes on tail. **Female –** Darker bars and less color on fins. Share burrow with alpheid shrimp. Sand and rubble areas in 5-55 m.
Asian Pacific: Indonesia and Philippines to S.W. Japan.

BIGFIN SHRIMPGOBY *Vanderhorstia macropteryx*
SIZE: to 11 cm (4 ¹/₄ in.) Gobies – Gobiidae
ID: Dark brown to bluish gray with orange and blue spots and lines on head; orange spots on body and dorsal fin, several thin orange bars behind pectoral fin, **blue lines on outer edge of anal fin and lower lobe of tail.** Share burrow with alpheid shrimp. Silty slopes in 20-40 m.
Asian Pacific: Flores in Indonesia, Philippines and S. Japan.

317

Gobies

GOLD-MARKED SHRIMPGOBY *Vanderhorstia auronotata*
SIZE: to 5 cm (2 in.)   Gobies – Gobiidae
ID: Pale bluish with about 6 narrow tan bars on forebody and about 10 bars behind, long lanceolate tail. **Female –** [bottom] Bluish tan with numerous dark tan bars on body. Sloping sand bottoms in 20-25 m.
Localized: Known only from Ambon, Indonesia and Philippines.

TAPERTAIL SHRIMPGOBY *Vanderhorstia attenuata*
SIZE: to 6 cm (2 1/4 in.)   Gobies – Gobiidae
ID: Pale bluish gray with 4 diffuse brown bars on body, often with accompanying narrow orange bars; small orange spots and lines on upper head and back. Share burrow with alpheid shrimp. Sand and rubble bottoms in 15-48 m.
Asian Pacific: Indonesia, Philippines and Solomon Is.

YELLOW-LINED SHRIMPGOBY *Vanderhorstia flavilineata*
SIZE: to 4.5 cm (1 3/4 in.)   Gobies – Gobiidae
ID: Blue gray shading to white below; **yellow midlateral stripe,** thinner yellow stripe on back and metallic blue stripe at base of dorsal fin, yellow spots on head. Share burrow with alpheid shrimp. Seaward sand slopes in 20-40 m.
Localized: Currently known only from Kimbe Bay on island of New Britain in Papua New Guinea.

CHEEK-STREAKED GOBY *Echinogobius hayashii*
SIZE: to 9 cm (3 1/2 in.)   Gobies – Gobiidae
ID: Elongate; whitish body with brownish spots and blotches, **light blue or pearly lines/streaks on head and front of body,** blue rimmed spots on tail. Solitary on open sand well away from reefs. Sheltered waters to 20 m.
Asian Pacific: West Papua in Indonesia, Micronesia and N. Australia.

YELLOWSTRIPE GOBY *Koumansetta hectori*
SIZE: to 5.5 cm (2 1/4 in.)   Gobies – Gobiidae
ID: Dark brown with 4 bright yellow stripes; black spot on 1st dorsal fin, yellow-edged black spot on 2nd dorsal fin and black spot on upper tail base. Solitary. Sand bottoms near base of reef formations in 3-30 m.
Indo-Asian Pacific: Red Sea to Andaman Sea, Brunei, Indonesia and Micronesia, north to S.W. Japan.

OLD GLORY *Koumansetta rainfordi*
SIZE: to 7.7 cm (3 in.)   Gobies – Gobiidae
ID: Charcoal-gray; 5 orange stripes, **row of white spots on upper back,** yellow-edged black spot on 2nd dorsal fin and black spot on upper tail base. Solitary or form small groups. Sand bottoms near base of reefs in 3-30 m.
West Pacific: Indonesia and Philippines, Micronesia to Great Barrier Reef and Fiji.

BANDED FLAT-HEAD GOBY *Callogobius hasseltii*
SIZE: to 7.5 cm (3 in.) Gobies – Gobiidae
ID: Tan to light brown; brown stripe from snout through eye to rear head, 3 broad brown bands below 1st dorsal fin, below 2nd dorsal and base of tail, **large black spot on upper front of tail.** Solitary. Sand and rubble of lagoons and inshore reefs to 15 m.
West Pacific: Indonesia, Philippines, Micronesia and Solomon Is. – S. Japan to Great Barrier Reef and Fiji.

SADDLED GOBY *Callogobius clitellus*
SIZE: to 5 cm (2 in.) Gobies – Gobiidae
ID: Whitish; brown stripe from snout to bar behind head, brown bar below 1st dorsal fin, band from 2nd dorsal fin to anal fin, skin ridges on head. Solitary. Hide in crevices and rubble. Coastal and seaward reefs in 3-22 m.
Asian Pacific: Vietnam to Indonesia, Philippines, Micronesia, Papua New Guinea and Solomon Is.

STELLAR GOBY *Callogobius stellatus*
SIZE: to 4.3 cm (1 3/4 in.) Gobies – Gobiidae
ID: White with prominent dark brown bands from below 1st and 2nd dorsal fins; dark bar across top of nape, small dark saddle between dorsal fins, narrow dark bar on tail base, **3 dark bands radiating from eye.** Coral rubble to 10 m.
Asian Pacific: Indonesia and Philippines.

SLIMY GOBY *Callogobius flavobrunneus*
SIZE: to 7 cm (2 3/4 in.) Gobies – Gobiidae
ID: Pale mottled brown with dark bands radiating from eye; dark blotchy bar below each dorsal fin and on tail base. Sand and rubble bottoms in 3-20 m.
Indo-Asian Pacific: Red Sea and E. Africa to Indonesia Philippines and Great Barrier Reef.

CENTRESCALE GOBY *Callogobius centrolepis*
SIZE: to 5.5 cm (2 1/4 in.) Gobies – Gobiidae
ID: Shades of brown with 4 dark brown blotchy body bars; **dorsal, anal and tail fins shades of brown with numerous dark spots and markings.** Solitary. Coral reefs in 10-30 m.
East Indo-Asian Pacific: Maldives to Indonesia, Philippines and Papua New Guinea.

OSTRICH GOBY *Callogobius maculipinnis*
SIZE: to 9 cm (3 1/2 in.) Gobies – Gobiidae
ID: Brown or yellowish brown with scattered pale brown spots; often 4 diffuse bars, **oblique brown and white bands on 1st and 2nd dorsal fins.**
Indo-West Pacific: Red Sea and E. Africa to Indonesia, Philippines, Micronesia, Papua New Guinea and Solomon Is. – S.W. Japan to N.W. and E. Australia and Fiji.

Gobies

CROSSHATCH GOBY *Amblygobius decussatus*
SIZE: to 8 cm (3 1/4 in.) Gobies – Gobiidae
ID: Pale bluish gray; **about 9 orange body bars intersect about 5 orange stripes**, 2 of the stripes extend onto head with dark borders, orange spot on tail base. Silt and mud and fine-sand bottoms of lagoons and sheltered bays in 3-25 m.
East Indo-Asian Pacific: Cocos-Keeling Is. to Indonesia, Micronesia and Solomon Is. – S.W. Japan to Great Barrier Reef.

BUAN GOBY *Amblygobius buanensis*
SIZE: to 8 cm (3 1/4 in.) Gobies – Gobiidae
ID: Pale gray with 3 reddish to dark stripes on upper body; 5 diffuse dark gray blotches on middle to lower side, **brown spot on tail base with white spot above and below.** Solitary or in pairs. Silty shores, often near mangroves to 4 m.
Asian Pacific: Indonesia, Philippines, Micronesia, Papua New Guinea and Solomon Is.

FRECKLED GOBY *Amblygobius stethophthalmus*
SIZE: to 8 cm (3 1/4 in.) Gobies – Gobiidae
ID: Tan; **white-edged dark stripe from snout through eye to forebody and a second below across gill cover,** several short black bar-like markings on back, black spot on tail base. Solitary or in pairs. Silty shores, often in turbid water to 5 m.
Asian Pacific: Andaman Sea to Indonesia and Philippines.

STRIPED MANGROVE GOBY *Amblygobius linki*
SIZE: to 6.5 cm (2 1/2 in.) Gobies – Gobiidae
ID: Three black and white stripes with crosshatcing of thin bars; pink markings on dorsal and anal fins. Solitary or pairs. Silty shores, often among mangroves or in brackish lakes to 3 m.
Asian Pacific: Vietnam, Indonesia and Philippines.

NOCTURN GOBY *Amblygobius nocturnus*
SIZE: to 7 cm (2 3/4 in.) Gobies – Gobiidae
ID: Pale gray with reddish tint; **orangish stripe from snout through eye fading onto body.** Solitary or in pairs. Silt and mud and fine-sand bottoms of lagoons and sheltered bays in 3-25 m.
Indo-Pacific: Persian Gulf to Indonesia, Philippines, Micronesia, Papua New Guinea, Solomon Is. and French Polynesia. – S.W. Japan to Australia.

SNOUTSPOT GOBY *Amblygobius esakiae*
SIZE: to 6.5 cm (2 1/2 in.) Gobies – Gobiidae
ID: Dull brownish to blue-gray; dark spots on snout, **broken black stripe behind eye and another from gill cover through pectoral base onto body.** In pairs. Silt or sand bottoms of estuaries and sheltered bays in 2-15 m.
Asian Pacific: Indonesia to Micronesia and Papua New Guinea.

SPHYNX GOBY   *Amblygobius sphynx*
SIZE: to 15 cm (6 in.)   Gobies – Gobiidae
ID: Yellowish or greenish brown; 5-6 dark brown bars interspersed with pairs of white markings. Solitary or in pairs. Sheltered sand bottoms or seagrass beds to 20 m.
Indo-Asian Pacific: S. Red Sea and E. Africa to Indonesia, Philippines, Micronesia and Papua New Guinea. – S.W. Japan to Australia.

HALFBANDED GOBY   *Amblygobius semicinctus*
SIZE: to 11 cm (4 1/4 in.)   Gobies – Gobiidae
ID: Greenish brown; 5 brown bars on side, **irregular white spots and lines on cheek and body,** dark spot above gill opening, dark spot at base of filamentous 1st dorsal fin, brown spot on upper tail and 1-2 dark spots on outer part of tail. Sand and rubble of reef flats and sheltered shores to 20 m.
Indian Ocean: E. Africa and Seychelles to Andaman Sea.

BANDED GOBY   *Amblygobius phalaena*
SIZE: to 15 cm (6 in.)   Gobies – Gobiidae
ID: Greenish brown with five dark body bars; pale-edged dark stripe through eye and another across cheek, **black spot on upper tail and 1st dorsal fin.** Solitary or in pairs. Sand and rubble bottoms of reef flats and sheltered shores to 20 m.
East Indo-Pacific: Cocos-Keeling Is. to Indonesia, Philippines, Micronesia and French Polynesia. – S.W. Japan to Australia.

Banded Goby – Variation
ID: Dark brown with only vague stripes on head and front of body and bars on body; black area on 1st dorsal fin. All variations may display a black spot behind upper rear gill. This variation most common in areas of black sand.

Banded Goby – Variation
ID: Whitish undercolor with 5 narrow white edged bars on body; darkish reticulated markings on back; darkish line from lower head to tail base.

SAND GOBY   *Arcygobius baliurus*
SIZE: to 5.5 cm (2 1/4 in.)   Gobies – Gobiidae
ID: Translucent with brown and white spotting; **paired brown spots interspersed with white spots on rear lower side.** Solitary or small groups. Sand of sheltered reefs in 3-15 m.
Indo-Pacific: E. Africa to Indonesia, Philippines, Micronesia and French Polynesia. – S.W. Japan to Australia.

Gobies

BLUEDOT GOBY   *Asterropteryx ensifera*
SIZE: to to 3.5 cm (1 1/2 in.)   Gobies – Gobiidae
ID: Dark brown (appears blackish underwater) with rows of small blue spots. Commonly form large aggregations that perch on, or hover above, extensive areas of rubble. Slopes exposed to currents on seaward reefs in 6-40 m.
Indo-Pacific: Seychelles to Indonesia, Philippines, Micronesia and French Polynesia. – S.W. Japan to Great Barrier Reef.

STARRY GOBY   *Asterropteryx semipunctata*
SIZE: to 5 cm (2 in.)   Gobies – Gobiidae
ID: Greenish brown; **dark brown blotches and scribbling, especially on back,** rows of small blue spots on head, body and anal fin. Solitary or form small groups. Algal-covered rocks and rubble of sheltered inshore reefs to 15 m.
Indo-Pacific: Red Sea and E. Africa to Indonesia and Micronesia to French Polynesia. – S.W. Japan to Great Barrier Reef.

BLACKFOOT GOBY   *Asterropteryx atripes*
SIZE: to 2.7 cm (1 in.)   Gobies – Gobiidae
ID: Dark brown shading to tan or whitish below; brilliant blue spots on head and body forming 3-4 rows, faint dark-edged pale spots on dorsal, anal and tail fins, dusky brown to black ventral fins. Sheltered bays on sand and rubble in 10-24 m.
Asian Pacific: Andaman Sea to Indonesia, Philippines and S.W. Japan.

TWINSPOT GOBY   *Asterropteryx bipunctata*
SIZE: to 4.2 cm (1 1/2 in.)   Gobies – Gobiidae
ID: Dusky pink to pale gray with bright orange spots; **white-edged dark blotch at base of 1st dorsal and another on tail base.** Solitary. Sandy ledges on steep lagoon and seaward reef slopes in 4-32 m.
West Pacific: Andaman Sea to Indonesia, Philippines, Micronesia, Papua New Guinea, Solomon Is. and Samoa.

STRIPED GOBY   *Asterropteryx striata*
SIZE: to 3 cm (1 1/4 in.)   Gobies – Gobiidae
ID: Brown upper body and white below; wide diffuse dark brown to blackish stripe from eye to tail, several rows of small blue spots on side. Solitary or form small groups. Rubble bottoms of sheltered coastal reefs in 10-30 m.
West Pacific: Sabah in Malaysia to E. Indonesia, Philippines, Micronesia, Papua New Guinea, Solomon Is. and Fiji.

Striped Goby – Dark Variation
ID: Entirely dark brown except rows of blue spots and pale belly; blue outer edge on anal fin.

SENOU'S GOBY *Asterropteryx senoui*
SIZE: to 3.3 cm (1 1/4 in.) Gobies – Gobiidae
ID: Pale gray with dense covering of variably sized brown spots on head and body; **narrow tapering dark bar below eye.** Sheltered shorelines on mud or sand and rubble bottoms in 15-65 m.
Asian Pacific: Halmahera, Raja Ampat, W. Papua in Indonesia to Philippines and S.W. Japan.

SPINY GOBY *Asterropteryx spinosa*
SIZE: to 4.3 cm (1 3/4 in.) Gobies – Gobiidae
ID: Pale gray with dense covering of variably sized brown spots; short bands on head and body, **broad dark bar with rounded tip below eye,** short dark vertical streak on tail base. Sheltered shorelines on mud or sand and rubble bottoms to 30 m.
Indo-West Pacific: Seychelles to Indonesia, Philippines, Micronesia and Solomon Is. – S.W. Japan to Great Barrier Reef and Fiji.

OVAL-SPOT GOBY *Asterropteryx ovata*
SIZE: to 3.6 cm (1 1/2 in.) Gobies – Gobiidae
ID: Pale gray to dusky pink with numerous orange spots on head and body; **dark oval spot with wide white rear border on tail base.** Solitary. Sand and mud in 15-40 m.
Asian Pacific: N. Sulawesi in Indonesia, Micronesia and S. Japan.

YOSHIGOUI'S GOBY *Ancistrogobius yoshigoui*
SIZE: to 5 cm (2 in.) Gobies – Gobiidae
ID: Grayish with dense brown mottling; row of 5 rectangular brown blotches between pair of narrow brown stripes on lower side, irregular orange to brown spots on head and forebody. Sand and mud among sheltered reefs in 5-30 m.
Asian Pacific: Indonesia, Philippines, Papua New Guinea and S.W. Japan.

ORANGETIP GOBY *Ancistrogobius yanoi*
SIZE: to 14.5 cm (1 3/4 in.) Gobies – Gobiidae
ID: Male – Pale with brown spots and blotches; **dark markings on upper iris,** brown bar below eye, white filament at tip of elongate 3rd dorsal spine with brown band below. Solitary. Sand, mud and rubble of sheltered reefs to 12 m.
Asian Pacific: Indonesia, Philippines, Micronesia, Papua New Guinea and Solomon Is. – S.W. Japan to Fiji.

DWARF GOBY *Pandaka pygmaea*
SIZE: to 1.2 cm (1/2 in.) Gobies – Gobiidae
ID: Translucent with scattered dark brown and white blotches; dark spot on upper pectoral fin base, often pepper-like spots on body. **Male –** Very tall dorsal fin pennant. Small groups among mangrove roots to 2 m.
Asian Pacific: Malaysia, Indonesia and Philippines.

Gobies

EAR-SPOT DWARFGOBY   *Eviota karaspila*
SIZE: to 2.3 cm (1 in.)   Gobies – Gobiidae
ID: Translucent body; reddish belly and white head, 2 black "ear" spots, red eyes with gold scribble markings. Solitary. Tide pools to exposed seaward reefs to 15 m.
Localized: Fiji

MELASMA DWARFGOBY   *Eviota melasma*
SIZE: to 3.3 cm (1 ¼ in.)   Gobies – Gobiidae
ID: Translucent with 8-9 brownish bars and narrow white spaces between; occasionally pair of dark brown "ear" spots above pectoral fins, gold ornate markings on iris. Solitary. Sponges or corals of seaward reefs slopes to 18 m.
West Pacific: Indonesia to Philippines, Micronesia. – S. Japan to Australia and Fiji.

SIGILLATA DWARFGOBY   *Eviota sigillata*
SIZE: to 2.5 cm (1 in.)   Gobies – Gobiidae
ID: Translucent with narrow reddish internal midlateral stripe; white stripe along spinal column broken by reddish bars that extend below midlateral stripe, white spots on head. Form groups on sand and rubble. Sheltered reefs in 3-20 m.
Indo-Asian Pacific: Madagascar to Indonesia, Philippines, Micronesia, Papua New Guinea. – S. Japan to Australia and Fiji.

STRIPED DWARFGOBY   *Eviota sebreei*
SIZE: to 3 cm (1 ¼ in.)   Gobies – Gobiidae
ID: Translucent with dark brown to blackish midlateral stripe with row of white to yellow marks on top; pale-edged black spot on tail base. Usually form small groups; rest on surface of large *Porites* coral heads. Lagoon and seaward reefs to 20 m.
Indo-West Pacific: Red Sea to Indonesia and Micronesia. – S.W. Japan to Australia and Fiji.

COMET DWARFGOBY   *Eviota cometa*
SIZE: to 2.4 cm (1 in.)   Gobies – Gobiidae
ID: Translucent body with broad reddish brown internal midlateral stripe; white spots from head to black spot on tail base, row of **white to yellow spots run along belly**. Sand near to coral reefs to 40 m.
West Pacific: Indonesia, Philippines, Papua New Guinea, Micronesia, Solomon Is. and Fiji.

RED-HEADED DWARFGOBY   *Eviota rubriceps*
SIZE: to 1.8 cm (¾ in.)   Gobies – Gobiidae
ID: Translucent body with **black sides and white belly**; red upper head striped in white. Solitary. Sand or rubble of sheltered reefs to 10 m.
Asian Pacific: Indonesia, Philippines, Papua New Guinea and S.W. Japan.

STORTHYNX DWARFGOBY *Eviota storthynx*
SIZE: to 2.8 cm (1 in.) Gobies – Gobiidae
ID: Translucent with white spots on dark brown lower body; red to brown spotting and blotches, 1 or 2 black "ear" spots, small dark spot at tail base. Solitary. Sand or rubble of sheltered reefs to 2 m.
Asian Pacific: W. Australia to Indonesia, Philippines, Micronesia and S.W. Japan.

ZEBRA DWARFGOBY *Eviota zebrina*
SIZE: to 2.4 cm (1 in.) Gobies – Gobiidae
ID: Translucent with grayish internal stripe broken by several white spots; short stripe of about 5 white spots behind pectoral fin, **red to black horseshoe-shaped marks on snout and another behind eye.** Dead coral algal-covered reefs in 3-15 m.
Indo-West Pacific: Red Sea and islands of W. Indian Ocean to West Papua in Indonesia, Micronesia, Australia and Fiji.

BROWN-BANDED DWARFGOBY *Eviota latifasciata*
SIZE: to 2 cm (³/₄ in.) Gobies – Gobiidae
ID: Translucent with 6-7 broad brown bars or blotches on side (narrow white bars may also be evident); white marks on upper head. Solitary. Exposed coastal reefs and outer slopes in 4-25 m.
East Indo-West Pacific: Christmas I. to Indonesia, Philippines, Papua New Guinea, Micronesia and Gilbert Is.

RED & WHITE-SPOTTED DWARFGOBY *Eviota prasites*
SIZE: to 3.5 cm (1¹/₂ in.) Gobies – Gobiidae
ID: Red with translucent back and underside; **row of white dashes on upper body,** red spots on back. Form small groups. Rubble bottoms of lagoon and coastal reefs in 5-20 m.
Indo-West Pacific: E. Africa to Indonesia, Philippines, Micronesia, Papua New Guinea and Solomon Is. – S. Japan to Australia, New Caledonia and Fiji.

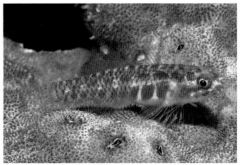

REDSPOTTED DWARFGOBY *Eviota rubrisparsa*
SIZE: to 3 cm (1¹/₄ in.) Gobies – Gobiidae
ID: Translucent with fine red spotting; head yellow or pale gray; 2-3 red or red-brown bars on side of belly, **dark iris with scrawled yellow markings.** Solitary or form small groups. Sand and rubble bottoms among reefs in 4-15 m.
Asian Pacific: Indonesia, Philippines, Papua New Guinea and Solomon Is.

TERRY'S DWARFGOBY *Eviota teresae*
SIZE: to 2.9 cm (1¹/₄ in.) Gobies – Gobiidae
ID: Translucent; white stripe from eye to rear belly, 3 red bars separated by pale bars on belly and white band on pectoral fin base, black and yellowish markings on red iris, peppered "ear" spot. Solitary or form groups. Lagoon and seaward reefs to 15 m.
West Pacific: Indonesia, Philippines and Micronesia. – S.W. Japan to Great Barrier Reef and Fiji.

325

Gobies

RUBBLE DWARFGOBY   *Eviota prasina*
SIZE: to 4 cm (1¹/₂ in.)   Gobies – Gobiidae
ID: Translucent green with 6-7 internal grayish brown bars; brown spots and/or bars on head, large black spot on upper tail base. Rubble and algal-covered dead reefs near shore to 5 m.
Indo-Asian Pacific: Red Sea and E. Africa to Indonesia, Philippines, Micronesia, Papua New Guinea and Solomon Is. – S. Japan to Great Barrier Reef and New Caledonia.

NEBULOUS DWARFGOBY   *Eviota nebulosa*
SIZE: to 2.4 cm (1 in.)   Gobies – Gobiidae
ID: Whitish with 6-9 dark brownish bars; short brown bar on top of head and upper back, large black spot on upper tail base. Shallow coral reefs to about 10 m.
Indo-West Pacific: E. Africa to Indonesia, Philippines, Micronesia, Papua New Guinea and Solomon Is. – S. Japan to Great Barrier Reef and Fiji.

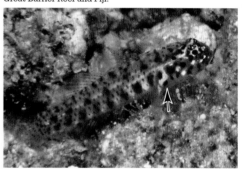

SPECKLED DWARFGOBY   *Eviota sparsa*
SIZE: to 3 cm (1¹/₄ in.)   Gobies – Gobiidae
ID: Translucent with 6 internal reddish brown bars that split on lower side; orange spots on head, dusky orange scale margins, **2 black bars separated by narrow whitish area behind eye.** Coral reefs in 2-18 m.
West Pacific: Indonesia, Philippines, Papua New Guinea, Solomon Is. – E. Australia to Fiji.

QUEENSLAND DWARFGOBY   *Eviota queenslandica*
SIZE: to 3 cm (1¹/₄ in.)   Gobies – Gobiidae
ID: Translucent with rows of dark spots and narrow brown scale margins; **pair of black spots on pectoral fin base,** black "ear" spot. Solitary or form small groups. Rubble bottoms of sheltered reefs to 6 m.
Asian Pacific: Andaman Sea to Indonesia, Philippines, Micronesia. – S. Japan to Great Barrier Reef and Vanuatu.

TWIN DWARFGOBY   *Eviota fallax*
SIZE: to 3 cm (1¹/₄ in.)   Gobies – Gobiidae
ID: Translucent with orange scale margins; internal orange stripe from head to tail interrupted by 7 white spots, **large orange spot behind eye,** white mark above pectoral fin, dark bars on upper iris. Solitary or loose groups. Dead reef and rubble in 10-20 m.
West Pacific: Indonesia to S. Japan, Micronesia, Papua New Guinea and Solomon Is.

WHITESPOTTED DWARFGOBY   *Eviota lachdeberei*
SIZE: to 2.6 cm (1 in.)   Gobies – Gobiidae
ID: Brown with pale underside; red iris, row of small white spots on upper back, **several white spots on lower front of body,** blackish spot on tail base. Solitary or form small groups in crevices. Silty inshore reefs to 20 m.
Asian Pacific: Indonesia, Philippines, Papua New Guinea and Solomon Is. to N. Australia.

NEON DWARFGOBY *Eviota atriventris*
SIZE: to 2.5 cm (1 in.) Gobies – Gobiidae
ID: Red midbody and translucent back and anterior body; pair of bright yellow stripes behind eye to midbody, belly often black with white stripe. Sheltered lagoon and coastal reefs in 3-20 m.
Asian Pacific: Brunei, Sabah in Malaysia, Indonesia, Philippines, Papua New Guinea and Solomon Is.

RAJA DWARFGOBY *Eviota raja*
SIZE: to 3 cm (1 1/4 in.) Gobies – Gobiidae
ID: Pair of red stripes bordering broad midlateral yellow stripe; narrow bar on tail base, white stripe on midsnout, white upper iris, blue streaks below eye and on gill cover. Small groups near or among branching corals. Sheltered reef slopes in 4-12 m.
Localized: Known only from Raja Ampat in Indonesia.

BLACKBELLY DWARFGOBY *Eviota nigriventris*
SIZE: to 2.4 cm (1 in.) Gobies – Gobiidae
ID: Red to dark brown sides with white to pale green back and belly; black blotch on tail base, black lips, pale upper iris. Form groups among branches of *Acropora* corals. Sheltered reefs in 4-20 m.
Asian Pacific: Andaman Sea to Indonesia, Philippines, Micronesia and Solomon Is. – S.W. Japan to Great Barrier Reef and Coral Sea.

PURPLE DWARFGOBY *Eviota dorsopurpurea*
SIZE: to 2.7 cm (1 in.) Gobies – Gobiidae
ID: Broad purplish brown stripe on lower body from snout to tail base and purplish pink on top of head and back. Form groups among branches of *Acropora* coral. Sheltered reefs in 8-30 m.
Localized: Papua New Guinea (Milne Bay Province, including D'Entrecasteaux Islands).

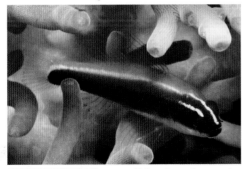

CREAMBACK DWARFGOBY *Eviota dorsogilva* complex
SIZE: to 2.3 cm (1 in.) Gobies – Gobiidae
ID: Broad red-brown to black stripe on lower body; top of head and back yellowish cream, dusky anal fin. Groups in branches of *Acropora* corals in 2-31 m. Distinguished from similar Blackbelly & Purple Dwarfgobies [above] by location.
West Pacific: Micronesia, Papua New Guinea and Solomon Is. – S.W. Japan to N. Great Barrier Reef, New Caledonia and Fiji.

TWOSTRIPE DWARFGOBY *Eviota bifasciata*
SIZE: to 3 cm (1 1/4 in.) Gobies – Gobiidae
ID: Translucent with pair of red stripes separated by wide white midlateral stripe; blackish blotch on lower tail base (occasionally on upper); white stripe on midsnout. Form groups. Hover above branching corals of sheltered reefs to 15 m.
Asian Pacific: Indonesia, Philippines, Micronesia, Papua New Guinea and Solomon Is. – S.W Japan to Great Barrier Reef.

Gobies

LARGETOOTH GOBY *Macrodontogobius wilburi*
SIZE: to 6.7 cm (2¹/₂ in.)   Gobies – Gobiidae
ID: Pale gray to tan with **about 7 darkish double bars on side;** row of dark black spots on back from nape to tail base, large 1st dorsal fin with filamentous tips on spines, about 4-5 dark band markings on anal fin. Solitary. Sand and silt bottoms to 10 m.
Indo-Pacific: Seychelles to Indonesia, Micronesia, Solomon Is. and Line Is. – S.W. Japan to Australia and Fiji.

Largetooth Goby – Female
ID: Pale tan with dense network of small brown spots; midlateral row of dark spots from pectoral fin to tail base, **dark patch below eye,** row of dark spots on back from nape to tail base.

BEAUTIFUL GOBY *Exyrias belissimus*
SIZE: to 16.5 cm (6¹/₂ in.)   Gobies – Gobiidae
ID: Brown with diffuse bars on side; large dorsal fins with orange to brown dash markings, scattered white spots on lower head and body. Solitary or form groups. Near reef overhangs on mud bottoms of silty coastal reefs and lagoons to 30 m.
Indo-West Pacific: E. Africa to Indonesia, Philippines, Papua New Guinea and Solomon Is. – S.W. Japan to Great Barrier Reef and Fiji.

Beautiful Goby – Variation
ID: With a yellow tint.

PUNTANG GOBY *Exyrias puntang*
SIZE: to 13 cm (5 in.)   Gobies – Gobiidae
ID: Male – Light brown with diffuse dark brown bars and scattered white spots; **large filamentous 1st dorsal fin with dark dashes.** Solitary or form groups. Mud and algal bottoms of mangrove shores, estuaries and brackish lakes to 5 m.
East Indo-Asian Pacific: Sri Lanka to Indonesia, Philippines and Solomon Is. – S.W. Japan to Great Barrier Reef and Fiji.

FERRARIS' GOBY *Exyrias ferrarisi*
SIZE: to 9.5 cm (3³/₄ in.)   Gobies – Gobiidae
ID: Brown to gray with brown spotting and lines; dark brown spot above pectoral fin, **4 large dark blotches on side,** tall sail-like 1st dorsal fin. Solitary or in pairs. Silty inshore reefs to 8 m.
Asian Pacific: Java, Bali and West Papua in Indonesia, north to Philippines.

FILAMENTED GOBY *Exyrias akihito*
SIZE: to 12 cm (4 3/4 in.) Gobies – Gobiidae
ID: Tan; narrow orange stripes, **row of 4 pairs of tiny brown spots on side,** orange spots on fins, tall large filamentous 1st dorsal fin. Solitary. Sand and light rubble areas near corals of sheltered reefs in 10-50 m.
Asian Pacific: Indonesia, Philippines and Papua New Guinea. – S.W. Japan to N. Great Barrier Reef.

THREADFIN SANDGOBY *Favonigobius reichei*
SIZE: to 4.5 cm (1 3/4 in.) Gobies – Gobiidae
ID: Elongate; **thread-like filament extends from 1st dorsal fin,** light tan with brown and white spotting. Usually form small groups. Sandy shores of protected bays and lagoons to 5 m.
Indo-Asian Pacific: E. Africa to Indonesia, Philippines, Micronesia. – S.W. Japan to Great Barrier Reef and Vanuatu.

BLOTCHED SANDGOBY *Fusigobius inframaculatus*
SIZE: to 6 cm (2 1/4 in.) Gobies – Gobiidae
ID: Male – Translucent with small orange spots; 4-5 large internal blackish blotches, **white dash followed by black spot on tail base,** long 1st dorsal spine. Solitary. Sand at base of coral overhangs of seaward reefs in 5-25 m.
Indo-West Pacific: Persian Gulf to Indonesia, Philippines and Solomon Is. – Taiwan to Great Barrier Reef and Fiji.

Blotched Sandgoby – Female
ID: Similar markings as male, but lack the long 1st dorsal spine.

SIGNALFIN SANDGOBY *Fusigobius signipinnis*
SIZE: to 6.3 cm (2 1/2 in.) Gobies – Gobiidae
ID: Translucent with tiny brown spots; red-brown iris, **black tip on foredorsal fin between 1st 3 spines** and darkish blotch below base. Solitary. Sand at base of coral formations of lagoon and seaward reefs in 3-30 m.
West Pacific: Indonesia, Philippines, Papua New Guinea and Solomon Is. – Taiwan to Great Barrier Reef and Fiji.

BLACKTIP SANDGOBY *Fusigobius melacron*
SIZE: to 4.5 cm (1 3/4 in.) Gobies – Gobiidae
ID: Translucent with numerous small yellowish brown or dusky orange spots; row of dark dashs at middle of body and another below dorsal fins, **black to light brown outer half of 1st dorsal fin.** Sand-rubble around reefs in 7-31 m.
West Pacific: Andaman Sea to Indonesia, Philippines, Micronesia and Solomon Is. – S.W. Japan to Great Barrier Reef and Fiji.

LARGE SANDGOBY   *Fusigobius maximus*
SIZE: to 8 cm (3 1/4 in.)   Gobies – Gobiidae
ID: Whitish with orangish spots; **black dash on 1st dorsal spine of rounded dorsal fin,** row of tiny white spots along base of dorsal fins, dusky spot on tail base. Solitary. Sand at base of corals of seaward reefs in 3-21 m.
Indo-Asian Pacific: Red Sea to Andaman Sea, Indonesia, Philippines and Papua New Guinea.

Large Sandgoby – Variation
ID: Translucent with pale orange spots; black dash on 1st dorsal spine.

SHOULDERSPOT SANDGOBY   *Fusigobius humeralis*
SIZE: to 4.4 cm (1 3/4 in.)   Gobies – Gobiidae
ID: Semitranslucent with numerous small dusky orange-brown spots on head and body; **pupil-sized black spot above pectoral fin base** and another on middle of tail base. Sand and rubble bottoms around reefs in 3-30 m.
East Indo-Pacific: Maldives to Indonesia, Philippines, Solomon Is. and French Polynesia. – S.W. Japan to New Caledonia.

PALE SANDGOBY   *Fusigobius pallidus*
SIZE: to 8 cm (3 1/4 in.)   Gobies – Gobiidae
ID: Translucent gray with numerous small dusky-edged pale orange spots on head, body, dorsal fin and tail; **small black spot between upper 1st and 2nd dorsal spines.** Sand and rubble bottoms around reefs in 12-35 m.
Indo-West Pacific: E. Africa to Indonesia, Philippines, Micronesia, Coral Sea and Fiji.

TWOSPOT SANDGOBY   *Fusigobius duospilus*
SIZE: to 5.7 cm (2 1/4 in.)   Gobies – Gobiidae
ID: Translucent gray with numerous orangish brown spots of varying size; **pair of black dash markings on 1st dorsal fin,** small dark blotch on base of tail. Solitary. Clean sand and rubble bottoms of seaward reefs to 46 m.
Indo-West Pacific: E. Africa to Indonesia, Philippines, Solomon Is. and Hawaii. – S. Japan to Australia and Fiji.

NEOPHYTE SANDGOBY   *Fusigobius neophytus*
SIZE: to 7 cm (2 3/4 in.)   Gobies – Gobiidae
ID: Translucent with numerous reddish brown spots of varying size; **short slanting bands below dorsal fins.** Solitary or form small groups. Sand and rubble areas of coastal reefs and lagoons to 10 m.
Indo-Pacific: E. Africa to Indonesia, Philippines, Solomon Is. and French Polynesia. – S.W. Japan to Australia.

ORANGESPOTTED SANDGOBY   *Istigobius rigilius*
SIZE: to 9.5 cm (3 ³/₄ in.)   Gobies – Gobiidae
ID: Translucent-whitish with numerous small orangish brown and white spots; **larger orangish brown double spots form broken midlateral line.** Solitary. Clean sand and rubble bottoms of lagoon and seaward reefs to 30 m.
West Pacific: Indonesia, Philippines, Micronesia and Solomon Is. – S.W. Japan to Great Barrier Reef and Fiji.

DECORATED SANDGOBY   *Istigobius decoratus*
SIZE: to 9 cm (3 ¹/₂ in.)   Gobies – Gobiidae
ID: Whitish to pale tan with faint brown lines and several rows of small brown spots; **row of brown streaks link paired white spots on lower side.** Solitary. Clean white to black sand of lagoon and seaward reefs to 25 m.
Indo-Asian Pacific: Red Sea and E. Africa to Indonesia, Philippines and Micronesia. – S.W. Japan to Australia and Fiji.

ORNATE SANDGOBY   *Istigobius ornatus*
SIZE: to 9 cm (3 ¹/₂ in.)   Gobies – Gobiidae
ID: Pale gray upper body with white below; numerous black streaks arranged in horizontal rows on side, **yellowish tip on 1st dorsal fin.** Solitary or form groups. Silt or mud bottoms of mangroves and protected bays to 2 m.
Indo-West Pacific: Red Sea and E. Africa to Indonesia, Philippines and Micronesia. – S.W. Japan to Australia and Fiji.

GOLDMANN'S SANDGOBY   *Istigobius goldmanni*
SIZE: to 6.7 cm (2 ³/₄ in.)   Gobies – Gobiidae
ID: Grayish upper body shading to white below with scale margins forming honeycomb pattern on upper body; **midlateral row of paired black spots on body from pectoral fin to tail base,** short black stripe behind eye. Sheltered sand to 5 m.
East Indo-West Pacific: India to Indonesia, Philippines, Micronesia and Solomon Is. – S.W. Japan to Australia and Fiji.

SHOULDERBAR GOBY   *Gnatholepis cauerensis*
SIZE: to 4.5 cm (1 ³/₄ in.)   Gobies – Gobiidae
ID: White with numerous horizontal rows of small close-set spots and blotches; thick bar through eye, **small orange "shoulder" spot.** Solitary or form small groups. Sand bottoms near rocky outcrops of sheltered reefs in 3-50 m.
Indo-Pacific: E. Africa to Indonesia, Micronesia, Hawaii and French Polynesia. – S.W. Japan to Australia and New Caledonia.

EYEBAR GOBY   *Gnatholepis anjerensis*
SIZE: to 9.5 cm (3 ³/₄ in.)   Gobies – Gobiidae
ID: White to pale gray with numerous horizontal rows of small spots and blotches; thin dusky bar through eye, **row of small white spots below dorsal fin.** Solitary or form groups. On sand of lagoons and seaward reefs to 46 m.
Indo-Pacific: Red Sea to Indonesia, Micronesia, Hawaii and French Polynesia. – S.W. Japan to New Caledonia.

Gobies

YELLOW CORALGOBY   *Gobiodon okinawae*
SIZE: to 3 cm (1¼ in.)   Gobies – Gobiidae
ID: Uniform bright yellow. Frequently perch on outer surface of corals such as tabletop *Acroporas* rather than among coral branches, like other species of *Gobiodon*. Usually form small groups. Sheltered reefs in 2-15 m.
East Indo-Asian Pacific: Coco-Keeling Is. to Indonesia, Philippines, Papua New Guinea and Solomon Is. – S. Japan to Australia.

LEMON CORALGOBY   *Gobiodon citrinus*
SIZE: to 6.6 cm (2½ in.)   Gobies – Gobiidae
ID: Variable color from yellow to yellowish green or black; **pair of blue to white bars through eye and 2nd pair across rear head.** Solitary. Usually rest on branching *Acropora* corals in sheltered lagoons to 15 m.
Indo-West Pacific: Red Sea and E. Africa, Indonesia and Philippines. – S. Japan to Great Barrier Reef and Fiji.

NEEDLESPINE CORALGOBY   *Gobiodon acicularis*
SIZE: to 4.6 cm (1¾ in.)   Gobies – Gobiidae
ID: Yellowish brown or black including fins; **long filamentous 1st dorsal spine.** Perch among branching corals in shallow protected reef areas to 5 m.
Asian Pacific: Raja Ampat in Indonesia to Micronesia, Papua New Guinea and Solomon Is. – S. Japan to Great Barrier Reef.

Needlespine Coralgoby – Black Variation
ID: Found among the branches of *Echinopora horrida, E. mammiformis, Pectinia* sp. and *Hydnophora rigida*.

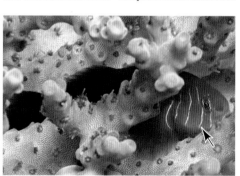

FIVE-LINED CORALGOBY   *Gobiodon quinquestrigatus*
SIZE: to 4.6 cm (1¾ in.)   Gobies – Gobiidae
ID: Dark brown body and lighter brown to **reddish head with 5 pale blue vertical lines. Juvenile –** Also vertical lines on body. Solitary or in pairs. Among branches of *Acropora* corals of lagoons and sheltered seaward reefs to 15 m.
Pacific: Brunei, Indonesia, Philippines, Micronesia and Solomon Is. to French Polynesia. – S. Japan to Australia.

WHITE-LINED CORALGOBY   *Gobiodon heterospilos*
SIZE: to 6.6 cm (2½ in.)   Gobies – Gobiidae
ID: Juvenile – White with a pair of black stripes; **black spots on head and tail. Adult –** Blackish without markings. Solitary or form groups. On wide variety of corals of sheltered lagoons and coastal reefs in 3-20 m.
Asian Pacific: Indonesia, Philippines, Papua New Guinea and Solomon Is. – S. Japan to Great Barrier Reef.

CERAM CORALGOBY *Gobiodon ceramensis*
SIZE: to 3.5 cm (1¹/₂ in.) Gobies – Gobiidae
ID: Black without distinctive markings. Solitary or in pairs. Among Pocilloporidae coral branches (frequently *Stylophora pistillata*) in lagoons and sheltered coastal reefs in 2-15 m.
Asian Pacific: Indonesia, Micronesia, Papua New Guinea and N. Great Barrier Reef.

UNICOLOR CORALGOBY *Gobiodon unicolor*
SIZE: to 3.5 cm (1¹/₂ in.) Gobies – Gobiidae
ID: Uniform dark brown to nearly black, including fins. Often associate with other *Gobiodon* species. Among branches of *Acropora* corals on shallow reefs in 2-8 m.
Asian Pacific: West Papua in Indonesia to Papua New Guinea and N. Australia.

RED-SPOTTED CORALGOBY *Gobiodon aoyagii*
SIZE: to 4.8 cm (2 in.) Gobies – Gobiidae
ID: Male – Green with red spots on head and body; red bar below eye and often 2-3 more on gill cover. **Two red spots on pectoral fin base.** Solitary or pairs among branches of *Acropora* corals. Reef crests and lagoons to 12 m.
Asian Pacific: Indonesia, Philippines and Micronesia. – S. Japan to Great Barrier Reef.

BROAD-BARRED CORALGOBY *Gobiodon histrio*
SIZE: to 4.8 cm (2 in.) Gobies – Gobiidae
ID: Male – Blue-green with about 5 red body stripes formed by lines and red spots; **dusky spot on upper rear gill cover,** 5 broad reddish bars on head. Solitary or pairs inside *Acropora* corals. Reef crests and lagoons to 15 m.
Indo-West Pacific: Red Sea to Indonesia, Philippines, Great Barrier Reef and Fiji.

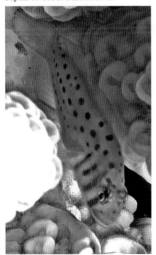

RED-STRIPED CORALGOBY
Gobiodon erythrospilus
SIZE: 4 cm (1¹/₂ in.)
Gobies – Gobiidae
ID: Yellowish green with uneven rows of red spots on body, 4-6 reddish bars on head and pectoral fin base. Solitary or pairs inside branches of *Acropora* corals. Reef crests and lagoons to 15 m.
West Pacific:
Indonesia, Philippines, Micronesia. – S. Japan to Great Barrier Reef and Fiji.

ELONGATE CORALGOBY *Gobiodon prolixus*
SIZE: to 3.8 cm (1¹/₂ in.) Gobies – Gobiidae
ID: Golden brown to orangish pink with 5 thin vertical pale blue lines on head and pectoral fin base. Solitary or pairs among branches of *Acropora* corals to 21 m.
East Indo-Pacific: Comoro Is. to Andaman Sea, Brunei, West Papua in Indonesia and French Polynesia. – S. Japan to N.E. Australia.

333

Gobies

MULTILINED CORALGOBY　　　　*Gobiodon rivulatus*
SIZE: to 4 cm (1 1/2 in.)　　　　　Gobies – Gobiidae
ID: Reddish brown, tan or greenish; 5 thin vertical lines on head to pectoral fin become less visible on body. Solitary or pairs among branches of *Acropora* corals to 15 m.
Indo-West Pacific: Red Sea and E. Africa to Andaman Sea, Brunei, Indonesia, Philippines, Micronesia, Papua New Guinea and Solomon Is. – S. Japan to Great Barrier Reef and Fiji.

Multilined Coralgoby – Color Variation

BLUEMAZE CORALGOBY　　　　*Gobiodon* sp.
SIZE: to 3.5 cm (1 1/2 in.)　　　　Gobies – Gobiidae
ID: Pinkish tan with **maze-like pattern of bluish lines on top of head** and similar markings on cheek and gill cover. Among branches of *Acropora* corals in 4-15 m.
Asian Pacific: Brunei to Sulawesi, Indonesia, Philippines and S. Japan.

BLACKFIN CORALGOBY　　　　*Paragobiodon lacunicolus*
SIZE: to 2.5 cm (1 in.)　　　　Gobies – Gobiidae
ID: Yellow to whitish often with reddish tints on head; fins black except occasionally pectoral fin translucent. Inhabit *Pocillopora damicornis* coral to 20 m.
Indo-Pacific: E. Africa to Indonesia, Philippines, Papua New Guinea, Solomon Is. to French Polynesia. – S.W. Japan to Australia.

GOLDEN CORALGOBY　　　*Paragobiodon xanthosoma*
SIZE: to 4 cm (1 1/2 in.)　　　　Gobies – Gobiidae
ID: Head, body and fins variable from bright yellow to medium green; bright red iris; head and nape covered with bristle-like papillae, often thin dark outer edge on dorsal fins. Among branches of *Seriatopora* corals, often *S. hystrix* in 2-10 m.
Indo-West Pacific: E. Africa to Indonesia, Philippines, Micronesia and Solomon Is. – S.W. Japan to Great Barrier Reef and Fiji.

REDHEAD CORALGOBY　　　*Paragobiodon echinocephalus*
SIZE: to 3.5 cm (1 1/2 in.)　　　　Gobies – Gobiidae
ID: Black body and fins with reddish orange head covered with tiny reddish bristles. Usually in pairs or small groups among branches of *Pocillopora*, *Seriatopora*, and *Stylophora* corals of lagoon and seaward reefs to 10 m.
Indo-Pacific: Red Sea to E. Africa Indonesia, Philippines, Micronesia and French Polynesia. – S.W. Japan to Australia.

334

WARTHEAD CORALGOBY   *Paragobiodon modestus*
SIZE: to 3.5 cm (1¹/₂ in.)   Gobies – Gobiidae
ID: Head and nape dark brownish orange shading to black body and fins; **narrow white outer edge on pectoral fin,** head covered with bristle-like papillae. Among branches of *Seriatopora* corals in 3-15 m.
Indo-Pacific: E. Africa to Andaman Sea, Indonesia, Micronesia, Solomon Is. and French Polynesia. – S. Japan to Great Barrier Reef.

BLACK CORALGOBY   *Paragobiodon melanosomus*
SIZE: to 3.5 cm (1¹/₂ in.)   Gobies – Gobiidae
ID: Entirely black; head and nape covered with bristle-like papillae. Among branches of *Seriatopora* corals in 4-15 m.
Asian Pacific: Brunei, Indonesia, Philippines, Micronesia, Papua New Guinea and Solomon Is. – S.W. Japan to Great Barrier Reef.

FRECKLED CORALGOBY   *Paragobiodon sp.*
SIZE: to 3.8 cm (1¹/₂ in.)   Gobies – Gobiidae
ID: Dark brown head, body and fins; bluish maze-like markings on head including nape, no markings on body or fins; head covered with bristle-like papillae. Among branches of *Seriatopora* corals in 3-15 m.
Localized: Currently known only from Raja Ampat, Indonesia.

DINAH'S GOBY   *Lubricogobius dinah*
SIZE: to 2.1 cm (³/₄ in.)   Gobies – Gobiidae
ID: Orange body, yellow-orange head and yellow fins; wide white stripe on back from nape to tail base (not always present). Sand and rubble bottoms, take refuge in dead shells, empty bottles, tunicates, etc. in 10-35 m.
Asian Pacific: Alor in Indonesia, Papua New Guinea and S.W. Japan.

ORNATE GOBY   *Lubricogobius ornatus*
SIZE: to 4 cm (1¹/₂ in.)   Gobies – Gobiidae
ID: Orange to yellow body and fins with pale bluish bands radiating from eye and inverted "Y" band originating on rear nape. Solitary. Reef, sand silt and rubble bottoms to 70 m.
Asian Pacific: Vietnam, Indonesia, Philippines, Papua New Guinea to N. Australia and Great Barrier Reef.

GOLDEN GOBY   *Lubricogobius exiguus*
SIZE: to 4 cm (1¹/₂ in.)   Gobies – Gobiidae
ID: Yellow to golden body, fins and green eyes. Often inside bottles, cans and coconut shell halves in sand and rubble areas in 15-70 m.
Asian Pacific: Indonesia, Philippines and Papua New Guinea. – S.W. Japan to N. Australia and New Caledonia.

Gobies

PALEBARRED REEFGOBY   *Priolepis pallidicincta*
SIZE: to 3.6 cm (1½ in.)   Gobies – Gobiidae
ID: Alternating brownish yellow and pale gray to whitish bars on head and body including 6 wide bars on body. Cryptic in crevices and caves to 56 m.
West Pacific: Brunei, Indonesia, Philippines, Papua New Guinea, Solomon Is. – Great Barrier Reef to Vanuatu and Fiji.

HALFBARRED REEFGOBY   *Priolepis semidoliata*
SIZE: to 3 cm (1¼ in.)   Gobies – Gobiidae
ID: Reddish with 3 bluish white bars on head below and behind eye and 2 bars joined to form an upside down "V" from nape to in front of pectoral fin; 5-7 faint bluish white bars on body, red spots on dorsal and tail fins. Coral reefs to 20 m.
Indo-Pacific: Red Sea to Andaman Sea, Indonesia, Philippines, Micronesia, Solomon Is. and Pitcairn Is. – S. Japan to Great Barrier Reef.

THREADFIN REEFGOBY   *Priolepis nuchifasciata*
SIZE: to 3.2 cm (1¼ in.)   Gobies – Gobiidae
ID: Reddish to mauve with **darkish red scale margins;** several dark edged pale bands on head and upper back, 2 bands on nape behind eye join to form a "Y" and a 3rd band behind splits to form and inverted "Y" in front of pectoral fin, elongate 2nd dorsal spine. Cryptic in crevices and caves in 5-30 m.
Asian Pacific: Indonesia, Philippines and Solomon Is. to Australia.

RIBBON REEFGOBY   *Priolepis vexilla*
SIZE: to 2.5 cm (1 in.)   Gobies – Gobiidae
ID: Reddish orange with **distinctly black scale margins;** dark edged pale bands on head and upper back below dorsal fins, 2 bands on nape behind eye join to form a "Y" and a 3rd band behind splits to form and inverted "Y" in front of pectoral fin, elongate 2nd dorsal spine. Solitary. Reef crevices in 2-20 m.
Asian Pacific: Sulawesi and Molucca Is., Indonesia and Philippines.

CONVICT REEFGOBY   *Priolepis cincta*
SIZE: to 5 cm (2 in.)   Gobies – Gobiidae
ID: Brown with **dark-edged whitish bars;** brown spotting on dorsal and tail fins. Solitary. Rocky crevices or among corals of coastal and outer reefs to 70 m.
Indo-West Pacific: E. Africa to Indonesia, Philippines, Micronesia, Papua New Guinea and Solomon Is. – S. Japan to Great Barrier Reef and Fiji.

NARROWBAR REEFGOBY   *Priolepis profunda*
SIZE: to 4.4 cm (1¾ in.)   Gobies – Gobiidae
ID: Male – Orange-brown with **narrow white or blue bars;** black spot on base of 1st dorsal fin, white margin on tail. **Female –** Rows of black spots on dorsal and tail fins. Solitary or pairs. Reef crevices in 20-114 m.
Asian Pacific: Andaman Sea to Philippines, Papua New Guinea and N.W. Australia.

CROSSROADS REEFGOBY *Priolepis compita*
SIZE: to 1.7 cm (³/₄ in.) Gobies – Gobiidae
ID: Yellowish brown with black-edged white lines radiating from eye; 3 black-edged white bars behind eye connected by black-edged white stripe, 3 black-edged white bars on forebody. Coral reefs in 10-30 m.
Indo-West Pacific: E. Africa to Indonesia, Great Barrier Reef and Fiji.

KAPPA REEFGOBY *Priolepis kappa*
SIZE: to 2.6 cm (1 in.) Gobies – Gobiidae
ID: Yellowish brown with dark brown scale edges; 5-6 narrow whitish bars on body, whitish lines radiate from eye including K-shaped configuration behind eye. Reef crevices to 19 m.
Indo-West Pacific: Comoro Is. to Indonesia, Philippines and Papua New Guinea. – Taiwan to N.E. Australia and Fiji.

BLACK-BARRED REEFGOBY *Priolepis nocturna*
SIZE: to 4.5 cm (1³/₄ in.) Gobies – Gobiidae
ID: Gray with black bars; semicircles of white spots on both dorsal fins and upper edge of tail. Solitary and cryptic. Hide inside coral crevices of seaward reef slopes in 8-30 m.
Indo-Pacific: Seychelles to Indonesia, Philippines, Papua New Guinea and French Polynesia.

BALI GOBY *Grallenia baliensis*
SIZE: to 2.5 cm (1 in.) Gobies – Gobiidae
ID: Elongate body with **protruding lower jaw;** translucent with numerous irregular brown bars, white speckles, tiny skin flaps (papillae) on cheek, back and lower body. Solitary. Open sand bottoms of sheltered coastal reefs in 5-15 m.
Localized: Known only from Bali in Indonesia.

SPIKEFIN GOBY *Discordipinna griessingeri*
SIZE: to 2.9 cm (1¹/₄ in.) Gobies – Gobiidae
ID: White with dark spots on head; wide spotted red upper borders on dorsal and tail fins and red border on pectoral and anal fins, elongate red spike-like 1st dorsal fin. Solitary. Mixed coral, sand and rubble bottoms in 2-45 m.
Indo-Pacific: Red Sea to Indonesia, E. Papua New Guinea and French Polynesia. – S.W. Japan to Great Barrier Reef.

WHISKERED GOBY *Discordipinna sp.*
SIZE: to 2.5 cm (1 in.) Gobies – Gobiidae
ID: Female – Cream to brown head and upper body; black to dark brown lower body with row of 6-10 white spots, **2 dark spots on upper tail. Male –** Similar, with tall 1st dorsal spine. Silty sand-rubble in 2-20 m.
West Pacific: E. Indonesia and Papua New Guinea. – S.W. Japan and Fiji.

337

FROGFACE MUDGOBY   *Oxyurichthys papuensis*
SIZE: to 17 cm (6³/₄ in.)   Gobies – Gobiidae
ID: Male – Pale brownish gray; large mouth; 4 brown blotches with smaller diffuse blotches between, dark marking on pectoral fin base, occasionally dark marking below eye, black spot on tail base. Burrows on silt or mud of sheltered shorelines to 50 m.
Indo-Asian Pacific: Red Sea to Indonesia, Philippines, Micronesia and Solomon Is. – S.W. Japan to Australia and New Caledonia.

THREADFIN MUDGOBY   *Oxyurichthys notonema*
SIZE: to 16 cm (6¹/₄ in.)   Gobies – Gobiidae
ID: Pale brownish gray to bluish gray with 4 large yellow-brown oval midlateral blotches and irregular bluish markings with orangish edges between; about 10 darkish saddles on back, bluish edge on anal fin. Burrows in silty sand in 10-25 m.
Indo-Pacific: Mozambique, Indonesia, Philippines and French Polynesia.

BLACKSPOT MUDGOBY   *Oxyurichthys zeta*
SIZE: to 8.5 cm (3¹/₄ in.)   Gobies – Gobiidae
ID: Shades of gray to brown; **large black spot on rear of 1st dorsal fin,** large diffuse brown blotches on side with blue streaks between, bluish above anal fin. Solitary. Soft bottoms of sheltered bays and estuaries in 5-20 m.
Asian Pacific: Japan to Palau, Indonesia, Papua New Guinea, and Solomon Is.

GREEN-SHOULDERED GOBY   *Acentrogobius caninus*
SIZE: to 14 cm (5¹/₂ in.)   Gobies – Gobiidae
ID: Pale gray with 5 diffuse brown saddles along back and 4-5 diffuse brown blotches along midside; scattered white spots on head and body, **greenish patch behind upper gill cover.**
Indo-West Pacific: Sri Lanka to Indonesia and Micronesia. – S. Japan to Australia and Fiji.

CENDERAWASIH GOBY   *Acentrogobius cenderawasih*
SIZE: to 5.5 cm (2¹/₄ in.)   Gobies – Gobiidae
ID: Pale bluish gray with row of 5 large brown spots on lower side; **small dark spot on upper and lower pectoral fin base,** dark spot on lower gill cover, brownish mottling on upper body. Build burrows in sand or silty slopes in 15-35 m.
Asian Pacific: West Papua in Indonesia to Philippines and Solomon Is.

SHADOW GOBY   *Yongeichthys nebulosus*
SIZE: to 18 cm (7 in.)   Gobies – Gobiidae
ID: Brownish to grayish with brown scribbling on head and body; 3 large dark brown blotches on side and tail base, filaments extend from 1st dorsal fin. Solitary; toxic skin. Sand or mud bottoms of coastal reefs and estuaries to 10 m.
Indo-Asian Pacific: Red Sea and E. Africa to Indonesia, Philippines, Micronesia and Solomon Is. – S. Japan to Australia and Fiji.

CANINE GOBY *Oplopomus caninoides*

SIZE: to 8.5 cm (3 1/4 in.)   Gobies – Gobiidae

ID: Grayish with tiny white and orange spots; widely spaced black spots on back, row of joined black spots/and or short dashes behind pectoral fin with small white spots between. Solitary. Sand and rubble bottoms in 4-20 m.

Indo-Asian Pacific: Red Sea and E. Africa to Indonesia, Philippines, Papua New Guinea, Solomon Is. and New Caledonia.

SPINECHEEK GOBY *Oplopomus oplopomus*

SIZE: to 8 cm (3 1/4 in.)   Gobies – Gobiidae

ID: Male – Pale gray with small blue and orange spots on body, blue streaks on side of head; blue-edged black spot on rear of 1st dorsal fin. **Female –** Lack black spot. Solitary or in pairs. Fine sand and silt bottoms of sheltered coasts to 12 m.

Indo-Pacific: Red Sea and E. Africa to Indonesia, Philippines, Micronesia and French Polynesia. – S.W. Japan to Australia.

YELLOW-LINED FAIRYGOBY *Tryssogobius flavolineatus*

SIZE: to 3.1 cm (1 1/4 in.)   Gobies – Gobiidae

ID: Translucent pale gray with yellow stripe from eye to below middle of 2nd dorsal fin; blue band below eye; yellow 1st dorsal fin with stripe on base, 2 thin stripes on bluish 2nd dorsal fin. Sheltered reefs with silty sand and rubble in 25-82 m.

Localized: Known only from Halmahera to West Papua, Indonesia and D'Entrecasteaux Is. in Papua New Guinea.

SARAH'S FAIRYGOBY *Tryssogobius sarah*

SIZE: to 3.3 cm (1 1/2 in.)   Gobies – Gobiidae

ID: Similar to members of dartfish family in appearance; pearl gray, pair of yellow stripes on 2nd dorsal fin, tall 1st dorsal fin, white ventral fins. Sand/rubble in areas of cool upwelling and currents in 42-75 m.

Asian Pacific: Indonesia, Papua New Guinea, Micronesia, and S.W. Japan.

SIGNAL GOBY *Signigobius biocellatus*

SIZE: to 7.3 cm (3 in.)   Gobies – Gobiidae

ID: Tan with brown blotches; twin **"eye" spots on dorsal fins,** brown bar below eye, blue-spotted black ventral and anal fins. Solitary or in pairs that share sand burrows. Silt or sand bottoms of sheltered coastal reefs and lagoons in 2-30 m.

Asian Pacific: Indonesia, Philippines, Micronesia, Great Barrier Reef and Vanuatu.

BRYOZOAN GOBY *Sueviota bryozophila*

SIZE: to 2.2 cm (3/4 in.)   Gobies – Gobiidae

ID: White with reddish spots and markings on back; brownish iris, 4 extended tubular nostrils on snout. Exclusively associate with white Lacy Bryozoan, *Triphyllozoon inornatum*, to 40 m.

Localized: Currently known only from Ambon, Alor and N. Sulawesi in Indonesia, but probably more wide spread.

RED-SPOTTED PYGMYGOBY *Trimma rubromaculatum*
SIZE: to 2.2 cm (1 in.)   Gobies – Gobiidae
IID: Pinkish with red spots and blotches; dotted with several white spots and streaks, **white stripe on center of snout.** Usually form small groups, often hover a few inches off the bottom. Rest on rubble at base of steep lagoon and seaward slopes in 20-35 m.
Asian Pacific: Indonesia, Philippines, Papua New Guinea and Solomon Is.

CAESIURA PYGMYGOBY *Trimma caesiura*
SIZE: to 3.5 cm (1¹/₂ in.)   Gobies – Gobiidae
ID: Red with dark brown scale margins forming netted pattern; **pale blue reticular pattern of interconnected lines on head and upper forebody.** Sand/rubble bottoms of caves and ledges in 5-30 m.
West Pacific: West Papua in Indonesia to Micronesia, Solomon Is. – S. Japan to N. Great Barrier Reef and Fiji.

LANTANA PYGMYGOBY *Trimma lantana*
SIZE: to 3.7 cm (1¹/₂ in.)   Gobies – Gobiidae
ID: Pale gray head and orange red body; large irregular red spots on head and forebody, **5 white spots/blotches along back to upper tail base,** row of 3-4 elongate white midlateral spots. Sand and rubble bottoms of caves and ledges in 5-30 m.
Asian Pacific: West Papua in Indonesia to Papua New Guinea and Solomon Is. – S. Japan to Australia.

RUBBLE PYGMYGOBY *Trimma naudei*
SIZE: to 3.5 cm (1¹/₂ in.)   Gobies – Gobiidae
ID: Bluish gray head with red spots and blotches; **red body with dark scale outlines,** 4-5 white spots on back and tail base, 3-4 white spots on side and row of whitish blotches on lower side. Sand/rubble bottoms of caves and ledges in 5-30 m.
Indo-Asian Pacific: Madagascar to Andaman Sea, West Papua in Indonesia, Philippines and north to S. Japan.

NOMURA'S PYGMYGOBY *Trimma nomurai*
SIZE: to 2.5 cm (1 in.)   Gobies – Gobiidae
ID: White head with narrow red bar behind eye; body mainly orange with row of small white spots along back and 5-6 white bars/spots along lower half of side. Solitary or loose groups. Rubble bottoms in 20-30 m.
Asian Pacific: Indonesia to S. Japan.

CANDYCANE PYGMYGOBY *Trimma cana*
SIZE: to 3.2 cm (1¹/₄ in.)   Gobies – Gobiidae
ID: Whitish or light gray with 6-7 bright red to orange bars encircling head and body. Solitary or form small groups. Perch on surface of hard corals. Steep outer reef slopes in 12-35 m.
West Pacific: Indonesia and Philippines to Micronesia, Papua New Guinea and Fiji.

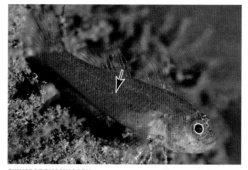

RINGEYE PYGMYGOBY *Trimma benjamini*
SIZE: to 3 cm (1¼ in.) Gobies – Gobiidae
ID: Red to orange with translucent fins; **thin white to lavender line encircling much of eye** often extends below eye across cheek. Solitary; rest on rocky bottoms. Steep seaward reef slopes in 4-35 m.
West Pacific: Indonesia, Philippines, Micronesia to Great Barrier Reef and Fiji.

SKINSPOT PYGMYGOBY *Trimma halonevum*
SIZE: to 3 cm (1¼ in.) Gobies – Gobiidae
ID: Light to dark salmon; **midlateral row of tiny dark brown spots;** a few scattered darkish spots on head and front of body. Solitary or form small groups. Sheltered coastal reefs and seaward slopes to 45 m.
East Indo-Asian Pacific: Maldives to Indonesia, Philippines, Papua New Guinea, Solomon Is. and Vanuatu.

CAVE PYGMYGOBY *Trimma taylori*
SIZE: to 2.5 cm (1 in.) Gobies – Gobiidae
ID: Reddish orange with violet highlights; **yellow bands on red iris. Male –** Elongate dorsal ray and yellow spotting on dorsal and anal fins. Form hovering aggregations in caves; mix with other *Trimma*. Along dropoffs in 15-50 m.
Indo-Pacific: Red Sea to Andaman Sea, Indonesia, Micronesia, Hawaii and French Polynesia. – S. Japan to Great Barrier Reef.

ERDMANN'S PYGMYGOBY *Trimma erdmanni*
SIZE: to 3.1 cm (1¼ in.) Gobies – Gobiidae
ID: Translucent pinkish on upper back; broad reddish orange midlateral stripe, whitish belly, **narrow blue stripe below eye.** Solitary. Rest on or hover just above bottom of outer reef dropoffs in 12-66 m.
Asia Pacific: Indonesia, Philippines, Papua New Guinea and Solomon Is.

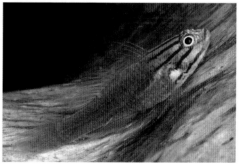

BLACKSPOT PYGMYGOBY *Trimma zurae*
SIZE: to 2 cm (¾ in.) Gobies – Gobiidae
ID: Reddish to golden to yellowish brown body; purplish head with pair of red bars below eye, **white-edged black spot on 1st dorsal fin.** Solitary. Caves and ledges of steep outer reef slopes in 10-40 m.
Localized: Philippines and N. Sulawesi in Indonesia.

STRIPEHEAD PYGMYGOBY *Trimma striatum*
SIZE: to 4 cm (1½ in.) Gobies – Gobiidae
ID: Maroon head and front of body with **red-orange stripes extending from head to anterior body;** pale yellow spotting on 2nd dorsal and tail fins. Solitary or form small groups. Rest on bottom of rocky crevices of lagoon and seaward reefs in 10-30 m.
East Indo-West Pacific: Maldives to Micronesia, Papua New Guinea and Solomon Is. – S. Japan to Great Barrier Reef and Fiji.

Gobies

OKINAWA PYGMYGOBY *Trimma okinawae*
SIZE: to 4.3 cm (1³/₄ in.) Gobies – Gobiidae

ID: Darkish undercolor with dense blotchy red to orange spotting on head and body. Form small groups. Inside caves and beneath ledges of seaward reef slopes in 8-35 m.

West Pacific: Vietnam to Indonesia, Philippines, Micronesia, Papua New Guinea and Solomon Is. – S. Japan to Australia and Fiji.

HARLOT PYGMYGOBY *Trimma fucatum*
SIZE: to 2.7 cm (1 in.) Gobies – Gobiidae

ID: Bluish gray gradating to reddish orange lower body; numerous reddish orange irregular spots and bars on head and upper body, reddish fins. Coral reef and boulders to 23 m.

Localized: E. Andaman Sea to Bali in Indonesia.

BIGEYE PYGMYGOBY *Trimma macrophthalmus*
SIZE: to 2.7 cm (1 in.) Gobies – Gobiidae

ID: Darkish undercolor with dense red to orange spotting; **pair of large red spots on pectoral fin base.** Small groups inside caves and beneath ledges. Seaward reef slopes in 5-35 m.

Indo-West Pacific: E. Africa to Indonesia, Philippines, Micronesia, Solomon Is. – S. Japan to Great Barrier Reef and Fiji.

FOUREYE PYGMYGOBY *Trimma hayashii*
SIZE: to 2.9 cm (1¹/₄ in.) Gobies – Gobiidae

ID: Purplish gray undercolor with orange spots on head and front of body becoming 3 stripes on rear body and tail base; streaks of orange on tail, **2 black spots ringed in blue on underside of head.** Crevices and caves of sheltered reefs in 2-20 m.

Asian Pacific: Indonesia, Philippines and Papua New Guinea to Micronesia, north to S. Japan.

PRINCESS PYGMYGOBY *Trimma marinae*
SIZE: to 2.7 cm (1 in.) Gobies – Gobiidae

ID: Translucent with reddish head and anterior body merging with **tapering red stripe extending to tail base;** elongate filament extends from 1st dorsal fin. Form hovering groups in caves and below ledges. Lagoon and seaward reefs in 9-26 m.

Asian Pacific: Indonesia, Philippines, Papua New Guinea and Solomon Is. – S.W. Japan to Great Barrier Reef.

NASAL PYGMYGOBY *Trimma nasa*
SIZE: to 2.8 cm (1 in.) Gobies – Gobiidae

ID: Pinkish head and translucent body; **pink midlateral stripe with thin yellow line (occasionally broken) above extending to rear body,** dark blotch at tail base. Forms schools around caves and overhangs in 5-41 m.

West Pacific: Indonesia, Philippines, Solomon Is. to Great Barrier Reef, Vanuatu, New Caledonia and Fiji.

PALE PYGMYGOBY *Trimma anaima*

SIZE: to 2 cm (³/₄ in.) Gobies – Gobiidae

ID: Translucent pinkish upper body and white lower head and belly; broad red to orange midlateral stripe from eye to tail base, light blue lines below eye and on snout and behind eye. Coral reef caves and crevices, often on dropoffs in 3-35 m.

East Indo-West Pacific: Maldives to Indonesia, Philippines, Papua New Guinea, Solomon Is., N. Great Barrier Reef and Fiji.

CHEN'S PYGMYGOBY *Trimma cheni*

SIZE: to 2.6 cm (1 in.) Gobies – Gobiidae

ID: Yellowish to pinkish head with 2 orange bars below eye and 2 yellowish bars on gill cover; bluish gray body with 5-6 poorly defined yellowish bars toward front and 3 poorly defined stripes on rear. Outer reef slopes in 10-52 m.

Asian Pacific: Indonesia, Philippines and Micronesia.

PORTHOLE PYGMYGOBY *Trimma finistrinum*

SIZE: to 3 cm (1 ¼ in.) Gobies – Gobiidae

ID: Yellow head and pinkish undercolor with **rows of bright yellow spots**; red rear tail base and white tail. Form hovering aggregations inside caves and beneath ledges of steep dropoffs of lagoons and outer slopes in 8-40 m.

Localized: Currently known only from Fiji.

BLOODSPOT PYGMYGOBY *Trimma haimassum*

SIZE: to 3.9 cm (1 ½ in.) Gobies – Gobiidae

ID: Pale yellowish to reddish brown or bluish gray; red to orange iris, **purplish stripe below eye extends onto gill cover**, dappled pattern of lavender markings on upper edge of eye between eyes and on snout. Ledges and caves on steep slopes (often orient belly to ceiling) in 15-70 m.

Asian Pacific: Indonesia and Philippines.

YELLOW & BLACK PYGMYGOBY *Trimma flavatrum*

SIZE: to 2.7 cm (1 in.) Gobies – Gobiidae

ID: Dusky yellow head and body with **yellow belly shading to blackish tail base**; transparent to whitish tail, yellow iris. Form schools in caves and beneath ledges (often orient belly to ceiling) in 10-30 m.

West Pacific: Indonesia, Philippines, Micronesia and Solomon Is. – S.W. Japan to Great Barrier Reef and Fiji.

FORKTAIL PYGMYGOBY *Trimma hoesei*

SIZE: to 2.5 cm (1 in.) Gobies – Gobiidae

ID: Pinkish orange with lavender or magenta markings on back and fins; **forked tail.** In pairs or form small groups. Swim upside-down beneath ledges and inside caves of steep outer reef slopes in 15-50 m.

East Indo-West Pacific: Chagos to Indonesia, Philippines and Solomon Is. – Great Barrier Reef to Fiji.

Gobies

YELLOWSKIN PYGMYGOBY *Trimma xanthochrum*
SIZE: to 3 cm (1¼ in.) Gobies – Gobiidae
ID: Pinkish orange with yellowish tints; white lower head extends to abdomen, brownish blotch on tail base, **broad yellow band on middle of dorsal fins,** yellow tail with whitish central stripe. Current-prone reefs in 12-70 m.
Localized: Raja Ampat, West Papua and Triton Bay in E. Indonesia.

MESHED PYGMYGOBY *Trimma agrena*
SIZE: to 3.8 cm (1½ in.) Gobies – Gobiidae
ID: Pale gray to mauve with orange scale margins; orange spots on head and **large dark spot on rear gill cover behind eye.** Reef crevices and caves in 2-20 m.
Asian Pacific: Sabah in Malaysia, Indonesia and Philippines.

YELLOWHEAD PYGMYGOBY *Trimma stobbsi*
SIZE: to 2.5 cm (1 in.) Gobies – Gobiidae
ID: **Yellow head,** mauve to grayish body; **small dark "ear" spot on upper rear corner of gill cover.** Solitary or form small groups. Inside caves and ledges of steep seaward slopes in 10-40 m.
Asian Pacific: Vietnam, Indonesia, Philippines, Papua New Guinea, Solomon Is., New Caledonia and Vanuatu.

MEANDER PYGMYGOBY *Trimma maiandros*
SIZE: to 3.1 cm (1¼ in.) Gobies – Gobiidae
ID: Bluish gray, violet, or pinkish with **orange U-shaped marking below eye;** short orange band behind eye, 2 rows of large orange blotches on side of body, lower row vertically elongate. Reef caves and crevices to 55 m.
East Indo-West Pacific: Cocos Keeling Is. to Indonesia, Micronesia and Solomon Is. – S. Japan to Great Barrier Reef and Fiji.

HONEYBEE PYGMYGOBY *Trimma anthrenum*
SIZE: to 2.6 cm (1 in.) Gobies – Gobiidae
ID: Entirely bright yellow to brownish yellow; **a blue and black triangular marking on the front of the iris and another on the rear;** darkish stripe on base of dorsal fins. Caves and recesses in reefs in 5-76 m.
West Pacific: Micronesia and Fiji.

REDEARTH PYGMYGOBY *Trimma milta*
SIZE: to 3 cm (1¼ in.) Gobies – Gobiidae
ID: Yellow to orange to reddish orange; occasionally front of head orange. Coral or rock walls or reef areas with mixed sand and rubble in 9-26 m.
Pacific: Indonesia, Philippines, Micronesia, Papua New Guinea, Solomon Is. to Hawaii and French Polynesia. – Taiwan to N. Great Barrier Reef.

TEVEGAE PYGMYGOBY *Trimma tevegae*
SIZE: to 3.5 cm (1¹/₂ in.) Gobies – Gobiidae
ID: Golden to reddish brown, dark tail base often with a white frontal bar, pale lower head and belly; white line through upper eye often with indistinct stripe behind. Form hovering aggregations in shaded areas of reefs in 8-40 m.
Asian Pacific: Indonesia and Philippines to Papua New Guinea and Solomon Is.

BLOTCH-TAILED PYGMYGOBY *Trimma caudomaculatum*
SIZE: to 3.5 cm (1¹/₂ in.) Gobies – Gobiidae
ID: Golden to reddish brown, dark tail base; **blue stripe from snout to dorsal fin** and another from eye to tail base; 3 small blue spots below eye and on gill cover; long 2nd dorsal spine. Form hovering aggregations in shaded areas of reefs in 8-40 m.
Asian Pacific: Maldives to Indonesia, Philippines and Solomon Is. – S.W. Japan to N.W. Australia.

YANO'S PYGMYGOBY *Trimma yanoi*
SIZE: to 2.6 cm (1 in.) Gobies – Gobiidae
ID: Translucent reddish to pink with 3 longitudinal rows of large pale rectangular orange blotches on side of body; red iris. Coral caves and crevices in 8-70 m.
Localized: Currently known only from West Papua in Indonesia, southern Luzon in Philippines and S.W. Japan.

FANG'S PYGMYGOBY *Trimma fangi*
SIZE: to 2.6 cm (1 in.) Gobies – Gobiidae
ID: Purplish gray with longitudinal rows of orange spots on side of body sometimes merging to form stripes toward rear; **dark spot behind upper rear gill cover and another on pectoral fin base,** orange spots or bands on head. Outer reef crevices and caves in 5-20 m.
Asian Pacific: Brunei, Indonesia and N. Palawan, Philippines.

BLACKNOSE DWARFGOBY *Sueviota atrinasa*
SIZE: to 2.5 cm (1 in.) Gobies – Gobiidae
ID: Translucent pale yellowish gray with about 7 internal brown body bars; dusky orange spots below eye and on gill cover, 3 brown to reddish bars on nape, **black nostrils.** Reef crevices in 10-30 m.
West Pacific: E. Indonesia and Philippines to W. Australia and Fiji.

LACHNER'S DWARFGOBY *Sueviota lachneri*
SIZE: to 2.4 cm (1 in.) Gobies – Gobiidae
ID: Translucent with 3 large orange to brown blotches behind pectoral fin; large dusky spots and bands on head, 2 large brown to reddish spots on pectoral fin base, small dark spots on dorsal and tail fin rays. Reef crevices in 3-50 m.
East Indo-West Pacific: Chagos to Indonesia, Philippines and Papua New Guinea. – S. Japan to Great Barrier Reef and Fiji.

345

BLUESTREAK GOBY *Valenciennea strigata*
SIZE: to 18 cm (7 in.) Gobies – Gobiidae

ID: White to pale gray with yellow snout and cheeks; bright blue stripe below eye and blue spots and/or bands below. Usually in pairs that share burrows. Sand and rubble areas on reef tops or sloping bottoms to 20 m.

Indo-Pacific: E. Africa to Andaman Sea, Indonesia, Philippines, Micronesia and French Polynesia. – S.W. Japan to Great Barrier Reef.

ORANGE-BARRED GOBY *Valenciennea decora*
SIZE: to 14 cm (5 1/2 in.) Gobies – Gobiidae

ID: Grayish white; **orange bar edged in blue on side behind pectoral fin,** often several additional bars on body and tail base, pearly spots and lines on head and orange margin on first dorsal fin, black inside mouth. Solitary or in pairs. Clean sandy bottoms of offshore reefs in 3-35 m.

West Pacific: Great Barrier Reef and Coral Sea to Fiji.

LONG-FINNED GOBY *Valenciennea longipinnis*
SIZE: to 15 cm (6 in.) Gobies – Gobiidae

ID: White to pale gray; **row of about 5 widely space midlateral spots,** pair of stripes on cheek and gill cover, several faint darkish stripes extend down back. In pairs. Fine sand of protected lagoons and coastal reefs to 4 m.

West Pacific: W. Indonesia to Micronesia. – S.W. Japan to Great Barrier Reef, New Caledonia and Fiji.

YELLOW FILAMENT GOBY *Valenciennea bella*
SIZE: to 9 cm (3 1/2 in.) Gobies – Gobiidae

ID: Pale gray with yellowish snout; 2-3 blue stripes on cheek, **long yellow filaments rays extend from foredorsal fin. Male –** Black throat. Solitary or in pairs. Mixed rubble and sand bottoms of coastal reefs in 12-35 m.

Asian Pacific: Indonesia and Philippines, north to S.W. Japan.

TWOSTRIPE GOBY *Valenciennea helsdingenii*
SIZE: to 9 cm (3 1/2 in.) Gobies – Gobiidae

ID: Pale gray with broad white midlateral stripe bordered with dark stripes; large black spot ringed with white on first dorsal fin. Usually in pairs. Mixed sand and rubble bottoms, at base of outer reef dropoffs in 5-40 m.

Indo-Pacific: Red Sea and E. Africa to Andaman Sea, Indonesia, Solomon Is. to French Polynesia. – S. Japan to Australia.

GREENBAND GOBY *Valenciennea randalli*
SIZE: to 10.8 cm (4 1/4 in.) Gobies – Gobiidae

ID: Pale gray to mauve; **white to blue-green stripe with orange edge below eye,** faint orange stripe along lower side to tail. Usually solitary. Silty bottoms of lagoons and sheltered seaward reefs in 8-35 m.

East Indo-Asian Pacific: Andaman Sea to Solomon Is. – S.W. Japan to Great Barrier Reef, New Caledonia and Fiji.

MURAL GOBY *Valenciennea muralis*
SIZE: to 11.5 cm (4 1/2 in.) Gobies – Gobiidae
ID: Pale yellowish gradating to blotchy gray on back and top of head; pair of pinkish red stripes from head to tail, black spot on rear 1st dorsal fin. Usually in pairs. Fine sand or silt bottoms of sheltered bays and lagoons to 15 m.
Asian Pacific: Andaman Sea to Indonesia, Micronesia and Solomon Is. – Philippines to Great Barrier Reef.

IMMACULATE GOBY *Valenciennea immaculata*
SIZE: to 12 cm (4 3/4 in.) Gobies – Gobiidae
ID: Yellowish white to pale mauve; pair of blue-edged orange stripes from snout and upper lip to tail base, and a 3rd stripe on back, **pointed lower lobe of tail.** Usually in pairs. Fine sand or mud bottoms of coastal reefs to 30 m.
Asian Pacific: Indonesia and Philippines. – Taiwan to Great Barrier Reef.

MUD GOBY *Valenciennea limicola*
SIZE: to 8.2 cm (3 1/4 in.) Gobies – Gobiidae
ID: Mauve to gray with pair of bright orange stripes from eye and mouth to tail; white streak below eye, orange streaks on **rounded tail.** Solitary or in pairs. Fine sand or mud bottoms of coastal reefs in 5-30 m.
West Pacific: Gulf of Thailand to Indonesia, Papua New Guinea and Fiji.

PARVA GOBY *Valenciennea parva*
SIZE: to 6.8 cm (2 1/2 in.) Gobies – Gobiidae
ID: White to pale gray; pair of **narrow orange stripes** from head to tail, blue/white stripe below eye, may display faint diffuse bars. In pairs or small groups. Clean sand around coral heads on coastal reefs to 20 m.
Indo-West Pacific: Seychelles to Indonesia, Philippines, Micronesia, Solomon Is. – S.W. Japan to Australia and Fiji.

ORANGE-DASHED GOBY *Valenciennea puellaris*
SIZE: to 14 cm (5 1/2 in.) Gobies – Gobiidae
ID: Pale gray with row of large orange spots on side; orange stripe or series of dashes below, smaller orange spots on back, blue streaks or spots on head. Usually in pairs. Sand bottoms in 8-25 m.
Indo-West Pacific: Red Sea and E. Africa to Indonesia, Philippines and Micronesia. – S. Japan to Australia, Vanuatu and Fiji.

Orange-dashed Goby – Variation
ID: Pale gray with scattered blue spots and dashes, especially on lower head; scattered orange spots on back, orange stripe from corner of mouth to tail, 7 orange bars across back extending to orange stripe.
Indian Ocean: Red Sea and Madagascar to Andaman Sea.

Gobies

SIXSPOT GOBY *Valenciennea sexguttata*
SIZE: to 14 cm (5 1/2 in.) Gobies – Gobiidae
ID: White to gray; 6 or more blue to bluish spots on cheek, **black tip on 1st dorsal fin.** Often in pairs. Share burrows under rocks of fine sand bottoms of shoreline reefs to 10 m.
Indo-Pacific: Red Sea and E. Africa to Indonesia, Philippines, Micronesia, Papua New Guinea, Solomon Is. and Line Is. – S.W. Japan to Great Barrier Reef and Fiji.

ALLEN'S GOBY *Valenciennea alleni*
SIZE: to 8.5 cm (3 1/4 in.) Gobies – Gobiidae
ID: Pale gray; **pair of blue-edged orange stripes on head,** the lower stripe extending to tail base, black tip on first dorsal fin. Solitary or in pairs. Shelter in sand burrows under rocks of silty coastal reefs in 2-10 m.
Localized: N. Australia from Shark Bay in the west to Great Barrier Reef.

WIDEBARRED GOBY *Valenciennea wardii*
SIZE: to 12.3 cm (5 in.) Gobies – Gobiidae
ID: Pale gray to whitish; 3 wide brown bars on side, blue band below eye, black pale-edged spot on 1st dorsal fin, narrow brown bar on tail base, brown outer half of tail. In pairs. Silty bottoms in 12-35 m.
Indo-Asian Pacific: Red Sea and E. Africa to Andaman Sea, Indonesia, Philippines. – S.W. Japan to Australia, New Caledonia.

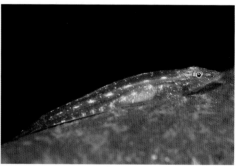

FLATHEAD GOBY *Phyllogobius platycephalops*
SIZE: to 3.5 cm (1 1/2 in.) Gobies – Gobiidae
ID: Translucent with scattered white flecks; wide flattened body. Usually in small groups, on and beneath flat leaf-like sponges, including *Phyllospongia*, in 2-18 m.
Indo-Asian Pacific: E. Africa to Sabah in Malaysia, Indonesia, Philippines, Papua New Guinea, Solomon Is. to Great Barrier Reef.

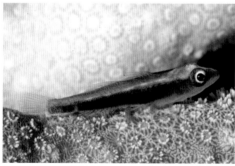

TRANSLUCENT CORAL GOBY *Bryaninops erythrops*
SIZE: to 2 cm (3/4 in.) Gobies – Gobiidae
ID: Translucent grayish with dark speckles red and white iris. On branching fire corals and *Porites* corals. Shallow lagoons and protected reefs to 10 m.
Indo-West Pacific: Red Sea to Philippines, Micronesia, Papua New Guinea and Solomon Is. – S.W. Japan to Great Barrier Reef and Samoa.

PINKEYE GOBY *Bryaninops natans*
SIZE: to 2.4 cm (1 in.) Gobies – Gobiidae
ID: Translucent body with bluish head and yellowish belly to above anal fin; **brilliant pink eyes.** Form groups that hover a short distance above *Acropora* corals in 12-25 m.
Indo-West Pacific: Red Sea to Philippines, Micronesia, Papua New Guinea, Solomon Is. to Cook Is. – S.W. Japan to N. Great Barrier Reef.

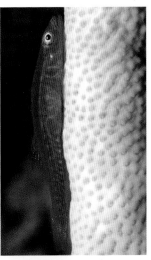

LARGE WHIP GOBY
Bryaninops amplus
SIZE: to 3.7 cm (1½ in.)
Gobies – Gobiidae
ID: Translucent reddish; blend in well with host gorgonian when viewed from above, iris has red outer and gold to white inner ring. On gorgonian sea whips and fans in areas of strong current in 5-30 m.
Indo-Pacific: Madagascar to Indonesia, Philippines, Micronesia, Papua New Guinea, Hawaii. – S. Japan to Great Barrier Reef and Vanuatu.

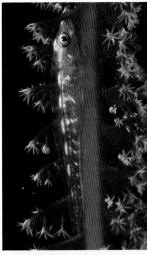

Large Whip Goby – Variation
ID: Occasionally with several irregular pale saddle markings on head and side.

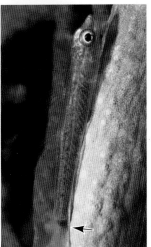

LOKI WHIP GOBY
Bryaninops loki
SIZE: to 3.7 cm (1¼ in.)
Gobies – Gobiidae
ID: Color variable according to host gorgonian, but usually with **darkish spot on lower tail base.** Found on gorgonian fans and whips including *Junceella, Ellisella* and *Subergorgia* in 6-45 m.
Indo-West Pacific: Chagos to Andaman, Indonesia, Micronesia, Solomon Is. – S. Japan to Great Barrier Reef, New Caledonia and Fiji.

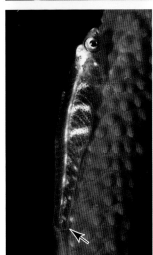

Loki Whip Goby – Variation
ID: Brown lower body and white bars, possibly associated with egg-guarding or other nuptial activity. **Note identifying spot on lower tail base.**

BLACK CORAL GOBY   *Bryaninops tigris*
SIZE: to 3 cm (1¼ in.)   Gobies – Gobiidae
ID: Elongate; translucent body with red to dark brown stripe (may be interrupted by white) from snout to lower tail base. Solitary. **Only on branches of black coral *Antipathes* in 15-53 m.**
East Indo-Pacific: Oman and Chagos to Indonesia, Philippines, Micronesia, Solomon Is., Hawaii and French Polynesia. – S.W. Japan to Great Barrier Reef and Fiji.

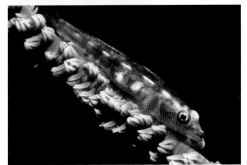

WIRE CORAL GOBY   *Bryaninops yongei*
SIZE: to 3.7 cm (1½ in.)   Gobies – Gobiidae
ID: Translucent upper body and brownish lower body; usually with a few diffuse bars. Solitary, **only on the wire coral, *Cirrhipathes anguina.*** Seaward and lagoon reefs in 3-45 m.
Indo-Pacific: Red Sea to Andaman Sea, Indonesia, Philippines, Micronesia, Papua New Guinea, Solomon Is., Hawaii and French Polynesia. – S. Japan to Great Barrier Reef.

349

Gobies

BLUECORAL GHOSTGOBY *Pleurosicya coerulea*
SIZE: to 2.2 cm (³/₄ in.) Gobies – Gobiidae
ID: Translucent bluish brown with narrow blue stripe from eye to tail base and another centered on back. Exclusively associate with Blue Coral, *Heliopora coerulea*, to 10 m.

Indo-West Pacific: Seychelles to Indonesia, Philippines, Micronesia, Papua New Guinea and Solomon Is. – S. Japan to Great Barrier Reef.

FOLDED GHOSTGOBY *Pleurosicya plicata*
SIZE: to 3 cm (1¹/₄ in.) Gobies – Gobiidae
ID: Translucent pale gray; sometimes with pale pink or reddish orange hue, often covering of pepper-like spots, dark, often reddish, stripe on side of snout, occasionally covers entire snout. Exclusively on sponges in 10-30 m.

East Indo-Asian Pacific: Oman to Indonesia, Philippines and Micronesia.

SOFT CORAL GHOSTGOBY *Pleurosicya boldinghi*
SIZE: 3.5 cm (1¹/₂ in.) Gobies – Gobiidae
ID: High body and pig-like snout profile; translucent tinted with color of host soft coral. Solitary or occasional in pairs. On *Dendronephthya* soft corals in 5-82 m.

Indo-Asian Pacific: E. Africa to Indonesia, Japan and Papua New Guinea.

LEATHER CORAL GHOSTGOBY *Pleurosicya muscarum*
SIZE: to 2.6 cm (1 in.) Gobies – Gobiidae
ID: Translucent reddish to grayish or bluish; red iris. Commensal with a variety of soft corals, usually prefer leather corals *Sinularia* and *Lobophytum* in 2-28 m.

Indo-West Pacific: E. Africa to Bali in Indonesia, Philippines and Micronesia. – S. Japan to Australia and Fiji.

BARREL SPONGE GHOSTGOBY *Pleurosicya labiata*
SIZE: to 3.5 cm (1¹/₂ in.) Gobies – Gobiidae
ID: Elongate flattened body; dusky with peppering of small brown spots, thin brown stripe on side of snout. Usually in groups. Rest in channels on outer surface of large barrel sponge, *Xestospongia testudinaria*, in 5-35 m.

East Indo-Asian Pacific: Sri Lanka to Brunei, Indonesia, Philippines, Papua New Guinea and N. Great Barrier Reef.

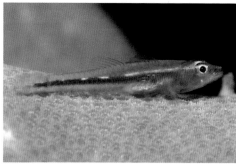

STONYCORAL GHOSTGOBY *Pleurosicya micheli*
SIZE: to 2.5 cm (1 in.) Gobies – Gobiidae
ID: Translucent with reddish brown midlateral stripe; white markings along spinal column. Solitary. On wide variety of hard corals of coral-rich reefs in 10-50 m.

Indo-Pacific: Seychelles to Indonesia, Philippines, Papua New Guinea, Solomon Is. to Hawaii and French Polynesia. – S.W. Japan to New Caledonia and Fiji.

COMMON GHOSTGOBY   *Pleurosicya mossambica*

SIZE: to 2.5 cm (1 in.)   Gobies – Gobiidae

ID: Highly variable markings and color from brown to red, greenish and translucent; consistently display dark speckling, often white dash markings along spinal column. Solitary. Seagrass beds to coral reefs in 2-30 m.

Indo-West Pacific: Red Sea and E. Africa to Andaman Sea, Indonesia, Micronesia, French Polynesia. – S. Japan to Australia, Fiji.

Common Ghostgoby – Variation

ID: Brown with dark grayish saddles and diffuse white blotches and markings. Note consistent dark speckling and white dash markings along spinal column.

Common Ghostgoby – Variation

ID: Not all variations display white dash markings along spinal column.

Common Ghostgoby – Variation

ID: All variations may perch on algae, seagrass, tunicates, sea pens, soft and hard corals and giant clams.

ELONGATE GHOSTGOBY   *Pleurosicya elongata*

SIZE: to 3.5 cm (1 1/2 in.)

ID: Elongate; protruding snout (similar to Wolfsnout Goby [next]); translucent body with tints similar to host sponge, internal brown dashed along vertebral column. On fan-shaped sponge *Ianthella basta.* Reefs in 10-40 m.

Asian Pacific: Brunei, E. Indonesia, Papua New Guinea, Solomon Is. and Great Barrier Reef.

WOLFSNOUT GOBY   *Luposicya lupus*

SIZE: to 3.5 cm (1 1/2 in.)   Gobies – Gobiidae

ID: Elongate; translucent body with dense covering of pepper-like spots, **prominent snout overhangs lower jaw.** Solitary. On sponges, especially *Phyllospongia foliascens,* to 10 m.

Indo-West Pacific: Red Sea and E. Africa to Indonesia and Papua New Guinea. – Great Barrier Reef to Fiji.

Small, Elongate Bottom Dwellers – Blennies

This ID Group consists of small, elongate, typically bottom-dwelling species in family Blenniidae.

FAMILY: Blennies – Blenniidae
Triplefins – Tripterygiidae
21 Genera – 107 Species Included

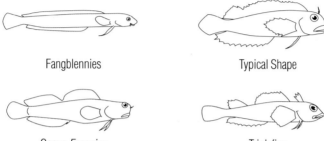

Fangblennies   Typical Shape

Genus *Ecsenius*   Triplefins

Because of similar elongate bodies and small size, species in the blenny and goby families, Gobiidae (previous ID Group), are often confused with one another. Blennies can be easily distinguished from gobies by noting their single, long continuous dorsal fins (except for a small group known as triplefins), ventral fins situated clearly in front of their pectoral fins, and their habit of resting on the bottom with curved bodies. Gobies have distinct, two-part dorsal fins and ordinarily perch with straight bodies.

Typical bottom-dwelling blennies also have blunt heads, long anal fins and fleshy, often branching appendages on their head, known as cirri. Most of these blennies have numerous, tiny comb-like teeth used for feeding on algae. Blennies classified in genera *Alticus*, *Entomacrodus* and *Istiblennius*, commonly known as rockskippers, inhabit tidal zones where they have the ability to jump between tide pools. Species in the rather large genus *Ecsenius*, generally inhabitants of coral-rich areas, are atypical for their members' limited distribution ranges. Possibly the most curious member of the blenny family is the Snake Blenny, *Xiphasia setifer*, whose length can approach 55 cm (21 in.), resulting in an appearance more characteristic of eels rather than blennies.

The Indo-Pacific is home to a large contingent of blennies, known as fangblennies, that spend much of their day swimming above the sea floor. This group, also commonly called sabretooth blennies, have two large, curved canine teeth in the front of their lower jaws which are used for defense. Members of genus *Plagiotremus* utilize their imposing canines to sever flesh from the fins of fishes for food. The Mimic Blenny, *Aspidontus taeniatus*, skillfully impersonates the Bluestreak Cleaner Wrasse, *Labroides dimidiatus*, allowing it to closely approach unsuspecting fishes before darting in to take a nip. The violent reactions of victims, who often vigorously chase the offending fangblennies, attest to the painful nature of these encounters. Members of genus *Meiacanthus* have venom glands associated with their canines. This defensive adaptation allows virtually unmolested access to open water where they feed on planktonic worms and crustaceans. When not feeding in the water column or on benthic crustaceans, fangblennies rest in wormholes with only their heads exposed.

Triplefins are in family Tripterygiidae. Because of the close relationship between the two families triplefins are occasionally called triplefin blennies. The triplefins' three dorsal fins instead of the blennies' single dorsal fin is the primary difference between the two families.

Small, Elongate Bottom Dwellers – Blennies

SLENDER SABRETOOTH BLENNY *Aspidontus dussumieri*

SIZE: to 12 cm (4 ³/₄ in.) Blennies – Blenniidae

ID: White with broad black stripe (sometimes segmented) on upper body; yellow dorsal and anal fins. Solitary and shy, retreat to safety of abandoned worm tubes when threatened. Lagoon and outer reefs to 20 m.

Indo-Pacific: Red Sea and E. Africa to Indonesia, Philippines, Micronesia and French Polynesia. – S. Japan to S.E. Australia.

FALSE CLEANERFISH *Aspidontus taeniatus*

SIZE: to 10.5 cm (4 ¹/₄ in.) Blennies – Blenniidae

ID: Light blue to white, may display yellowish upper body; black stripe enlarges from snout to tail. Solitary; mimic Bluestreak Cleaner Wrasse, can best be differentiated from the wrasse by the underslung mouth and occasional presence in hiding holes with heads exposed. Coastal, lagoon and outer reefs to 25 m.

Indo-Pacific: Red Sea and E. Africa to French Polynesia.

YELLOWTAIL FANGBLENNY *Meiacanthus atrodorsalis*

SIZE: to 11 cm (4 ¹/₄ in.) Blennies – Blenniidae

ID: Blue-gray head and forebody shading to pale yellow behind, often dark stripe on dorsal fin base; **blue-edged black band from eye to front of dorsal fin.** Solitary or in pairs. Coastal, lagoon and seaward reefs to 30 m.

West Pacific: Indonesia, Philippines, Micronesia, Papua New Guinea and Solomon Is. – S.W. Japan to Australia, New Caledonia and Fiji.

SMITH'S FANGBLENNY *Meiacanthus smithi*

SIZE: to 8 cm (3 ¹/₄ in.) Blennies – Blenniidae

ID: Pale gray; **black line from eye to dorsal fin** where it joins broad black stripe, black stripes and margins on tail. Solitary or in pairs. Coastal, lagoon and outer reefs to 10 m.

East Indian Ocean: Maldives and India to Sumatra and N.W. Bali in Indonesia.

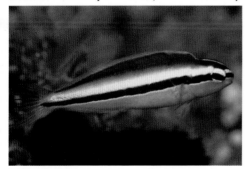

TWIN FANGBLENNY *Meiacanthus geminatus*

SIZE: to 6.5 cm (2 ¹/₂ in.) Blennies – Blenniidae

ID: Gray back and dorsal fin and yellow belly; wide white stripe above black midlateral stripe. Mimic the Yellowbelly Cardinalfish. Solitary. Sheltered coastal reefs to 15 m.

Localized: Sabah in Malaysia to Sulu Archipelago and Palawan in Philippines.

BUNDOON FANGBLENNY *Meiacanthus bundoon*

SIZE: to 8 cm (3 ¹/₄ in.) Blennies – Blenniidae

ID: Brown to gray to black; tapering yellowish green to yellow or tan stripe from head to tail base, black borders on clear to white tail. Solitary or in pairs. Coral reefs in 5-20 m.

Localized: Fiji and Tonga.

Blennies

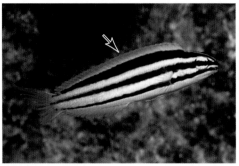

BLACKSTRIPE FANGBLENNY   *Meiacanthus vittatus*
SIZE: to 6.5 cm (2 1/2 in.)   Blennies – Blenniidae
ID: Pale gray upper body and whitish below; **black midlateral stripe with white border above.** Mainly lagoon reefs to 20 m.
Localized: Papua New Guinea including the island of New Britain.

HAIRYTAIL FANGBLENNY   *Meiacanthus crinitus*
SIZE: to 6.5 cm (2 1/2 in.)   Blennies – Blenniidae
ID: White with 3 black stripes; **narrow black edge on dorsal fin.** Solitary or in pairs. Sheltered coastal reefs to 20 m.
Asian Pacific: N. Sulawesi to West Papua in Indonesia, New Guinea and Solomon Is.

SULU FANGBLENNY   *Meiacanthus abditus*
SIZE: to 11 cm (4 1/4 in.)   Blennies – Blenniidae
ID: White with 3 black stripes; **mainly blackish dorsal fin with thin white outer edge and thin white stripe on base.** Form aggregations among gorgonians and black coral. Coastal reefs and steep slopes in 5 - 40 m.
Asian Pacific: Sabah in Malaysia to Philippines and Indonesia.

DOUBLEPORE FANGBLENNY   *Meiacanthus ditrema*
SIZE: to 6.5 cm (2 1/2 in.)   Blennies – Blenniidae
ID: White with silver reflection; 2 narrow black stripes, the **lower stripe splits before crossing head.** Form aggregations among branches of gorgonians or black coral. Sheltered coastal reefs in 3-20 m.
West Pacific: Indonesia, Philippines, Papua New Guinea and Solomon Is. – S.W. Japan to Australia and Fiji.

BLUEFIN FANGBLENNY   *Meiacanthus cyanopterus*
SIZE: to 6.4 cm (2 1/4 in.)   Blennies – Blenniidae
ID: Three black stripes from head to tail base with narrower white stripes between and white underside; **black stripe on upper dorsal fin with bluish stripe below.** Outer reef slopes in 40-65 m.
Localized: Bali to Alor in Indonesia.

KOMODO FANGBLENNY   *Meiacanthus abruptus*
SIZE: to 4.5 cm (1 3/4 in.)   Blennies – Blenniidae
ID: Pale yellowish to bright yellow with black stripe from snout to tail base and another from nape to upper margin of tail base. Solitary or small groups. Sheltered bays, rocky and mangrove shores to 5 m.
Localized: Bali to Komodo in Indonesia.

STRIPED FANGBLENNY   *Meiacanthus grammistes*
SIZE: to 10 cm (4 in.)   Blennies – Blenniidae
ID: White with yellowish head and upper body with 3 black stripes; **black spots on tail and tail base.** Act as a model for mimic Shorthead Fangblenny [below] and juvenile Bridled Monocle Bream. Solitary or in pairs. Lagoons and outer slopes to 20 m.
Asian Pacific: Indonesia, Micronesia, Papua New Guinea and Solomon Is. – S.W. Japan to Great Barrier Reef.

LINED FANGBLENNY   *Meiacanthus lineatus*
SIZE: to 9.5 cm (3 3/4 in.)   Blennies – Blenniidae
ID: Bright yellow with white belly; 3 black stripes on head and body, **black submarginal stripe on yellow dorsal fin.** Solitary or in pairs. Lagoons and seaward reefs in 3-18 m.
Localized: Great Barrier Reef.

CANARY FANGBLENNY   *Meiacanthus oualanensis*
SIZE: to 10 cm (4 in.)   Blennies – Blenniidae
ID: Yellow with greenish tints on dorsal, anal and tail fins. Very similar to Fiji Fangblenny, but has broader body and holds dorsal fin erect less. Mimicked by the Fiji Fangblenny, and the yellow Fijian variation of the juvenile Bridled Monocle Bream. Solitary or form small groups. Coral reefs to about 20 m.
Localized: Known only from Fiji.

SPOTFIN FANGBLENNY   *Adelotremus deloachi*
SIZE: to 7.5 cm (3 in.)   Blennies – Blenniidae
ID: White with dark stripe running from snout down back of body. **Male –** [pictured] Large rounded dorsal with black ocellated spot rimmed in blue toward front. Solitary. Photographed specimen living in sand burrow at 17 m.
Localized: Known only from N. Sulawesi, Bali and Komodo in Indonesia.

SHORTHEAD FANGBLENNY   *Petroscirtes breviceps*
SIZE: to 13 cm (5 in.)   Blennies – Blenniidae
ID: Variable, commonly white to yellow with 3 blackish stripes. Solitary or pairs; mimic Striped Fangblenny [above]. Shelter and nest in abandoned worm tubes, shells or small-necked bottles. Coastal and lagoon reefs, weed and sand areas to 15 m.
Indo-Asian Pacific: E. Africa to Indonesia, Micronesia and Solomon Is. – S.W. Japan to N.W. Australia and New Caledonia.

BULBSNOUT FANGBLENNY   *Petroscirtes thepassii*
SIZE: to 7.1 cm (2 3/4 in.)   Blennies – Blenniidae
ID: Mottled greenish brown, usually darker back; **elongate bulbous snout.** Solitary or small groups. Weedy habitats with seagrass, algae and *Sargassum* around coastal and lagoon reefs to 5 m.
Asian Pacific: Ternate and Halmahera in Indonesia, Micronesia, N. Papua New Guinea and Solomon Is.

Blennies

HIGHFIN FANGBLENNY *Petroscirtes mitratus*
SIZE: to 7.7 cm (3 in.) Blennies – Blenniidae
ID: Mottled shades of greenish brown; **tall front dorsal fin.** Solitary or small groups. Weedy habitats including beds of *Sargassum*, often cluster around encrusted mooring lines in coastal and lagoon reefs to 5 m.
Indo-West Pacific: Red Sea and E. Africa to Indonesia, Philippines and Micronesia. – S.W. Japan to Great Barrier Reef and Fiji.

SMOOTH FANGBLENNY *Petroscirtes xestus*
SIZE: to 7.5 cm (3 in.) Blennies – Blenniidae
ID: White; densely mottled and spotted, **thin dark brown stripe extends from eye to tail base.** Solitary. Sand, coral rubble and weed flats to 5 m.
Indo-Pacific: E. Africa to Andaman Sea, Indonesia, Philippines, Micronesia, Papua New Guinea, Solomon Is. and French Polynesia. – Great Barrier Reef to Fiji.

VARIABLE FANGBLENNY *Petroscirtes variabilis*
SIZE: to 7.5 cm (3 in.) Blennies – Blenniidae
ID: Variable olive to greenish upper body and whitish to pale greenish yellow below to entirely yellow; densely mottled and spotted, often large dark blotches on back. Solitary. Weedy areas of coastal and lagoon reefs to 5 m.
East Indo-West Pacific: Sri Lanka to Indonesia, Philippines, Papua New Guinea. – Taiwan to Great Barrier Reef and Fiji.

BLUESTRIPED FANGBLENNY *Plagiotremus rhinorhynchos*
SIZE: to 12 cm (4 3/4 in.) Blennies – Blenniidae
ID: Adult – Yellowish orange; **pair of neon-blue stripes from snout to tail.** Solitary. Feed on scales of other fishes, which they aggressively attack. Coastal, lagoon and outer reefs to 40 m.
Indo-Pacific: E. Africa to Andaman Sea, Indonesia, Philippines, Micronesia to French Polynesia. – S. Japan to Australia.

Bluestriped Fangblenny – Variation
ID: Dark blue to black with pair of neon-blue stripes from snout to tail, translucent fines. Mimic juvenile Bluestreak Cleaner Wrasse. Feed on scales of other fishes, which they aggressively attack. Occasionally bite divers. Often reside in small holes with their heads and forebodies extended.

Bluestriped Fangblenny – Variation
ID: Dark blue to black with pair of neon-blue stripes from snout to tail. Brownish body and head, reddish orange dorsal and anal fins; yellow tail.

Small, Elongate Bottom Dwellers – Blennies

BICOLOR FANGBLENNY *Plagiotremus laudandus*
SIZE: to 7.5 cm (3 in.) Blennies – Blenniidae
ID: Blue-gray head and forebody shading to yellow rear and tail; dark stripe on dorsal fin. Solitary; similar to Yellowtail Fangblenny, but **lack bar from eye to dorsal fin.** Feed on scales of other fishes, which they aggressively attack. Coastal and seaward reefs to 30 m.
West Pacific: Indonesia, Philippines and Micronesia. – S. Japan to Australia and Fiji.

FIJI FANGBLENNY *Plagiotremus flavus*
SIZE: to 7 cm (2 3/4 in.) Blennies – Blenniidae
ID: Yellow with greenish tints on dorsal and anal fins. Similar to and believed to mimic Canary Fangblenny, but has more slender body and usually hold dorsal fin in erect position. Feed on scales of other fishes, which they aggressively attack. Solitary or form small groups. Coral reefs to about 20 m.
Localized: Fiji and Tonga.

IMPOSTER FANGBLENNY *Plagiotremus phenax*
SIZE: to 5 cm (2 in.) Blennies – Blenniidae
ID: Pale gray with black submarginal stripe on dorsal fin. Solitary. Feed on scales of other fishes, which they aggressively attack. Coastal, lagoon and outer reefs to 10 m.
East Indian Ocean: Maldives, Sri Lanka to Andaman Sea and S. Sumatra in Indonesia.

PIANO FANGBLENNY *Plagiotremus tapeinosoma*
SIZE: to 12 cm (4 3/4 in.) Blennies – Blenniidae
ID: Yellow back and tail; wide **black stripe on upper side from snout to tail base formed by numerous short bars.** Solitary. Feed on scales of other fishes, which they aggressively attack. Lagoon and seaward reefs to 20 m.
Indo-Pacific: Red Sea and E. Africa to Indonesia, Micronesia and French Polynesia. – S. Japan to Great Barrier Reef.

Snake Blenny
ID: Often seen with heads and forebodies extending from sand burrows. When the fordorsal fin is extended, it is brownish with 2-3 blue stripes and a black spot with blue edging.

SNAKE BLENNY *Xiphasia setifer*
SIZE: to 55 cm (22 in.) Blennies – Blenniidae
ID: Extremely elongate resembling an eel; prominent dorsal fin black with blue margin, alternating dark and light brown banded body. Solitary. Sand or mud bottoms near reefs in 2-20 m.
Indo-Pacific: Red Sea and E. Africa to Indonesia, Micronesia and French Polynesia. – S. Japan to Australia and Vanuatu.

Blennies

REDSPOTTED BLENNY *Blenniella chrysospilos*
SIZE: to 14 cm (5 1/2 in.)   Blennies – Blenniidae
ID: Tan to nearly white with red to dusky H-shaped bars on side; **numerous red spots on head and forebody,** branched cirri above eyes, red and bluish streaks through eye.
Indo-Pacific: Red Sea and E. Africa to Andaman Sea, Indonesia, Micronesia and French Polynesia. – S.W. Japan to Australia.

Redspotted Blenny
ID: Often occupy holes with only heads exposed along shallow surge-swept reef crests from near surface to 3 m.

LINED-TAIL BLENNY *Blenniella caudolineata*
SIZE: to 10 cm (4 in.)   Blennies – Blenniidae
ID: Male – Grayish with about 9 brownish or olive bars on side; several dark stripes on middle of body, pale yellow crest on head. **Female –** 3-4 dark lines on side breaking into spots and dashes toward tail. Shallow exposed shoreline and tidepools to 3 m.
Pacific: Indonesia, Japan to Solomon Is. and French Polynesia.

BLUEDASHED BLENNY *Blenniella paula*
SIZE: to 13 cm (5 in.)   Blennies – Blenniidae
ID: Pale to medium gray, olive or brown; 8 H-shaped dark bars each containing one or more pale blue dark edged dashes or spots. Solitary; rest on bottom or in holes with only head protruding. Intertidal reef flats with cracks and reefs to 5 m.
Pacific: Micronesia, Great Barrier Reef, New Caledonia to French Polynesia.

MANGROVE OYSTER BLENNY *Omobranchus obliquus*
SIZE: to 7 cm (2 3/4 in.)   Blennies – Blenniidae
ID: White to pale brown or gray undercolor with **narrow brown bars across throat** and slanting brown bands and chevron markings on side. Solitary or small groups. Coral, rock and rubble bottoms of coastal reefs and estuaries to 3 m.
East Indo-West Pacific: Nicobar Is. to Indonesia, Philippines, Micronesia and Solomon Is. – S. Japan to Australia and Samoa.

ELONGATE OYSTER BLENNY *Omobranchus elongatus*
SIZE: to 6.3 cm (2 1/2 in.)   Blennies – Blenniidae
ID: Gray with dark blotches and oblique bars on side; **white diagonal line behind eye,** pale-edged dark spot near rear gill cover. Solitary or form small groups. Rocky reefs and estuaries that support oyster growth to 2 m.
Indo-West Pacific: E. Africa to Andaman Sea, Indonesia, Philippines and Solomon Is. – S. Japan to Australia and Fiji.

Small, Elongate Bottom Dwellers – Blennies

BROWN CORAL BLENNY *Atrosalarias fuscus*
SIZE: to 14.5 cm (5 3/4 in.)   Blennies – Blenniidae
ID: Adult – Entirely dark brown to nearly black with clear tail except Great Barrier Reef and Palau populations, which have a yellow tail. Solitary among branches of live and dead corals. Lagoons to sheltered outer reef slopes in 2-12 m.
Indo-Pacific: Red Sea and E. Africa to Micronesia and French Polynesia. – S.W. Japan to Australia and New Caledonia.

Brown Coral Blenny – Juvenile
SIZE: to 7.5 cm (3 in.)
ID: Yellow to orange body and fins except for clear tail.

EARED BLENNY *Cirripectes auritus*
SIZE: to 9 cm (3 1/2 in.)   Blennies – Blenniidae
ID: Brown to pinkish red; **black "ear" spot and small dark spots on side,** fringing cirri on nape. Solitary. Coastal fringing reefs and outer slopes, usually less than 10 m, but may reach 20 m.
Indo-Pacific: E. Africa to Indonesia, Philippines, Micronesia, Papua New Guinea, Solomon Is. and Line Is. – Taiwan to Australia.

ZEBRA BLENNY *Cirripectes quagga*
SIZE: to 9 cm (3 1/2 in.)   Blennies – Blenniidae
ID: Color variable, but commonly uniform dark brown with small white spots; often show 12-14 grayish bars on body, sometimes red to yellow rear body. Shallow reefs to 10 m.
Indo-Pacific: E. Africa to Hawaii and French Polynesia. – Taiwan to Great Barrier Reef. Mainly oceanic islands. Only Asian Pacific record from Molucca Is. in Indonesia.

BARRED BLENNY *Cirripectes polyzona*
SIZE: to 8.5 cm (3 1/4 in.)   Blennies – Blenniidae
ID: Adult – Usually gray undercolor, occasionally greenish; about 12 brown body bars and spot above pectoral fin; reddish reticulations and bands on head.. Solitary. Algal ridges and reef crests in surge channels of outer reefs to 3 m.
Indo-Pacific: E. Africa to Indonesia, Micronesia, Solomon Is. and Line Is. – S. Japan to Great Barrier Reef and Fiji.

Barred Blenny – Female
SIZE: to 4 cm (1 1/2 in.)
ID: Dark gray back, midlateral stripe, white lower body with pale bars; **brown head with white spot between and behind eyes and white bar below on upper jaw extending above.**

Blennies

CHESTNUT BLENNY   *Cirripectes castaneus*
SIZE: to 11.5 cm (4 1/2 in.)   Blennies – Blenniidae
ID: Male – Brownish green with reddish brown bands on head and bars on side; ocellated "ear" spot, fringe-like cirri on nape. **Female –** Dark brown with network of pale olive polygons on side. Surge zones of outer reef crests to about 3 m.

Indo-West Pacific: Red Sea and E. Africa to Andaman Sea, Micronesia and Solomon Is. – S. Japan to Australia and Fiji.

LADY MUSGRAVE BLENNY   *Cirripectes chelomatus*
SIZE: to 12.3 cm (4 3/4 in.)   Blennies – Blenniidae
ID: Dark brown with fine red dots on head and body; dark iris, fringe-like cirri on nape. Solitary or form small groups. Surge zones of outer reef crests to 10 m.

West Pacific: E. Papua New Guinea and Great Barrier Reef to Fiji.

FILAMENTOUS BLENNY   *Cirripectes filamentosus*
SIZE: to 9 cm (3 1/2 in.)   Blennies – Blenniidae
ID: Male – Dark brown; **thin red line markings on snout and cheek,** fringe-like cirri on nape. **Female –** Dark brown with vague red marking on head. Solitary or form small loose groups. Coral and rocky reefs to 16 m, but usually shallow.

Indo-Pacific: Red Sea and E. Africa to Andaman Sea, Indonesia, Philippines, Papua New Guinea, Solomon Is. – Taiwan to Australia.

SPRINGER'S BLENNY   *Cirripectes springeri*
SIZE: to 8 cm (3 1/4 in.)   Blennies – Blenniidae
ID: Dark brown with **bright red spots on head** and forebody; occasionally with spotting over entire body, fringe-like cirri on nape. Solitary or small loose groups. Upper edge of outer reef slopes to 18 m, but usually shallow.

Asian Pacific: Indonesia, Philippines, Papua New Guinea, and Solomon Is.

RED-SPECKLED BLENNY   *Cirripectes variolosus*
SIZE: to 8 cm (3 1/4 in.)   Blennies – Blenniidae
ID: Male – Dark reddish brown with red spots and short dash markings on head; **dark around eye with silver iris,** fringe-like cirri on nape. Solitary; often seek shelter in holes, fissures. Living corals on exposed seaward reefs to 5 m.

Pacific: Bonin Is. in S. Japan to Palau in Micronesia, Johnston Atoll and French Polynesia.

RED-STREAKED BLENNY   *Cirripectes stigmaticus*
SIZE: to 12 cm (4 3/4 in.)   Blennies – Blenniidae
ID: Male – Dark brown with red to brown network of reticulated lines on head and forebody with thin broken lines behind; fringe-like cirri on nape. Solitary or loose groups. Upper edge of seaward reef slopes to 20 m, but usually shallow.

Indo-West Pacific: E. Africa to Indonesia, Philippines, Micronesia, Papua New Guinea, Solomon Is. and Fiji.

IMITATOR BLENNY *Cirripectes imitator*
SIZE: to 11.3 cm (4 1/2 in.) Blennies – Blenniidae
ID: **Male –** Dark brown with tips of anterior dorsal fin spines and upper edge of tail yellow. Shallow rock and reefs.
North Asian Pacific: Philippines, Taiwan and S.W. Japan.

Imitator Blenny – Female
ID: Dark brown with network of gray polygons on head and body; yellow tips on anterior dorsal fin spines.

AUSTRALIAN CORALBLENNY *Ecsenius australianus*
SIZE: to 5 cm (2 in.) Blennies – Blenniidae
ID: Reddish brown; **short dark brown stripe with pale margins behind eye,** double row of narrow rectangular white markings on side. Solitary or form small groups. Coral areas in lagoons and on outer slopes in 3-22 m.
Localized: S.E. Papua New Guinea, N. Great Barrier Reef and Coral Sea.

LUBBOCK'S CORALBLENNY *Ecsenius lubbocki*
SIZE: to 4 cm (1 1/2 in.) Blennies – Blenniidae
ID: Shades of brown with dark bars and pale stripes; bright gold iris, **dark margin on gill cover,** row of white streaks on back. Solitary or in pairs. Coastal reefs encrusted with coralline algae in 3-12 m.
East Indian Ocean: E. Andaman Sea and N.W. Sumatra in Indonesia.

MONOCLE CORALBLENNY *Ecsenius monoculus*
SIZE: to 6.4 cm (2 1/2 in.) Blennies – Blenniidae
ID: Pale brown with irregular rows of pale yellow to white spots on back and side; **black saddle on tail base with white blotch in front and behind.** Solitary or small, loose groups. Algal-covered, steep-walled gutters in 2-10 m.
Asian Pacific: Molucca Is. and West Papua in E. Indonesia, north to Philippines and S. China Sea.

SPOTTED-ROW CORALBLENNY *Ecsenius collettei*
SIZE: to 4 cm (1 1/2 in.) Blennies – Blenniidae
ID: Brown to gray-brown; double row of black spots on sides and often another at base of dorsal fin, yellow speckling on iris. Solitary or form small groups on coral outcroppings. Silty coastal reefs and lagoons to 10 m.
Localized: N. Papua New Guinea.

Blennies

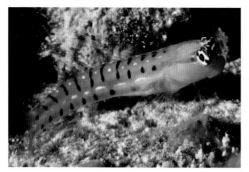

TIGER CORALBLENNY   *Ecsenius tigris*
SIZE: to 4 cm (1 ¹/₂ in.)   Blennies – Blenniidae
ID: Tan; narrow black bars or spots on upper body and row of black spots behind belly, row of small white spots on upper back, white dashes form midlateral line. Solitary or form small groups. Lagoon reefs and channels to 10 m.
Localized: Coral Sea off E. Australia, including Osprey Reef.

SQUARESTRIPED CORALBLENNY   *Ecsenius fourmanoiri*
SIZE: to 4 cm (1 ¹/₂ in.)   Blennies – Blenniidae
ID: Whitish with 3 brown to black stripes from head to tail base, middle stripe extends from eye and bottom stripe from chin, 2 rows of white squarish spots between black stripes. Solitary. On and around coral reefs to 25 m.
West Pacific: New Caledonia and Vanuatu to Fiji.

FIJI CLOWN CORALBLENNY   *Ecsenius fijiensis*
SIZE: to 4 cm (1 ¹/₂ in.)   Blennies – Blenniidae
ID: Pale reddish brown front body shading to yellow from midbody to tail; short white stripe above eye and another below, about 8 dark tear-shaped spot bars from midbody to tail base. Solitary, perch on coral outcroppings. Coral reefs in 3-25 m.
Localized: Known only from Fiji.

Fiji Clown Coralblenny – Variation
ID: Reddish brown; reddish brown stripe behind eye to center of body, double row of rectangular to rounded bluish spots. Distinguished from similar Australian Coralblenny [previous page] by location. Solitary or from small groups. Coral areas in lagoons and on outer slopes in 3-22 m.

CLOWN CORALBLENNY   *Ecsenius axelrodi*
SIZE: to 5.8 cm (2 ¹/₄ in.)   Blennies – Blenniidae
ID: Orange with dark upper head; ocellated spot above pectoral fin base, **orange strip on pectoral fin base,** spots on back become bars toward rear of body, 2 pale body stripes. Solitary or small groups. Outer reef slopes in 8-30 m.
Asian Pacific: N. Sulawesi and West Papua in Indonesia, Papua New Guinea and Solomon Is.

Clown Coralblenny – Gray Variation
ID: Gray with yellow head and pair of yellow stripes through eye; 3 black stripes and dusky spots on side. Also a yellow variation which may display striped or barred pattern. All variations have distinctive ocellated spot and **orange strip on pectoral fin base.**

Small, Elongate Bottom Dwellers – Blennies

BATH'S CORALBLENNY *Ecsenius bathi*
SIZE: to 4.4 cm (1 ³/₄ in.) Blennies – Blenniidae
ID: Gray with yellow head; pair of yellow stripes through eye, 3 black body stripes. Solitary or form small groups on coral outcroppings, commonly around sponges and tunicates. Coastal reefs and outer slopes in 3-25 m.
Localized: Bali, Molucca Is., Sulawesi, Halmahera to West Papua in E. Indonesia.

Bath's Coralblenny – Orange-striped Variation
ID: Alternating orange and gray stripes are intersected by narrow brown bars forming rectangular "windows"; pair of white stripes through eye.
Localized: Sabah in Malaysia to West Papua in E. Indonesia. (Does not overlap range of Twocoat Coralblenny [next].)

TWOCOAT CORALBLENNY *Ecsenius dilemma*
SIZE: to 4.2 cm (1³/₄ in.) Blennies – Blenniidae
ID: Alternating orange and gray stripes are intersected by narrow dark bars on back forming rectangular "windows". Also has black striped variation same as Bath's Coralblenny [previous]. Coral reefs in 5-25 m, usually below 10 m.
Localized: Philippines. (Does not overlap range of Bath's Coralblenny [previous]).

THREE-LINED CORALBLENNY *Ecsenius trilineatus*
SIZE: to 3.4 cm (1¹/₂ in.) Blennies – Blenniidae
ID: Shades of brown; 3 narrow dark brown body stripes with double row of white spots between, yellow-rimmed pupil with spoke-like marks on iris. Solitary or form small groups on coral outcroppings. Sheltered and seaward reefs in 2-10 m.
Localized: Brunei and Sabah in Malaysia to Molucca Is. and West Papua in Indonesia to Papua New Guinea and Solomon Is.

WHITE-LINED CORALBLENNY *Ecsenius taeniatus*
SIZE: to 4.8 cm (2 in.) Blennies – Blenniidae
ID: Yellowish brown head and dark brown body, narrow white stripe on back and wider white midlateral stripe, pair of yellow stripes on iris. Solitary or form small groups. Fringing reefs to 8 m.
Localized: Milne Bay in S.E. Papua New Guinea, including D'Entrecasteaux Is.

STRIPED CORALBLENNY *Ecsenius prooculis*
SIZE: to 5.3 cm (2 in.) Blennies – Blenniidae
ID: Pale brown to pale gray with 3 black body stripes; bright yellow marks on iris. Solitary or small loose groups. Sheltered lagoon reefs to 10 m.
Localized: N. Papua New Guinea and Solomon Is.

Blennies

BICOLOR CORALBLENNY *Ecsenius bicolor*
SIZE: to 10 cm (4 in.) Blennies – Blenniidae
ID: The most common form of this highly variable species has dark head and forebody and bright yellow-orange rear body; long straight cirri. Solitary; often use abandoned worm tubes for burrows. Coastal, lagoon and outer reefs to 25 m.
East Indo-Central Pacific: Maldives to Micronesia, Phoenix Is. in Central Pacific. – S.W. Japan to S. Great Barrier Reef and Fiji.

Bicolor Coralblenny – White-belly Variation
ID: Dark gray upper front body abruptly gradating to yellow toward rear; white belly. Different variations can often be identified by the presence of **an arched line behind and below the eye.**

Bicolor Coralblenny – Brown Variation
ID: Brown body with blue highlights on dorsal and tail fins.

FOURLINE CORALBLENNY *Ecsenius aequalis*
SIZE: to 3 cm (1 1/4 in.) Blennies – Blenniidae
ID: Pale gray to tan with 4 longitudinal black lines; 1st at base of dorsal fin, a 2nd on upper back and 2 on side of body (uncommonly lines broken or absent). Coral reefs in 2-11 m.
Localized: S.E. Papua New Guinea to Great Barrier Reef and Coral Sea.

LINED CORALBLENNY *Ecsenius lineatus*
SIZE: to 7 cm (2 3/4 in.) Blennies – Blenniidae
ID: Gray shading to dirty yellow upper body and light blue-gray below; **black stripe from eye to tail composed of black rectangular segments.** Solitary or form small groups. Seaward reef slopes to 28 m.
Indo-Asian Pacific: Mauritius and Reunion Is. to W. Australia, Sabah in Malaysia, Indonesia, Philippines and S.W. Japan.

KURT'S CORALBLENNY *Ecsenius kurti*
SIZE: to 4.5 cm (1 3/4 in.) Blennies – Blenniidae
ID: Pale gray; 4 thin black stripes on side break into spots on rear body, dark spoke-like lines on iris. Solitary or form small groups on coral outcroppings, often on species of *Porites*. Coastal fringing reefs and lagoons to 10 m.
Localized: Palawan, Cuyo and Calamianes Is. in W. Philippines.

MIDAS CORALBLENNY *Ecsenius midas*

SIZE: to 13 cm (5 in.) Blennies – Blenniidae

ID: Highly variable in both color and pattern, but commonly in shades of yellow/gold, mauve and/or brown, colors may be mixed or uniform; occasionally display irregular bars, **dark spot in front of anus in all variations.** Seaward reefs to 30 m.

Indo-Pacific: Red Sea and E. Africa to Indonesia, Micronesia, Solomon Is. to French Polynesia. – Philippines to Great Barrier Reef.

Midas Coralblenny – Variation

ID: All variations have the ability to rapidly change color and patterns; often uniformly gold (source of common name) or brown. Very similar to yellow variation of Bicolor Fangblenny distinguished by smaller eye located further back on head. Form plankton feeding aggregations, often mixing with and mimicking Anthias.

Midas Coralblenny – Nuptial Male

ID: Large male displaying nuptial color pattern with dark head and forebody and barred body. Nesting male occupy holes in rocky reefs. Dart out to display for passing female.

BLUEBELLY CORALBLENNY *Ecsenius caeruliventris*

SIZE: to 3 cm (1 1/4 in.) Blennies – Blenniidae

ID: Bicolored head with pair of white stripes through eye; **blue belly,** 3-4 darkish saddles on back [sometime not apparent]. Solitary or form small loose groups. Coral reefs to 20 m.

Localized: Togean and Sangihe Is. off northern Sulawesi in Indonesia.

TWINSPOT CORALBLENNY *Ecsenius bimaculatus*

SIZE: to 4 cm (1 1/2 in.) Blennies – Blenniidae

ID: Shades of brown with pale lower head and belly; **pair of black spots on the upper belly,** pair of white stripes through eye, white midlateral stripe from eye to midbody. Solitary or form small groups. Coastal reefs and outer slopes to 15 m.

Asian Pacific: Sabah in Malaysia to Indonesia and Philippines.

SHIRLEY'S CORALBLENNY *Ecsenius shirleyae*

SIZE: to 5 cm (2 in.) Blennies – Blenniidae

ID: Light brown with pale lower head and belly; pair of yellow stripes through eye, white midlateral stripe from eye to midbody. Solitary or form small groups on coral outcroppings. Coastal reefs and outer slopes in 3-12 m.

Localized: Java, S. Sulawesi, Lesser Sunda Is. and east to Flores in Indonesia.

365

Blennies

SADLE CORALBLENNY *Ecsenius sellifer*
SIZE: to 5 cm (2 in.) Blennies – Blenniidae
ID: Brown; **several separate ocellated spots on upper rear body,** several pale narrow stripes, black marks on edge of gill cover. Solitary or form small groups. Fringing reefs to 8 m.
Localized: Palau in Micronesia, Trobriand Is. off S.E. Papua New Guinea to Solomon Is.

OCULAR CORALBLENNY *Ecsenius oculus*
SIZE: to 6 cm (2¼ in.) Blennies – Blenniidae
ID: Brown with dark brown bars; **several pairs of ocellated spots on rear body,** several pale wavy broken stripes. Solitary or form small groups. Coastal and lagoon reefs to 15 m.
North Asian Pacific: Philippines to S. Taiwan and S.W. Japan.

FIJI SPOTTED CORALBLENNY *Ecsenius pardus*
SIZE: to 6 cm (2½ in.) Blennies – Blenniidae
ID: Brown with black stripe behind eye and combination of fine white lines and variable sized white-edged black spots, including 4-5 large ones on midside. Solitary or small, loose groups. Wave-exposed reefs in 2-10 m.
Localized: Fiji.

EYESPOT CORALBLENNY *Ecsenius ops*
SIZE: to 5.5 cm (2¼ in.) Blennies – Blenniidae
ID: Blue-gray head, brown body, occasionally tan to white lower body; white to bluish white line below eye, brilliant gold iris, **small black spot behind eye.** Solitary or form small groups. Sheltered coastal reefs to 15 m.
Localized: Java Sea to Banggai Is. E. of Sulawesi and Flores in Indonesia.

TAILSPOT CORALBLENNY *Ecsenius stigmatura*
SIZE: to 5.8 cm (2¼ in.) Blennies – Blenniidae
ID: Shades of brown to orange with pale lower head; **white-edged black spot on tail base,** yellow iris with outer rim of orange, white dark-edged line below eye. Solitary or form small groups. Coastal reefs and lagoons in 2-30 m.
Localized: Molucca and Raja Ampat in Indonesia and S. Philippines.

TRICOLOR CORALBLENNY *Ecsenius tricolor*
SIZE: to 4.8 cm (2 in.) Blennies – Blenniidae
ID: Blue-gray front body, yellow-orange rear body; white stripe from below eye to midside, stripe on head is edged with blue above and orange below. Small loose groups. Seaward and sheltered reefs in 5-30 m.
Localized: Sabah in Malaysia and Palawan and Calamianes Is. in W. Philippines.

Small, Elongate Bottom Dwellers – Blennies

SCHROEDER'S CORALBLENNY *Ecsenius schroederi*
SIZE: to 5 cm (2 in.) Blennies – Blenniidae
ID: Pale brown to gray with wide dusky bars; white stripe from eye to above pectoral fin where it continues as a row of dashes, 2nd thin broken stripe below dorsal fin. Solitary. Sheltered coastal and lagoon reefs in 2-12 m.
Localized: N.W. Australia to Molucca Is. and West Papua in E. Indonesia.

YAEYAMA CORALBLENNY *Ecsenius yaeyamaensis*
SIZE: to 6 cm (2¼ in.) Blennies – Blenniidae
ID: Shades of brown; **2 dark dashes behind eye,** black margin on cheek, "Y" marking on pectoral fin base, numerous white spots and blotches on side. Solitary or small groups. Boulder habitats in 2-15 m.
East Indo-Asian Pacific: Sri Lanka to Indonesia, Philippines and Micronesia. – S.W. Japan to N. Australia and Vanuatu.

SPOTTED CORALBLENNY *Ecsenius stictus*
SIZE: to 5.5 cm (2¼ in.) Blennies – Blenniidae
ID: Light gray; faint white midlateral stripe with dark borders, several rows of faint dark lines and spots, "Y" marking on pectoral fin base, dark margin on gill cover. Solitary. Reef flats and lagoons to 8 m.
Localized: Great Barrier Reef.

PICTUS CORALBLENNY *Ecsenius pictus*
SIZE: to 5 cm (2 in.) Blennies – Blenniidae
ID: Pale brown head, dark brown body, **yellowish tail base;** numerous thin white stripes with white spots on upper stripe and midlateral stripe. Solitary or in pairs. Coastal and outer reef slopes in 11-40 m.
Asian Pacific: Sulawesi to Flores and West Papua in Indonesia and Philippines to Solomon Is.

BLACK CORALBLENNY *Ecsenius namiyei*
SIZE: to 10 cm (4 in.) Blennies – Blenniidae
ID: Blotchy brown to charcoal with either pale gray or yellow tail; usually pale curving line behind eye. Solitary. Perch on sponges and corals. Lagoon and seaward reefs to 30 m.
Asian Pacific:Bali, Sabah in Malaysia to Molucca Is. and West Papua in E. Indonesia, Philippines, Papua New Guinea andSolomon Is. – S.W. Japan to Australia.

Black Coralblenny – Phase
ID: When resting on the bottom or inside a hole often display white line and spot markings on head, commonly including a wavy line from conter of mouth to behind eye. Identification confirmed by 4 cirri between eyes.

367

Blennies

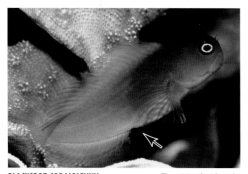

BLACKSPOT CORALBLENNY   *Ecsenius lividanalis*
SIZE: to 4.7 cm (2 in.)   Blennies – Blenniidae
ID: Variable from entirely yellow-orange to purplish gray-brown front body and yellow-orange behind; **dark spot around anus** [not visible in photo]. Form small groups. Lagoons and sheltered coastal reefs to 12 m.
Asian Pacific: N.W. Australia to E. Indonesia, Papua New Guinea and Solomon Is.

Blackspot Coralblenny – Yellow-fin Variation
ID: Purplish gray head and body with yellow dorsal fin.

Blackspot Coralblenny – Yellow Variation
ID: Yellow body and fins except for clear tail.

EYELINED CORALBLENNY   *Ecsenius melarchus*
SIZE: to 6.2 cm (2 1/2 in.)   Blennies – Blenniidae
ID: Light brown head, dark brown body; bright gold iris, **black-edged pale gray to white stripe from lower eye to pectoral fin base with orange stripe just below**. Solitary or form small groups. Sheltered coastal reefs in 1-30 m.
Localized: Java Sea in Indonesia, north to Sabah in Malaysia and W. Philippines.

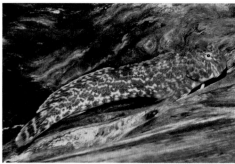

BANDA CORALBLENNY   *Ecsenius bandanus*
SIZE: to 4.1 cm (1 1/2 in.)   Blennies – Blenniidae
ID: Brown on back shading to white on lower side; black stripe through eye extending to rear head and golden stripes on upper eye. Solitary or small groups on coral outcrops. Coastal reefs and outer slopes in 2-15 m.
Localized: Banda Is., Molucca Is. and West Papua in Indonesia.

AMBON BLENNY   *Paralticus amboinensis*
SIZE: to 15 cm (6 in.)   Blennies – Blenniidae
ID: White undercolor covered with clusters of gray-brown spots; broad, leaf-like tentacle above eye; gray bars below eye. Solitary. Shoreline reefs or among mangroves, often on submerged logs to 1 m.
Localized: Sabah in Malaysia to N. Papua New Guinea.

DELICATE BLENNY *Glyptoparus delicatulus*
SIZE: to 4.5 cm (1 ³/₄ in.) Blennies – Blenniidae
ID: Female – Tan to pale gray; covered with scattered white spots and rows of brown spots on body, **brown to green bar wraps under chin and 2 elongate markings behind.** In pairs or form small groups. Dead reefs with algal turf in 2-8 m.
Indo-West Pacific: E. Africa to Indonesia, Micronesia, Papua New Guinea and Solomon Is. – S.W. Japan to Great Barrier Reef and Fiji.

Delicate Blenny – Male
ID: Tan to pale gray; covered with scattered white spots and rows of brown spots on body, **brown to green to blue bar wraps under chin and 2 elongated markings behind,** several narrow brown bars on snout.

LINED ROCKSKIPPER *Istiblennius lineatus*
SIZE: to 14 cm (5 ¹/₂ in.) Blennies – Blenniidae
ID: Male – Pale yellowish to whitish with vertical black lines on head; 5 sets of double bars on side, blade-like crest. **Female –** Body lines usually darker. Usually in groups that graze algae from rocks. Rocky shores and intertidal zone exposed to waves to 2 m.
East Indo-Pacific: Maldives to Indonesia, Philippines, Micronesia and Pitcairn Is. – S. Japan to Australia.

STREAKY ROCKSKIPPER *Istiblennius dussumieri*
SIZE: to 12.5 cm (5 in.) Blennies – Blenniidae
ID: Male – Grayish upper body, white below; 8-9 bars branching ventrally on side, about 5 squarish white patches on upper back, red spots and short lines on head, yellow streaks on medial fins. Solitary or form groups. Sheltered rocky shores to 2 m.
Indo-West Pacific: E. Africa to Indonesia, Philippines, Micronesia and Papua New Guinea. – Taiwan to E. Australia and Fiji.

BEAUTIFUL ROCKSKIPPER *Istiblennius bellus*
SIZE: to 16.5 cm (6 ¹/₂ in.) Blennies – Blenniidae
ID: Male – Charcoal-gray with alternating light and dark bars; large sail-like skin crest on top of head, 2 unbranched cirri. Solitary. Rocky shores and intertidal flats exposed to strong waves to 3 m.
Scattered: E. Africa to Fiji. Probably absent in Asian Pacific.

Beautiful Rockskipper – Female
ID: Gray upper body, pale gray below; numerous small black spots, may display several gray bars on rear body.

369

Blennies

WHITE-SPOTTED BLENNY *Salarias alboguttatus*
SIZE: to 6.5 cm (2 1/2 in.) Blennies – Blenniidae
ID: Gray to brown with numerous scattered white spots on head; **rows of white spots and dash lines on side,** 7-8 irregular dusky bars, unbranched tentacle above eye. Solitary. Lagoon and sheltered coastal reefs to 8 m.
West Pacific: Philippines and Micronesia to Great Barrier Reef, New Caledonia and Fiji.

FINE-SPOTTED BLENNY *Salarias guttatus*
SIZE: to 14 cm (5 1/2 in.) Blennies – Blenniidae
ID: Greenish brown; heavily mottled and spotted, irregular whitish area along midside, about 7 white-spotted gray saddles on back, **large dark-edged white spot in front of lower pectoral fin base.** Solitary on dead coral. Sheltered reefs to 5 m.
Asian Pacific: Indonesia, Philippines, Papua New Guinea, Solomon Is. and Great Barrier Reef.

PATZNER'S BLENNY *Salarias patzneri*
SIZE: to 5 cm (2 in.) Blennies – Blenniidae
ID: Covered with numerous white spots of unequal size; 8 pairs of irregular olive-brown bars, large white spot on breast, unbranched cirri above eye. Solitary; perch on coral. Sheltered shoreline reefs to 5 m.
Asian Pacific: Indonesia, Philippines and Papua New Guinea.

STARRY BLENNY *Salarias ramosus*
SIZE: to 5 cm (2 in.) Blennies – Blenniidae
ID: Shades of brown; covered with numerous small white spots, highly branched tentacle above eye. **Juvenile –** White spots larger and more obvious and of varying sizes. Solitary. Inhabit rock or coral outcroppings. Sheltered areas with mixed sand and weed bottoms to 5 m.
Asian Pacific: N.W. Australia, Indonesia and Philippines.

SEGMENTED BLENNY *Salarias segmentatus*
SIZE: to 6.8 cm (2 3/4 in.) Blennies – Blenniidae
ID: Whitish undercolor, darkish head with white spots; 3 rows of large spots form 13-14 bars, several large white spots in front of pectoral fin. Solitary; on coral outcroppings. Sheltered reefs, often in turbid water to 8 m.
West Pacific: Sabah in Malaysia and Sulawesi in Indonesia to Micronesia, Solomon Is. and Fiji.

JEWELLED BLENNY *Salarias fasciatus*
SIZE: to 14 cm (5 1/2 in.) Blennies – Blenniidae
ID: White undercolor; 8 dark bars with several white circular to oval spots, numerous oval spots between bars, thin wavy dark stripes on body. Solitary. Coral rocks or mixed sand and weed bottoms to 5 m.
Indo-West Pacific: Red Sea and E. Africa to Indonesia, Philippines, Micronesia and Solomon Is. S.W. Japan and Australia to Fiji

CERAM BLENNY   *Salarias ceramensis*
SIZE: to 14 cm (5 1/2 in.)   Blennies – Blenniidae
ID: Complex pattern of variable-sized dark-edged pale spots on body; dark spotting on fins, large individuals often display several large dark patches on upper side. Solitary. Coral-rich areas of lagoons or sheltered coastal reefs in 2-12 m.
Asian Pacific: Indonesia, Philippines, Papua New Guinea and Solomon Is.

OBSCURE BLENNY   *Salarias obscurus*
SIZE: to 13 cm (5 in.)   Blennies – Blenniidae
ID: Generally similar in shape and behavior to Ceram Blenny [previous], but uniformly dark brownish without dark spotting on pectoral and tail fins. Solitary; perch on corals. Sheltered shoreline reefs, frequently in turbid water to 6 m.
Localized: Sabah in Malaysia, West Papua in Indonesia, Palawan Province in W. Philippines, including Cuyo and Calamianes Is.

LEOPARD BLENNY   *Exallias brevis*
SIZE: to 14.5 cm (5 3/4 in.)   Blennies – Blenniidae
ID: Male – Large with high arched back; pale gray to red with netted pattern of red spots, branched cirri above eye and fringe-like cirri across nape. **Female –** Brown spots. Solitary; among branches of *Pocillopora* corals. Coral-rich reefs 3-20 m.
Indo-Pacific: Red Sea and E. Africa to Indonesia, Micronesia, Hawaii and French Polynesia. – S.W. Japan to Australia and Fiji.

Leopard Blenny – Juvenile
SIZE: to 7.5 cm (3 in.)
ID: Whitish with large brownish spots with dark centers on body and fins; polygonal brown spots on head.

THROATSPOT BLENNY   *Nannosalarias nativitatis*
SIZE: to 4.2 cm (1 1/2 in.)   Blennies – Blenniidae
ID: Brown to gray upper body, white below; white dots on head and body, about 8 wide dark diffuse bars on lower body, **dark reddish band behind eye;** one or two dark spots on side of throat. Outer reefs with some surge in 2-12 m.
West Pacific: Andaman Sea to Philippines, Micronesia and Solomon Is. – S.W. Japan to Great Barrier Reef and Fiji.

TRIPLESPOT BLENNY   *Crossosalarias macrospilus*
SIZE: to 8.2 cm (3 1/2 in.)   Blennies – Blenniidae
ID: Shades of brown with numerous spots and ovals; midlateral stripe formed by dark brown rectangular blotches, **large brown spot in front of dorsal fin,** 2 ocellated spots on throat. Solitary. Lagoon and outer reefs to 25 m.
West Pacific: Andaman Sea, Indonesia, Philippines, Micronesia and Solomon Is. – S.W. Japan to Great Barrier Reef and Fiji.

Triplefins

HIGHHAT TRIPLEFIN   *Enneapterygius tutuilae*
SIZE: to 2.8 cm (1 1/4 in.)   Triplefins – Tripterygiidae
ID: Highly variable from reddish browns to brown to nearly all white; tall whitish first dorsal fin, 2nd and 3rd dorsal fins usually translucent; usually white saddle or bar behind 2nd dorsal fin. Solitary or loose groups. Tidepools to outer slopes to 55 m.
Indo-Pacific: Red Sea and E. Africa to Indonesia, Micronesia, Solomon Is. and French Polynesia. – Taiwan to Great Barrier Reef.

Highhat Triplefin – White Variation
ID: Translucent white with scattered dark markings; all variations commonly have a black or dark spot in upper 2nd dorsal fin; usually row of darkish bars with white between on lower body.

Highhat Triplefin – Variation
ID: This variation has no obvious dark spot on the 2nd dorsal fin, but clearly shows the distinctive tall whitish 1st dorsal fin. Also, note the whitish "V" shaped marking behind the 2nd dorsal fin and white spot and bar markings along the lower body.

PALE-SPOTTED TRIPLEFIN   *Enneapterygius pallidoserialis*
SIZE: to 3.6 cm (1 1/2 in.)   Triplefins – Tripterygiidae
ID: **Female –** Translucent; dark brown saddles with dashes and blotches below, several white spots and streaks. **Male –** Dark brown or black. Solitary or in pairs. Coral reefs to 8 m.
Asian Pacific: Philippines, Taiwan, Japan, Chuuk (Truk) in Micronesia to Vanuatu.

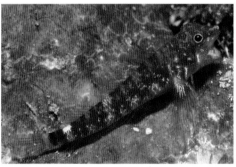

YELLOWEYE TRIPLEFIN   *Enneapterygius flavoccipitis*
SIZE: to 3.2 cm (1 1/4 in.)   Triplefins – Tripterygiidae
ID: **Female –** Tan to red with fine speckling; 6-7 darkish bars (may be poorly defined), **last bar black with white saddle in front. Male –** Dark gray with large yellow patch on lower head and pectoral fin base. Lagoon and outer reefs to 25 m.
Asian Pacific: Indonesia, Philippines, Papua New Guinea and Solomon Is. – S.W. Japan to Great Barrier Reef and Vanuatu.

PHILIPPINE TRIPLEFIN   *Enneapterygius philippinus*
SIZE: to 3.8 cm (1 1/2 in.)   Triplefins – Tripterygiidae
ID: **Male –** Green head shading to black rear body and tail. **Female –** Light green; dorsal fins translucent without black pigment. Solitary or in pairs. Reef flats and tide pools to 4 m.
Indo-West Pacific: Madagascar to Andaman Sea, Indonesia, Philippines, Papua New Guinea and Solomon Is. – S.W. Japan to Australia, New Caledonia and Fiji.

BLACK AND RED TRIPLEFIN *Enneapterygius similis*
SIZE: to 3.9 cm (1 1/2 in.) Triplefins – Tripterygiidae
ID: **Male –** Red forebody and black rear body including tail; **irregular white spot below and behind eye.** Reef crests with some wave action and algal growth to 13 m.
Asian Pacific: Sabah in Malaysia, Indonesia and Philippines. – Great Barrier Reef to Vanuatu and New Caledonia.

Black and Red Triplefin – Female
ID: Translucent grayish to yellowish green with red scale margins; numerous white flecks on body; white snout tip.

UMPIRE TRIPLEFIN *Enneapterygius rhabdotus*
SIZE: to 3.2 cm (1 1/4 in.) Triplefins – Tripterygiidae
ID: **Female –** Yellowish with 3 double bars divided by thin gray line and wide black bar on tail base. **Male –** Blackish fins and body to rear of 2nd dorsal fin; black band from rear of 3rd dorsal fin to anal fin, yellow tail. Coral and rocky reefs in 1-8 m.
Pacific: Gulf of Thailand to Taiwan, Philippines, Micronesia, Vanuatu, Fiji and French Polynesia.

NEGLECTED TRIPLEFIN *Helcogramma desa*
SIZE: to 5.5 cm (2 1/4 in.) Triplefins – Tripterygiidae
ID: **Nuptial Male –** Black lower head to pectoral fin base; blue stripe from upper lip passes below eye onto gill cover, several blue spots on black pectoral fin base. Commonly gray to yellowish with about 5 sets of slanted double bars on side. Rocky boulder areas and surge channels to 7 m.
Localized: Cuyo Is. and Palawan in Philippines and Vietnam.

RANDALL'S TRIPLEFIN *Helcogramma randalli*
SIZE: to 4.7 cm (2 in.) Triplefins – Tripterygiidae
ID: **Male –** Reddish with large irregular brown spots/blotches scattered over body; black lower head; blue to white stripe from upper lip passing below eye onto gill cover, few blue to white spots on black pectoral fin base. Exposed coasts in areas with boulders in 2-12 m.
Localized: Bali to Alor in Indonesia.

Randall's Triplefin – Female
ID: White and reddish brown mottling on back; two rows of partially merged reddish brown spots on midbody from behind head to tail base, yellowish pectoral fin.

Triplefins

STRIPED TRIPLEFIN　　　　　*Helcogramma striata*
SIZE: to 5 cm (2 in.)　　　　Triplefins – Tripterygiidae
ID: Red with whitish lower body; 3 white or bluish white stripes, bright yellow iris. Solitary or small groups. Coastal, lagoon and outer reefs to 20 m.

West Pacific: Andaman Sea to Indonesia, Philippines, Micronesia, Papua New Guinea and Solomon Is. – S. Japan to Australia and Fiji.

SCARF TRIPLEFIN　　　　　*Helcogramma trigloides*
SIZE: to 4.8 cm (2 in.)　　　Triplefins – Tripterygiidae
ID: Female – Translucent grayish with series of poorly defined brown diagonal H-shaped bars on side; brown markings on head including dark upper lip, translucent fins. Coral and rock bottoms to 6 m.

Asian Pacific: Andaman Sea, Taiwan, Indonesia, Philippines, Papua New Guinea, Solomon Is. and Vanuatu.

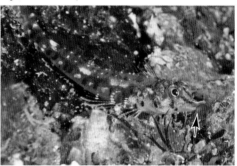

WHITESPOT TRIPLEFIN　　　*Helcogramma albimacula*
SIZE: to 4.2 cm (1 3/4 in.)　　Triplefins – Tripterygiidae
ID: Nuptial Male – Dark brown head and body with about 5 narrow whitish bands on body; bright red iris, bright red rear dorsal and tail fins. **Female –** Reddish head and forebody. **Male –** Head with brown, white and yellow mottling. Exposed coasts in areas with rocky boulders and surge channels to 6 m.

Localized: Apo I. and Batangas Province, Luzon, Philippines.

RHINOCEROS TRIPLEFIN　　*Helcogramma rhinoceros*
SIZE: to 3.7 cm (1 1/2 in.)　　Triplefins – Tripterygiidae
ID: Male – Translucent brown to gray with several large white internal patches along vertebrae; blue band below eye, midlateral row of brown blotches, yellow lower head, **extended snout.** Solitary or pairs. Boulder and coral to 6 m.

Asian Pacific: Andaman Sea, Indonesia and Philippines, Solomon Is., Vanuatu and New Caledonia.

Rhinoceros Triplefin – Female

ID: Translucent brown to gray; series of wide brown dash markings along vertebral column with white dashes between, short pointed snout.

Rhinoceros Triplefin – Nuptial Male

ID: Black beneath head from tip of long pointed snout to pectoral fin; bright blue line above.

LITTLE HOODED TRIPLEFIN *Helcogramma chica*
SIZE: to 4 cm (1 ½ in.) Triplefins – Tripterygiidae
ID: First dorsal fin shorter than others, blunt snout, and relatively colorless. **Male** – Black lower head. Solitary or small groups algal-covered rocky surfaces. Intertidal zone and outer reef slopes in 1-32 m.
East Indo-Pacific: Andaman Sea to Marshall Is., Phoenix Is., Vanuatu and Fiji.

HELEN'S TRIPLEFIN *Ceratobregma helenae*
SIZE: to 4.7 cm (1 ¾ in.) Triplefins – Tripterygiidae
ID: Reddish brown head; 14-16 dark bands on side [poorly defined in photo], orange scale margins; sometimes orange with a few white scale centers. Solitary. Lagoon and outer reef slopes to 37 m.
West Pacific: Christmas I. to Indonesia, Philippines and Micronesia. – S.W. Japan to Great Barrier Reef and Fiji.

SCALYFIN TRIPLEFIN *Norfolkia brachylepis*
SIZE: to 7.3 cm (3 in.) Triplefins – Tripterygiidae
ID: Pale gray with about 5 ragged dark bands extending from dorsal fins onto body; white and black bands on anal fin, ragged dark bars on tail; tall 1st dorsal fin. Coral reef recesses and crevices in 1-25 m.
Indo-Asian Pacific: Red Sea and E. Africa to Indonesia, Philippines and Micronesia. – S. Japan to Australia and Fiji.

LARGEMOUTH TRIPLEFIN *Ucla xenogrammus*
SIZE: to 5.7 cm (2 ¼ in.) Triplefins – Tripterygiidae
ID: Male – Translucent with alternating red and white vertebrae; large mouth extends to below eye, projecting lower jaw. Solitary. Perch on coral heads in lagoon and outer reefs in 2-41 m.
Indo-West Pacific: Andaman Sea to Indonesia, Philippines, Micronesia, Solomon Is. – S.W. Japan to Great Barrier Reef and Fiji.

Largemouth Triplefin – Female
ID: Translucent with light reddish mottling.

Largemouth Triplefin – Nuptial Male
ID: Dark brown bars on side of body, and brown to black band on 2nd and 3rd dorsal fins.

Odd-shaped Bottom Dwellers
Frogfishes – Scorpionfishes & Lionfishes – Stonefishes – Waspfishes – Flatheads – Flounders & Soles – Others

This ID Group consists of fishes that normally rest on the bottom and do not have typical fish-like shapes.

FAMILY: Frogfishes (Anglerfishes) – Antennariidae
5 Genera – 16 Species Included

Typical Shape

Frogfishes, also known as anglerfishes, are globular in shape with large, extremely upturned mouths which can be opened to the width of their bodies to engulf prey. Their pectoral and ventral fins have evolved into webbed, hand-like appendages, which they use to grasp, perch or "walk." The small circular gill openings are located behind and/or below the pectoral fins. The first dorsal spine, located on the snout, has evolved into a thin, stalk-like structure (illicium) tipped with a lure (esca), which is wiggled energetically, much like a casting rod, to attract prey. Escas vary from realistic fish-like and shrimp-like shapes to nondescript tufts. Masters at camouflage, frogfishes can slowly change colors to match various backgrounds. At rest they often look like sponges or clumps of algae making the stationary, ambush predators difficult to sight. Subtle differences in body markings of several similar-appearing frogfishes often make identification difficult.

FAMILY: Scorpionfishes & Lionfishes – Scorpaenidae
13 Genera – 37 Species Included

Scorpionfishes – Typical Shape

Lionfishes – Typical Shape

Scorpionfishes derived their name from venomous fin spines common to most species. Poison is produced by glands embedded in long grooves on both sides of the spines. The pain from wounds varies from uncomfortable to intense, and occasionally requires medical attention. Immersing the affected area in hot water offers some relief. Many of these solitary, ambush predators have stout bodies adorned with skin flaps and tassels that augment their abilities to change colors for better camouflage.

The visually flamboyant group of scorpionfishes, commonly known as lionfishes, are renown for their dramatically elongate, feather-like pectoral and dorsal fins. Lionfishes typically hover just above the bottom or nestle in crevices during the day. At night they become active hunters searching the bottom for crustaceans and small fishes. Most species of lionfishes can be easily identified; however, several of the similar-appearing, bottom-dwelling scorpionfishes present quite a problem differentiating between species.

FAMILY: Stonefishes – Synanceiidae
5 Genera – 9 Species Included

Stonefishes

Genus *Inimicus*

The two stonefishes, in genus *Synanceia,* produce the most deadly fish venom known. The neurotoxin produced in glands at the base of the dorsal fin is injected through hollow spines. The deadly pair have poorly defined globular bodies that can exceed a foot in length. Most victims are shoeless waders in shallow tropical waters who fail to see the motionless fishes expertly camouflaged as algae-covered stones.

All fins and spines with bulging eyes and upturned snouts, members of genus *Inimicus,* commonly known as devilfishes, are among the most bizarre fishes inhabiting shallow seas. The bottom-oriented predators, also producers of powerful, but much less potent venom, spend much of the time buried beneath the sand. The three similar-appearing species are best distinguished by the patterns displayed on the inner surface of their spread pectoral fins.

FAMILY: Waspfishes – Tetrarogidae
4 Genera – 7 Species Included

Typical Shape

Waspfishes can be distinguished from the similar-appearing Leaf Fishes in the scorpionfish family by dorsal fins that begin above or in front of the eyes. The small ambush predators sway back and forth on the sand mimicking leaf debris as they wait for unsuspecting fish and crustaceans to venture close.

FAMILY: Flatheads – Platycephalidae
6 Genera – 11 Species Included

Typical Shape

Flatheads are close relatives of scorpionfishes with dorsally-compressed bodies, two separate dorsal fins, and as their name implies, flattened heads. While resting on the bottom the ambush predators resemble miniature crocodiles. Their distinctive eyes are draped with an elaborate curtain of tassels that expand or contract as the intensity of sunlight dictates. The tasseled-structure also helps disguise the eyes' location from prey.

FAMILY: Flounders – Bothidae/Pleuronectidae/Paralichthyidae & Soles – Soleidae

12 Genera – 16 Species Included

Flounders – Typical Shape

Soles – Typical Shape

Members of the Order Pleuronectiformes, the flatfishes, include flounders, soles and several other flat forms of fishes from temperate climes. Flatfishes have been placed in their respective families primarily by the location of their eyes. Bothidae are mainly shallow-water tropical species with both eyes on their left sides; righteye flounders, Pleuronectidae, typically live in temperate or deep waters, and Soleidae, the soles, have eyes on their left side and slightly out of alignment. A few species from both families have "reversed" individuals whose eyes are located on the contrasting side.

To acquire their flattened bodies, flatfishes pull off one of the animal kingdom's most astonishing feats of developmental biology. While adrift in their pelagic larval stage, flounders have typical bilateral fish-shaped bodies, properly aligned fins and pigmented eyes – one on each side of the head. Weeks later, sometime before settling to the sea floor, muscles, skin, blood vessels and bones inexorably shift into the flattened shape of thumbnail-sized benthic juveniles. During the metamorphosis, one eye migrates across the head until next to the other, swimbladders disappear, the dorsal and anal fins line oval bodies, and a lone pectoral fin, often quite long on the males, extends from the center of their backs. The thin, pancake profile not only hides them from predators and prey but also allows them to bury quickly in the sand, leaving only their stalked, independently functioning, 180-degree-rotating, periscopic eyes exposed. But, without question, the tropical flounders' best defensive strategy is rapid adaptive camouflage. Once settled, after moving to a new location, visual cues from the immediate surroundings are transmitted via nerves to thousands of irregular-shaped cells in the skin, known as chromatophores, which change body patterns within seconds to match their environment.

FAMILY: Others

Toadfishes – Batrachoididae

Batfishes – Ogcocephalidae

Sea Moths – Pegasidae

Flying Gurnards – Dactylopteridae

Velvetfishes – Aploactinidae

Coral Crouchers – Caracanthidae

Stargazers – Uranoscopidae

Clingfishes – Gobiesocidae

ID: Highly variable colors, most commonly in shades of tan and brown, but yellow, orange, pink, red, green, white and black. Colors can slowly change taking a few days to several weeks. May display pepper like spots and vague variably-sized spots.

GIANT FROGFISH   *Antennarius commerson*

SIZE: to 45 cm (18 in.)   Frogfishes – Antennariidae

ID: Large; typically smooth skin but occasionally have a few warty projections and rough-textured striations, long thin rod tipped with tiny white lure. Solitary, frequently on sponges. Coastal, lagoon and outer reefs to 50 m.

Indo-Pacific: Red Sea and E. Africa to Indonesia, Philippines, Micronesia, Solomon Is. and Hawaii. – S. Japan to Australia.

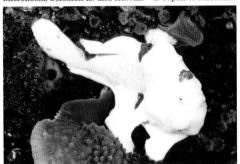

Giant Frogfish – White Phase
ID: In all phases may develop scab-like patches.

Giant Frogfish – Brown Mottled Phase
ID: Occasionally acquire mottled colors with rough striations. Occasionally develop fringe-like projections further enhancing their camouflage.

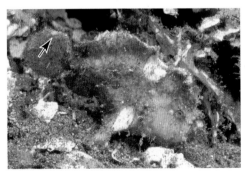

RANDALL'S FROGFISH   *Antennarius randalli*

SIZE: to 4.6 cm (1 3/4 in.)   Frogfishes – Antennariidae

ID: Small; highly variable in shades of yellow to red, brown and black, **small to large white spots generally behind eye, above pectoral fin and on upper edge of tail,** short rod with split filamented lure. Solitary. Seaward reefs in 8-31 m.

Pacific: Indonesia, Philippines, Micronesia and Papua New Guinea to Hawaii and Easter Is. – S. Japan to Fiji

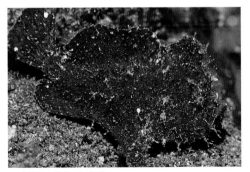

Randall's Frogfish – Dark Phase
ID: Color changeable to match surroundings. This dark individual photographed on black sand at Bali, Indonesia. Note consistent white spot markings, commonly a 2nd white spot on lower edge of tail opposite the upper spot.

Frogfishes

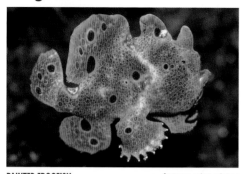

PAINTED FROGFISH   *Antennarius pictus*
SIZE: to 21 cm (8 1/4 in.)   Frogfishes – Antennariidae
ID: Color and markings highly variable, often have pale-edged dark spots of varying sizes; frequently three spots on tail but spots can be scattered across body, skin smooth, but may have variable-sized rough patches. Solitary. Coral and rocky reefs to 75 m.
Indo-Pacific: E. Africa to Indonesia, Philippines, Papua New Guinea, Micronesia and French Polynesia. – S. Japan to Australia.

Painted Frogfish – Small Juvenile
SIZE: to 2.5 cm (1 in.)
ID: Small size. Wide range of colors from white and yellow to dark purple and black. Usually display three spots on mid-tail. Often in open sand and gravel.

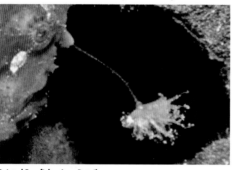

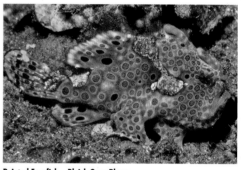

Painted Frogfish – Lure Detail
ID: Long thin rod with translucent crustacean-like lure. Lures are often damaged by prey. Lures grow back, but seldom as ornate as originals.

Painted Frogfish – Bluish Gray Phase
ID: Often develop large scab-like patches on body and pale-edged dark spots may be spread across body, head and fins.

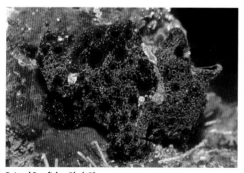

Painted Frogfish – Black Phase
ID: The black phase is most commonly encountered in areas of black volcanic sand.

Painted Frogfish – Ocellated Spots
ID: Have the ability to change to almost any color including white, black, red, pink, blue, purple, orange, yellow and brown.

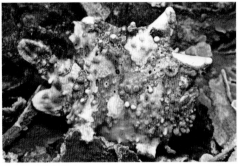

WARTY FROGFISH
SIZE: to 11 cm (4 1/4 in.)

Antennarius maculatus
Frogfishes – Antennariidae

ID: **Numerous knob-like warts;** variable color, usually with reddish brown patches (commonly with patch extending from eye and angling up to 3rd dorsal fin). Solitary. Frequently among algae, sponges and soft corals of coastal reefs to 15 m.

Indo-Pacific: Mauritius to Indonesia, Philippines, Micronesia, Solomon Is. and French Polynesia. – S.W. Japan to Australia.

Warty Frogfish – Small Juvenile
SIZE: to 2 cm (1 in.)

ID: Small size. Usually white or yellow with red or orange patch markings similar to adults. An uncommon juvenile variation is dark, most commonly black, with orange spots covering the body, occasionally with blue edges on fins.

Warty Frogfish – Lure Detail

ID: Long thin rod with translucent brown to grey shrimp or fish-shaped lure, occasionally eye-like spot and shrimp-like legs with white tips. Lures are occasionally damaged by prey. Lures grow back, but seldom as ornate as originals.

Warty Frogfish – Variation

ID: Colors include white, pink, red, yellow, greenish yellow, and shades of brown and black; often have red to orange on leading edges of 3 broad dorsal fins. May have nearly wartless skin and/or display numerous dark spots over body, head and fins.

BANDFIN FROGFISH
SIZE: to 9 cm (3 1/2 in.)

Antennatus tuberosus
Frogfishes – Antennariidae

ID: Usually pale gray to yellowish with dark brown or reddish brown reticular pattern on body; often with lavender to pink and white patches on head, **dark central bar on tail.** Solitary. Occasionally in hard coral branches to 73 m, but usually less than 10 m.

Indo-Pacific: E. Africa to Hawaii and French Polynesia.

Bandfin Frogfish – Subadult

ID: Dark brown body with lavender head. **Juvenile –** White without markings except dark central bar on tail and dark inner-half of anal fin, similar markings on adult and subadult.

Frogfishes

STRIATED FROGFISH *Antennarius striatus*
SIZE: to 22 cm (8 ³/₄ in.) Frogfishes – Antennariidae
ID: Variable colors usually with **dark zebra-like banding,** spots and patches; often have covering of filamentous skin appendages used for camouflage. Frequently on sand or mud bottoms to 219 m.
Indo-Pacific: Circumtropical.

Striated Frogfish – Hairy Variation
ID: Filamentous skin appendages are extremely well developed in individuals inhabiting bottoms with filamentous algae and soft coral. **Distinctive worm-like lure** on tip of short stocky rod. Flexible lure generally has 2 or 3 fleshy extentions.

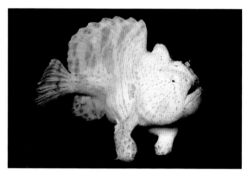

Striated Frogfish – Variation
ID: Uncommon reddish variation with white filamentous skin appendages are extremely well developed in some individuals.

Striated Frogfish – Variation
ID: White individual without filaments encountered on black sand bottom.

HISPID FROGFISH *Antennarius hispidus*
SIZE: to 19 cm (7 ¹/₂ in.) Frogfishes – Antennariidae
ID: Distinctive marble-sized **pom-pom-like lure of short white filaments;** scattered filamentous skin appendages, variable beige, yellow, orange or white, often display zebra-like banding. Solitary or occasionally in groups. Often among leaf debris of coastal reefs to 69 m.
Indo-West Pacific: E. Africa to Taiwan, Australia and Fiji.

BANDTAIL FROGFISH *Antennarius dorehensis*
SIZE: to 6 cm (2 ¹/₄ in.) Frogfishes – Antennariidae
ID: Small; color highly variable, but usually drab shades, **often band on tail.** short stocky rod with tapering unbranched fleshy lure. Solitary and cryptic. Wedge in cracks and crevices of sheltered coral reefs, usually in less than 1 m.
Indo-Pacific: E. Africa to Indonesia, Papua New Guinea and Solomon Is. to French Polynesia. S.W. Japan to Australia.

FRECKLED FROGFISH *Antennarius coccineus*
SIZE: to 12 cm (4 3/4 in.) Frogfishes – Antennariidae
ID: Lack tail base (dorsal and anal fins end at start of tail); short rod half height of adjacent dorsal ray with white-tufted lure, darkish blotch below rear dorsal fin. Reef crevices or rubble of coastal, lagoon and outer reefs to 75 m, but usually shallow.
Indo-Pacific: Red Sea and E. Africa to Indonesia, Philippines, Micronesia and C. America. – S.W. Japan to Australia.

Freckled Frogfish – Variation
ID: Colors variable, usually mottled red to pink, yellowish, or reddish brown; darkish blotch below rear dorsal fin often indistinct.

SPOTFIN FROGFISH *Antennarius nummifer*
SIZE: to 12.5 cm (5 in.) Frogfishes – Antennariidae
ID: Variable colors from white to yellow, orange, red, brown or tan with contrasting blotches; large pale-ringed dark spot on base of rear dorsal fin. Inshore and offshore reefs to 176 m.
Indo-West Pacific: Red Sea to Micronesia – S.W. Japan to N.E. Australia and Samoa.

Spotfin Frogfish – Lure and 2nd Dorsal Spine Detail
ID: Lure and 2nd dorsal spine are about the same length; lure tip bulbous with filaments; **2nd dorsal spine slender with large fleshy base and covered with clusters of short filaments.**

ROSY FROGFISH *Antennarius rosaceus*
SIZE: to 5.8 cm (2 1/4 in.) Frogfishes – Antennariidae
ID: Tan to brown or reddish with mottling; usually large pale edged dark spot below dorsal fin; lure long tipped with tuft of filaments; **2nd dorsal spine long and slender, may have long filaments.** On reefs to 130m, most common 30-40m.
Indo-West Pacific: Red Sea to Micronesia. – S.W. Japan to N.E. Australia and Samoa.

OCELLATED FROGFISH *Nudiantennarius subteres*
SIZE: to 7.8 cm (3 in.) Frogfishes – Antennariidae
ID: Highly variable colors; usually large pale-ringed dark spot on rear dorsal fin that extends onto back, occasionally have other large ringed spots; lure short about half the length of second dorsal spine; 2nd dorsal spine long, rod-like, often with short filaments. Shallow sand and rubble areas to 128 m.
Localized: E. Indonesia to Philippines.

383

Frogfishes – False Scorpionfishes – Lionfishes

SPOT-TAIL FROGFISH   *Lophiocharon trisignatus*
SIZE: to 18 cm (7 in.)   Frogfishes – Antennariidae
ID: Shades of brown to green; curved lure, darkish reticulations on head, dark-edged white spots on tail. Large (2 mm) eggs attach to side of males [pictured]. Solitary. Often under wharf pilings or among debris of coastal reefs in 2-20 m.
Asian Pacific: Singapore, Indonesia, Philippines and Papua New Guinea to N. Australia and Great Barrier Reef.

MARBLE-MOUTH FROGFISH   *Lophiocharon lithinostomus*
SIZE: to 12 cm (4 ³/₄ in.)   Frogfishes – Antennariidae
ID: Mottled green, often resembles algal-cover rock; **rows of 2-4 dark spots on membrane between each pair of tail fin rays.** Often clusters of large eggs are attached to side of males. Solitary. Rubble areas around shallow coastal reefs in 1-10 m.
Asian Pacific: Sabah in Malaysia, Indonesia and Philippines.

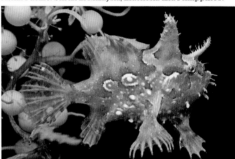

BUTLER'S FROGFISH   *Tathicarpus butleri*
SIZE: to 14 cm (5 ¹/₂ in.)   Frogfishes – Antennariidae
ID: Long rod tipped with hair-like filaments; long separate 2nd dorsal spine, **transparent dorsal and fin membranes,** skin covered with random patches of short filaments, variable colors. Weeds and rubble of coastal reefs in 7-146 m, but usually below 25 m.
Localized: Molucca Is. in Indonesia, Papua New Guinea to N. Australia and Great Barrier Reef.

SARGASSUMFISH   *Histrio histrio*
SIZE: to 14 cm (5 ¹/₂ in.)   Frogfishes – Antennariidae
ID: Shades of brown to yellow with skin flaps; random lines, spots and dusky blotches. Usually solitary; several individuals may inhabit same float of *Sargassum* seaweed near shore to open ocean.
Circumtropical: All tropical seas except E. Pacific.

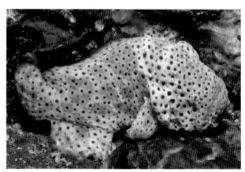

CRYPTIC FROGFISH   *Histiophryne cryptacanthus*
SIZE: to 9.6 cm (3 ³/₄ in.)   Frogfishes – Antennariidae
ID: Lure inconspicuous; dorsal spines embedded in fleshy hump; no tail base, highly variable coloration that blends with surroundings. Solitary. Often on sponges in reef crevices of coastal reefs in 4-130 m.
Asian Pacific: Two populations: W. Papua in Indonesia, Taiwan and Papua New Guinea, also S. and W. Australia.

BEARDED FROGFISH   *Histiophryne pogonius*
SIZE: to 6.5 cm (2 ¹/₂ in.)   Frogfishes – Antennariidae
ID: Whitish gray with numerous dark spots encircled with white rings giving a sponge-like appearance.
Asian Pacific: Indonesia, Philippines and Papua New Guinea.

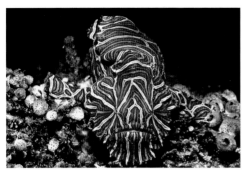

PSYCHEDELIC FROGFISH   *Histiophryne psychedelica*

SIZE: to 11 cm (4 1/4 in.)   Frogfishes – Antennariidae

ID: Yellowish tan to peach to red with complex pattern of swirling white stripes and bands radiating from eyes and continuing onto body and tail. Solitary; often in tight recesses, crevices and holes in small coral heads and rubble in 5-10 m.

Localized: Known only from Ambon and Bali in Indonesia.

FALSE SCORPIONFISH   *Centrogenys vaigiensis*

SIZE: to 25 cm (10 in.) False Scorpionfishes – Centrogennidae

ID: Mottled brown with numerous dark blotches; dark bands on fins except dorsal which has spots and long spines, **nostrals fringed by flaps between eyes.** Resembles a scorpionfish, possibly to discourage predators. Solitary or groups. Estuaries and silty coastal reefs in 2-10 m.

Asian Pacific: Indonesia and S. Japan to Australia.

SHORTFIN LIONFISH   *Dendrochirus brachypterus*

SIZE: to 15 cm (6 in.)   Lionfishes – Scorpaenidae

ID: Red to brown and yellow with fan-like non-filamentous pectoral fins marked with 6-10 dark bands on males (4-6 on females); short skin flap below each eye. Solitary or form small groups. Sand of coastal reefs and lagoons in 2-80 m.

Indo-West Pacific: Red Sea and E. Africa to Indonesia, Philippines, Micronesia and Solomon Is. – S. Japan to Australia and Fiji.

Shortfin Lionfish – Yellow Variation

ID: Usually shades of red to brown, but on rare occasion bright yellow.

ZEBRA LIONFISH   *Dendrochirus zebra*

SIZE: to 20 cm (8 in.)   Lionfishes – Scorpaenidae

ID: White with wide brown body bars; **whitish fan-like pectoral fins with radiating brownish streaks** and short ray filaments. Solitary or form small groups; nocturnal feeders. Coral and rock of sheltered reefs to 75 m.

Indo-West Pacific: E. Africa to Indonesia, Philippines, Micronesia and Solomon Is. – S. Japan to Australia and Fiji.

TWINSPOT LIONFISH   *Dendrochirus biocellatus*

SIZE: to 12 cm (4 3/4 in.)   Lionfishes – Scorpaenidae

ID: Brown body with wide bars and 2-3 wide dark bands; whitish fan-like non-filamentous pectoral fins, pair of eye spots on rear dorsal fin, long tentacle-like skin flap below each eye. Solitary. Caves and under ledges of coral-rich areas to 40 m.

Indo-Pacific: E. Africa to Indonesia, Philippines, Micronesia, Solomon Is. and French Polynesia. - S. Japan to N.W. Australia.

Lionfishes

RED LIONFISH *Pterois volitans*

SIZE: to 38 cm (15 in.)   Lionfishes – Scorpaenidae

ID: Numerous reddish brown to nearly black bands with white lines between; long feather-like pectoral fin rays with light and dark bands, dark spotted dorsal, anal and tail fins. Solitary. Coastal, lagoon and seaward reefs to 50 m.

East Indo-Pacific: Cocos-Keeling Is. to French Polynesia. – S. Japan to Australia. Absent Andaman Sea.

Red Lionfish – Young Juvenile

ID: Translucent body with narrow black bands; pectoral fin rays with reddish brown to pink ocellated spots. As they mature pectoral and dorsal fins become more elongate and body bands, which may be reddish brown to black, widen. Older juveniles and adults may be solitary or form small groups; during day often inside caves, under ledges or around wreckage.

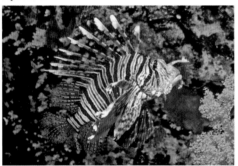

INDIAN LIONFISH *Pterois miles*

SIZE: to 31 cm (12 in.)   Lionfishes – Scorpaenidae

ID: Visually identical to Red Lionfish (previous), distinguished by location and having one less ray in rear dorsal and anal fins and slightly shorter pectoral fin rays. Solitary or form small groups. Coastal, lagoon and seaward reefs in 2-50 m.

Indian Ocean: Red Sea and E. Africa to Andaman Sea and Sumatra in Indonesia.

SPOTFIN LIONFISH *Pterois antennata*

SIZE: to 20 cm (8 in.)   Lionfishes – Scorpaenidae

ID: Pale with numerous red-brown bands of varying width; whitish translucent fan-like **pectoral fins with a few large spots** and long filamentous rays. Solitary or form small groups in caves and crevices. Coastal, lagoon and outer reefs to 50 m.

Indo-Pacific: E. Africa to Indonesia, Philippines, Micronesia, Solomon Is. and French Polynesia. – S. Japan to Australia.

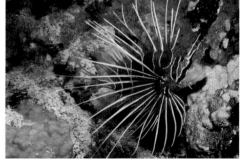

FALSE MOMBASA LIONFISH *Pterois paucispinula*

SIZE: to 18.6 cm (7 1/4 in.)   Lionfishes – Scorpaenidae

ID: Pale with numerous red-brown bands of varying width; **whitish fan-like pectoral fins with semicircular bands formed of large spots** and long filamentous rays. Solitary. Soft coral, sponge and rubble of sheltered reefs in 10-50 m.

Asian Pacific: Indonesia, Philippines and Papua New Guinea. – S.W. Japan to Australia.

CLEARFIN LIONFISH *Pterois radiata*

SIZE: to 18 cm (7 in.)   Lionfishes – Scorpaenidae

ID: Wide brown bars with thin white lines; horizontal white-edged band on tail base, **long filamentous pectoral fin rays with connecting translucent tissue only near base.** Solitary. Coastal and offshore reefs in 3-20 m.

Indo-Pacific: Red Sea and E. Africa to Indonesia, Philippines, Micronesia and French Polynesia. – S.W. Japan to New Caledonia.

CLEARTAIL LIONFISH *Pterois russelli*
SIZE: to 27 cm (10 ¹/₂ in.)   Lionfishes – Scorpaenidae
ID: Numerous brown to reddish brown bars of variable widths on head and body, separated by narrower white bars; median fins clear without dark spots. Solitary. Mud or fine sand bottoms of estuaries, bays and coastal waters in 3-50 m.
Indo-Asian Pacific: E. Africa to Indonesia, Philippines and N. Australia. Absent Papua New Guinea and Solomon Is.

Cleartail Lionfish – Juvenile
ID: Young often with strong reddish hue on pectoral fins, which are more feather-like in appearance than in adult. Usually seen in same areas as adults, often over open sand.

LONGSPINE LIONFISH *Pterois andover*
SIZE: to 23 cm (9 in.)   Lionfishes – Scorpaenidae
ID: Similar to Red Lionfish [previous page] except has longer dorsal spines with pennant-like flaps; 13 instead of 14 pectoral rays, sparsely scattered small black spots on rear dorsal, anal and tail fins. Mud and sand of coastal waters to 50 m.
Localized: Sulawesi to West Papua in Indonesia and Milne Bay in Papua New Guinea.

GURNARD LIONFISH *Parapterois heturura*
SIZE: to 38 cm (15 in.)   Lionfishes – Scorpaenidae
ID: White with brown bars; **thread-like filaments on dorsal spines** and outermost rays of tail, non-filamentous fan-like **pectoral fins with fine blue bands** on interior surface. Solitary, sometimes partially buried. Mud and sand in 3-300 m.
Asian Pacific: Bali, Alor, Ambon in Indonesia and Philippines, north to S. Japan.

BLEEKER'S LIONFISH *Ebosia bleekeri*
SIZE: to 22 cm (8 ³/₄ in.)   Lionfishes – Scorpaenidae
ID: Pale with 8 brown bars and spots between; whitish fan-like pectoral fins marked with semicircular bands and no ray filaments, reddish dorsal spines. Solitary or small groups. Sand and mud slopes in 10-85 m, often in areas of cool upwellings.
Asian Pacific: S. Japan to E. Australia, rare in Indonesia, Philippines and Papua New Guinea.

PYGMY LIONFISH *Brachypterois serrulata*
SIZE: to 15.5 cm (6 in.)   Lionfishes – Scorpaenidae
ID: Pale brown with 4 irregular dark brown bars on side, including narrow bar at tail base; pale-edged dark spot on gill cover, banding on pectoral fins. Solitary. Mud bottoms of estuaries and offshore trawling grounds in 6-82 m.
Indo-Asian Pacific: E. Africa to Indonesia, Philippines and S. Japan.

Scorpionfishes

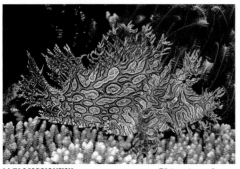

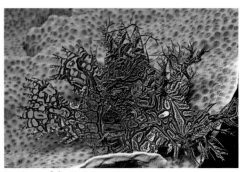

LACY SCORPIONFISH *Rhinopias aphanes*
SIZE: to 24 cm (9 1/2 in.) Scorpionfishes – Scorpaenidae
ID: **Maze-like pattern of dark lined oblong designs;** long highly branched cirrus above each eye, fins have intricate lace-like structure, especially pectoral and tail fins. Solitary or pairs. Seaward reefs in 5-30 m.
Asian Pacific: S. Japan to Papua New Guinea, Great Barrier Reef and New Caledonia. Absent Indonesia and Philippines.

Lacy Scorpionfish – Variation
ID: The three genus members presented on this page are, at times, difficult to distinguish. All three have up-turned mouths, are variable in color, markings and ornamentation and "walk" using pectoral and ventral fins. Can often be differentiated by location: The Lacy is absent from Indonesia and Philippines, but present in Papua New Guinea, which lacks populations of the Weedy or Paddle-flap.

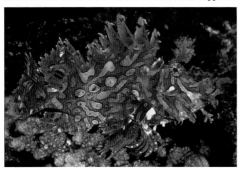

WEEDY SCORPIONFISH *Rhinopias frondosa*
SIZE: to 20 cm (8 in.) Scorpionfishes – Scorpaenidae
ID: Colors, markings and ornamentation variable; branched cirrus above eye can be complex or simple, fins may have lace-like structure or nearly solid, body patterns, if present, tend to be circular. Seaward reefs in 10-297 m.
Indo-Asian Pacific: E. Africa to Indonesia, Philippines and Caroline Is. in Micronesia, north to Japan. Absent Papua New Guinea.

Weedy Scorpionfish – Variation
ID: Most frequently confused with Lacy [previous], but body patterns of the Weedy are typically more circular and less complex; branched filaments on fins, snout and lower jaw generally shorter and less ornate. Tail and pectoral fins tend to be solid although occasionally somewhat lacy in structure. Unlike the others, the cirri above eyes of Paddle-flap [next] are solid, and when present, filaments are short.

PADDLE-FLAP SCORPIONFISH *Rhinopias eschmeyeri*
SIZE: to 23 cm (9 in.) Scorpionfishes – Scorpaenidae
ID: Body and fins generally smooth, unpatterned and solid color, but occasionally blotched; **flat paddle-like, generally unfilamented, flap above each eye,** often series of fleshy tabs along middle of rear body.
Indo-Asian Pacific: Seychelles and Mauritius to Vietnam, Indonesia and Philippines. – Japan to Australia.

Paddle-flap Scorpionfish – Variation
ID: Color highly variable including red, pink, orange, yellow, blue and lavender to purplish; occasionally randomly blotched. Solitary. Coral reefs and rubble bottoms in 18-55 m.

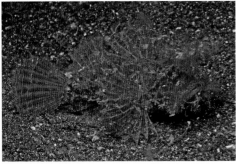

AMBON SCORPIONFISH *Pteroidichthys amboinensis*
SIZE: to 12 cm (4³/₄ in.)   Scorpionfish – Scorpaenidae
ID: Commonly in shades of brown; **extremely long cirrus above each eye,** fan-like pectoral fins, usually numerous skin flaps and filaments on head, jaw, body and fins. Solitary. Algae, sand and mud bottoms in 3-40 m.
Indo-Asian Pacific: Red Sea to S. Japan, Indonesia, Philippines and Papua New Guinea.

Ambon Scorpionfish – Variation
ID: The length of skin flaps and filaments is highly variable, may be absent in habitats without filamentous algae.

MOZAMBIQUE SCORPIONFISH *Parascorpaena mossambica*
SIZE: to 12 cm (4³/₄ in.)   Scorpionfishes – Scorpaenidae
ID: Small with **well developed cirri above eyes;** blotchy shades of white to brown and lavender, no dark spot on dorsal fin. Solitary. Inside crevices or on sand and rubble bottoms of reef flats, lagoons and channels to 20 m.
Indo-Pacific: E. Africa to Indonesia, Philippines, Papua New Guinea, Solomon Is. and French Polynesia. – S.W. Japan to Australia.

PAINTED SCORPIONFISH *Parascorpaena picta*
SIZE: to 15 cm (6 in.)   Scorpionfishes – Scorpaenidae
ID: Mottled red to brown to greenish brown; large rounded head with short blunt snout, reddish bands in iris, cirri above eyes not always present, **bands on lips.** Solitary. Inside crevices of coastal reefs to 15 m.
East Indo-West Pacific: Sri Lanka to Indonesia, Philippines, Papua New Guinea and Solomon Is. – Taiwan to N. Australia and Fiji.

FILAMENTOUS SCORPIONFISH *Hipposcorpaena filamentosus*
SIZE: to 5 cm (2 in.)   Scorpionfishes – Scorpaenidae
ID: Colors variable from whitish to reddish brown often with yellow or white highlights; 2 dark bars on side below 1st dorsal fin and other below 2nd dorsal fin. Silty inshore reefs in 10 to 20 m. Known from only a few specimens.
Asian Pacific: Indonesia, Philippines and Papua New Guinea.

DECOY SCORPIONFISH *Iracundus signifer*
SIZE: to 13 cm (5 in.)   Scorpionfishes – Scorpaenidae
ID: Red to brown; **unique 1st dorsal fin mimics shape of small fish when extended,** fin used as lure. Solitary. Seaward reefs in 10-50 m.
Indo-Pacific: Oceanic islands from Mauritius to Hawaii and Pitcairn Is. in southeastern Pacific. Rare in Asian Pacific.

Scorpionfishes

GUAM SCORPIONFISH *Scorpaenodes guamensis*
SIZE: to 12.5 cm (5 in.) Scorpionfishes – Scorpaenidae
ID: Relatively elongate body; blotchy shades of brown with **distinct dark spot on gill cover.** Solitary and cryptic; nocturnal feeders. Rubble, rocks and coral crevices of reef flats and lagoons to 12 m.
Indo-Pacific: Red Sea and E. Africa to Indonesia, Micronesia, Solomon Is. and French Polynesia. – S. Japan to Australia.

HAIRY SCORPIONFISH *Scorpaenodes hirsutus*
SIZE: to 5.5 cm (2 1/4 in.) Scorpionfishes – Scorpaenidae
ID: Blotchy red and brown; **pair of dark bands radiating from lower eye,** golden iris with numerous black marks. Solitary, rarely in open except at night. Caves and crevices of lagoon and outer reef slopes in 5-50 m.
Indo-Pacific: Red Sea and E. Africa to Indonesia, Philippines, Micronesia, Hawaii and Pitcairn Is. – S.W. Japan to Australia.

CHEEKSPOT SCORPIONFISH *Scorpaenodes evides*
SIZE: to 10.5 cm (4 in.) Scorpionfishes – Scorpaenidae
ID: Mottled brown to reddish brown with variable brown markings on head; **dark spot on lower edge of gill cover.** Solitary and nocturnal. Caves and crevices or among rocks of seaward reefs to 40 m, often in areas with cool upwellings.
Indo-Pacific: Red Sea and E. Africa to Hawaii and French Polynesia. – Rare in Asian Pacific.

BLOTCHFIN SCORPIONFISH *Scorpaenodes varipinnis*
SIZE: to 7.5 cm (3 in.) Scorpionfishes – Scorpaenidae
ID: Blotchy red to reddish brown with **white bar followed by red bar tail base edged in black.** Solitary, in open at night. Reef crevices of reef flats, lagoons and seaward reefs to 200 m.
Indo-Asian Pacific: Red Sea and E. Africa to Indonesia, Philippines, Micronesia, Papua New Guinea and Solomon Is. – S. Japan to Australia and New Caledonia.

SHORTFIN SCORPIONFISH *Scorpaenodes parvipinnis*
SIZE: to 13 cm (5 in.) Scorpionfishes – Scorpaenidae
ID: Blotchy shades of red to brown; **very short dorsal fin** (longest spine usually less than the eye diameter). Solitary. Reef crevices of coastal, lagoon and outer reefs to 46 m.
Indo-Pacific: Red Sea and E. Africa to Indonesia, Philippines, Micronesia, Papua New Guinea, Solomon Is., Hawaii and French Polynesia. – S.W. Japan to Australia.

DWARF SCORPIONFISH *Scorpaenodes kelloggi*
SIZE: to 4.8 cm (2 in.) Scorpionfishes – Scorpaenidae
ID: Reddish brown, often with 4 diffuse body bars; dark bar on rear of tail base, **dark-edged white bands radiate from eye.** Solitary. Coral and rubble in 1-24 m.
Indo-Pacific: E. Africa to Indonesia, Philippines, Micronesia and French Polynesia.

BARCHIN SCORPIONFISH *Sebastapistes strongia*

SIZE: to 9.5 cm (3 3/4 in.)   Scorpionfishes – Scorpaenidae

ID: Red to brownish with white blotches and bars; often broad zone of white across nape to pectoral fin, **red and white bands on lower jaw**. Solitary and cryptic, in open only at night. Sheltered coastal reefs, reef flats and lagoons to 37 m.

Indo-Pacific: Red Sea and E. Africa to Indonesia, Philippines, Micronesia, Papua New Guinea, Solomon Is. and French Polynesia.

MAURITIUS SCORPIONFISH *Sebastapistes mauritiana*

SIZE: to 9.5 cm (3 3/4 in.)   Scorpionfishes – Socrpaenidae

ID: **Deep pit on top of head behind eyes** and pair of strong bony ridges associated with spine between eyes; color variable, usually red to greenish brown with white blotches. Solitary and cryptic. Exposed reef flats and lagoon margins to 5 m.

Indo-Pacific: E. Africa to Indonesia, Philippines, Micronesia, Papua New Guinea and French Polynesia. – S.W. Japan to Australia.

YELLOW-SPOTTED SCORPIONFISH *Sebastapistes cyanostigma*

SIZE: to 7 cm (2 3/4 in.)   Scorpionfishes – Scorpaenidae

ID: Pink to reddish body with numerous tiny white spots, yellow blotches and yellowish fins. Solitary or from small groups. Branches of *Pocillopora* coral heads in surge areas of outer reefs in 2-15 m.

Indo-West Pacific: Red Sea and E. Africa, Indonesia, Micronesia, Solomon Is. and Line Is. – S.W. Japan to Australia and Fiji.

DEVIL SCORPIONFISH *Scorpaenopsis diabolus*

SIZE: to 23.4 cm (9 1/4 in.)   Scorpionfishes – Scorpaenidae

ID: Pronounced hump on back; highly variable color, but generally drab shades blending with surroundings, **wide dark band on tail**. Solitary or in pairs. Rubble or weed bottoms of coastal, lagoon and seaward reefs in 2-70 m.

Indo-Pacific: Red Sea and E. Africa to Indonesia, Micronesia, Philippines, Solomon Is., French Polynesia. – S. Japan to Australia.

Devil Scorpionfish – Phase

ID: Colors can vary dramatically according to surroundings; some of most colorful examples display patches of pink, red, orange and blue are found on coralline algal bottoms.

Devil Scorpionfish – Brown Phase

ID: Inner surface of pectoral fin is brilliant yellow and orange with isolated black spots, similar Flasher Scorpionfish [next page] has broad black border; length of snout about 1.3 – 1.5 eye diameter, similar Flasher's snout about equal to diameter of eye.

Scorpionfishes

FLASHER SCORPIONFISH *Scorpaenopsis macrochir*
SIZE: to 13.6 cm (5 1/4 in.)   Scorpionfishes – Scorpaenidae
ID: Similar to Devil Scorpionfish [previous], but much smaller maximum size, shorter snout and less pronounced hump on back; color highly variable. Solitary or in pairs. Rubble, weed and rocks of coastal reefs to 80 m.
Indo-Pacific: Mauritius to Indonesia, Philippines, Micronesia, Solomon Is. and French Polynesia. – S.W. Japan to Australia.

Flasher Scorpionfish – Orange Phase
ID: All species of *Scorpaenopsis* at times display colors that effectively mimic sponge growth. Tolerate brackish environments near river mouths; also occur on deep offshore trawling grounds.

Flasher Scorpionfish – Gray-brown Phase
ID: Inner surface of pectoral fin is brilliant yellow-orange with isolated black spots and wide black border. Similar Devil Scorpionfish [previous] lacks this border.

BANDTAIL SCORPIONFISH *Scorpaenopsis neglecta*
SIZE: to 17 cm (6 3/4 in.)   Scorpionfishes – Scorpaenidae
ID: Similar to Flasher Scorpionfish [previous], but bony ridge above eye serrated; head profile not as steep and snout more humped, colors highly variable. Solitary. Open sand and mud bottoms of seaward reefs to 40 m.
East Indo-Asian Pacific: India and E. Andaman Sea to Indonesia and Philippines. – S. Japan to Australia.

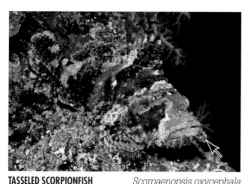

TASSELED SCORPIONFISH *Scorpaenopsis oxycephala*
SIZE: to 36 cm (14 1/4 in.)   Scorpionfishes – Scorpaenidae
ID: Long humped snout; eye cirri absent in adults, **prominent skin tassels on lower head,** small scales (60-65 in lateral row on side). Highly variable color and markings. Solitary. Coral and rocky bottoms to 43 m.
Indo-Asian Pacific: Red Sea and E. Africa to Indonesia, Philippines, Micronesia and Taiwan (absent Papua New Guinea).

Tasseled Scorpionfish – Variation
ID: Variable colors and patterns depending on surroundings.

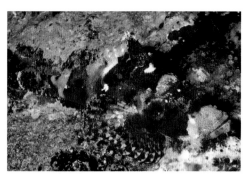

POSS'S SCORPIONFISH *Scorpaenopsis possi*

SIZE: to 19.4 cm (7 1/2 in.)   Scorpionfishes – Scorpaenidae

ID: Similar to Tasseled Scorpionfish and Papuan Scorpionfish [previous and next], but shorter snout and only 17 pectoral rays (versus 19 or 20); eye cirri absent or shorter than eye diameter, usually drab shades. Coral reefs to 40 m.

Indo-Pacific: Red Sea and E. Africa to Indonesia, Philippines, Micronesia, Papua New Guinea to Cook Is. – S.W. Japan to Australia.

Poss's Scorpionfish – Juvenile

SIZE: to 9 cm (3 1/2 in.)

ID: Young frequently display white patches on nape and below eye. Solitary. Reef crevices and under ledges during day. Mainly seaward reefs and passages.

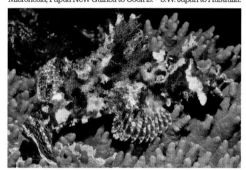

PAPUAN SCORPIONFISH *Scorpaenopsis papuensis*

SIZE: to 19.5 cm (7 1/2 in.)   Scorpionfishes – Scorpaenidae

ID: Very similar to Tasseled Scorpionfish [previous page], but larger scales (48-55 in lateral on side); best distinguished by location except in overlapping range of Indonesia, Palau and Philippines. Solitary. Seaward reefs to 40 m.

Pacific: Indonesia, Philippines, Micronesia, Papua New Guinea, Solomon Is. to French Polynesia. – S.W. Japan to Great Barrier Reef.

Papuan Scorpionfish – Young

SIZE: to 14 cm (5 1/2 in.)

ID: Young usually have long cirri above eyes and exaggerated facial skin appendages; color often mottled brownish red, but is highly variable and can be changed to match surroundings.

RAGGY SCORPIONFISH *Scorpaenopsis venosa*

SIZE: to 36 cm (4 1/4 in.)   Scorpionfishes – Scorpaenidae

ID: Long humped snout; eye cirri absent in adults, **prominent skin tassels on lower head,** small scales (60-65 in lateral row on side). Highly variable color and markings. Solitary. Coral and rocky bottoms to 43 m.

Asian Pacific: Red Sea and E. Africa to Indonesia, Philippines, Micronesia and Taiwan (absent Papua New Guinea).

BLUNTSNOUT SCORPIONFISH *Scorpaenopsis obtusa*

SIZE: to 10 cm (4 in.)   Scorpionfishes – Scorpaenidae

ID: Colors variable; blunt face, wide alternating color bands on tail. Known from only a few specimens collected in less than 15 m.

Asian Pacific: Indonesia, Philippines and Papua New Guinea.

Scorpionfishes – Waspfishes – Longfin Waspfishes

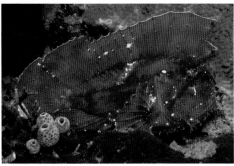

LEAF SCORPIONFISH   *Taenianotus triacanthus*
SIZE: to 10 cm (4 in.)   Scorpionfishes – Scorpaenidae
ID: Strongly compressed head and body; tall dorsal fin begins behind eyes, leafy appendages above eyes, color highly variable. Solitary or in pairs. Coastal, lagoon and outer reefs to 134 m.
Indo-Pacific: E. Africa to Indonesia, Philippines, Solomon Is., Micronesia and French Polynesia. – S.W. Japan to Australia and Fiji.

Leaf Scorpionfish – Yellow Variation
ID: Color variations include white, pink, yellow, tan, brown and black with mottling.

SPINY WASPFISH   *Ablabys macracanthus*
SIZE: to 9 cm (3 1/2 in.)   Waspfishes – Tetrarogidae
ID: Light to dark brown; compressed body, smooth-edged dorsal fin that begins above eye forming a triangular sail-like projection. Solitary or in pairs, sway with water motion. Sand or mud bottoms of coastal reefs in 1-20 m.
East Indo-Asian Pacific: Maldives to Andaman Sea, Indonesia and Philippines.

Spiny Waspfish – Pale Variation
ID: Light brown to nearly white with dark brown mask. Distinguished from similar Cockatoo Waspfish [next] dorsal fin count (15 vs 17).

COCKATOO WASPFISH   *Ablabys taenianotus*
SIZE: to 12 cm (4 3/4 in.)   Waspfishes – Tetrarogidae
ID: Similar to Spiny Waspfish [previous], difficult to distinguish except for higher number of dorsal spines (17 vs 15). Solitary or in pairs. Sand, mud, rubble and weed bottoms of coastal reefs to 20 m.
West Pacific: Andaman Sea to Indonesia, Philippines, Micronesia and Solomon Is. – S. Japan to Australia and Fiji.

Cockatoo Waspfish – Variation
ID: Dark brown with white "face" is common. All variations may also have one or more white spots on side as displayed on previous variation.

LONGSPINE WASPFISH   *Paracentropogon longispinis*
SIZE: to 11 cm (4 1/4 in.)   Waspfishes – Tetrarogidae
ID: Shades of reddish brown with white spots and blotches on body; **alternating dark and light diagonal banding on dorsal fin,** dorsal fin with V-shaped notches between spines. Solitary. Silt and sand bottoms of coastal reefs in 7-30 m.
East Indo-Asian Pacific: India to Indonesia, Philippines, north to Taiwan.

Longspine Waspfish – Variation
ID: Mottled reddish brown with only a few white spots and a white "face." This species is frequently found among weed beds during day, but hunts in open for small crustaceans at night.

BANDTAIL WASPFISH   *Paracentropogon zonatus*
SIZE: to 5.5 cm (2 1/4 in.)   Waspfishes – Tetrarogidae
ID: Color variable, usually mottled brown to reddish brown, often whitish face; white blotch on upper side, **irregular whitish bar under rear 2nd dorsal fin** and another on tail. Rocky bottoms and coral reef crevices in 10-40 m.
Asian Pacific: Indonesia and Philippines.

Bandtail Waspfish – Variation
ID: Often reddish brown with darker brown blotches. Color pattern of this variation blends with surroundings.

WHITEFACE WASPFISH   *Richardsonichthys leucogaster*
SIZE: to 7 cm (2 3/4 in.)   Waspfishes – Tetrarogidae
ID: Blotchy shades of red to brown; dorsal fin deeply incised between each spine, **row of rear-pointing spines on gill cover.** Solitary. Mud and sand of coastal reefs in 3-18 m.
Indo-Asian Pacific: E. Africa to Andaman Sea, Indonesia, Philippines and N. Australia.

LONGFIN WASPFISH   *Apistus carinatus*
SIZE: to 16.5 cm (6 1/2 in.)   Longfin Waspfishes – Apistidae
ID: Shades of brown; **elongate pale-edged black area on spinous dorsal fin,** large wing-like pectoral fins, chin "whiskers". Solitary or form groups, sometimes partly bury. Sand or mud bottoms in 10-45 m.
Indo-Asian Pacific: Red Sea and E. Africa to Malaysia, Indonesia and Philippines. – S. Japan to Australia.

Stonefishes

REEF STONEFISH *Synanceia verrucosa*

SIZE: to 38 cm (15 in.)   Stonefishes – Synanceiidae

ID: Globular, appear as an algae-covered stone; prominent warts and skin flaps, eyes far apart with deep pit between. Solitary or small groups, venomous fin spines deadly. Often bury. Reef flats and outer slopes to 20 m.

Indo-Pacific: Red Sea and E. Africa to Indonesia, Philippines, Micronesia and French Polynesia. – S.W. Japan to Australia.

Reef Stonefish – Variation

ID: An ambush feeder that typically modifies color and skin flaps to blend with surrounds; however, it may display brilliant shades of red, lavender or orange. Dorsal, ventral and anal fin spines are highly venomous, wounds can be fatal.

ESTUARINE STONEFISH *Synanceia horrida*

SIZE: to 30 cm (12 in.)   Stonefishes – Synanceiidae

ID: Globular and poorly defined, appear as algae-covered stone; prominent warts, **bony ridge above and between eyes.** Solitary, venomous spines deadly. Often partially bury in sand and rubble of coastal reefs and estuaries to 40 m.

East Indo-Asian Pacific: India to Andaman Sea, Indonesia, Philippines, Papua New Guinea, Solomon Is. and Australia.

Estuarine Stonefish – Variation

ID: An ambush feeder that typically modifies color and skin flaps to blend with surrounding; however, it may display shading of red, yellow and orange. Dorsal, ventral and anal fin spines are highly venomous, wounds can be fatal.

ROUGH-HEAD STINGFISH *Minous trachycephalus*

SIZE: to 11.2 cm (4 1/2 in.)   Stonefishes – Synanceiidae

ID: Variable shades of gray and brown; **reddish hue on cheek and pectoral fin base** and dark spot behind jaw, faint bands on front of pectoral fins; 1st dorsal fin spine short. Sand and mud bottoms in 10-50 m.

East Indo-Asian Pacific: Sri Lanka to Indonesia, Philippines and Taiwan.

MANY-BARBED STINGFISH *Choridactylus multibarbus*

SIZE: to 12 cm (4 3/4 in.)   Stonefishes – Synanceiidae

ID: Large protruding eyes and blunt snout; 3 lowest pectoral fin rays unattached from fin are used for "walking", variable colors, may be marbled or blotched. Solitary. Sand and mud bottoms of coastal waters in 10-40 m.

Asian Pacific: E. Andaman to Indonesia and Philippines.

Odd-Shaped Bottom Dwellers

SPINY DEVILFISH　　　　　*Inimicus didactylus*
SIZE: to 18 cm (7 in.)　　Stonefishes – Synanceiidae
ID: Upturned snout and bulbous eyes set on top of head; fan-like pectoral fins, dorsal fin with isolated spiky spines except 1st 3 connected with membrane. Solitary or in pairs. Sheltered inshore areas of sand, rubble, silt and mud 5-55 m.
Asian Pacific: Thailand and Vietnam to Indonesia, Philippines, Micronesia, Solomon Is., New Caledonia and Vanuatu.

Spiny Devilfish – Variation
ID: Color highly variable, often drab taking on the appearance of sand where they spend much of the time buried with only their venomous dorsal fins exposed. Move along surface of sand using two separated lower spines of pectoral fins.

Spiny Devilfish – Variation
ID: Colors often bright including yellow, pink, red, lavender and orange.

Spiny Devilfish – Variation
ID: Flare pectoral fins when disturbed. **Large dark semicircle around base of spread pectoral fin followed by a wide pale band and a colorful outer band frequently with white markings.**

CALEDONIAN DEVILFISH　　*Inimicus caledonicus*
SIZE: to 22 cm (8 ¾ in.)　　Stonefishes – Synanceiidae
ID: Similar to Spiny Devilfish [previous] and Spotted Devilfish [next]; best **distinguished by pale to yellow inner surface of pectoral fin marked with dark band through middle and dark area around base.** Solitary. Sand, rubble and mud in 15-60 m.
Asian Pacific: Andaman Sea, Great Barrier Reef and New Caledonia.

SPOTTED DEVILFISH　　　*Inimicus sinensis*
SIZE: to 18 cm (7 in.)　　Stonefishes – Synanceiidae
ID: Similar to Spiny Devilfish and Caledonian Devilfish [previous]; best **distinguished by dark inner surface of pectoral fin with large pale to yellow spots (no bands of color).** Solitary. Sand, rubble and mud in 5-90 m.
East Indo-Asian Pacific: India to W. Australia, Java in Indonesia and Philippines, north to Taiwan and S. China.

Flatheads

CROCODILE FLATHEAD   *Cymbacephalus beauforti*
SIZE: to 47 cm (19 in.)   Flatheads – Platycephalidae
ID: Elongate flattened crocodile-like snout and head; earthtone camouflage markings often with colorful highlights. Solitary; can slowly change colors to blend with surroundings. Sand, rubble and seagrass near mangroves and reefs to 12 m.
Asian Pacific: Indonesia, Philippines, Micronesia, Papua New Guinea and Solomon Is. to New Caledonia.

Crocodile Flathead – Black Variation
ID: Although Crocodile Flatheads typically depend on camouflaged patterns to match the bottom, they are commonly sighted in a black variation.

SMALLEYED FLATHEAD   *Cymbacephalus bosschei*
SIZE: to 44 cm (17 in.)   Flatheads – Platycephalidae
ID: Pale brown with numerous small dark brown spots; dorsal fin spines marked with alternating black and white rings, **distinctive elongate dark blotches on upper and lower margins of tail.** Solitary. Sand bottoms near reefs in 10-50 m.
Asian Pacific: Singapore to Indonesia, Philippines, Papua New Guinea and N. Australia.

FRINGE-EYED FLATHEAD   *Cymbacephalus nematophthalmus*
SIZE: to 58 cm (23 in.)   Flatheads – Platycephalidae
ID: Brown with 4-5 poorly defined dark saddles on back; **dark area toward front of 1st dorsal fin,** fins with dark diagonal bands, long cirrus centered above each eye. Solitary. Sand, rocks, alage and mangroves to 1-5 m.
Asian Pacific: Andaman Sea to Malaysia, Indonesia, Philippines, Papua New Guinea and Solomon Is. to Australia.

LONGSNOUT FLATHEAD   *Thysanophrys chiltonae*
SIZE: to 22 cm (8 3/4 in.)   Flatheads – Platycephalidae
ID: Pale grayish brown to tan with darker brown blotches and scattered white spots; 4-5 indistinct brown saddles on rear half of body, brown bar below eye, narrow space between eyes. Solitary. Sandy fringe of coral reefs in 5-38 m.
Indo-Pacific: Red Sea and E. Africa to Indonesia, Micronesia, Solomon Is. and French Polynesia. – S.W. Japan to Australia.

CELEBES FLATHEAD   *Thysanophrys celebica*
SIZE: to 15 cm (6 in.)   Flatheads – Platycephalidae
ID: Mottled tan with irregular dark brown saddles on upper half of body including **broad bar below 1st dorsal fin and smaller wedge-shaped saddle below 2nd dorsal fin;** dark bar below eye. Solitary. Sandy fringe of coral reefs in 10-45 m.
Indo-Asian Pacific: E. Africa to Indonesia, Philippines, Papua New Guinea and Solomon Is. – Taiwan to Great Barrier Reef.

FRINGELIP FLATHEAD   *Sunagocia otaitensis*
SIZE: to 25 cm (10 in.)   Flatheads – Platycephalidae
ID: Flat body mottled tan with fine brown spots mixed with white spotting; **brown and white bars on lower lip,** white fins have black spotting. Solitary, usually bury in sand. Sandy fringe of coral reefs in 3-40 m.
Indo-Pacific: E. Africa to Indonesia, Philippines, Micronesia, Solomon Is. and French Polynesia. – S. Japan to Great Barrier Reef.

BROADHEAD FLATHEAD   *Sunagocia arenicola*
SIZE: to 37 cm (15 in.)   Flatheads – Platycephalidae
ID: Mottled and flecked with shades of brown and gray; commonly display six indistinct dark bars and pale blotches, **wide head with bluntish snout,** no dark black markings on fins. Often bury in sand. Sand around reefs in 3-30 m.
Indo-West Pacific: E. Africa to Indonesia, Philippines, Micronesia and Papua New Guinea. – S.W. Japan to Great Barrier Reef and Fiji.

THORNY FLATHEAD   *Rogadius pristiger*
SIZE: to 21 cm (8 1/4 in.)   Flatheads – Platycephalidae
ID: Brown with 3-4 dark brown saddles on back; **dark 1st dorsal fin with white and black bands on spines,** pectoral fin has white base with dark spots and white margin. Solitary, bury during day. Sand of coastal waters in 10-25 m.
Indo-Asian Pacific: Red Sea and Madagascar to Indonesia, Philippines, Papua New Guinea and New Caledonia.

BLACK-BANDED FLATHEAD   *Rogadius patriciae*
SIZE: to 27 cm (10 3/4 in.)   Flatheads – Platycephalidae
ID: Mottled light brown to tan with dark spotting on pectoral, dorsal and anal fins; ventral fin dusky with narrow white margin, **tail white with blackish blotchs along upper edge and dark bands below.** Coastal sand bottoms in 10-100 m.
Asian Pacific: Indonesia and S.W. Japan to Great Barrier Reef and New Caledonia.

WELANDER'S FLATHEAD   *Rogadius welanderi*
SIZE: to 13 cm (5 in.)   Flatheads – Platycephalidae
ID: Light brown with faint saddles; 1st dorsal fin blackish with white and black bands, black spots on 2nd dorsal fin, bi-lobed iris lappet, **snout long compared to similar Spiny Flathead [next].** Bury in sand near sheltered coral reefs in 5-40 m.
Indo-West Pacific: Reunion I. and N.W. Australia to Indonesia and Micronesia.

SPINY FLATHEAD   *Onigocia spinosa*
SIZE: to 13 cm (5 in.)   Flatheads – Platycephalidae
ID: Gray to reddish brown with 3-4 dark brown cross bands on back and sides; **prominent pale band between dorsal fins and another on tail base,** short snout compared to Welander's Flathead [previous]. Solitary. Sand of coastal reefs in 5-250 m.
Asian Pacific: N.W. Australia to Indonesia, Philippines and S. Japan.

BROADBAND FLATHEAD
Onigocia pedimacula

SIZE: to 11 cm (4 1/4 in.) Flatheads – Platycephalidae

ID: Mottled gray-brown with about 4-6 brown to olive bands across back and sides; a broad band between rear margin of head and start of 2nd dorsal fin and narrow band on base of tail. Solitary. Sand and rubble near reefs in 10-110 m.

Indo-West Pacific: E. Africa to Indonesia, Philippines, Solomon Is. to Great Barrier Reef and Fiji.

JAPANESE FLATHEAD
Inegocia japonica

SIZE: to 25 cm (10 in.) Flatheads – Platycephalidae

ID: Elongate and flattened head and snout; gray-brown back, frequently with 5-6 faint blackish saddles, **dusky area along midside,** dark spots on dorsal, pectoral and tail fins. Solitary. Sand and mud bottoms of coastal waters in 5-85 m.

East Indo-Asian Pacific: India and Sri Lanka to Papua New Guinea. – S. Japan to E. Australia.

SPOTTED FLATHEAD
Cociella punctata

SIZE: to 35 cm (14 in.) Flatheads – Platycephalidae

ID: Brownish gray with numerous small dark spots; **broad dark margin on 1st dorsal fin,** pale tail with dark spots or horizontal streaks. Solitary. Sand or mud bottoms of coastal waters in 3-250 m.

Indo-Asian Pacific: Red Sea and E. Africa to Indonesia, Philippines, Solomon Is. – Taiwan to Australia and Vanuatu.

TRIPLEFIN VELVETFISH
Neoaploactis tridorsalis

SIZE: to 4.5 cm (1 3/4 in.) Velvetfishes – Aploactinidae

ID: Variable coloration, but usually mottled dark gray; **irregular white spots and blotches on pectoral fins,** usually whitish band behind eye, sometimes red to orange head. Solitary. Sand and rubble bottoms, frequently near reefs, in 3-4 m.

Asian Pacific: Indonesia to Great Barrier Reef and New Caledonia.

PHANTOM VELVETFISH
Paraploactis obbesi

SIZE: to 5 cm (2 in.) Velvetfishes – Aploactinidae

ID: Compressed body; dark brown to orange, dorsal fin begins in front of eyes, no scales, **bony knobs on head (instead of spines).** Solitary, well camouflaged. Sand and rubble strewn bottoms of coastal reefs in 10-30 m.

Asian Pacific: Indonesia and Philippines.

GHOST VELVETFISH
Cocotropus larvatus

SIZE: to 7 cm (2 3/4 in.) Velvetfishes – Aploactinidae

ID: Variable from dark brown to tan or white with vague markings; round tail, straight almost vertical front edge of foredorsal fin to snout. Coral reefs and rubble in 4-40 m.

Asian Pacific: Andaman Sea to Indonesia, Papua New Guinea, Micronesia to S.W. Japan.

HELMUT GURNARD   *Dactyloptena orientalis*
SIZE: to 40 cm (16 in.)   Flying Gurnards – Dactylopteridae
ID: Elongate rigid body with antenna-like dorsal fin ray above head; large wing-like pectoral fins with protruding filamentous tips. Solitary, "walk" on bottom with finger-like ventral rays. Sandy areas near reefs to 68 m.
Indo-Pacific: Red Sea and E. Africa to Indonesia, Philippines, Micronesia and French Polynesia. – S. Japan to N. Australia.

Helmut Gurnard – Juvenile
SIZE: to 10 cm (4 in.)
ID: Wing-like pectorals usually marked with bright ocellated spot near base that becomes a dark blotch with age; pectoral fins of both young and adult are typically marked with electric blue wavy lines and spots. When alarmed, fully extend pectoral fins and rapidly swim away.

DRAGON SEA MOTH   *Eurypegasus draconis*
SIZE: to 9.2 cm (3 1/2 in.)   Sea Moths – Pegasidae
ID: Hard bony carapace; long narrow flattened snout and wing-like pectoral fins often with pale border, brown with netted reticulations. "Walk" with finger-like ventral fins. Solitary or in pairs. Sand, rubble and seagrass bottoms to m.
Indo-Pacific: Red Sea and E. Africa to Indonesia, Philippines, Papua New Guinea to French Polynesia. – S. Japan to Australia.

SLENDER SEA MOTH   *Pegasus volitans*
SIZE: to 20 cm (8 in.)   Sea Moths – Pegasidae
ID: Slender with long tapered snout and thin uniformly tapered body; occasional with speckling and dark blotches and dark and pale bands on tail. Solitary or in pairs. Sand, rubble and seagrass bottoms of coastal reefs to 73 m.
Indo-Asian Pacific: E. Africa to Indonesia, Philippines, Papua New Guinea. – S. Japan to Australia.

THREESPINE TOADFISH   *Batrachomoeus trispinosus*
SIZE: to 31 cm (12 1/4 in.)   Toadfishes – Batrachoididae
ID: Brown to gray with dark bands; scorpionfish-like body, but soft scaleless appearance and few spines and skin flaps on head. Solitary. Wedge in reef crevices, often in open at night to 36 m.
Asian Pacific: E. Andaman Sea to Indonesia, Papua New Guinea and Australia.

HUTCHIN'S TOADFISH   *Halophryne hutchinsi*
SIZE: to 17.6 cm (7 in.)   Toadfishes – Batrachoididae
ID: Pale brownish with about five irregular bars from lower dorsal fin to lower side; two dark bands extend from lower front and lower rear eye. Silty reefs in 3-15 m.
Localized: Raja Ampat in Indonesia to southern Luzon in Philippines.

Stargazers – Tongue Sole – Halibuts – Lefteye Flounders

WHITEMARGIN STARGAZER *Uranoscopus sulphureus*
SIZE: to 35 cm (14 in.) Stargazers – Uranoscopidae
ID: Whitish with brown pigmentation and diffuse brown spots on back; large black spot on 1st dorsal fin, white margins on all fins; large, rounded head with upturned mouth and small eyes. Solitary. Sand bottoms of coastal waters in 5-150 m.
Indo-West Pacific: Red Sea to Indonesia, Micronesia, Coral Sea and Fiji.

Whitemargin Stargazer
ID: Members of this family are ambush predators that often bury in the sand with only their eyes, mouth and a small part of their head exposed as they wait for prey to pass by.

RETICULATE STARGAZER *Uranoscopus bicinctus*
SIZE: to 35 cm (14 in.) Stargazers – Uranoscopidae
ID: White to pale gray with **reddish brown to charcoal spots, blotches and reticulations;** 2 wide somewhat indistinct bars on side, large black spot on 1st dorsal fin, large rounded head. Bury in sand. Coastal waters in 5 to 100 m.
Asian Pacific: Indonesia, Philippines and Micronesia.– S.W. Japan to Australia.

ESTUARY BATFISH *Halieutaea indica*
SIZE: to 9 cm (3 1/2 in.) Batfishes – Ogcocephalidae
ID: Tan with pair of large dark diffuse areas on either side of back; bright yellow pectoral fin tips; flattened disk-shaped body, fringe of hair-like filaments on edge of head and body. Solitary. Soft bottoms of coastal waters in 1-50 m.
Localized: Indonesia.

SPECKLED TONGUESOLE *Cynoglossus puncticeps*
SIZE: to 25 cm (10 in.) Tongue Sole – Cynoglossidae
ID: Light to dark brown elongate body marked with fine speckling to dark blotching; fins with thin to broad bands. Solitary and nocturnal. Sand bottoms in 3-30 m.
East Indo-Pacific: India to N. Australia to Philippines and Taiwan.

PACIFIC HALIBUT *Psettodes erumei*
SIZE: to 60 cm (2 ft.) Halibuts – Psettodidae
ID: Light to dark brown with scattering of white or dark speckles; occasionally with broad, dark bands [pictured], large tooth-lined mouth. Solitary and often nocturnal. Sand bottoms in 1-100 m.
Indo-Pacific: E. Africa to Andaman Sea, Indonesia, Philippines and Papua New Guinea. – S.W. Japan to Australia.

PEACOCK FLOUNDER *Bothus mancus*
SIZE: to 48 cm (19 in.)   Lefteye Flounders – Bothidae
ID: Gray to brown with **blue spots and circles;** eyes widely spaced and lower eye in front of upper eye, male has elongate pectoral fin rays. Solitary on sandy bottoms or rocky surfaces to 84 m.
Indo-Pacific: Red Sea and E. Africa to Indonesia, Philippines, Micronesia, Hawaii and Pitcairn Is. – S.W. Japan to Australia.

LEOPARD FLOUNDER *Bothus pantherinus*
SIZE: to 39 cm (15 in.)   Lefteye Flounders – Bothidae
ID: Brown to gray with dark-edged pale spots and florets; large diffuse dark blotch on middle of rear body, eyes about equal to one eye diameter apart and lower eye nearly aligns with upper, male has elongate pectoral fin rays. Solitary. Sand bottoms to 60 m.
Indo-Pacific: Red Sea and E. Africa to Indonesia, Micronesia, Hawaii and French Polynesia. – S. Japan to Australia.

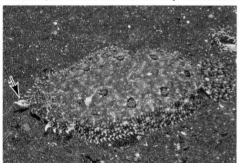

ANGLER FLOUNDER *Asterorhombus intermedius*
SIZE: to 15 cm (6 in.)   Lefteye Flounders – Bothidae
ID: Blotchy pale brown with several large dark edged spots; **1st dorsal fin ray with large fluffy tip used as luring device** (half the length of lure on Longlure Flounder [next]). Solitary. Sand near coastal coral reefs in 1-96 m.
Indo-West Pacific: Red Sea and E. Africa to Indonesia, Philippines and Solomon Is.– S.W. Japan to Great Barrier Reef and Fiji.

LONGLURE FLOUNDER *Asterorhombus filifer*
SIZE: to 13 cm (5 in.)   Lefteye Flounders – Bothidae
ID: Blotchy pale brown with several large dark edged spots; **long tapering white 1st spine with several brown bands, used as a luring device** (twice the length of the Angler Flounder [previous]). Solitary. Sand near coastal reefs in 3-57 m.
Indo-Pacific: E. Africa to Indonesia, Philippines, Great Barrier Reef, Hawaii and French Polynesia.

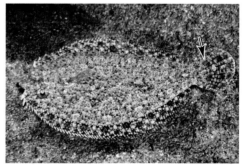

TWOSPOT DWARF FLOUNDER *Engyprosopon grandisquama*
SIZE: to 15 cm (6 in.)   Lefteye Flounders – Bothidae
ID: Brown to gray with numerous scattered dark and light spots and rings; **pair of large prominent black spots on tail with white spot between.** Solitary. Mud and sand bottoms of coastal seas in 7-100 m.
Indo-Asian Pacific: E. Africa to Indonesia, Philippines, Papua New Guinea, Solomon Is. – S Japan to Australia and New Caledonia.

LARGESCALE DWARF FLOUNDER *Engyprosopon macrolepis*
SIZE: to 7.5 cm (3 in.)   Lefteye Flounders - Bothidae
ID: Densely mottled with shades of brown, black and white; **pair of large blackish spots near tail base** and central white spot. Solitary. Sand and mud bottoms, occasionally near reefs, in 3-91 m.
Indo-Asian Pacific: Red Sea and Maldives to Philippines, Papua New Guinea and New Caledonia.

THREESPOT FLOUNDER *Samariscus triocellatus*

SIZE: to 7 cm (2 3/4 in.) Righteye Flounders – Pleuronectidae

ID: Elongate oval body; mottled with irregular light and dark brown markings and **2-3 dark-edged ocelli along midline of body.** Solitary, frequently under ledges. Sand areas of lagoon and seaward reefs in 5-30 m.

Indo-Pacific: E. Africa to Indonesia, Philippines, Micronesia, Hawaii and French Polynesia. – S. Japan to Australia.

COCKATOO FLOUNDER *Samaris cristatus*

SIZE: to 22 cm (8 3/4 in.) Righteye Flounders – Pleuronectidae

ID: Brown, frequently with darker brown stripes and white spots on fins; **elongate white dorsal fin rays above head spread across sand when threatened.** Solitary. Sand and silt bottoms of sheltered bays and estuaries in 5-70 m.

Asian Pacific: E. Africa to Indonesia and Philippines. – Taiwan to Australia and New Caledonia.

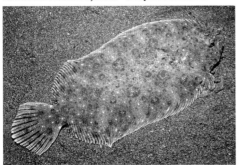

JAVAN FLOUNDER *Pseudorhombus javanicus*

SIZE: to 39 cm (15 in.) Sand Flounders – Paralichthyidae

ID: Blotchy shades of brown with small white spots and variably sized yellow-brown rings; brown spots on rear dorsal, anal and tail fins, occasionally display two dark blotches on lateral line. Coastal sand and mud bottoms in 5-25 m.

East Indo-Asian Pacific: E. India to Andaman Sea, Indonesia and Philippines.

OCELLATED FLOUNDER *Pseudorhombus dupliciocellatus*

SIZE: to 40 cm (16 in.) Sand Flounders – Paralichthyidae

ID: Shades of brown with reticulum of lines, rings and spots; **2-4 pairs of large black spots ringed with white dots.** Solitary. Sand and mud bottoms of coastal seas commonly in 50-150 m, but occasionally in shallows.

Asian Pacific: Andaman Sea to Indonesia and Philippines (primarily on continental shelfs). – S. Japan to Australia.

AMBON SOLE *Aseraggodes suzumotoi*

SIZE: to 9 cm (3 1/2 in.) Soles – Soleidae

ID: Shades of brown with numerous, variable-sized, irregular, dark-edged white spots and vermiculations, including several that form doughnut-shaped rings. Occasionally in aggregations. Sand flats, coastal sand slopes and estuaries in 4-26 m.

Localized: Bali, Sulawesi and Molucca Is. in Indonesia.

STRANGE SOLE *Aseraggodes xenicus*

SIZE: to 6.8 cm (2 3/4 in.) Soles – Soleidae

ID: Light brown with numerous variable-sized irregular, dark-edged white spots and vermiculations. Pattern variable; white markings sometimes vague or obscure. Sand bottoms including reef flats and tide pools to 10 m.

Asian Pacific: Andaman Sea to Indonesia, Philippines Micronesia and S.W. Japan.

WHITE SOLE *Aseraggodes albidus*
SIZE: to 3.5 cm (1¹/₂ in.) Soles – Soleidae
ID: White with no markings; dorsal, anal and tail fins translucent. Known and scientifically described from a single specimen collected on dark silty sand in 15 m depth.
Localized: Known only from Lembeh Strait in N. Sulawesi, Indonesia.

BANDED SOLE *Soleichthys heterorhinos*
SIZE: to 14 cm (5 ¹/₂ in.) Soles – Soleidae
ID: Variable; white to tan and brown with numerous dark brown lines or narrow bars across body and fins, **black outer border on pale tail fin.** Sand and coral areas of coastal reefs to 10 m.
Indo-West Pacific: Red Sea to Indonesia, Philippines, Micronesia, Papua New Guinea, Solomon Is. – S. Japan to Australia and Fiji.

Banded Sole – Variation
ID: Pale yellowish tan with nearly black, thin, uneven and often wavy bars across fins.

SPOTTED-TAIL SOLE *Zebrias zebra*
SIZE: to 26 cm (10 in.) Soles – Soleidae
ID: Pale brown with 20-25 dark-edged brown cross-bars more or less arranged in pairs; tail black with pale yellow spots, encircled with white. **encircling fins with thin alternating dark and pale rays.** Solitary and nocturnal. Sand and silt bottoms of sheltered bays, estuaries and coastal in 5-40 m.
Asian Pacific: Indonesia and Philippines to Japan.

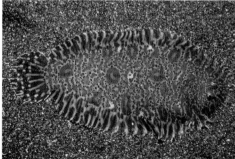

UNICORN SOLE *Aesopia cornuta*
SIZE: to 20 cm (8 in.) Soles – Soleidae
ID: About 12-16 brown and pale bars across head across body; **encircling fins with wide alternating black and white bands,** black tail with spots and lines, occasionally about four whitish spots between head and tail. Sand and mud bottoms in 3-100 m.
Indo-Asian Pacific: E. Africa to Papua New Guinea. – S. Japan to N. E. Australia.

CARPET SOLE *Liachirus melanospilos*
SIZE: to 15 cm (6 in.) Soles – Soleidae
ID: Body shades of brown with dark scattering of irregular brown blotches and circles and **several irregular dark-edged brown blotches;** encircling fins with alternating dark and light bands. Solitary. Silt or sand bottoms of estuaries and coastal sand slopes in 4-40 m.
Asian Pacific: Indonesia and Philippines, north to Japan.

Soles – Clingfishes – Coral Crouchers

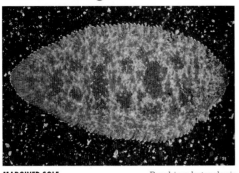

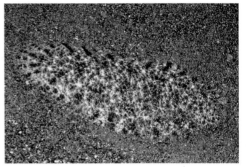

MARGINED SOLE — *Brachirus heterolepis*
SIZE: to 38 cm (15 in.) — Soles – Soleidae
ID: Blotchy brown over tan to white undercolor; **edging of white tips on fin rays around head and body,** clusters of hair-like skin flaps on upper surface of body. Solitary and nocturnal. Sand patches of coastal waters in 3-30 m.
Localized: Sumatra to West Papua in Indonesia.

BLOTCHED SOLE — *Brachirus marmoratus*
SIZE: to 35 cm (14 in.) — Soles – Soleidae
ID: Mottled tan to light gray with dark brown to blackish blotches; scattering of small white spots over body, **series of dark blotches encircling body.** Sand bottoms in 2-15 m.
Localized: Java, Bali, Flores and Sulawesi in Indonesia.

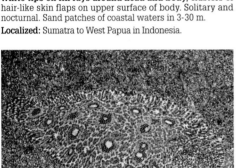

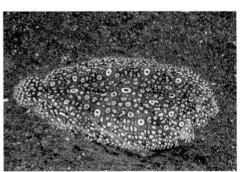

HOOK-NOSE SOLE — *Heteromycteris hartzfeldii*
SIZE: to 13 cm (5 in.) — Soles – Soleidae
ID: Light tan to dark brown with maze of whitish reticulations; **black blotches with numerous pale dots and occasionally large white spots inside.** Sand bottoms on coastal slopes in 2-15 m.
Asian Pacific: Indonesia and Philippines.

PEACOCK SOLE — *Pardachirus pavoninus*
SIZE: to 30 cm (12 in.) — Soles – Soleidae
D: Gray to brownish with **sprinkling of gold spots** interspersed with numerous larger irregular pale spots (often containing one or more small black dots) and donut-shaped circles. Sand or silty bottoms in 1-40 m.
East Indo-West Pacific: Sri Lanka to Andaman Sea, Indonesia and Solomon Is. – Japan to Great Barrier Reef and Fiji.

URCHIN CLINGFISH — *Diademichthys lineatus*
SIZE: to 5 cm (2 in.) — Clingfishes – Gobiesocidae
ID: Red to dark red-brown; pair of white to yellow stripes from head to tail, long spatulate snout and elongate body. Solitary, shelter among spines of sea urchins or in branching corals. Coral reefs to 20 m.
Indo-West Pacific: Mauritius to Indonesia, Philippines, Solomon Is. – S.W. Japan to N. Australia, New Caledonia and Fiji.

MIMIC CLINGFISH — Undescribed
SIZE: to 3 cm (1 1/4 in.) — Clingfishes – Gobiesocidae
ID: Dark brown to black with 3 wide white stripes from snout to tail base. Believed to bury in rubble during much of day. Shelter among *Diadema* sea urchins toward late afternoon. Rubble slopes from 5-15 m.
Asian Pacific: Bali, Flores, Alor and Lembeh Strait in Indonesia, also Manus in Papua New Guinea.

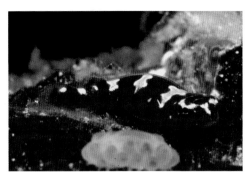

RED CLINGFISH *Pherallodus indicus*

SIZE: to 3.5 cm (1 1/2 in.) Clingfishes – Gobiesocidae

ID: Purple to red with white markings on upper head and irregular white to goldish bars of varying size across back. Sometimes commensal with sea urchins. Coral reefs to 5 m.

Pacific: Indonesia to Japan and French Polynesia.

CRINOID CLINGFISH *Discotrema crinophilum*

SIZE: to 3 cm (1 1/4 in.) Clingfishes – Gobiesocidae

ID: Dark brown to dark red to nearly black with white to yellow stripe on back from nose to tail and stripe from eye to tail on each side; often row of dark spots between stripes. Usually commensal with crinoids. Coral reefs in 5-20 m.

West Pacific: Indonesia, Philippines, S.W. Japan to New Caledonia, Vanuatu and Fiji.

DOUBLESTRIPED CLINGFISH *Lepadichthys lineatus*

SIZE: to 2.8 cm (1 in.) Clingfishes – Gobiesocidae

ID: Wide stripe on side from eye to tail, another below on side of belly, a 3rd stripe extends down center of back; spots between stripes, display color of host crinoid. Solitary or in pairs. Coral reefs in 5-25 m.

Indo-Asian Pacific: Red Sea to Indonesia and Papua New Guinea.

ONELINE CLINGFISH
Discotrema monogrammum

SIZE: to 2.6 cm (1 in.)

Clingfishes-Gobiesocidae

ID: Black to pale reddish brown; wide white to yellow stripe runs along each side of body from snout through eye to tail. Typically display color of host crinoid. Solitary or in pairs. Coral reefs in 8-20 m.

Asian Pacific: Indonesia, Philippines, Papua New Guinea and Great Barrier Reef.

SPOTTED CROUCHER
Caracanthus maculatus

SIZE: to 5 cm (2 in.)

Coral Crouchers – Caracanthidae

ID: Uniform light gray, often with pink tints, and numerous small red to dark maroon spots; body covered with fuzzy appearing papillae. Inhabit branches of *Acropora, Stylophora* and *Pocillopora* corals. Coral reefs in 3 - 15 m.

East Indo - Pacific: Cocos-Keeling Is. to Indonesia, Micronesia and French Polynesia. – S.W. Japan to Great Barrier Reef.

GRAY CORAL CROUCHER *Caracanthus unipinna*

SIZE: to 5 cm (2 in.) Coral Crouchers – Caracanthidae

ID: Uniform shades of brown to gray; body covered with fuzzy appearing papillae. Inhabit branches of *Acropora, Stylophora* and *Pocillopora* corals. Coral reefs in 1 - 15 m.

Indo-Asian Pacific: E. Africa to Indonesia, Philippines, Micronesia, Papua New Guinea, Solomon Is. and French Polynesia. – S.W. Japan to Australia and New Caledonia.

IDENTIFICATION GROUP 17
Odd-shaped Swimmers
Boxfishes – Goatfishes – Sweepers – Triggerfishes – Filefishes – Puffers – Porcupinefishes – Others

This ID Group consists of swimming fishes that do not have typical fish-like shapes.

FAMILY: Boxfishes – Ostraciidae
4 Genera – 11 Species Included

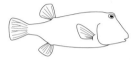

Typical Shape

Boxfishes, also commonly known as trunkfishes, are protected by a square, triangular or rounded bony carapace formed by a series of polygonal armor plates. A sharp spine over each eye of fishes in genus *Latoria* provides the common name cowfishes. Boxfishes have small protruding mouths used for feeding during the day on a variety of attached bottom invertebrates, including sponges, tunicates, and algae. These relatively slow swimmers move with a sculling action of their dorsal, anal and pectoral fins. The broom-like tail is only brought into play when a burst of speed is required.

FAMILY: Goatfishes – Mullidae
3 Genera – 21 Species Included

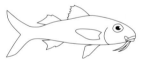

Typical Shape

Two food-searching, chemosensory appendages (barbels) extending from the chin give goatfishes their common family name. When not in use scouring the sand for worms, crustaceans, brittle stars and small fishes, barbels slip under the lower gill covers. Certain species feed during the day, others at night, and a few both day and night. During the day the nighttime feeders often form aggregations in open water or near the protection of the reef. Several goatfishes alter their normal color patterns dramatically when resting on the bottom or while attending cleaning stations.

FAMILY: Sweepers – Pempheridae
2 Genera – 5 Species Included

Typical Shape

Sweepers, also commonly known as bullseyes, are easily recognized by their thin, hatchet-shaped bodies and tendency to gather in aggregations within caves and the deep shadows of the reef's structure during the day. At dusk they disperse to feed on current-borne zooplankton at the reef's periphery.

FAMILY: Triggerfishes – Balistidae
9 Genera – 20 Species Included

Typical Shape   Typical Shape   Typical Shape

    Triggerfishes have relatively large, laterally-compressed bodies with two-part dorsal fins. The first stout, elongate spine of the first dorsal fin can be held erect and locked in place by a rigid second spine, "trigger" that must be depressed to lower the first spine. The mechanism can be used as a formidable defense against predators or to lock the fish firmly inside a crevice. Triggerfishes are the largest demersal egg-layers. Several of the larger species blow shallow depressions in the sand for nests. Deposited eggs form a translucent gelatinous mass. Some species, including the Titan Triggerfish, attack and may seriously nip divers approaching their nesting areas.

FAMILY: Filefishes – Monacanthidae
14 Genera – 25 Species Included

Typical Shape   Typical Shape

    Filefishes and closely related triggerfishes are included together in superfamily Balistoidea. When raised, the filefishes' first elongate dorsal spine resembles a woodworker's rattail file, which most species can lock in place. Many triggerfishes and filefishes can quickly alter their color and pattern to match their surroundings.

FAMILY: Puffers – Tetraodontidae
5 Genera – 27 Species Included

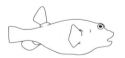 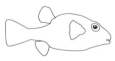

Typical Shape   Tobies – Typical Shape

    Puffers are odd-shape fishes with tough, scaleless skin that, when molested, have the ability to greatly expand the size of their bodies by drawing water into the ventral portion of their stomachs. The family is divided into two subfamilies: larger puffers in Tetraodontinae and the smaller puffers, known as tobies, in Canthigasterinae. The family produces a potent neurotoxin (tetrodotoxin) in their tissue that is potentially deadly if ingested.

FAMILY: Porcupinefishes – Diodontidae
4 Genera – 7 Species Included

Typical Shape

Like puffers, porcupinefishes, also have the ability to inflate their bodies with water. But, unlike the closely related puffers, porcupinefishes are covered with short spines. These spines stand permanently erect in species of *Chilomycterus* and *Cyclichthys*, commonly known as burrfishes. The spines of those in genus *Diodon* lay flat unless the body is inflated. Members of the family should be treated with respect because of their ability to inflict nasty bites with a plated mouth structure easily capable of crushing the shells of gastropods.

FAMILY: Others

Trumpetfishes – Aulostomidae

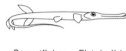

Cornetfishes – Fistulariidae

Flashlightfishes – Anomalopidae

Shrimpfishes
Centriscidae

Eel-tailed Catfishes – Plotosidae

Cobias – Rachycentridae

Remoras – Echeneidae

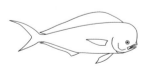

Molas – Molidae

Dolphinfishes – Coryphaenidae

LONGHORN COWFISH   *Lactoria cornuta*

SIZE: to 46 cm (18 in.)   Boxfishes – Ostraciidae

ID: **Pair of long horns in front of eyes,** a second pair on lower rear body; gray to brown, olive or yellow with blue or white spots. Tail can open wide into fan shape but usually closed. Solitary. Sand and mud bottoms in 1-50 m.

Indo-Pacific: Red Sea and E. Africa to Indonesia, Philippines, Micronesia to French Polynesia. – S. Japan to Australia.

Longhorn Cowfish – Juvenile

SIZE: to 7.5 cm (3 in.)

ID: Yellow with whitish spots.

THORNBACK COWFISH   *Lactoria fornasini*

SIZE: to 15 cm (6 in.)   Boxfishes – Ostraciidae

ID: Pale reddish to yellowish brown with blue spots and scrawl markings; pair of horns in front of eyes, a second pair on lower rear body and a singe thorn on midback. Solitary. Sand, rubble and weed bottoms of coastal, lagoon and outer reefs to 30 m.

Indo-Pacific: E. Africa to Indonesia, Philippines, Micronesia, Solomon Is., Hawaii and French Polynesia. – Japan to Australia.

Thornback Boxfish – Small Juvenile

SIZE: 2-3 cm ($^3/_4$ -1$^1/_4$ in.)

ID: Often orange to yellow-orange with some dark or blue spots; often display dark hexagonal carapace plates on sides. The snout of this species angles down at about a 45-degree angle in contrast to the nearly horizontal bottom of the carapace.

ROUNDBELLY COWFISH   *Lactoria diaphana*

SIZE: to 25 cm (10 in.)   Boxfishes – Ostraciidae

ID: **Rounded belly;** mottled shades of white, brown and gray with honeycomb net pattern, pairs of short horns in front of eyes and lower rear body and a single thorn on back. Solitary. Coastal and outer reefs to 50 m.

Indo-Pacific: E. Africa to Indonesia, Philippines, Micronesia, Papua New Guinea, Hawaii and Panama.

SMALLSPINE TURRETFISH   *Tetrosomus concatenatus*

SIZE: to 22 cm (8$^3/_4$ in.)   Boxfishes – Ostraciidae

ID: Triangular carapace with **pair of thorn-like spines on raised middorsal ridge;** gray to yellowish brown with dark network of hexagons and small blue spots. Solitary. Seagrass and weed bottoms of coastal reefs to 20 m.

Indo-Asian Pacific: E. Africa to Andaman Sea, Indonesia, Philippines and Papua New Guinea. – Japan to Australia.

Boxfishes

HUMPBACK TURRETFISH *Tetrosomus gibbosus*
SIZE: to 22 cm (8 ³/₄ in.) Boxfishes – Ostraciidae
ID: Triangular carapace; pale gray with light to dark brown network of hexagons, **tall dorsal ridge and single spine at tip**, may display brown blotches. Solitary. Seagrass and weed bottoms of coastal reefs to 20 m.
Indo-Asian Pacific: Red Sea and E. Africa to Indonesia and Solomon Is. – S. Japan to Great Barrier Reef and New Caledonia.

Humpback Turretfish – Postlarval Juvenile
ID: Large spines along dorsal and ventral ridges.

SPOTTED BOXFISH *Ostracion meleagris*
SIZE: to 15 cm (6 in.) Boxfishes – Ostraciidae
ID: Male – Black to dark brown back with white spots; blue head and sides with bright orange body spots. Solitary or in pairs. Coastal, lagoon and outer reefs to 30 m.
Indo-Pacific: E. Africa to Andaman Sea, Indonesia, Philippines, Micronesia, Papua New Guinea, Solomon Is., Hawaii, French Polynesia. – S. Japan to Great Barrier Reef and New Caledonia.

Spotted Boxfish – Female
ID: Dark brown to blackish with numerous white spots.

SOLOR BOXFISH *Ostracion solorensis*
SIZE: to 11 cm (4 ¹/₄ in.) Boxfishes – Ostraciidae
ID: Male – Black back with light blue maze pattern; head below eyes and sides bluish to black with pale blue lines and spots with black borders. Solitary or pairs. Seaward reefs to 20 m.
West Pacific: Christmas I. to Andaman Sea, Indonesia, Philippines, Micronesia, Papua New Guinea, Solomon Is. – S. Japan to N.W. Australia and N. Great Barrier Reef and Fiji.

Solor Boxfish – Female
ID: Black back with pale brown maze pattern, black upper side and yellowish brown below including head with dark spots and pale maze line markings.

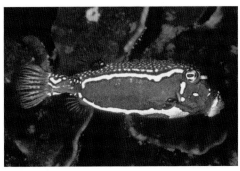

WHITLEY'S BOXFISH   *Ostracion whitleyi*

SIZE: to 15.5 cm (6 in.)   Boxfishes – Ostraciidae

ID: Male – Bluish gray with dark-edged white margin along side; white spots on back. Solitary or in pairs. Rock and coral bottoms of lagoon and seaward reefs in 3-27 m.

Central Pacific: Hawaii and Johnston I. to French Polynesia. Rare except at Marquesas Is.

Whitley's Boxfish – Female

ID: Dark brown back and snout with white spots, pale brown lower body with brown spots and markings; white midlateral stripe from snout tip to tail base.

YELLOW BOXFISH   *Ostracion cubicus*

SIZE: to 45 cm (18 in.)   Boxfishes – Ostraciidae

ID: Large Adult – Bump on snout tip; brownish purple with indistinct spots, often yellow crinkled line markings on head and yellowish tail base. Solitary. Coastal, lagoon and outer reefs to 35 m.

Indo-Pacific: Red Sea and E. Africa to Indonesia, Philippines, Micronesia, Solomon Is. and French Polynesia. – Japan to Australia.

Yellow Boxfish – Intermediate Adult

SIZE: to 30 cm (12 in.)

ID: Bump on snout tip; tan to yellowish brown with dark ringed bluish spots, may have black to blue or yellow crinkled line markings on head and around pectoral fin base.

Yellow Boxfish – Young Adult

SIZE: 9-15 cm (3 1/2 - 6 in.)

ID: Small bump on snout tip; yellowish brown to olive with black spots on head and a few dark ringed bluish spots on body.

Yellow Boxfish – Juvenile

SIZE: 3-8 cm (1 1/4 - 3 1/4 in.)

ID: Absent or only a hint of bump on snout tip; bright yellow with black spots on head and body.

NASAL BOXFISH *Ostracion nasus*

SIZE: to 30 cm (12 in.) Boxfishes – Ostraciidae

ID: Pale gray to yellowish with brown spots, usually one per hexagonal plate of carapace. Solitary. Sheltered bays exposed to heavy silt deposition, and mud bottoms of coastal trawling grounds in 3-40 m.

East Indo-Asian Pacific: Sri Lanka to Indonesia, Philippines, Micronesia and Papua New Guinea. – S. Japan to Australia

Nasal Boxfish – Juvenile

SIZE: 2-4 cm ($^3/_4$ -1$^1/_2$ in.)

ID: Pale gray to pale brown with numerous black spots; often pale hexagonal outline of carapace plates visible.

SHORTNOSE BOXFISH *Ostracion rhinorhynchos*

SIZE: to 35 cm (14 in.) Boxfishes – Ostraciidae

ID: Gray to pale brown with dark spots on back to tail; darkish outlines of hexagonal carapace plates on sides, **large conical protruding snout.** Solitary or in pairs. Sand and rubble patches near reefs in 3-35 m.

East Indo-Asian Pacific: Sri Lanka to Indonesia, Philippines, Papua New Guinea and Micronesia. – Japan to Australia.

Shortnose Boxfish – Juvenile

SIZE: 3-10 cm (1$^1/_4$ -4 in.)

ID: Small bump on snout tip; whitish to pale gray or brown with dark brown spots mainly concentrated on back and tail base.

BICOLOR GOATFISH *Parupeneus barberinoides*

SIZE: to 25 cm (10 in.) Goatfishes – Mullidae

ID: Head and front of body dark reddish brown with 2 diagonal whitish bands; rear white and yellow with black spot below rear dorsal fin. Adults solitary; juveniles often form schools. Rubble, weed and coral reefs to 15 m.

West Pacific: Indonesia, Philippines, Micronesia, Papua New Guinea and Solomon Is. – S. Japan to Australia and Fiji.

DASH-DOT GOATFISH *Parupeneus barberinus*

SIZE: to 53 cm (21 in.) Goatfishes – Mullidae

ID: White with pale gray to yellow on upper back; **narrow black stripe extends from lip to below 2nd dorsal fin,** large black spot on tail base. Solitary or form small groups. Sand and rubble bottoms near reefs to 100 m.

Indo-Pacific: E. Africa to Indonesia, Philippines, Micronesia, Solomon Is., French Polynesia. – S. Japan to Australia and Vanuatu.

MANYBAR GOATFISH *Parupeneus multifasciatus*
SIZE: to 30 cm (12 in.) Goatfishes – Mullidae
ID: Light gray to brownish, purplish or red; 3-4 alternating black and white bars of variable width, black band behind eye. Solitary. Coral reefs and adjacent sand and rubble to 140 m.
East Indo-Pacific: Cocos-Keeling Is. to Indonesia, Philippines, Micronesia, Hawaii, Pitcairn Is. – S. Japan to Great Barrier Reef.

Manybar Goatfish – Variation
SIZE: 3-10 cm (1$\frac{1}{4}$-4 in.)
ID: One of the most vairable goatfish, often changing from a well defined barred pattern to one with rather indistinct or hazy bars. Colors can be modified from grayish brown to reddish brown; white areas can also acquire colors of adjacent bars.

SIDESPOT GOATFISH *Parupeneus pleurostigma*
SIZE: to 33 cm (13 in.) Goatfishes – Mullidae
ID: Yellowish to purplish gray to light red; white oval patch preceded by large black patch on midbody, blackish under 2nd dorsal fin. Solitary; feed during day. Sand and rubble bottoms near coral reefs in 1-75 m.
Indo-Pacific: E. Africa to Indonesia, Philippines, Micronesia, Solomon Is., Hawaii and French Polynesia. – S. Japan to Australia.

LONGBARBEL GOATFISH *Parupeneus macronemus*
SIZE: to 32 cm (12$\frac{1}{2}$ in.) Goatfishes – Mullidae
ID: Yellowish to grayish white; wide black stripe from eye to below 2nd dorsal fin, large black spot at middle of tail base, **black stripe along base of 2nd dorsal.** Solitary or form small groups. Sand, rubble and weed bottoms in 3-40 m.
Indo-Asian Pacific: Red Sea and E. Africa to Indonesia and Philippines.

INDIAN GOATFISH *Parupeneus indicus*
SIZE: to 35 cm (14 in.) Goatfishes – Mulliade
ID: Bluish white to grayish or brownish; **bright yellow oval patch on midbody,** black spot on tail base, often dark bridle from snout to behind eye. Rest in aggregations during day. Coastal reefs, lagoons and outer slopes to 113 m.
Indo-West Pacific: E. Africa to Indonesia, Philippines, Micronesia, Papua New Guinea, Solomon Is. – S. Japan to Australia and Fiji.

BLACKSPOT GOATFISH *Parupeneus spilurus*
SIZE: to 32 cm (12$\frac{1}{2}$ in.) Goatfishes – Mullidae
ID: Alternating white and reddish brown to yellowish brown stripes; **white patch behind 2nd dorsal fin precedes a black saddle spot on tail base.** Solitary; rest on bottom much of day, active at night. Sandy areas and reefs to 30 m.
Asian Pacific: Indonesia and Philippines. – S. Japan to W. Australia, Great Barrier Reef and New Caledonia.

Goatfishes

CARDINAL GOATFISH *Parupeneus ciliatus*
SIZE: to 38 cm (15 in.) Goatfishes – Mullidae
ID: Light red or purplish to yellowish; pair of white bands extend from eye to below 2nd dorsal fin base, often display darkish saddle on upper tail base. Solitary. Lagoons, seaward reefs and seagrass beds to 40 m.
Indo-Pacific: E. Africa to Indonesia, Philippines, Micronesia, Solomon Is. and French Polynesia. – S. Japan to Australia.

Cardinal Goatfish – Phase
ID: Have ability to rapidly change color; the red phase is quite common. Often display a white patch behind 2nd dorsal fin.

CINNABAR GOATFISH *Parupeneus heptacanthus*
SIZE: to 36 cm (14 in.) Goatfishes – Mullidae
ID: Brownish yellow to light red; lavender to iridescent blue stripes around eye and on upper sides, **small dark spot on midside**. Solitary. Turbid areas on silty sand or weedy bottoms in 15-100 m.
Indo-West Pacific: Red Sea and E. Africa to Indonesia, Philippines, Micronesia and Papua New Guinea. – S. Japan to Australia and Fiji.

GOLDSADDLE GOATFISH *Parupeneus cyclostomus*
SIZE: to 50 cm (20 in.) Goatfishes – Mullidae
ID: Highly changeable, various combinations of purple, brown, gray, green and yellow; blue line markings around eye. often a yellow saddle is apparent on upper tail base. Coastal, lagoon and outer reefs in 2-125 m.
Indo-Pacific: Red Sea and E. Africa to Indonesia, Philippines, Solomon Is. and French Polynesia. – S.W. Japan to Australia.

Goldsaddle Goatfish – Phase
ID: Mixed tan, purple and yellow combination, note bright yellow saddle behind 2nd dorsal fin. Usually solitary, but occasionally in pairs.

Goldsaddle Goatfish – Phase
ID: Yellow to gold without markings except some blue lines around eye. Unlike most goatfishes that feed predominantly on small sand or rubble-dwelling invertebrates, Goldsaddles generally swim rapidly about the bottom feeding primarily on small fishes that are frightened from holes in the reefs by their probing barbels.

INDIAN DOUBLEBAR GOATFISH *Parupeneus trifasciatus*
SIZE: to 30 cm (12 in.) Goatfishes – Mullidae
ID: White often with purplish or yellow tints with orange spots on scales; black patch around eye, pair of large black bars below dorsal fins the **1st bar extending past the pectoral base.** Solitary. Coastal, lagoon and outer reefs to 80 m.
Indian Ocean: Red Sea and Andaman Sea to Sumatra, Java and Bali in Indonesia.

YELLOWFIN GOATFISH *Mulloidichthys vanicolensis*
SIZE: to 38 cm (15 in.) Goatfishes – Mullidae
ID: Bluish white with yellowish back, yellow fins; yellow stripe from eye to tail base. Form stationary daytime aggregations, but solitary when feeding on sand-dwelling animals at night. Coastal reefs, lagoons and outer slopes to 113 m.
Indo-Pacific: Red Sea and E. Africa to Indonesia, Philippines, Micronesia, Solomon Is., Hawaii and Easter I. – S. Japan to Australia.

RED GOATFISH *Mulloidichthys pfluegeri*
SIZE: to 48 cm (19 in.) Goatfishes – Mullidae
ID: Orange-red to pink; display red head and 3-4 dark red bars on side when feeding, fade when not feeding. Solitary or in pairs. Sand flats near coral reefs in 15-110 m.
Indo-Pacific: Primarily around oceanic islands from Réunion to Micronesia, Hawaii and Marquesas Is. in French Polynesia.

DOUBLEBAR GOATFISH *Parupeneus crassilabris*
SIZE: to 38 cm (15 in.) Goatfishes – Mullidae
ID: White to purple or yellow with orange spots on scales; black patch around eye, pair of large black bars or spots, one below each dorsal fin the **1st extending no further than pectoral base.** Solitary. Coastal, lagoon and outer reefs to 80 m.
West Pacific: Indonesia, Philippines, Micronesia, Papua New Guinea andSolomon Is. – S.W. Japan to Australia and Fiji.

MIMIC GOATFISH *Mulloidichthys mimicus*
SIZE: to 30 cm (12 in.) Goatfishes – Mullidae
ID: Yellow with 4 blue stripes. Effective mimic of Bluestripe Snapper with which it aggregates during day; disperses at night to feed. Rocky reefs and coral areas to 35 m.
Central Pacific: Line, Phoenix and Marquesas Is.

YELLOWSTRIPE GOATFISH *Mulloidichthys flavolineatus*
SIZE: to 40 cm (16 in.) Goatfishes – Mullidae
ID: Silvery white with indistinct yellow stripe; can rapidly display or fade a dark spot on yellow stripe. Form stationary daytime aggregations. Sandy areas of sheltered reefs and outer slopes in 1 to 76 m.
Indo-Pacific: Red Sea and E. Africa to Indonesia, Philippines, Micronesia, Solomon Is. and Hawaii – S.W. Japan to Australia, Fiji.

Goatfishes – Cornetfishes – Trumpetfishes – Shrimpfishes

BANDTAIL GOATFISH — *Upeneus taeniopterus*
SIZE: to 36 cm (14 in.) — Goatfishes – Mullidae
ID: Silvery with **pair of orange-yellow stripes (lower stripe faint);** about 10-12 black bands on tail. Usually form small groups. Sand bottoms near rock or coral reefs to 25 m.
Indo-Pacific: E. Africa to Micronesia, Hawaii and Tuamotu Is. in French Polynesia, primarily oceanic islands.

MOLUCCA GOATFISH — *Upeneus moluccensis*
SIZE: to 22 cm (8 ³/₄ in.) — Goatfishes – Mullidae
ID: Silvery pink or whitish with **single golden-yellow stripe from eye to upper base of tail;** upper lobe of tail with 6-7 red cross bars, lower tail lobe reddish with dark streak across back edge. Usually occurs in aggregations. Muddy bottoms in 10-80 m.
Indo-Asian Pacific: Red Sea and E. Africa to Indonesia, Philippines and Solomon Is. – S. Japan to Australia and New Caledonia.

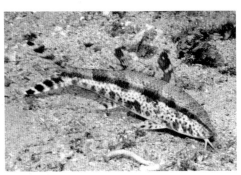

FRECKLED GOATFISH — *Upeneus tragula*
SIZE: to 30 cm (12 in.) — Goatfishes – Mullidae
ID: Whitish or tan with heavy mottling of brown blotches and spots; dark yellowish to brown stripe from snout to tail, black bands on tail. Solitary or form groups. Sand and rubble to 25 m.
Indo-Asian Pacific: E. Africa to Indonesia, Philippines, Micronesia, Papua New Guinea, Solomon Is. – Japan to Australia, New Caledonia.

Freckled Goatfish – Red Phase
ID: Can rapidly change to tan or red with dark red stripe from snout to tail. Change to red phase when being cleaned or sleeping at night.

STRIPED GOATFISH — *Upeneus vittatus*
SIZE: to 28 cm (11 in.) — Goatfishes – Mullidae
ID: Silvery with greenish or yellowish tints; **4 brassy yellow stripes** and black bands on dorsal and tail fins. Solitary or form groups. Silt or mud bottoms in 5-100 m.
Indo-Pacific: Red Sea and E. Africa to Andaman Sea, Indonesia, Philippines, Micronesia, Papua New Guinea, Solomon Is., Hawaii, French Polynesia. – S. Japan to Australia.

CORNETFISH — *Fistularia commersonii*
SIZE: to 150 cm (5 ft.) — Cornetfishes – Fistulariidae
ID: Silvery with pale olive back; elongated snout, body and whip-like tail filament, may display blue midlateral stripe from head to tail base. Solitary or form schools. Virtually all habitats to 128 m.
Indo-Pacific: Red Sea to French Polynesia including all of the Asian Pacific.

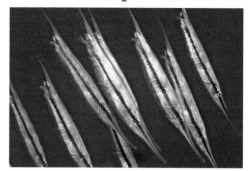

TRUMPETFISH *Aulostomus chinensis*
SIZE: to 80 cm (2 ¹/₂ ft.) Trumpetfishes – Aulostomidae
ID: Elongate body with trumpet-like snout; commonly gray to reddish brown with whitish stripes, also all yellow, blackish tail base with white spots, yellow tail with 2 black spots. Solitary. Inshore and seaward reefs to 122 m.
Indo-Pacific: Red Sea to French Polynesia including all of the Asian Pacific.

RIGID SHRIMPFISH *Centriscus scutatus*
SIZE: to 14 cm (5 ¹/₂ in.) Shrimpfishes – Centriscidae
ID: Very similar to Hinged Shrimpfish [next], but **1st dorsal spine extending from rear body is rigid and straight lacking a hinge,** midlateral stripe brown to red. Form head down schools. Sheltered reefs in 2-15 m.
Indo-Asian Pacific: Red Sea to Indonesia, Micronesia and Solomon Is. – S. Japan to Great Barrier Reef and New Caledonia.

HINGED SHRIMPFISH
Aeoliscus strigatus
SIZE: to 15 cm (6 in.)
Shrimpfishes – Centriscidae
ID: Silvery with brown midlateral stripe; thin elongate snout and flattened body, **1st dorsal spine extends from rear body with hinge allowing the rear half to angle in varying directions.** Head down schools. Reefs to 42 m.
Asian Pacific: Andaman Sea to Indonesia, Philippines, Papua New Guinea. – S. Japan to Great Barrier Reef and New Caledonia.

SMOOTH SHRIMPFISH
Centriscus cristatus
SIZE: to 30 cm (12 in.)
Shrimpfishes – Centriscidae
ID: Silvery white with yellowish brown midlateral stripe that runs from elongate snout to tail base; flattened body slightly wider than other shrimpfish, **1st dorsal spine rigid, straight and lacks a hing.** Solitary or form small groups. Seagrass beds to 10 m.
Asian Pacific: Indonesia, Philippines, Papua New Guinea to Australia and New Caledonia.

WHITELIPPED EEL CATFISH *Paraplotosus albilabris*
SIZE: to 130 cm (4 ¹/₄ ft.) Eel-tailed Catfishes – Plotosidae
ID: Brown to gray to blackish occasionally with dark bands or mottling; elongate body with eel-like rear dorsal, tail and anal fins joined. Seldom larger than 16 in. Solitary or small groups; nocturnal. Clear to turbid coastal flats and reefs 1-10 m.
Asian Pacific: Singapore to Indonesia, Philippines, Papua New Guinea and N. Australia.

STRIPED CATFISH *Plotosus lineatus*
SIZE: to 32 cm (13 in.) Eel-tailed Catfishes – Plotosidae
ID: Black to brown with white belly; pair of narrow white stripes extend from head to tail, 4 pairs of barbels around mouth. Young form tightly pack feeding schools that often contain hundreds of fish. Sand and seagrass near reefs to 35 m.
Indo-West Pacific: Red Sea and E. Africa to Indonesia, Philippines, Micronesia and Solomon Is. – S. Japan to Australia and Fiji.

COBIA *Rachycentron canadum*
SIZE: to 2 m (6 ½ ft.) Cobias – Rachycentridae
ID: Silver to dark brown; often dusky midbody stripe, lower jaw protrudes, tail forked, long dark pectoral fins. Coastal reefs in 5-40 m, but generally pelagic.
Circumtropical: Widespread in tropical seas, including Atlantic and Indo-Pacific.

REMORA *Remora* sp.
SIZE: to 40 cm (16 in.) Remoras – Echeneidae
ID: Elongate body with suction disc on top of head; dusky white to nearly black often with white speckles. Swim with or attached to turtles, sharks, mantas and other large fishes, occasionally free-swimming. May attempt to attach to divers. If attached, push forward to release. Pelagic, occasionally near reefs to 50 m.
Circumtropical.

SHARKSUCKER *Echeneis naucrates*
SIZE: to 90 cm (3 ft.) Remoras – Echeneidae
ID: Elongate body with suction disc on top of head; pale gray to nearly black with white edged black stripe from head to tail. Swim with or attached to sharks, turtles, mantas and other large fishes, occasionally free-swimming. May attempt to attach to divers. If attached, push forward to release.
Circumtropical.

Sharksucker – Variation
ID: Pale gray with dusky stripe through eye. Suction disc clearly visible on top of head of free-swimming individuals.

TWOFIN FLASHLIGHTFISH *Anomalops katoptron*
SIZE: to 16 cm (6 ¼ in.) Flashlightfishes – Anomalopidae
ID: Dark gray with black head and fins with pale margins; oval-shaped light organ below eye, **2 dorsal fins.** Form aggregations. Outer reef slopes with caves in 1-40 m.
Pacific: Indonesia and Philippines to French Polynesia. – S. Japan to Australia.

ONEFIN FLASHLIGHTFISH *Photoblepharon palpebratum*
SIZE: to 10 cm (4 in.) Flashlightfishes – Anomalopidae
ID: Dark brown to nearly black; crescent-shaped light organ below eye, **single dorsal fin.** Form aggregations on outer reef slopes with caves 1-50 m; during moonless nights; usually below 15 m.
Pacific: Indonesia, Philippines, Micronesia to French Polynesia, south to Great Barrier Reef.

GOLDEN SWEEPER   *Parapriacanthus ransonneti*
SIZE: to 10 cm (4 in.)   Sweepers – Pempheridae
ID: Golden brown translucent body with yellowish head; cardinalfish-like appearance, but with single dorsal fin. Form huge tightly packed schools in caves, under ledges or under *Acropora* plate corals. Coastal, lagoon and outer reefs in 3-30 m.
West Pacific: Indonesia to Micronesia, New Caledonia and Fiji. – Japan to Australia

Golden Sweeper – Close up
ID: Family Pempheridae, primarily restricted to the Indo-Pacific and consisting of about 25 members, many are difficult to distinguish to species. Except for two Atlantic species, family members are restricted to the Indo-Pacific. As night approaches they disperse from their daytime shelters to feed on zooplankton.

COPPER SWEEPER   *Pempheris oualensis*
SIZE: to 22 cm (8 3/4 in.)   Sweepers – Pempheridae
ID: Copper-brown; leading edge and tip of 1st dorsal fin blackish, **black spot on pectoral fin base** (no black margin on anal fin or tail). Form aggregations in caves or under ledges. Lagoon and seaward reefs to 35 m.
East Indo-Pacific: Cocos-Keeling Is. to Micronesia and French Polynesia. – S.W. Japan to Great Barrier Reef.

VANIKORO SWEEPER   *Pempheris vanicolensis*
SIZE: to 18 cm (7 in.)   Sweepers – Pempheridae
ID: Copper-brown; broad black tip on 1st dorsal fin, **black margin on anal fin** and tail. Form aggregations in caves or under ledges. Lagoon and seaward reefs to 25 m.
Indo-West Pacific: Red Sea and E. Africa to Philippines, Vanuatu and Fiji.

DUSKY SWEEPER   *Pempheris adusta*
SIZE: to 17 cm (6 3/4 in.)   Sweepers – Pempheridae
ID: Copper-brown with silvery to brassy reflections; black tip on 1st dorsal fin and **black stripe along base of anal fin**, steep curve in lateral line below 1st dorsal fin. Form aggregations in caves or under ledges. Lagoon and seaward reefs to 30 m.
Indo-Asian Pacific: East Africa to east New Guinea.

SILVER SWEEPER   *Pempheris schwenkii*
SIZE: to 15 cm (6 in.)   Sweepers – Pempheridae
ID: Silvery, occasionally coppery; purplish tinting on back and head, often lavender and greenish iridescence, in coppery phase distinguished from similar species by only 3-4 scale rows above lateral line. Form schools in protected areas. Lagoon and seaward reefs to 40 m.
Indo-West Pacific: Africa to Fiji. – Philippines to Great Barrier Reef.

Triggerfishes

ROUGH TRIGGERFISH *Canthidermis maculata*
SIZE: to 50 cm (20 in.) Triggerfishes – Balistidae
ID: Elongate body; pale gray to blackish often with white spots, large dark diffuse spot around pectoral fin base. White longitudinal groove in front of eye. Form schools. Steep outer reef slopes in 15-55 m. Often near large floating objects in open sea.
Circumtropical.

TITAN TRIGGERFISH *Balistoides viridescens*
SIZE: to 75 cm (2 1/2 ft.) Triggerfishes – Balistidae
ID: Dark body with yellow-green to blue crosshatches; yellow-green snout and cheek, whitish rear body and tail base, **dark "moustache" band above mouth.** Solitary; nesting females will attack divers. Lagoon and outer reefs in 3-50 m.
Indo-Pacific: Red Sea and E. Africa to Indonesia, Philippines, Micronesia, Solomon Is. and French Polynesia. – S. Japan to Australia.

CLOWN TRIGGERFISH *Balistoides conspicillum*
SIZE: to 50 cm (20 in.) Triggerfishes – Balistidae
ID: Black undercolor with **large white spots on lower body;** orange lips, yellowish band across top of snout, pale patch with dark spots on back. Solitary. Clear waters and coral-rich areas of outer reef slopes to 75 m.
Indo-West Pacific: E. Africa to Indonesia, Philippines, Micronesia, Solomon Is. – S. Japan to Australia and Fiji.

Clown Triggerfish – Juvenile
SIZE: 4-8 cm (1 1/2 - 3 1/4 in.)
ID: Black with large white spots; yellowish to bright yellow snout with orange lips, white band across top of snout, bright yellow patch on back. Young inhabit caves below 20 m on steep slopes, but may be shallower on rare occasions.

STARRY TRIGGERFISH *Abalistes stellatus*
SIZE: to 50 cm (20 in.) Triggerfishes – Balistidae
ID: Grayish undercolor with yellow to yellow-brown network; **dark upper back with 3-4 white blotches,** yellow-brown spots on head and body; narrow tail base. Solitary. Muddy or fine silt and sand bottoms in 4-120m.
Indo-West Pacific: Red Sea and E. Africa to Indonesia, Philippines, Papua New Guinea, Solomon Is. – S. Japan to N. Australia and Fiji.

Starry Triggerfish – Juvenile
SIZE: to 5 cm (2 in.)
ID: White with yellow spots; black back with 4 white blotches extending from behind eye to tail base. Juveniles inhabit isolated patch reefs in silty areas.

BLUE TRIGGERFISH   *Pseudobalistes fuscus*
SIZE: to 55 cm (22 in.)   Triggerfishes – Balistidae
ID: Blue to bluish gray often with yellowish scale spots; pale blue to muddy red margins on all fins. Solitary; nesting females very aggressive and may attack divers. Sheltered inner reefs and seaward slopes to 50 m.
Indo-Pacific: Red Sea and E. Africa to Indonesia, Micronesia, Solomon Is. and French Polynesia. – S. Japan to Australia and New Caledonia.

Blue Triggerfish – Juvenile
SIZE: 8-15 cm (3¹/₄-6 in.)
ID: Yellow-orange with maze of iridescent blue lines, blue dorsal, anal and tail fins with yellow-orange spots.

YELLOWMARGIN TRIGGERFISH   *Pseudobalistes flavimarginatus*
SIZE: to 60 cm (2 ft.)   Triggerfishes – Balistidae
ID: Tan body with dark spots and crosshatch pattern, pale orangish snout and cheeks; yellow to orange margins on median fins. Solitary; nest in sand and rubble channels. Coastal reefs, lagoons and sheltered outer reefs in 2-50 m.
Indo-Pacific: Red Sea and E. Africa to Indonesia, Micronesia, Solomon Is. and French Polynesia. – S. Japan to Great Barrier Reef.

Yellowmargin Triggerfish – Subadult
SIZE: 8-20 cm (3¹/₄-8 in.)
ID: Tan with yellowish tints; variable-sized dark spots on body and dusky gray dorsal, anal and tail fins with yellow margins.

Yellowmargin Triggerfish – Small Juvenile
SIZE: 3-5 cm (1¹/₄-2 in.)
ID: White back with yellow below; 4 black bars or saddles on back; scattered spots on body. Solitary; very shy, retreat into reef when approached.

ORANGE-LINED TRIGGERFISH   *Balistapus undulatus*
SIZE: to 30 cm (12 in.)   Triggerfishes – Balistidae
ID: Dark green to brown with diagonally curved orange bands; **large black spot on tail base.** Solitary; feed on coral, algae, sponges, worms, crabs, urchins and fishes. Coral-rich areas of lagoons and outer reefs in 2-50 m.
Indo-Pacific: Red Sea and E. Africa to Indonesia, Philippines, Micronesia, Solomon Is., French Polynesia. – S. Japan to Australia.

Triggerfishes

BRIDLED TRIGGERFISH *Sufflamen fraenatum*
SIZE: to 38 cm (15 in.) Triggerfishes – Balistidae
ID: Changeable light to dark brown; median fins darker, narrow **pale yellow to pink band under chin.** Solitary. Seaward reefs, over open bottoms with sand and rubble patches in 8-186 m.
Indo-Pacific: Red Sea and E. Africa to Indonesia, Micronesia, Hawaii and French Polynesia. – S. Japan to Australia.

Bridled Triggerfish – Juvenile
SIZE: 3-8 cm (1¼-3¼ in.)
ID: White with dark brown back; wavy horizontal brown lines on body.

FLAGTAIL TRIGGERFISH *Sufflamen chrysopterum*
SIZE: to 22 cm (8¾ in.) Triggerfishes – Balistidae
ID: Dark brown, changeable to yellowish brown; bluish chin and belly, **narrow bluish to yellow or orange bar from lower rear corner of eye to front of pectoral fin,** white edged yellow-brown tail. Solitary. Lagoon and seaward reefs in 2-30 m.
Indo-West Pacific: E. Africa to Indonesia, Philippines, Micronesia, Papua New Guinea and Solomon Is. – S. Japan to Australia and Fiji.

Flagtail Triggerfish – Small Juvenile
SIZE: 3-8 cm (1¼-3¼ in.)
ID: White with brown back, dark horseshoe-shaped marking on tail. Usually seen in areas of mixed coral and rubble; quickly retreat to rocky holes when threatened.

SCYTHE TRIGGERFISH *Sufflamen bursa*
SIZE: to 24 cm (9½ in.) Triggerfishes – Balistidae
ID: Gray to brown with white chin and belly; **yellow or brown scythe-shaped band behind eye with a 2nd band across rear gill cover,** thin white line from mouth to anal fin. Solitary. Seaward reefs in mixed coral, sand and rubble in 3-90 m.
Indo-Pacific: E. Africa to Indonesia, Philippines, Micronesia, Hawaii and French Polynesia. – S. Japan Great Barrier Reef.

Scythe Triggerfish – Phase
ID: Has the ability to rapidly change scythe-markings from bright yellow-orange to dark brown, also may darken or lighten or change shades of body color.

BLACKPATCH TRIGGERFISH *Rhinecanthus verrucosus*
SIZE: to 23 cm (9 in.)   Triggerfishes – Balistidae
ID: Brownish upper body, white below; tapering dark bar through eye, **large black oval patch on lower body,** 3 rows of black dots on tail base. Solitary or form loose groups. Sheltered areas of mixed coral and rubble or seagrass to 20 m.
Indo-Asian Pacific: Seychelles to Indonesia, Philippines, Micronesia, Solomon Is. – S. Japan to Great Barrier Reef, Vanuatu.

LAGOON TRIGGERFISH *Rhinecanthus aculeatus*
SIZE: to 25 cm (10 in.)   Triggerfishes – Balistidae
ID: Snout and back shades of tan with white below; yellow-orange band from snout intersects black bar below eye, **black midbody patch with black bands extending to anal fin.** Solitary or form groups. Lagoon and reef flats to 4 m.
Indo-Pacific: E. Africa to Indonesia, Philippines, Micronesia, Papua New Guinea, Hawaii and French Polynesia. – S. Japan to Australia.

HALFMOON TRIGGERFISH *Rhinecanthus lunula*
SIZE: to 28 cm (11 in.)   Triggerfishes – Balistidae
ID: White with gray back and upper snout; **yellowish stripe from snout to rear body with interrupting black spot below eye,** black ring on tail base, black 1st dorsal fin, black spot over anus. Solitary. Seaward reefs in 10-25 m.
South Pacific: Great Barrier Reef and New Caledonia to Fiji and French Polynesia.

WEDGETAIL TRIGGERFISH *Rhinecanthus rectangulus*
SIZE: to 25 cm (10 in.)   Triggerfishes – Balistidae
ID: Light brown snout and back, white below; black band through eye enlarges and runs to anal fin, **black triangular mark on tail base.** Solitary or form groups. Surge-affected reef flats and seaward reefs to 12 m.
Indo-Pacific: Red Sea and E. Africa to Indonesia, Philippines, Micronesia, Solomon Is., Hawaii. – S.W. Japan to E. Australia.

CROSSHATCH TRIGGERFISH *Xanthichthys mento*
SIZE: to 22 cm (8 3/4 in.)   Triggerfishes – Balistidae
ID: **Male –** Yellow-gold with black scale margins forming crosshatch pattern, bright red tail; blue lines on cheek. **Female –** Yellow tail; red rear dorsal and anal fin margins (yellow on males). Form groups. Seaward reefs above dropoffs in 10-100 m.
Pacific: Japan and Micronesia to French Polynesia (most common around subtropical islands including Hawaii and Galapagos).

BLUELINE TRIGGERFISH *Xanthichthys caeruleolineatus*
SIZE: to 42 cm (17 in.)   Triggerfishes – Balistidae
ID: Light brown upper body, grayish white below with a **metallic blue stripe separating the two areas.** Solitary or form groups. Outer reef slopes, usually below 25 m.
Indo Pacific: Mainly oceanic islands W. Indian Ocean to Hawaii, Cook Is. and French Polynesia. Absent Asian Pacific.

GILDED TRIGGERFISH   *Xanthichthys auromarginatus*
SIZE: to 22 cm (8 ³/₄ in.)   Triggerfishes – Balistidae
ID: Male – Steel blue with white scale spots; **large dark blue patch on lower head,** yellow margins on dorsal, anal and tail fins. Form groups that feed on zooplankton above bottom. Outer reef slopes in 15-140 m, usually below 20 m.
Indo-Pacific: Mauritius to Micronesia, Hawaii and French Polynesia; mainly islands. – S.W. Japan to Great Barrier Reef.

Gilded Triggerfish – Female
ID: Steel blue with white scale spots forming horizonal rows; **maroon stripe on base of dorsal and anal fins and margin on tail.** On steep slopes, often on the upper edge of dropoffs below 20 m.

INDIAN TRIGGERFISH   *Melichthys indicus*
SIZE: to 24 cm (9 ¹/₂ in.)   Triggerfishes – Balistidae
ID: Dark grayish black with jet black fins; white band at base of dorsal and anal fins, **white margin on tail.** Solitary or form groups. Coral-rich areas of outer reefs slopes in 2-20 m.
Indo-Asian Pacific: Red Sea and E. Africa to Andaman Sea, Sumatra and Bali in Indonesia.

BLACK TRIGGERFISH   *Melichthys niger*
SIZE: to 32 cm (12 ¹/₂ in.)   Triggerfishes – Balistidae
ID: Black with bluish scale margins; pale blue to white band at base of dorsal and anal fins, may display blue markings on upper head and yellow marking on cheek. Solitary or form loose groups. Outer reefs in 2-75 m.
Circumtropical.

PINKTAIL TRIGGERFISH   *Melichthys vidua*
SIZE: to 34 cm (13 ¹/₂ in.)   Triggerfishes – Balistidae
ID: Brown with yellowish snout and pectoral fins; white dorsal and anal fins with black margins, **white tail with wide pink margin.** Solitary or form loose groups; feed on algae, mixed invertebrates and fishes. Outer reefs in 4-60 m.
Indo-Pacific: E. Africa to Indonesia, Micronesia, Hawaii and French Polynesia. – S. Japan to Great Barrier Reef and New Caledonia.

REDTOOTH TRIGGERFISH   *Odonus niger*
SIZE: to 40 cm (16 in.)   Triggerfishes – Balistidae
ID: Dark blue to purplish body; pale blue head, red teeth usually visible, 2 blue lines extend from eye toward mouth, lunate tail with long lobes. Form plankton-feeding aggregations above reef, retreat to recesses when threatened. Outer reefs slopes in 5-40 m.
Indo-Pacific: E. Africa to Indonesia, Micronesia, Solomon Is., French Polynesia. – S. Japan to Great Barrier Reef and New Caledonia.

UNICORN FILEFISH *Aluterus monoceros*
SIZE: to 75 cm (2 1/2 ft.) Filefishes – Monacanthidae
ID: Pale gray to brownish often with silvery highlights and faint spotting; profile distinctly concave below small mouth, snout and nape long and convex. Solitary. Commonly in open water, occasionally over outer reefs to 1- 80 m.
Circumtropical.

Unicorn Filefish – Reticulated Phase
ID: May rapidly change from nearly solid color to a reticulated pattern of pale lines and gray blotches. Juveniles [pictured above] often associate with floating debris such as *Sargassum* and jellyfish.

SCRAWLED FILEFISH *Aluterus scriptus*
SIZE: to 75 cm (2 1/2 ft.) Filefishes – Monacanthidae
ID: Mottled to uniform shades of gray to brown to olive, can quickly change colors and markings; irregular blue spot and line markings and black spots, tall slender 1st dorsal spine. Solitary. Coastal, lagoon and outer reefs in 2-80 m.
Circumtropical.

BROOM FILEFISH *Amanses scopas*
SIZE: to 19 cm (7 1/2 in.) Filefishes – Monacanthidae
ID: Dark brown with black tail; several blackish bars on midside, occasionally pale central area. **Male –** Patch of long black spines in front of tail base. **Female –** Patch of short bristles in front of tail base. Solitary or in pairs. Seaward reefs in 3-18 m.
Indo-Pacific: Red Sea to Indonesia, Philippines, Micronesia, Solomon Is., French Polynesia. – S. Japan to S. Great Barrier Reef.

LEAFY FILEFISH *Chaetodermis penicilligerus*
SIZE: to 18 cm (7 in.) Filefishes – Monacanthidae
ID: Brown to brownish yellow; shaggy, **covered with numerous skin-flaps,** wavy brown lines on side, black dots on transparent dorsal, anal and tail fins. Solitary. Weed-covered bottoms of coastal reefs in 2-25 m.
Asian Pacific: Sabah in Malaysia, Indonesia, Philippines and Papua New Guinea. – S. Japan to Great Barrier Reef.

BEARDED FILEFISH *Anacanthus barbatus*
SIZE: to 35 cm (14 in.) Filefishes – Monacanthidae
ID: Elongate with long dark barbel (beard) extending from chin; shades of brown with netted pattern of dark markings and dusky bars. Solitary or small groups in mixed sand-weed areas. Coastal reefs and mangrove estuaries in 3-20 m.
East Indo-Asian Pacific: India to N.W. Australia and Indonesia.

427

Filefishes

BARRED FILEFISH *Cantherhines dumerilii*
SIZE: to 38 cm (15 in.) Filefishes – Monacanthidae
ID: Brownish gray to blue-gray; **faint dark bars on rear body,** yellow iris, 4 yellowish spines on tail base, black line marking above pectoral fin base. Usually in pairs; feed on live corals. Coastal, lagoon and outer reefs to 35 m.
Indo-Pacific: E. Africa to Indonesia, Philippines, Micronesia, Papua New Guinea, Hawaii, French Polynesia. – S. Japan to Australia.

Barred Filefish – Juvenile
SIZE: 3-8 cm (1¼ - 3¼ in.)
ID: Gray with white spots; yellow iris, yellow fins and black line marking above pectoral fin base.

SPECTACLED FILEFISH *Cantherhines fronticinctus*
SIZE: to 24 cm (9½ in.) Filefishes – Monacanthidae
ID: Yellowish brown with some dark spots and blotches; **black band between eyes,** white band on tail base. Solitary; shy remain close to cover. Outer reefs to 40 m, usually below 15 m.
Indo-Asian Pacific: E. Africa to Indonesia, Philippines, Micronesia, Papua New Guinea and Solomon Is. – S. Japan to Australia.

WIRENET FILEFISH *Cantherhines pardalis*
SIZE: to 25 cm (10 in.) Filefishes – Monacanthidae
ID: Bluish gray to bluish brown with bluish stripes on head and pale blue netted pattern on body; usually white spot on upper tail base. Solitary. Outer reefs in 2-20 m.
Indo-Pacific: Red Sea and E. Africa to Indonesia, Philippines, Micronesia, Papua New Guinea, Solomon Is. and French Polynesia. – S. Japan to Australia.

WAXY FILEFISH *Cantherhines cerinus*
SIZE: to 12 cm (4¾ in.) Filefishes – Monacanthidae
ID: Juvenile – [pictured] Yellow with narrow brown bar behind eye to lower gill cover; few indistinct dark blotches and bands on body. **Adult** – Yellowish gray-brown; markings similar. Outer reefs in 15-28 m.
Localized: Known only from Luzon and Calamianes Is., Philippines.

FAN-BELLIED FILEFISH *Monacanthus chinensis*
SIZE: to 38 cm (15 in.) Filefishes – Monacanthidae
ID: Pale brown with medium brown blotches covered with dark brown spots; **large fan-shaped ventral skin flap,** concave snout. Solitary. Sheltered coastal reefs and rocky areas, frequently in weed beds in 5-50 m.
West Pacific: Sabah in Malaysia, Indonesia, Philippines and Papua New Guinea. – S. Japan to E. Australia, Fiji. Absent Solomon Is.

RHINO FILEFISH   *Pseudalutarius nasicornis*
SIZE: to 18 cm (7 in.)   Filefishes – Monacanthidae
ID: Male – Elongate with **dorsal spine in front of eye;** white to bluish white with 2 dark orangish brown stripes from head to tail, often small gold spots. **Female** – Body less deep; lack gold spots. Solitary or small groups. Sheltered coasts in 2-55 m.
Indo-Asian Pacific: E. Africa to Indonesia, Philippines and Micronesia. – S. Japan to E. Australia and New Caledonia.

PUFFER FILEFISH   *Brachaluteres taylori*
SIZE: to 5 cm (2 in.)   Filefishes – Monacanthidae
ID: White to gray or yellowish green; brown to dark gray horizontal lines on body with rows of brown to white spots between, circular head and body with only snout slightly protruding. Solitary. Bottom growth of coastal reefs in 8-35 m.
Asian Pacific: Indonesia, Philippines, Micronesia, Papua New Guinea and E. Australia.

MINUTE FILEFISH   *Rudarius minutus*
SIZE: to 3 cm (1¼ in.)   Filefishes – Monacanthidae
ID: Tiny; pale gray with numerous pale brown spots and fine speckling. **Male** – Dark spot with pale rim above anal-fin base. Solitary or form small groups among gorgonian, soft coral or fire coral branches. Coastal reefs and lagoons in 2-15 m.
Asian Pacific: Sabah in Malaysia, Indonesia, Micronesia, Papua New Guinea and Great Barrier Reef.

DIAMOND FILEFISH   *Rudarius excelsus*
SIZE: to 2.5 cm (1 in.)   Filefishes – Monacanthidae
ID: Tiny with a somewhat diamond-shaped body; green to brown with numerous small skin flaps. **Male** – Long bristles on side of tail base. Weedy bottoms near reefs in 8-25 m.
Asian Pacific: Sabah in Malaysia, Indonesia and Papua New Guinea. – S.W. Japan to Great Barrier Reef.

MIMIC FILEFISH   *Paraluteres prionurus*
SIZE: to 10 cm (4 in.)   Filefishes – Monacanthidae
ID: White with brown to black spots; 2 dark brown saddles across back, yellow tail and highlights on fins. Mimic Black-saddled Toby (p. 435) which lacks file-like 1st dorsal fin. Solitary or small groups. Seaward reefs to 25 m.
Indo-West Pacific: E. Africa to Indonesia, Philippines, Micronesia and Solomon Is. – S. Japan to Great Barrier Reef and Fiji.

Mimic Filefish – Juvenile
SIZE: 3 - 5 cm (1¼ - 2 in.)
ID: Dark saddle markings more pronounced and extend onto belly.

Filefishes

STRAPWEED FILEFISH   *Pseudomonacanthus macrurus*

SIZE: to 23 cm (9 in.)   Filefishes – Monacanthidae

ID: Highly variable color and markings, mottled shades of yellow to brown, greenish; **head and body covered with numerous dark spots,** tan and grayish, usually darkish blotch below and behind eye. Solitary. Reef flats, lagoons and weed or seagrass beds to 10 m.

Asian Pacific: Sabah in Malaysia, Indonesia and Philippines.

Strapweed Filefish – Phase

ID: Colors can be rapidly change, intensified or pale; usually a blue spotted border on extendible belly appendage.

Strapweed Filefish – Juvenile

SIZE: 5-7.5 cm (2-3 in.)

ID: Mottled shades of light tan, pale and dark spots; note dark blotch behind and below eye.

BLACKBAR FILEFISH   *Pervagor janthinosoma*

SIZE: to 14 cm (5 1/2 in.)   Filefishes – Monacanthidae

ID: Brown to olive; blue spots on dorsal and anal fins, orangish tail, dark bar above pectoral fin base. Solitary and secretive. Lagoon and seaward reefs to 20 m.

Indo-West Pacific: E. Africa to Indonesia, Philippines, Micronesia, Papua New Guinea and Solomon Is. – S.W. Japan to Great Barrier Reef, New Caledonia and Fiji.

BLACKHEADED FILEFISH   *Pervagor melanocephalus*

SIZE: to 11 cm (4 1/4 in.)   Filefishes – Monacanthidae

ID: Dark brown head, orange body with red or orange tail; **orange spots on snout,** dark bar above pectoral fin base. Solitary or in pairs; remain close to shelter. Seaward reefs in 3-40 m.

West Pacific: Indonesia, Philippines, Micronesia, Papua New Guinea and Solomon Is. – S.W. Japan to Great Barrier Reef to Fiji.

ORANGETAIL FILEFISH   *Pervagor aspricaudus*

SIZE: to 12 cm (4 3/4 in.)   Filefishes – Monacanthidae

ID: Blue-gray head and front of body with **numerous tiny black spots;** orange rear body and tail. Solitary. Corals and rubble areas of lagoons and outer reef slopes to 25 m.

Indo-Pacific: Mauritius to Japan, Micronesia and Hawaii, south to Great Barrier Reef, New Caledonia; primarily oceanic. Absent from much of Asian Pacific.

YELLOWEYE FILEFISH *Pervagor alternans*
SIZE: to 16 cm (6 1/4 in.) Filefishes – Monacanthidae
ID: Dark brown with black horizontal lines; **orange iris and adjacent ring around eye,** orange tail markings with a black submarginal band. Solitary. Coral and rocky reefs to 15 m.
Asian Pacific: Micronesia to Great Barrier Reef and New Caledonia.

BLACK-LINED FILEFISH *Pervagor nigrolineatus*
SIZE: to 10 cm (4 in.) Filefishes – Monacanthidae
ID: Brown with **white stripe from snout to dorsal fin base;** frequently curved white band below eye to midbody. Solitary and secretive; often around soft corals. Lagoon and outer reefs in 3-25 m.
Asian Pacific: Sabah in Malaysia, Indonesia, Philippines, Micronesia, Papua New Guinea and Solomon Is.

SHORTSNOUT FILEFISH *Paramonacanthus curtorhynchos*
SIZE: to 11 cm (4 1/4 in.) Filefishes – Monacanthidae
ID: Tan to yellowish with tiny brown spots and **dark brown rectangular marking below beginning of soft dorsal fin.** Solitary. Sandy or weed-covered bottoms of coastal reefs in 1-20 m.
East Indo-West Pacific: India to Indonesia, Philippines, Micronesia, Papua New Guinea, and Solomon Is. to Great Barrier Reef and Fiji.

Shortsnout Filefish – Juvenile
SIZE: 3-4.5 cm (1 1/4 - 1 3/4 in.)
ID: White to tan with wide brown stripe from eye to upper base of tail interspersed with a dark rectangular marking below beginning of soft dorsal fin, brown midlateral stripe. Can rapidly change, darken or pale both colors and markings.

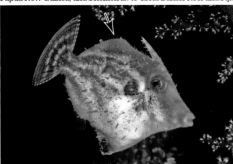

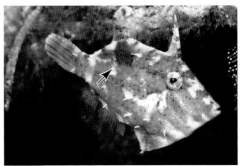

FAINTSTRIPE FILEFISH *Paramonacanhus pusillus*
SIZE: to 18.5 cm (7 1/4 in.) Filefishes – Monacanthidae
ID: Grayish upper body with 3-4 broad diffuse brown stripes; white head and lower body, **base of 2nd dorsal fin elevated.** Sand and mud bottoms in 20-40 m, occasionally in *Sargassum* rafts near surface.
Asian Pacific: Red Sea to Indonesia, Philippines, Papua New Guinea and Solomon Is. – Japan to N.W. Australia.

WHITEBAR FILEFISH *Paramonacanthus choirocephalus*
SIZE: to 15 cm (6 in.) Filefishes – Monacanthidae
ID: White to gray with brown to black blotches; **usually dark blotch below 2nd dorsal fin,** deep diamond shaped body. Solitary or form groups; often shelter among crinoid arms. Sheltered coastal reefs to 25 m.
Asian Pacific: Indonesia, Philippines, Papua New Guinea to Australia. Absent from Solomon Is.

BRISTLE-TAILED FILEFISH
Acreichthys tomentosus

SIZE: to 11.5 cm (4 1/2 in.)   Filefishes – Monacanthidae

ID: Mottled green, yellow, brown or white; **diffuse white stripe through eye,** numerous skin flaps on body. Solitary. Sheltered reefs frequently in seagrass beds to 15 m.

East Indo-West Pacific: Sri Lanka to Indonesia, Philippines, Micronesia and Solomon Is. – S.W. Japan to Australia, New Caledonia.

RADIAL FILEFISH
Acreichthys radiatus

SIZE: to 7 cm (2 3/4 in.)   Filefishes – Monacanthidae

ID: Brown and white bands radiate from eye onto head and forebody; irregular brown and white bars on body and tail base. Often concealed among tentacles of *Xenia* soft coral. Coastal and lagoon reefs in 2-12 m.

Asian Pacific: Indonesia, Philippines and Papua New Guinea. – S.W. Japan to Great Barrier Reef and New Caledonia.

LONGNOSE FILEFISH
Oxymonacanthus longirostris

SIZE: to 10 cm (4 in.)   Filefishes – Monacanthidae

ID: Blue-green with rows of orange spots; black spot on tail, elongated snout with small upturned mouth. In pairs or small groups among *Acropora* coral branches on which it feeds. Lagoons and seaward reefs to 35 m.

Indo-West Pacific: E. Africa to Indonesia, Philippines, Micronesia and Solomon Is. – S.W. Japan to Australia and Fiji.

STAR PUFFER
Arothron stellatus

SIZE: to 90 cm (3 ft.)   Puffers – Tetraodontidae

ID: Pale gray with dense covering of black spots on head, body and fins; larger black spots or irregular blotches around pectoral fin base. Solitary. Lagoon and seaward reefs in 3-58 m.

Indo-Pacific: Red Sea to Indonesia, Philippines, Micronesia, Papua New Guinea, Solomon Is., French Polynesia. – S. Japan to Australia.

Star Puffer – Older Juvenile

SIZE: 4-10 cm (1 1/2 - 4 in.)

ID: Gray to light gray to yellowish tan with numerous black spots; yellow to orange patches and **curved blackish bands on belly,** which disappear with age. Solitary in shallow sand, rubble or weedy areas near protected inner reefs.

Star Puffer – Young Juvenile

SIZE: 2.5-3 cm (1 - 1 1/4 in.)

ID: Orange covered with curving black line markings. Solitary in shallow sand, rubble and weedy areas near protected inner reefs.

BLUE-SPOTTED PUFFER   *Arothron caeruleopunctatus*

SIZE: to 80 cm (2 1/2 ft.)   Puffers – Tetraodontidae

ID: Yellowish brown back, bluish below and covered with small blue spots; alternating light and dark rings around eye, **black patch with white spots on pectoral-fin base.** Solitary. Seaward reef slopes in 2-45 m.

East Indo-West Pacific: Maldives to Indonesia, Micronesia, Papua New Guinea, Solomon Is. – Japan to Australia, New Caledonia, Fiji.

Blue-Spotted Puffer – Variation

ID: Brown with only a few blue spots, except on fins; alternating light and dark rings around eye, network of line markings on back, black patch with white spots on pectoral fin base, occasionally black-edged white spots over body.

RETICULATED PUFFER   *Arothron reticularis*

SIZE: to 43 cm (17 in.)   Puffers – Tetraodontidae

ID: Brown with white spots on back and rear body and tail; white bands encircle eye and pectoral fin base, **white curved bands run from head to pectoral fin.** Solitary. Mangrove estuaries, tidal creeks and open sand or mud bottoms to 20 m.

East Indo-West Pacific: India to Indonesia, Philippines and Papua New Guinea. – S.W. Japan to Australia and Fiji.

WHITE-SPOTTED PUFFER   *Arothron hispidus*

SIZE: to 48 cm (19 in.)   Puffers – Tetraodontidae

ID: Gray to greenish brown upper body, paler below covered with white spots; large white-edged black spot around pectoral fin base. Solitary. Mixed coral, sand, rubble and weed bottoms in 1-50 m.

Indo-Pacific: Red Sea and E. Africa to Indonesia, Micronesia, Papua New Guinea, Hawaii, French Polynesia. – S.W. Japan to Australia.

MAP PUFFER   *Arothron mappa*

SIZE: to 60 cm (2 ft.)   Puffers – Tetraodontidae

ID: Gray with black maze markings, white to yellowish belly; **spoke-like lines radiate from eye,** irregular black botch around pectoral fin base and another on belly. Solitary. Lagoon and seaward reefs in 4-30 m.

Indo-West Pacific: E. Africa to Indonesia, Philippines, Micronesia, Papua New Guinea and Solomon Is. – S.W. Japan to Australia, Fiji.

Map Puffer – Young Juvenile

SIZE: 2.5-6 cm (1-2 1/4 in.)

ID: Gray to shades of brown with lighter, nearly white, blotches; black honeycomb markings on body and head, several light and dark lines radiate from eye, irregular black blotch around pectoral fin base. Solitary, usually in shallow rocky areas.

Puffers

IMMACULATE PUFFER   *Arothron immaculatus*
SIZE: to 28 cm (11 in.)   Puffers – Tetraodontidae
ID: Brown shading to white underside; large yellowish brown to black blotch around pectoral fin base, black-edged yellowish tail. Solitary. Coastal mangroves, seagrass and weedy and silt bottoms near reefs in 3-30 m.
Indo-Asian Pacific: Red Sea and E. Africa to W. Indonesia, Philippines and S.W. Japan.

Immaculate Puffer – Variation
ID: Occasionally gray blotches and mottling.

STRIPED PUFFER   *Arothron manilensis*
SIZE: to 31 cm (12 1/2 in.)   Puffers – Tetraodontidae
ID: Brown to greenish gray with **dark brown to yellowish stripes;** large dark spot around pectoral fin base. Coastal mangroves, seagrass beds, weedy areas and sand and silt bottoms near reefs in 2-20 m.
West Pacific: Indonesia, Philippines, Papua New Guinea and Solomon Is. – S.W. Japan to E. Australia to Fiji.

GUINEAFOWL PUFFER   *Arothron meleagris*
SIZE: to 32 cm (12 1/2 in.)   Puffers – Tetraodontidae
ID: Black covered with small white spots. Solitary; feed mainly on live tips of branching corals. Coral-rich areas and rocky bottoms of coastal, lagoon and outer reefs to 15 m.
Indo-Pacific: E. Africa to S.E. Indonesia, Hawaii, and Pitcairn Is. – S. Japan to Australia. Most common around oceanic islands.

Guineafowl Puffer – Variation
ID: Occasionally bright yellow-gold, may display a few scattered black spots.

Guineafowl Puffer – Juvenile
SIZE: 2.5 - 3 cm (1 - 1 1/4 in.)
ID: Black with bright orange-yellow spots.

BLACKSPOTTED PUFFER
Arothron nigropunctatus
SIZE: to 30 cm (12 in.)   Puffers – Tetraodontidae
ID: Highly variable in wide combination of colors, most commonly pale gray; **consistent markings include black lips,** black pectoral fin base and a few scattered black spots. Solitary. Coral-rich areas of lagoons to seaward reefs in 3-25 m.
Indo-West Pacific: E. Africa to Indonesia, Philippines, Micronesia, Papua New Guinea and Solomon Is. – S.W. Japan to E. Australia, Fiji.

Blackspotted Puffer – Variation
ID: Feed heavily on living corals, occasionally on other cnidarians, crustaceans and molluscs.

Blackspotted Puffer – Variation
ID: Many variations display a white bar across snout and often a dark 2nd dorsal fin.

BICOLORED TOBY
Canthigaster smithae
SIZE: to 9 cm (3 1/2 in.)   Puffers – Tetraodontidae
ID: Brown back with white below; **brown to yellowish stripe on lower side from chin to lower tail base,** yellow iris with blue "spokes," dark margins on tail. Solitary. Seaward reefs slopes in 20-40 m.
Indian Ocean: E. Africa and islands of W. Indian Ocean to Andaman Sea and W. Sumatra in Indonesia.

CROWN TOBY
Canthigaster axiologus
SIZE: to 13.5 cm (5 1/4 in.)   Puffers – Tetraodontidae
ID: White with dark brown bar between eyes and 3 wedge-shaped; **dark brown saddles on back outlined with yellow-orange spots.** Similar Black-saddled Toby [next], saddles extend onto side. Solitary. Sand and rubble of seaward reefs in 10-80 m.
West Pacific: Indonesia to Philippines, Micronesia, Papua New Guinea and Solomon Is. – S.W. Japan to Australia and Fiji.

BLACK-SADDLED TOBY
Canthigaster valentini
SIZE: to 10 cm (4 in.)   Puffers – Tetraodontidae
ID: White with light brown spots and **4 dark brown to blackish saddles, the middle two extend onto lower side.** Mimicked by Mimic Filefish. Solitary or form small groups. Lagoon and seaward reefs to 50 m.
Indo-Pacific: Red Sea to Indonesia, Philippines, Micronesia, Papua New Guinea, Solomon Is., French Polynesia. – S. Japan to Australia.

435

Puffers

AMBON TOBY *Canthigaster amboinensis*
SIZE: to 15 cm (6 in.) Puffers – Tetraodontidae
ID: Brown to orange-brown with bluish hue on underside; dark brown to blue spots and bands on head, small blue and blackish spots on body. Solitary. Shallow outer reefs affected by surge to 10 m.
Indo-Pacific: E. Africa to Indonesia, Philippines, Micronesia, Solomon Is. and French Polynesia. – S. Japan to S.E. Australia.

Ambon Toby – Juvenile
SIZE: 4-8 cm (1 1/2 - 3 1/4 in.)
ID: Brown; small white spots on cheek and larger white spots on body, **white lines radiate from eye and wrap around snout.**

LANTERN TOBY *Canthigaster epilampra*
SIZE: to 11 cm (4 1/4 in.) Puffers – Tetraodontidae
ID: Darkish upper back white below; blue streaked yellow tail, yellowish around eye with blue radiating lines, blue lines on snout, tiny blue spots on body. Solitary, usually in caves and under ledges. Steep outer reefs slopes in 9-60 m.
Pacific: Indonesia, Philippines, Papua New Guinea, Solomon Is., French Polynesia. – S.W. Japan to Australia and New Caledonia.

HONEYCOMB TOBY *Canthigaster janthinoptera*
SIZE: to 9 cm (3 1/2 in.) Puffers – Tetraodontidae
ID: Brown with dense covering of white spots; occasionally with lines radiating from eyes. Solitary or in pairs; usually close to shelter or in caves and crevices. Lagoon and seaward reefs to 30 m.
Indo-Pacific: Red Sea to Indonesia, Philippines, Micronesia, Papua New Guinea, Solomon Is., French Polynesia. – S. Japan to Australia.

LEOPARD TOBY *Canthigaster leoparda*
SIZE: to 7 cm (2 3/4 in.) Puffers – Tetraodontidae
ID: Pale brown; blue lines on snout and nape, blue-ringed brown spot on upper nape, **clusters of brown blotches** and small blue spots on side. Solitary or in pairs; usually in caves or under ledges. Steep outer reefs in 30-50 m.
Asian Pacific: Indonesia, Philippines and Micronesia.

TYLER'S TOBY *Canthigaster tyleri*
SIZE: to 8 cm (3 1/4 in.) Puffers – Tetraodontidae
ID: Pale brown to orange; blue lines on snout and nape, **large brown spots cover body.** Solitary; usually in caves or under ledges. Steep outer reef slopes in 8-40 m.
Indian Ocean: Tanzania, Comores, Mauritius, Christmas I. to Molucca Is. in E. Indonesia.

FINGERPRINT TOBY   *Canthigaster compressa*
SIZE: to 11 cm (4 1/4 in.)   Puffers – Tetraodontidae
ID: Brownish with pale belly; **orangish tail with thin bluish bars,** numerous wavy white to blue to green lines cover body, ocellated black spot below dorsal fin base. Silty bays and harbors, often around wharf pilings in 2-25 m.
Asian Pacific: Indonesia to Micronesia, Papua New Guinea and Solomon Is. – S. Japan to Great Barrier Reef and Vanuatu.

WHITEBELLY TOBY   *Canthigaster bennetti*
SIZE: to 10 cm (4 in.)   Puffers – Tetraodontidae
ID: Brown upper body, whitish below; **blue lines radiate from eye often interspersed with orange,** blue-edged black spot at base of dorsal fin. Solitary or form groups. Sand and rubble bottoms of sheltered coastal reefs, reef flats and lagoons to 15 m.
Indo-West Pacific: E. Africa to Indonesia, Philippines, Micronesia, Solomon Is. and Hawaii. – S. Japan to Great Barrier Reef.

INDIAN TOBY   *Canthigaster petersii*
SIZE: to 9 cm (3 1/2 in.)   Puffers – Tetraodontidae
ID: Brown, covered with white, blue or greenish spots, similar colored lines on snout and back; ocellated spots on dorsal fin base. Solitary. Lagoon and seaward reefs in 1-25 m.
Indian Ocean: E. Africa to Andaman Sea.

SHY TOBY   *Canthigaster ocellicincta*
SIZE: to 6.5 cm (2 1/2 in.)   Puffers – Tetraodontidae
ID: Brown; **2 dark brown to black saddles with white band between extend onto belly,** black spot on dorsal fin base, bluish line markings on head, dark spots on body. Solitary; in caves and crevices. Steep seaward reef slopes in 20-53 m.
West Pacific: Indonesia, Philippines, Micronesia, E. Papua New Guinea, Solomon Is. – S.W. Japan to Great Barrier Reef and Fiji.

PAPUAN TOBY   *Canthigaster papua*
SIZE: to 9 cm (3 1/2 in.)   Puffers – Tetraodontidae
ID: Reddish brown; covered with white to blue to green spots including tail, white to blue to green lines on snout and back, ocellated black spot on dorsal fin base, **orange around mouth.** Solitary or in pairs. Coral reefs to 35 m.
Asian Pacific: Indonesia, Philippines, Micronesia, Papua New Guinea, Solomon Is. to Great Barrier Reef and New Caledonia.

SOLANDER'S TOBY   *Canthigaster solandri*
SIZE: to 10.5 cm (4 1/4 in.)   Puffers – Tetraodontidae
ID: Similar to Indian Toby [above left] but generally with fewer pale spots and belly orange; **yellow-orange on tail.** Solitary. Lagoon and seaward reefs in 1-55 m.
Pacific: S.W. Japan, Micronesia (except Palau) and E. Australia to French Polynesia.

SHORTFIN PUFFER *Torquigener brevipinnis*
SIZE: to 14 cm (5 1/2 in.) Puffers – Tetraodontidae
ID: Brown back with numerous whitish spots, white to tan below; four bars below eye extend from mouth to pectoral fin. Form small groups. Shallow rubble, sand, weed and seagrass areas to 100 m.
Asian Pacific: S. Japan and Indonesia to Great Barrier Reef and New Caledonia.

MILKSPOTTED PUFFER *Chelonodontops patoca*
SIZE: to 33 cm (13 in.) Puffers – Tetraodontidae
ID: Whitish to brown undercolor with large oval white spots; broad dark band behind pectoral fin another under dorsal fin, white to pale yellow underside. Solitary. Estuaries and mangrove areas, occasionally around sheltered inshore coral reefs to 5 m.
East Indo-Asian Pacific: Arabian Gulf to Indonesia, Philippines, Papua New Guinea and Solomon Is. – S.W. Japan to Australia.

SILVER PUFFER *Lagocephalus sceleratus*
SIZE: to 85 cm (2 3/4 ft.) Puffers – Tetraodontidae
ID: Elongate bright silver; back with greenish to bluish tinting and dark spots or blotches. Usually form pelagic schools to 100 m; occasionally over shallow sand areas around reefs and may rest on bottom. Can be aggressive and bite divers.
Indo-Pacific: Red Sea and E. Africa to French Polynesia. – S. Japan to Australia.

SPOTTED BURRFISH *Chilomycterus reticulatus*
SIZE: to 70 cm (2 1/4 ft.) Porcupinefishes – Diodontidae
ID: Brown to gray with white underside; numerous fixed triangular spines, body and fins covered with black spots, 3 dusky body bars and dusky bar under eye. Solitary. Rocky and coral reefs in 3-25 m.
Circumglobal: Rare in tropical waters, more common in subtropical and warm temperate seas.

YELLOW-SPOTTED BURRFISH *Cyclichthys spilostylus*
SIZE: to 28 cm (11 in.) Porcupinefishes – Diodontidae
ID: Brown to gray with white underside; numerous fixed triangular spines with a yellow or dark spot around base, occasionally 3 large dusky blotches on side. Solitary. Coral, sand and rubble and weed bottoms in 3-90 m.
Indo-Pacific: Red Sea to Indonesia, Philippines and Galapagos Is.

ORBICULAR BURRFISH *Cyclichthys orbicularis*
SIZE: to 14 cm (5 1/2 in.) Porcupinefishes – Diodontidae
ID: Brown to red-brown; several large darkish gray to red-brown blotches, stout fixed spike-like spines, unspotted yellowish translucent fins. Solitary. Sand and rubble bottoms in 5-30 m.
Indo-Asian Pacific: Red Sea to Andaman Sea, Indonesia, Philippines, Micronesia and Papua New Guinea. – Japan to Great Barrier Reef.

PORCUPINEFISH *Diodon hystrix*
SIZE: to 71 cm (2 1/4 ft.)   Porcupinefishes – Diodontidae
ID: Yellow-brown to brown, olive or gray with white underside; numerous short to medium movable spines, **numerous small black spots on head, body and fins.** Solitary, except when courting. Coral and rocky reefs to 50 m.
Circumglobal: Tropical, subtropical and warm temperate seas.

BALLOONFISH *Diodon holocanthus*
SIZE: to 38 cm (15 in.)   Porcupinefishes – Diodontidae
ID: Light gray-brown with white underside; numerous long moveable spines, **covered with small black spots except fins,** brown bar under eye, several brown blotches on back. Solitary. Coral reefs and open sand to 100 m.
Circumglobal: Tropical, subtropical and warm temperate seas.

BLACK-BLOTCHED PORCUPINEFISH *Diodon liturosus*
SIZE: to 50 cm (20 in.)   Porcupinefishes – Diodontidae
ID: Brown with numerous short to long movable spines; **large dark brown to black blotches with white margins on back** and around and below eye. Solitary. Coral reefs to 90 m.
Indo-Pacific: Red Sea and E. Africa to Indonesia, Philippines, Micronesia to Solomon Is. and French Polynesia. – S. Japan to Australia.

BLACKLIP PORCUPINEFISH *Lophodiodon calori*
SIZE: to 30 cm (12 in.)   Porcupinefishes – Diodontidae
ID: Brown with white underside; short movable spines on head and belly and fixed spines on back and sides, **black lips,** blackish bar below eye, 2-3 dusky dark patches on lower side. Solitary. Sand and rubble bottoms in 10-50 m.
Indo-Asian Pacific: E. Africa to S. Indonesia and N.W. Australia.

OCEAN SUNFISH *Mola mola*
SIZE: to 420 cm (14 ft.)   Molas – Molidae
ID: Broad oval body with long dorsal and anal fins and no tail; silver to gray to gray-brown gradating to whitish belly. Solitary or in pairs. Generally oceanic, often near surface, occasionally near reefs to be cleaned.
Circumgobal: Tropical and temperate seas.

DOLPHINFISH *Coryphaena hippurus*
SIZE: to 162 cm (5 1/4 ft.)   Dolphinfishes – Coryphaenidae
ID: Bluish silver with yellow tinting; long continuous dorsal fin from above eye to tail base, long ventral fins and large forked tail. **Male** – Large blunt head. **Female** – Torpedo-shaped head. Form groups; pelagic, often under *Sargassum* floats.
Circumtropical.

Pipefishes & Seahorses

This ID Group consists of fishes with long tubular snouts and elongate bodies encased in bony ring-like body segments.

FAMILY: Ghost Pipefishes – Solenostomidae
Single Genus – 6 Species Included

Ornate Ghost Pipefishes

Ghost pipefishes must rank among the most exotic creatures inhabiting the Earth's shallow seas. Unfortunately, both the classification and natural history of these close relatives of pipefishes and seahorses lack adequate investigation. At present, only four forms are considered to be valid species. Adding confusion, each form appears to have several distinct variations. The little presently known about the natural history of the genus is briefly discussed in the Ornate Ghost Pipefish gallery on the facing page.

FAMILY: Pipefishes & Seahorses – Syngnathidae
Pipefish Subfamily – Syngnathinae, 14 Genera – 36 Species Included
Seahorse Subfamily – Hippocampinae, Single Genus – 12 Species Included

---

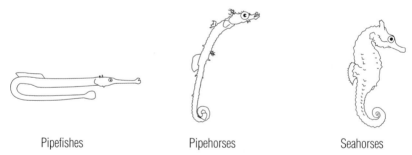

| Pipefishes | Pipehorses | Seahorses |

Pipefishes and seahorses have slender bodies fashioned by encircling bony segments, elongate snouts, and a single spineless dorsal fin. The heads of seahorses angle down from the body's main axis, and their bodies end with curved prehensile tails. Pipefishes have basically straight bodies generally culminating with small, rounded, flat tails. Both groups are carnivorous feeding on tiny crustaceans, which are sucked whole into their mouths. Seahorses generally attach to a holdfast near the bottom where they blend expertly with their surroundings. Pipefishes characteristically slip over the bottom searching for prey; however, species in genus *Doryrhamphus*, commonly known as flagtail pipefishes, hover in pairs or small groups inside crevices or under overhangs where they act as cleaners, removing small parasites from client fishes.

The family's reproductive behavior equals the members' peculiar appearance. An extended, harmonized courtship dance culminates with intertwined bodies. Female seahorses deposit from 50 to 500 eggs into an enclosed abdominal pouch of males where fertilization occurs. The eggs of pipefishes are attached to an external abdominal patch on males where the mass is easily observed. Highly developed offspring hatch after an extended incubation period.

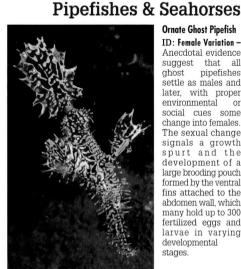

Ornate Ghost Pipefish

ID: Female Variation – Anecdotal evidence suggest that all ghost pipefishes settle as males and later, with proper environmental or social cues some change into females. The sexual change signals a growth spurt and the development of a large brooding pouch formed by the ventral fins attached to the abdomen wall, which many hold up to 300 fertilized eggs and larvae in varying developmental stages.

ORNATE GHOST PIPEFISH *Solenostomus paradoxus*
SIZE: to 11 cm (4 1/4 in.) Ghost Pipefishes – Solenostomidae
ID: Male – Short skin filaments on snout and body and fins with jagged edges giving "spiky appearance." Solitary, small groups or more commonly in male/female pairs; usually remain in restricted home range. Coastal, lagoon and outer reefs in 4-35 m.

Indo-West Pacific: Red Sea and E. Africa to Indonesia, Philippines, Micronesia, Solomon Is., E. Australia and Fiji.

Ornate Ghost Pipefish

ID: Female Variation – It is believed that Ghost Pipefishes have an annual cycle, with the majority of their life spent in the pelagic as larvae before settling to the bottom and becoming sexually mature.

Ornate Ghost Pipefish

ID: Female Variation – Frequently hover head down among arms of crinoids, black corals, gorgonians and soft corals. Feed primarily on tiny crustaceans, which are snapped from open water.

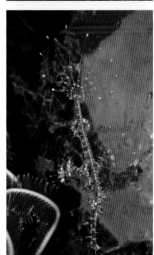

Ornate Ghost Pipefish

ID: Male Variation – Highly variable colors usually dependent on habitat, such as host crinoids.

Ornate Ghost Pipefish

ID: Juvenile – Recently settled individuals have large wispy transparent tails with silver to white tips on the tail and dorsal fin spines. As they mature, in the first few days after settling, the large wispy tails gradually become smaller and acquire pigmentation.

Ghost Pipefishes – Pipefishes

HALIMEDA GHOST PIPEFISH   *Solenostomus halimeda*

SIZE: to 6.4 cm (2 ¹/₂ in.)   Ghost Pipefishes – Solenostomidae

ID: Female – Green to whitish gray; fin lobes rounded resembling the leaf-like segments of *Halimeda* algae where they shelter. Solitary or in pairs. Sheltered coastal and lagoon reefs in 3-23 m.

East Indo-West Pacific: Maldives, N.W. Australia, Indonesia, Papua New Guinea, Micronesia and Fiji.

Halimeda Ghost Pipefish – Male Variation

ID: In areas with stands of dead algal patches, change color to camouflage with surroundings.

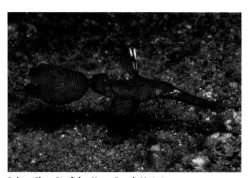

ROBUST GHOST PIPEFISH   *Solenostomus cyanopterus*

SIZE: to 15 cm (6 in.)   Ghost Pipefishes – Solenostomidae

ID: Female/Male Pair – Red, green, purplish, yellow or brown, often with darker mottling; long narrow dorsal fin, fan-like ventral and tail fins, **short tail base.** Solitary or in pairs. Coastal and lagoon reefs to 10 m.

Indo-West Pacific: Red Sea to Indonesia, Philippines, Micronesia, Papua New Guinea, Solomon Is. – S. Japan to Australia and Fiji.

Robust Ghost Pipefish – Young Female Variation

ID: A relatively rare red variation.

ROUGHSNOUT GHOST PIPEFISH

Solenostomus paegnius

SIZE: to 12 cm (4 ³/₄ in.)

Ghost Pipefishes – Solenostomidae

ID: Shades of brown, light tan and green with skin filaments on snout and body (often long and hair-like). Solitary or in pairs. Sand and rubble areas often with filamentous algae to 10 m.

Indo-West Pacific: E. Africa to Indonesia, Solomon Is. – S. Japan to E. Australia and Fiji.

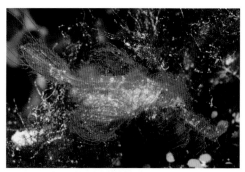

Roughsnout Ghost Pipefish – Variation

ID: Extent of "hairy" growth variable, occasionally individuals have their outline entirely obscured. Hairy variety more common on and around objects covered with filamentous algae, usually in sand or rubble areas. Possibly a separate species. This variation is also commonly known as the Irish Setter, Rufus or Hairy Ghost Pipefish.

VELVET GHOST PIPEFISH
Solenostomus sp.
SIZE: to 5 cm (2 in.)
Ghost Pipefishes –
Solemidae
ID: Classification
uncertain; possibly a
variation of the Robust
Ghost Pipefish.
Asian Pacific:
Indonesia.

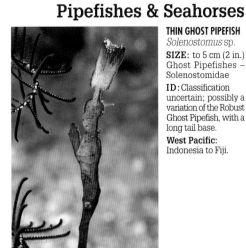

THIN GHOST PIPEFISH
Solenostomus sp.
SIZE: to 5 cm (2 in.)
Ghost Pipefishes –
Solenostomidae
ID: Classification
uncertain; possibly a
variation of the Robust
Ghost Pipefish, with a
long tail base.
West Pacific:
Indonesia to Fiji.

SHORTPOUCH PYGMY PIPEHORSE *Acentronura breviperula*
SIZE: to 5 cm (2 in.)   Pipefishes – Syngnathidae
ID: Brown to nearly black with white speckling; slightly
expanded seahorse-like "chest," pale snout, skin flaps of
various sizes and shapes on head and body. Solitary or in pairs.
Sparse seagrass near sheltered reefs to 20 m.
Asian Pacific: Indonesia, Philippines, Papua New Guinea to
Fiji, but possibly more widespread.

Shortpouch Pygmy Pipehorse – Variation
ID: The species are typically light to dark brown, however
often acquire the colors of their habitat. Can have smooth
head and body with few filaments.

RUMENGAN'S PIPEHORSE *Kyonemichthys rumengani*
SIZE: to 2.7 cm (1 in.)   Pipefishes – Syngnathidae
ID: Tiny and hairlike in brown shades that match
surroundings; a long, single red or brown filament extends
from top of head, center of upper back and apex of trunk. Often
in pairs. Attach to growth on walls and on silt bottoms in 15-20 m.
West Pacific: Indonesia, Philippines, Papua New Guinea
and Fiji.

LYNNE'S PIPEFISH *Festucalex rufus*
SIZE: to 6 cm (2¼ in.)   Pipefishes – Syngnathidae
ID: Pink overall with some scattered white flecks. Found
with small thin pink sponge clusters that grow on steep
slopes and walls below 15 m.
Asian Pacific: Indonesia and Papua New Guinea.

Pipefishes

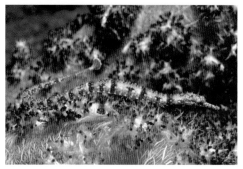

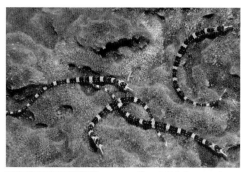

BROKEN-BANDS PIPEFISH   *Corythoichthys benedetto*
SIZE: to 7 cm (2³/₄ in.)   Pipefishes – Syngnathidae
ID: Tan; 12 dark encircling bars with a short non-encircling bar between, 2 rows of white spots run length of body, head and snout spotted. In pairs. Sheltered coastal and seaward reefs in 5-20 m.
Asian Pacific: Andaman Sea, Indonesia, Papua New Guinea and Great Barrier Reef.

BROWN-BANDED PIPEFISH   *Corythoichthys amplexus*
SIZE: to 9.5 cm (3³/₄ in.)   Pipefishes – Syngnathidae
ID: Whitish to pale gray; broad reddish brown bars with tiny white spots encircle body. Solitary or form loose groups. Lagoon and seaward reefs in 3-20 m.
Indo-West Pacific: Seychelles to Indonesia, Philippines, Papua New Guinea, Solomon Is. to Australia and Fiji.

MESSMATE PIPEFISH   *Corythoichthys intestinalis*
SIZE: to 17 cm (6³/₄ in.)   Pipefishes – Syngnathidae
ID: Whitish to pale yellowish with dark wavy or reticulated line stripes (absent on back); diffuse dark body bars. Usually in pairs, but occasionally form small aggregations. Coastal and lagoon reefs with corals and sponges in 3-12 m.
West Pacific: Indonesia, Philippines, Micronesia, Papua New Guinea, Australia and Fiji.

REEFTOP PIPEFISH   *Corythoichthys haematopterus*
SIZE: to 20 cm (8 in.)   Pipefishes – Syngnathidae
ID: Whitish with dark wavy line stripes forming netted pattern; **white-edged reddish tail**, about 15 wide dark bars often with pale centers. Solitary, in pairs or small groups. Sand, rubble and weeds in 2-15 m.
Indo-West Pacific: E. Africa to Indonesia. – S. Japan to N. Australia, Vanuatu and Fiji.

NETWORK PIPEFISH   *Corythoichthys flavofasciatus*
SIZE: to 18 cm (7 in.)   Pipefishes – Syngnathidae
ID: Pale yellowish undercolor with line stripes; black stripes on head and about 20 dusky bands encircle body. Pacific and Indian Ocean variations may prove to be separate species. Solitary, in pairs or groups. Sheltered reefs in 2-25 m.
Indo-Pacific: Red Sea and Madagascar to French Polynesia. – S. Japan to Australia.

SCHULTZ'S PIPEFISH   *Corythoichthys schultzi*
SIZE: to 15 cm (6 in.)   Pipefishes – Syngnathidae
ID: Pale tan with numerous dark-edged orange to brown rectangular spots (**lack white speckles**); widely spaced narrow whitish bars encircle body, **long snout**. Solitary or in pairs. Sand and rubble around reefs in 2-15 m.
Indo-West Pacific: Red Sea and E. Africa to Indonesia, Solomon Is. – S. Japan to N. Australia, Great Barrier Reef and Fiji.

BLACK-BREASTED PIPEFISH　*Corythoichthys nigripectus*
SIZE: to 11 cm (4¹/₄ in.)　　Pipefishes – Syngnathidae
ID: Pale gray undercolor; 4-5 wide lavender to orange body bars before dorsal fin, **orange gill cover,** black mark on breast. Solitary or in pairs. Among coral and algal patches primarily in clear water outer reefs in 5-30 m.
Pacific: Indonesia, Micronesia to New Caledonia, Fiji and French Polynesia.

ORANGE-SPOTTED PIPEFISH　*Corythoichthys ocellatus*
SIZE: to 11 cm (4¹/₄ in.)　　Pipefishes – Syngnathidae
ID: Tan with numerous dark-edged orange rectangular spots and white speckles; widely spaced narrow whitish bars encircle body, long snout. Solitary or in pairs. Sand, rubble and weed bottoms around reefs to 12 m.
West Pacific: Indonesia, Philippines, Papua New Guinea, Solomon Is., Great Barrier Reef and Fiji.

BARRED PIPEFISH　*Choeroichthys cinctus*
SIZE: to 4.5 cm (1³/₄ in.)　　Pipefishes – Syngnathidae
ID: White to yellow with dark body bars, bars on tail formed by spots; short with body and tail base about equal in length. Solitary. Tidal estuaries, algae and seagrass beds (often near reefs) and reefs exposed to freshwater discharge in 10-40 m.
West Pacific: Indonesia, Papua New Guinea and Solomon Is. to N. Great Barrier Reef, Vanuatu and Fiji.

SHORT-BODIED PIPEFISH　*Choeroichthys brachysoma*
SIZE: to 6.5 cm (2¹/₂ in.)　　Pipefishes – Syngnathidae
ID: **Male –** Tan with scattered small white spots; short with wide body, tail base about equal in length to body. **Female –** Slender body with 2 rows of black spots on side. Solitary. Lagoon, seaward reefs and seagrass areas to 20 m.
Indo-Pacific: Red Sea and E. Africa to Indonesia, Philippines, Micronesia and French Polynesia. – S. Japan to Australia.

ESTUARY PIPEFISH　*Hippichthys cyanospilos*
SIZE: to 16 cm (6¹/₄ in.)　　Pipefishes – Syngnathidae
ID: Yellow to greenish or nearly black; white and dark bars on lower half of snout, 12-15 whitish saddles or spots on back and midside. Solitary. Tidal estuaries, algae and seagrass beds and reefs exposed to freshwater discharge to 4 m.
Indo-West Pacific: Red Sea and E. Africa to Indonesia, Philippines, New Guinea, Solomon Is. – S.W. Japan to N. Australia and Fiji.

SHORT-TAILED PIPEFISH　*Trachyrhamphus bicoarctatus*
SIZE: to 40 cm (16 in.)　　Pipefishes – Syngnathidae
ID: Slender with tiny tail, **head often raised with a bent neck posture;** shades of green, brown, white and yellow with fine speckling and usually pale saddles. Solitary. Sand, rubble, and weeds of sheltered coastal reefs and lagoons to 25 m.
Indo-Asian Pacific: Red Sea and E. Africa to Indonesia, Micronesia, Solomon Is. – S. Japan to N. Australia, New Caledonia.

Pipefishes

HONSHU PIPEFISH   *Doryrhamphus japonicus*
SIZE: to 8 cm (3 ¼ in.)   Pipefishes – Syngnathidae
ID: Orange-brown; black margined narrow blue strip from snout to tail base, **circular tail with 3 defined orange spots.** Solitary or in pairs; a cleaner. Often near sponges and *Diadema* urchins in coastal, lagoon and outer reefs to 30 m.
Asian Pacific: Indonesia to Papua New Guinea, north to S. Japan.

BLUESTRIPE PIPEFISH   *Doryrhamphus excisus*
SIZE: to 7 cm (2 ½ in.)   Pipefishes – Syngnathidae
ID: Orange-brown; wide blue stripe from snout to tail base, **circular tail with orange fan marking on base with single spot** behind. Solitary or in pairs; a cleaner. Ceiling of caves or under ledges in coastal, lagoon and outer reefs to 45 m.
Asian Pacific: Indonesia, Philippines, Papua New Guinea, to S.W. Japan.

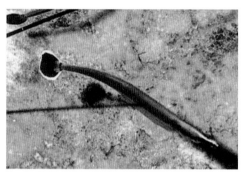

FLAGTAIL PIPEFISH   *Doryrhamphus negrosensis*
SIZE: to 5 cm (2 in.)   Pipefishes – Syngnathidae
ID: Blue to blue-gray; pale stripe on top of snout and head, **blackish tail with orange base and white margin.** Solitary, in pairs or small groups. Inside reef crevices or shelter among sea urchin spines of protected reefs to 20 m.
Asian Pacific: Sabah in Malaysia, Philippines, N. Papua New Guinea and Micronesia.

CLEANER PIPEFISH   *Doryrhamphus janssi*
SIZE: to 13 cm (5 in.)   Pipefishes – Syngnathidae
ID: Orange midbody with blue head and rear body; **black tail with white margin and center spot.** Solitary or in pairs; an active cleaner. Reef crevices of primarily seaward reef slopes in 5-35 m.
Asian Pacific: Andaman Sea to Indonesia, Philippines, Micronesia, Solomon Is., W. Australia and Great Barrier Reef.

MANY-BANDED PIPEFISH   *Dunckerocampus multiannulatus*
SIZE: to 17 cm (6 ¾ in.)   Pipefishes – Syngnathidae
ID: Numerous alternating narrow red to blackish and white to pinkish bars encircle body; **white-edged red tail.** Solitary, in pairs or small aggregations. Crevices of lagoon and seaward slopes to 45 m.
Indian Ocean: Red Sea and E. Africa to Andaman Sea and Sumatra in Indonesia.

ORANGE-BANDED PIPEFISH   *Dunckerocampus pessuliferus*
SIZE: to 16 cm (6 ¼ in.)   Pipefishes – Syngnathidae
ID: Numerous alternating narrow red to black and yellow-orange bars encircle body; **red tail with yellow central spot and white upper margin.** Solitary or in pairs. Isolated coral patches on sand and mud slopes of coastal reefs in 15-35 m.
Asian Pacific: N.W. Australia, Sulawesi, Bali and West Papua in Indonesia and Sulu Archipelago in Philippines.

RINGED PIPEFISH *Dunckerocampus dactyliophorus*
SIZE: to 18 cm (7 in.)   Pipefishes – Syngnathidae
ID: Red to maroon and white alternating bars encircle body; **red tail with white borders and central white spot.** Solitary, in pairs or small aggregations. Caves or ledges of coastal, lagoon and outer reefs to 55 m.
Indo-Pacific: Red Sea and E. Africa to Indonesia, Philippines, Micronesia, Solomon Is. to French Polynesia. – S. Japan to Australia.

BROAD-BANDED PIPEFISH *Dunckerocampus boylei*
SIZE: to 16 cm (6¼ in.)   Pipefishes – Syngnathidae
ID: Red to maroon and white alternating bars with thin black margins encircle body; white margined red tail. **Lack central white spot on tail like Ringed Pipefish** [previous]. Solitary or pairs. Caves and crevices of seaward reefs in 20-40 m.
Indo-Asian Pacific: Known only from Red Sea, Mauritius, Bali and West Papua in Indonesia.

NAIA PIPEFISH *Dunckerocampus naia*
SIZE: to 15 cm (6 in.)   Pipefishes – Syngnathidae
ID: Numerous dark brown to reddish and pale yellow bands; **red tail with white borders and white spot on base.** Solitary or pairs under ledges and in caves in 15-40 m.
West Pacific: Indonesia, S. Japan, Micronesia, Solomon Is. and Fiji.

SAMOAN PIPEFISH *Halicampus mataafae*
SIZE: to 13 cm (5 in.)   Pipefishes – Syngnathidae
ID: Very short snout; skin flap appendages (most noticeable on head), tiny tail, brown to nearly black with narrow whitish saddles along back. Solitary. Inside crevices and under rocks and debris of coastal, lagoon and outer reefs to 15 m.
Indo-West Pacific: Red Sea and E. Africa to Indonesia, Philippines, Micronesia. – S.W. Japan to N. Great Barrier Reef Fiji and Samoa.

WINGED PIPEFISH *Halicampus macrorhynchus*
SIZE: to 16 cm (6¼ in.)   Pipefishes – Syngnathidae
ID: Variably colored, but commonly shades of red, yellow, green or brown; **8-10 pairs of skin flaps extend from each side of back** may resemble small "wings;" long snout. Solitary or in pairs. Sand and weeds of coastal and lagoon reefs in 4-25 m.
Indo-West Pacific: Red Sea to Indonesia, Philippines, Papua New Guinea, Solomon Is., Great Barrier Reef and Fiji.

BROCK'S PIPEFISH *Halicampus brocki*
SIZE: to 12 cm (4¾ in.)   Pipefishes – Syngnathidae
ID: Whitish unmarked body, occasionally brown; branched filaments on head and skin flaps on body, brown variation has about 15 white saddles on back. Solitary and cryptic. Coral and algae patches of coastal reefs in 3-45 m.
Asian Pacific: Indonesia, Philippines to Micronesia. – S.W. Japan to Great Barrier Reef and W. Australia.

447

Pipefishes – Seahorses

DOUBLE-ENDED PIPEFISH   *Syngnathoides biaculeatus*
SIZE: to 28 cm (11 in.)   Pipefishes – Syngnathidae
ID: Mottled green; **rear body tapers to a tailless point.**
Solitary, in pairs or groups. Within seagrasses and floating
or attached *Sargassum* algae of sheltered coastal reefs and
lagoons to 10 m.
Indo-West Pacific: Red Sea and E. Africa to Indonesia and
Solomon Is. – S. Japan to Great Barrier Reef and Fiji.

PYGMY PIPEFISH   *Micrognathus pygmaeus*
SIZE: to 6 cm (2 1/4 in.)   Pipefishes – Syngnathidae
ID: Snout short; skin flaps absent except for tiny flaps on
head, tiny tail, brown with about 10 narrow whitish saddles
on back and upper side. Solitary and cryptic. Inside caves
and crevices of coastal, lagoon and outer reefs to 10 m.
West Pacific: West Papua in Indonesia, Micronesia, Papua
New Guinea, Solomon Is. to Society Is.

SOFT CORAL PIPEFISH   *Siokunichthys breviceps*
SIZE: to 15 cm (6 in.)   Pipefishes – Syngnathidae
ID: Pale gray to whitish with numerous diffuse brown bars
on head, trunk and tail. Soft corals in 3-12 m.
Indo-Asian Pacific: Mozambique, Indonesia, Philippines and
Great Barrier Reef.

BRUCE'S PIPEFISH
Bulbonaricus brucei
SIZE: to 4.5 cm (1 3/4 in.)
Seahorses –
Syngnathidae
ID: Toothpick size
green body with
splotches of mauve,
pair of thin blue-
green stripes extend
down back with
wider white stripes
below; black upper
head with triangular
white patch between
eyes. Solitary or pairs
live on surface of
Galaxea spp. coral
from 3 to 20 m.
**East Africa-Asian
Pacific:** Known from
Tanzania and Triton
Bay in West Papua,
Indonesia.

**Braun's Pughead Pipefish
Variation –**
ID: Dark toothpick
size reddish brown
body with random
red splotches near
the head; white
encircles mouth and
forms oval patch
between eyes.

BRAUN'S PUGHEAD PIPEFISH   *Bulbonaricus brauni*
SIZE: to 5.5 cm (2 in.)   Pipefishes – Syngnathidae
ID: Toothpick size bright red body covered with numerous
tiny white specks, white face and crown of head. Solitary or
pairs live on surface of Galaxea spp. coral from 10 to 20 m.
Asian Pacific: Known from W. Australia, Sumatra, N.E.
Sulawesi, West Papua Indonesia, Taiwan, S. Japan and Palau.

BARRED XENIA PIPEFISH   *Siokunichthys bentuviai*
SIZE: to 6.5 cm (2 1/2 in.)   Pipefishes – Syngnathidae
ID: Whitish to light green with numerous thin close-set reddish bars on body; **broad reddish brown band extends from side of snout through eye to nape.** In and around *Xenia* soft corals to 10 m.
Indo-Pacific: Red Sea to N. Sulawesi in Indonesia.

SPOTTED XENIA PIPEFISH   *Siokunichthys herrei*
SIZE: to 7.8 cm (3 1/4 in.)   Pipefishes – Syngnathidae
ID: Whitish to tan with prominent **dark brown spots on top and side of head;** diffuse dark stripe on side of snout. In and around *Xenia* soft corals to 10 m.
Localized: N. Sulawesi in Indonesia and Philippines.

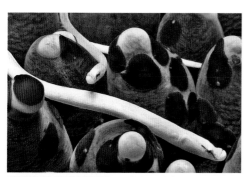

MUSHROOM CORAL PIPEFISH   *Siokunichthys nigrolineatus*
SIZE: to 8 cm (3 1/4 in.)   Pipefishes – Syngnathidae
ID: White and **unmarked except for thin dark band through eye;** worm-like body, short upturned snout. Solitary or small groups. Among tentacles of mushroom corals (primarily *Heliofungia actiniformis*) of coastal reefs and lagoons in 8-20 m.
Asian Pacific: Indonesia, Philippines and Papua New Guinea.

THORNY SEAHORSE
Hippocampus histrix
SIZE: to 15 cm (6 in.)
Seahorses – Syngnathidae
ID: **Spines protrude from head and body ridges;** elongate snout usually with a few white bars, variable shades of red, brown, yellow or green, often with gold speckling. Solitary. Sponges, gorgonians, soft corals and weeds of coastal reefs in 15-40 m.
Asian Pacific: S. Japan and Indonesia to Coral Sea.

COMMON SEAHORSE
Hippocampus kuda
SIZE: to 25 cm (10 in.)
Seahorses – Syngnathidae
ID: Dusky brown to blackish; covered with tiny black and white spots, females sometimes yellowish with several dark spots, **back-swept relatively smooth crown.** Solitary. Inhabit seagrass and mangroves to 15 m.
Asian Pacific: Philippines and N. Sulawesi in Indonesia.

ZEBRA-SNOUT SEAHORSE
Hippocampus barbouri
SIZE: to 15 cm (6 in.)
Seahorses – Syngnathidae
ID: Tan to pale yellow with black speckles; **profuse line markings on snout,** spiny crown on top of head, pair of spines above eyes and thick spine in front of eyes. Solitary; cling to living corals. Shallow coastal reefs in 5-12 m.
Asian Pacific: Philippines and N. Sulawesi in Indonesia.

Seahorses

GREAT SEAHORSE   *Hippocampus kelloggi*
SIZE: to 28 cm (11 in.)   Seahorses – Syngnathidae
ID: Tan to orange or red often with horizontal white lines of tiny dots, high plate in front of crown on top of head; long slightly backward pointing rounded cheek spines. Soft bottoms and areas with gorgonians in 20 to 120 m.
Indo-West Pacific: E. Africa and Red Sea to Philippines, Indonesia and Vanuatu – S. Japan to Great Barrier Reef.

THREE-SPOT SEAHORSE   *Hippocampus trimaculatus*
SIZE: to 17 cm (6 ³/₄ in.)   Seahorses – Syngnathidae
ID: Orange, tan to black, low bony crown on top of head; often with black spots on 1ˢᵗ, 4ᵗʰ and 7ᵗʰ trunk rings, may have brown and white zebra striping; hooked cheek and eye spines. Sand and grass inside estuaries, mangroves and gorgonians in 10 to 40 m.
East Indo-Asian Pacific: India to Indonesia and Philippines– Japan to Australia.

SATOMI'S PYGMY SEAHORSE   *Hippocampus satomiae*
SIZE: to 1.4 cm (¹/₂ in.)   Seahorses – Syngnathidae
ID: Tiny; white to tan with black spot immediately in front of eye, blotchy red markings on gill cover, body and tail. Reef overhangs, associated with soft corals, small sea fans and hydroids in 10-20 m.
Localized: Sabah in Malaysia, Lembeh Strait and Togean Is. in Sulawesi in Indonesia.

TIGERTAIL SEAHORSE
Hippocampus comes
SIZE: 16 cm (6 ¹/₄ in.)
Seahorses – Syngnathidae
ID: Large Adults –
Frequently bright yellow occasionally with dark blotches. **Males –** Often black with yellow blotches; darkish bars often encircle rear body. In pairs; cling to soft corals, in open at night. Reefs rich with soft corals to 20 m.
East Indo-Asian Pacific: Andaman Sea to South China Sea, Indonesia and Philippines.

WALEA PYGMY SEAHORSE   *Hippocampus waleananus*
SIZE: to 1.8 cm (³/₄ in.)   Seahorses – Syngnathidae
ID: Tiny; pinkish to yellowish with red to brown blotches. Occurs on *Nephea* soft coral in 5-20 m.
Localized: Known only from Tomini Bay, Togean Is., in N. Sulawesi in Indonesia.

COLEMAN'S PYGMY SEAHORSE   *Hippocampus colemani*
SIZE: to 2.2 cm (1in.)   Seahorses – Syngnathidae
ID: Tiny; smooth skin texture, protruding belly, and whitish color with widely scattered, faint reddish spots. Solitary or small groups. Among algae in 8-15 m.
Localized: S.W. Japan and Milne Bay Province in Papua New Guinea.

PYGMY SEAHORSE
Hippocampus bargibanti
SIZE: to 2 cm (³/₄ in.)
Seahorses – Syngnathidae

ID: Tiny; pinkish with large orange to red warts and spots, encircling bands on tail. Solitary or form small groups. Sea fans of similar color especially *Muricella* spp. of reefs and slopes in 15-40 m.

Asian Pacific: Sabah in Malaysia, Indonesia, Philippines, Papua New Guinea. – S. W. Japan to Great Barrier Reef, New Caledonia.

Pygmy Seahorse – Variation

ID: Frequently adapt color of host seafan.

DENISE'S PYGMY SEAHORSE   *Hippocampus denise*
SIZE: to 2.4 cm (1 in.)   Seahorses – Syngnathidae

ID: Tiny; variable from whitish, yellow, orange, pink or combination of red and white, scattered tubercles closely match expanded feeding polyps of host gorgnian and can be contracted when polyps are withdrawn.

Asian Pacific: Indonesia, Papua New Guinea, Solomon Is., Micronesia and Vanuatu.

Denise's Pygmy Seahorse – Variation

ID: Inhabit the blades of gorgonian sea fans *Annella*, *Echinogorgia* and *Muricella* to 12-70 m.

PONTOH'S PYGMY SEAHORSE   *Hippocampus pontohi*
SIZE: to 1.6 cm (⁵/₈ in.)   Seahorses – Syngnathidae

ID: Tiny and thin; white with yellow to orange wash on back and crown, reddish filaments on back and occasionally crown, can he brown with red highlights. Solitary, but usually in small groups, tend to move about. Live on hydroids and algae, mimic dead segments of *Halimeda* algae in 5-20 m.

West Pacific: E. Indonesia and Papua New Guinea to Fiji.

Pontoh's Pygmy Seahorse – Dark Variation

ID: Brown to reddish brown hue with red highlights. Formerly classified as *H. severni*.

Eels

This ID Group consists of fishes with long snake-like bodies.

FAMILY: Morays – Muraenidae
10 Genera – 42 Species Included

Typical Shape

Morays have no pectoral or ventral fins; their dorsal, tail and anal fins form a single, long continuous fin that begins behind the head, encircles the tail and extends midway down the belly. Their elongate, scaleless bodies are coated with a clear, protective mucous layer.

Morays constantly open and close their mouths, a behavior often perceived as a threat, but in reality the action simply moves water through their gills for respiration. By nature morays are not aggressive, although they can inflict a nasty bite if molested, and will swim off the bottom to greet approaching divers in areas where fish feeding occurs. During the day, most species are reclusive and tend to hide in dark recesses. Normally they are sighted with only their heads extending from holes.

Species in genera *Gymnothorax* and *Enchelycore* have long pointed teeth for feeding on fishes and octopuses. In most cases octopuses with less than the prescribed eight legs have been victims of moray encounters. Those species with blunt crushing teeth, such as members of genus *Echidna,* prey on crustaceans, primarily crabs. Translucent, ribbon-like larval eels, known as leptocephali, have a lengthy pelagic stage. Once settled to the sea floor younger morays tend to be more slender than older adults.

FAMILY: Snake Eels – Ophichthidae
7 Genera – 15 Species Included

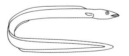

Typical Shape

Most species of snake eels are virtually without fins and strongly resemble snakes. Their pointed snouts and sharp bony tails allow them to burrow forward or backward beneath the sediment where they spend most of their lives. During the day they are often seen with only their heads extending from the bottom. A few species prowl in the open at night.

FAMILY: Conger Eels & Garden Eels – Congridae
4 Genera – 13 Species Included

Conger Eels – Typical Shape

Garden Eels – Typical Shape

Like morays, the dorsal, anal and tail fins of conger eels join to form a single continuous fin that encircles most of their bodies; but, unlike morays, most conger eels have pectoral fins. The nocturnal predators primarily feed on crustaceans and fishes.

Those species in subfamily Heterocongrinae, commonly known as garden eels, are typically pencil thin with reduced or absent pectoral fins and small upturned mouths. They form small to large colonies on sand plains and slopes where they extend their front bodies from burrows to pick plankton from the currents.

DRAGON MORAY
Enchelycore pardalis

SIZE: to 92cm (3 ft.)
Morays – Muraenidae

ID: Pale brownish yellow to orange with numerous dark-ringed white spots; curved jaws with many dagger-like teeth, pair of long tubular nostrils above eyes. Solitary. Inhabit reef crevices of outer reefs in 15-50 m.

Indo-Pacific: Mauritius to Hawaii and French Polynesia. Rare in Asian Pacific. Mainly oceanic islands.

HOOKJAW MORAY   *Enchelycore bayeri*
SIZE: to 70 cm (2¼ ft.)   Morays – Muraenidae
ID: Uniform brown; curved or "hooked" jaws with many dagger-like teeth. Solitary. Inhabit reef crevices of exposed reef flats and outer reef slopes in 1-64 m.

East Indo-Pacific: Maldives to Indonesia, Philippines, Micronesia, Papua New Guinea, Solomon Is. and French Polynesia. – S.W. Japan to Australia and New Caledonia.

BENTJAW MORAY   *Enchelycore schismatorhynchus*
SIZE: to 120 cm (4 ft.)   Morays – Muraenidae
ID: Light tan to gray with white margin on fins; curved or "hooked" jaws with many dagger-like teeth. Solitary, usually in open only at night. Inhabit crevices of lagoon and outer reefs in 5-35 m.

East Indo-Pacific: Chagos to Indonesia, Philippines, Papua New Guinea and French Polynesia, north to S.W. Japan.

VIPER MORAY   *Enchelynassa canina*
SIZE: to 154 cm (5 ft.)   Morays – Muraenidae
ID: Reddish brown to gray; **"wrinkled skin" with darkish lines in grooves,** curved jaws with many dagger-like teeth. Solitary and nocturnal. During day inhabit deep reef recesses of reefs and reef flats in areas of surge to 15 m.

Indo-Pacific: Mauritius to Hawaii, Pitcairn Is. and Panama. Rare in Asian Pacific; inhabit shorelines in Micronesia.

BARRED MORAY *Echidna polyzona*
SIZE: to 72 cm (2¹/₂ ft.) Morays – Muraenidae
ID: Lightly mottled shades of brown with 25-30 indistinct often encircling bars; **dark patch on corner of jaw.** Solitary. Inhabit crevices of reef flats and lagoons in 2-20 m.
Indo-Pacific: Red Sea and E. Africa to Indonesia, Philippines, Micronesia, Papua New Guinea, Solomon Is., Hawaii and French Polynesia. – S.W. Japan to Australia.

Barred Moray – Small Adult
SIZE: to 50 cm (20 in.)
ID: With age bars become progressively obscure and white undercolor becomes lightly mottled shades of brown.

Barred Moray – Juvenile
SIZE: to 30 cm (12 in.)
ID: White with 25-30 dark brown bars most encircling body and fins, dark patch extends from dark bar over corner of mouth.

FINESPECKLED MORAY *Echidna delicatula*
SIZE: to 44 cm (17 in.) Morays – Muraenidae
ID: Tan to dark brown with fine yellowish speckles; speckles of young [pictured] join to form reticulations. Solitary and cryptic. Inhabit cracks, crevices and deep recesses of inshore reefs and reef flats to 15 m.
East Indo-West Pacific: Sri Lanka to Indonesia, Philippines, Micronesia. – S. Japan to Fiji.

SNOWFLAKE MORAY
Echidna nebulosa
SIZE: 75 cm (2¹/₂ ft.)
Morays – Muraenidae
ID: Whitish with pattern of large black blotches containing yellow spots and numerous small blackish spots and scribble markings between. Solitary, often in open. Reef flats and rocky shorelines in 1-18 m.
Indo-Central Pacific: Red Sea and E. Africa to Indonesia, Philippines, Micronesia, Papua New Guinea, Solomon Is., Hawaii. – S.W. Japan to Australia.

LARGEHEAD SNAKEMORAY *Uropterygius macrocephalus*
SIZE: to 45 cm (18 in.) Morays – Muraenidae
ID: Brown with dense network of lichen-like white to yellowish blotches; no fins (except at tail tip). Solitary and secretive, rarely in open during day. Inhabit reef crevices of lagoon and seaward reefs in 1-14 m.
Indo-Central Pacific: Seychelles to Indonesia, Philippines, S. Japan, Micronesia, Papua New Guinea, Solomon Is. and Hawaii.

MOON SNAKEMORAY *Uropterygius kamar*
SIZE: to 37 cm (15 in.) Morays – Muraenidae
ID: Reddish brown to dark brown with irregular white spots or whitish mottling; white to **pale and dark bars on lower jaw.** Coral reefs and rubble bottoms in 3-55 m.
Indo-Pacific: E. Africa to Indonesia, Micronesia and Pitcairn Is.

YELLOWFIN SNAKEMORAY *Uropterygius xanthopterus*
SIZE: to 60 cm (2 ft.) Morays – Muraenidae
ID: Reddish brown to brown with whitish to pale gray blotches roughly arranged in three rows the length of body; pale dots on head and forebody, bright yellow tail tip. Cryptic, lagoon and seaward reefs in 2-56 m.
East Indo-Pacific: Maldives to Indonesia, Philippines, Micronesia, Solomon Is. and French Polynesia.

BROWN SNAKEMORAY *Uropterygius concolor*
SIZE: to 30 cm (12 in.) Morays – Muraenidae
ID: Uniform brown head and body without markings. Shallow reefs and mangrove estuaries in 1-10 m.
Indo-Pacific: Red Sea and E. Africa to Indonesia, Philippines, Micronesia, Papua New Guinea, Solomon Is., and French Polynesia. – S.W. Japan to Australia.

WHITE-EYED MORAY *Gymnothorax thyrsoideus*
SIZE: to 65 cm (2¼ ft.) Morays – Muraenidae
ID: Whitish or pale yellow-brown with dense scattering of small brown spots; **purplish gray head with distinctive white iris.** Solitary or small groups. Inhabit crevices of shallow reef flats to 7 m.
East Indo-Pacific: Maldives to Indonesia, Philippines, Micronesia, Solomon Is. and French Polynesia. – S.W. Japan to N. Australia.

MASKED MORAY *Gymnothorax breedeni*
SIZE: to 65 cm (2¼ ft.) Morays – Muraenidae
ID: Brown with tan flecks; black blotchy band from eye to behind mouth and black blotch over gill opening. Solitary, can be aggressive toward divers. Inhabit reef crevices of outer reef slopes in 4-25 m.
Indo-Pacific: E. Africa to Christmas I., Micronesia and French Polynesia, mainly oceanic islands. Rare Asian Pacific.

CHLAMYDATUS MORAY *Gymnothorax chlamydatus*
SIZE: to 60 cm (2 ft.) Morays – Muraenidae
ID: White with 13 black encircling bars; small black spots and blotches on head and white spaces between bars. Solitary, enter sand burrows tail first, head often exposed. Inhabit sand areas in 5-30 m.
Asian Pacific: Indonesia, Philippines, Taiwan and S.W. Japan.

Morays

LATTICETAIL MORAY   *Gymnothorax buroensis*
SIZE: to 39 cm (15 in.)   Morays – Muraenidae
ID: Small; light brown head and front body becoming dark brown over rear with small black spots and white flecks, yellow fin margins. Solitary. Inhabit coral branches, crevices and algal beds of lagoon and seaward reef slopes in 2-24 m.
Indo-Pacific: Red Sea to Indonesia, Philippines, Micronesia, Solomon Is., Hawaii, French Polynesia. – S.W. Japan to Australia.

PALECHIN MORAY   *Gymnothorax herrei*
SIZE: to 30 cm (12 in.)   Morays – Muraenidae
ID: Brown with **whitish lower head** and pale tail tip. Solitary or form groups in reef crevices; engage in mass spawning with multiple males entwined around a single female. Coastal reefs and tide pools to 2 m, often in turbid water.
Indo-Asian Pacific: Red Sea to Indonesia, Philippines, Papua New Guinea and Great Barrier Reef.

SMALLSPOT MORAY   *Gymnothorax microstictus*
SIZE: to 30 cm (12 in.)   Morays – Muraenidae
ID: Reddish brown upper head and body, whitish lower head and belly; dense covering of somewhat darker spots on head, body and fins. Coral reefs in 5-30 m.
Asian Pacific: Indonesia, Philippines, Papua New Guinea, Solomon Is. and Vanuatu.

WHITE-MARGINED MORAY   *Gymnothorax albimarginatus*
SIZE: to 105 cm (3 1/2 ft.)   Morays – Muraenidae
ID: Brown with distinctive **white margin on fins;** white spot surrounding each sensory pore on jaws, pale iris. Solitary, possibly has venomous bite. Inhabit reef crevices of lagoons and outer reefs in 5-25 m.
Pacific: Indonesia to Hawaii and French Polynesia. – Taiwan and S. Japan and to New Caledonia and Fiji.

YELLOW-HEADED MORAY   *Gymnothorax rueppellii*
SIZE: to 80 cm (2 1/2 ft.)   Morays – Muraenidae
ID: White to light grayish brown with 16-21 dark brown bars encircling body and fins; top of head yellow to brownish yellow, **small dark spot on rear corner of mouth.** Solitary, nocturnal. Inhabit reef crevices of lagoon and outer reefs in 1-30 m.
Indo-Pacific: Red Sea and E. Africa to Indonesia, Micronesia, Hawaii and French Polynesia. – S.W. Japan to Australia.

ENIGMATIC MORAY   *Gymnothorax enigmaticus*
SIZE: to 58 cm (23 in.)   Morays – Muraenidae
ID: Whitish with brown mottling and approximately 20 dark bars encircling body and fins; **bars often darker on dorsal fin than body. Juveniles –** Bright white with black encircling bars. Solitary, in open at night. Lagoon and outer reefs in 1-15 m.
Indo-West Pacific: E. Africa to Indonesia, Philippines, Micronesia, Papua New Guinea. – S.W. Japan to Great Barrier Reef and Fiji.

YELLOWMOUTH MORAY   *Gymnothorax nudivomer*
SIZE: to 100 cm (3¼ ft.)   Morays – Muraenidae
ID: Light brown head and front body shading to dark brown toward rear with numerous small white spots; inside of mouth yellow, dark blotch on gill opening. Solitary, produces skin toxin. Inhabit crevices of mainly outer reefs in 5-165 m.
Indo-Pacific: Red Sea and E. Africa to Micronesia and French Polynesia. – S.W. Japan to Great Barrier Reef. Rare Asian Pacific.

WHITEMOUTH MORAY   *Gymnothorax meleagris*
SIZE: to 100 cm (3¼ ft.)   Morays – Muraenidae
ID: Dark brown with numerous close-set white spots; inside of mouth and tip of tail white. Solitary, head frequently protrudes from reef recess. Lagoon and outer reefs in 1-36 m.
Indo-Pacific: E. Africa to Indonesia, Philippines, Micronesia, Papua New Guinea, Solomon Is., Hawaii and French Polynesia. – S. Japan to Great Barrier Reef.

YELLOWMARGIN MORAY   *Gymnothorax flavimarginatus*
SIZE: to 120 cm (4 ft.)   Morays – Muraenidae
ID: Yellowish orange body densely mottled with dark brown, dark purplish brown front of head; black blotch over gill opening, orange iris. Solitary and curious, head frequently protrudes from reef recess. Lagoon and outer reefs in 1-150 m.
Indo-Pacific: Red Sea to Indonesia, Micronesia, Solomon Is., Hawaii and French Polynesia. – S.W. Japan to Great Barrier Reef.

GIANT MORAY   *Gymnothorax javanicus*
SIZE: to 250 cm (8¼ ft.)   Morays – Muraenidae
ID: Tan to brown and greenish with irregular dark brown spots on head, body and fins; black blotch on gill opening. Solitary, most common large moray. Inhabit reef holes of lagoon and outer reefs in 1-46 m.
Indo-Pacific: Red Sea to Indonesia, Micronesia, Solomon Is., Hawaii and French Polynesia. – S.W. Japan to Great Barrier Reef.

SPOTTED MORAY   *Gymnothorax isingteena*
SIZE: to 100 cm (3¼ ft.)   Morays – Muraenidae
ID: Whitish with **irregular black spots occasionally joining to form circular markings.** Solitary, heads often protrude from crevices. Inhabit coral outcroppings of coastal reefs and outer slopes in 3-30 m.
Asian Pacific: Sabah in Malaysia to Indonesia and S. Japan.

BLACKSPOTTED MORAY   *Gymnothorax favagineus*
SIZE: to 180 cm (6 ft.)   Morays – Muraenidae
ID: White to yellow with rounded black leopard-like spotting; young have larger and fewer spots, large adults have smaller spots that form honeycomb-like pattern. Solitary. Inhabit reef crevices of lagoon and outer reefs in 1-50 m.
Indo-West Pacific: E. Africa to Indonesia, Philippines and Papua New Guinea. – S.W. Japan to Great Barrier Reef and Fiji.

FIMBRIATED MORAY   *Gymnothorax fimbriatus*
SIZE: to 80 cm (2 ½ ft.)   Morays – Muraenidae
ID: Tan to grayish brown, head with yellow hue; highly variable pattern of widely spaced irregular dark brown to black spots, partial bars and bands. Solitary. Reef crevices, often beneath debris on outer reefs in 7-50 m.
Indo-Pacific: Seychelles to Indonesia, Philippines, Micronesia, Solomon Is. and French Polynesia. – S.W. Japan to Australia.

WHITELIP MORAY   *Gymnothorax chilospilus*
SIZE: to 30 cm (12 in.)   Morays – Muraenidae
ID: Mottled reddish brown and white with **irregular dark brown bars and blotches;** several sensory pores on jaws frequently surrounded by large white spots. Solitary. Inhabit crevices and fissures of lagoon and outer reefs slopes in 1-45 m.
Indo-Asian Pacific: Seychelles to Indonesia, Philippines, Solomon Is. – S.W. Japan to Australia.

RETICULATED MORAY   *Gymnothorax richardsonii*
SIZE: to 34 cm (13 in.)   Morays – Muraenidae
ID: White to tan with dense brown reticulum on head, body and fins. Solitary. Inhabit reef crevices and rubble of shallow reef flats and estuaries in 1-12 m.
Indo-Pacific: E. Africa to Indonesia, Philippines, Micronesia, Papua New Guinea, Solomon Is. and French Polynesia. – S.W. Japan to Australia and Fiji.

AUSTRALIAN MORAY   *Gymnothorax cribroris*
SIZE: to 47 cm (19 in.)   Morays – Muraenidae
ID: Pale brown with darker irregular netted pattern on body and fins; **dark irregular, but well defined spots behind eye.** Solitary. Inhabit crevices and recesses during day, hunt in open at night. Tide pools, lagoons, bays and fore-reef slopes to 15 m.
Localized: West Australia and Great Barrier Reef.

INDIAN MUD MORAY
Gymnothorax tile
SIZE: to 60 cm (2 ft.)
Morays –
Muraenidae

ID: Mottled brown with whitish flecks especially evident on head. Solitary. Near debris on muddy bottoms and estuaries and lower portion of rivers in 1-10 m.

Indo-Asian Pacific: India to Singapore, Indonesia and Philippines.

UNDULATED MORAY   *Gymnothorax undulatus*
SIZE: to 120 cm (4 ft.)   Morays – Muraenidae
ID: Pale brown to yellowish brown to brown with close-set irregular dark brown blotches and small spots, blotches often merge to form irregular bars on rear body and fins. Solitary. Inhabit reef flats, lagoons and outer slopes to 50 m.
Indo-Pacific: Red Sea to Indonesia, Micronesia, Solomon Is., Hawaii, French Polynesia. – S.W. Japan to Great Barrier Reef.

BARRED-FIN MORAY   *Gymnothorax zonipectis*
SIZE: to 47 cm (19 in.)   Morays – Muraenidae
ID: Tan with 2-4 longitudinal rows of brown blotches on body and bands on fins; **white bands and blotches on lips,** dark brown band at rear edge of eye. Solitary. Inhabit reef crevices, mainly of outer slopes in 4-40 m.

Indo-Pacific: E. Africa to Indonesia, Philippines, Micronesia, Solomon Is. and French Polynesia. – S.W Japan to Australia.

SLENDERTAIL MORAY   *Gymnothorax gracilicauda*
SIZE: to 32 cm (13 in.)   Morays – Muraenidae
ID: Whitish to pale brown with over 30 **dark bars which do not extend over the dorsal fin.** Solitary, hide inside crevices and reef recesses during day. Lagoon reefs and fore-reef slopes to 20 m.

Indo-Pacific: E. Africa to Indonesia, Philippines, Micronesia, Solomon Is. to Hawaii and French Polynesia. – S. Japan to Australia.

PEPPERED MORAY   *Gymnothorax pictus*
SIZE: to 120 cm (4 ft.)   Morays – Muraenidae
ID: White to light gray with dense dark brown to blackish spotting. **Small Juveniles –** Relatively large spots in about 3 longitudinal rows. Solitary, inhabit crevices in rock; feed mainly on crabs. Shallow reef flats to 3 m.

Indo-West Pacific: E. Africa to Indonesia, Philippines, Micronesia, Solomon Is., Hawaii and E. Pacific. – S.W. Japan to Australia and Fiji.

DWARF MORAY   *Gymnothorax melatremus*
SIZE: to 30 cm (12 in.)   Morays – Muraenidae
ID: Three color patterns: all brown, brown with dark brown mottling [pictured], bright yellow; all may have dark mark on gill opening. Solitary and cryptic, seldom seen by divers.

Indo-Pacific: E. Africa to Indonesia, Philippines, Micronesia, Papua New Guinea, Solomon Is., Hawaii and Pitcairn Is. – S.W. Japan to Australia.

TIGER SNAKE MORAY   *Scuticaria tigrina*
SIZE: to 120 cm (4 ft.)   Morays – Muraenidae
ID: Yellowish to reddish tan with large rough edged spots and scattered smaller spots between; nearly cylindrical with no fins (except at tail tip). Solitary. Inhabit ledges and crevices of lagoon and seaward reefs in 8-25 m.

Indo-Pacific: E. Africa to Hawaii, French Polynesia and Panama. Only Asian Pacific records Indonesia and Philippines.

OKINAWAN SNAKE MORAY   *Scuticaria okinawae*
SIZE: to 93 cm (3 ft.)   Morays – Muraenidae
ID: Gray-brown without markings; nearly cylindrical with no fins (except at tail tip). Solitary. Inhabit reef crevices of rocky shores and mainly seaward reefs in 5-30 m.

Indo-North Pacific: Mauritius to Indonesia, S. Japan and Hawaii.

ZEBRA MORAY *Gymnomuraena zebra*
SIZE: to 100 cm (3 ¼ ft.) Morays – Muraenidae
ID: Dark brown with numerous narrow white bars encircling head, body and fins. Solitary, often in open; feed mainly on crabs. Inhabit reef crevices and under ledges of exposed reef flats and seaward slopes in 1-40 m.
Indo-Pacific: Red Sea to Indonesia, Philippines, Micronesia, Solomon Is., Hawaii and French Polynesia. – S.W. Japan to Australia.

GIANT ESTUARINE MORAY *Strophidon sathete*
SIZE: to 375 cm (12 ¼ ft.) Morays – Muraenidae
ID: Very large; grayish brown without markings, small canine teeth in 2 rows along side of upper and front of lower jaws. **Young –** White stripe on dorsal fin. Solitary. Inhabit burrows in mud bottoms of estuaries and river mouths to 15 m.
Indo-West Pacific: Red Sea to Indonesia, Micronesia, Papua New Guinea, Solomon Is. – S.W. Japan to Great Barrier Reef to Fiji.

RIBBON EEL *Rhinomuraena quaesita*
SIZE: to 130 cm (4 ¼ ft.) Morays – Muraenidae
ID: Male – Brilliant blue body with yellow dorsal fin, snout, lower jaw and eye; **large fan-shaped nostrils and chin barbels.** Solitary or pairs. Inhabit sandy burrows of coastal, lagoon and seaward reefs in 1-67 m.
Indo-West Pacific: Madagascar to Indonesia, Philippines, Micronesia and French Polynesia. – S.W. Japan to Great Barrier Reef and Fiji.

Ribbon Eel – Female
SIZE: 85-120 cm (2 ¾-4 ft.)
ID: This species is a protandrous hermaphrodite (males changing into females). At approximately 85 cm males begin to develop female sex organs and change color becoming yellowish blue to entirely yellow. Females are less common.

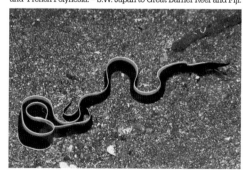

Ribbon Eel – Juvenile
SIZE: to 65 cm (2 ¼ ft.)
ID: Jet-black body with yellow dorsal fin; flared translucent nostrils, white stripe on lower jaw. Rarely sighted outside protection of crevices.

WHITE RIBBON EEL *Pseudechidna brummeri*
SIZE: to 103 cm (3 ½ ft.) Morays – Muraenidae
ID: Long laterally compressed body; pale gray, brown or green, small dark spots on head, narrow white margin on fins. Solitary and cryptic, rarely in open except at night. Sand and rubble bottoms of sheltered coastal reefs and reef flats in 1-8 m.
Indo-West Pacific: E. Africa to Indonesia, Micronesia, Papua New Guinea. – S.W. Japan to Great Barrier Reef, Fiji and Cook Is.

REPTILIAN SNAKE EEL   *Brachysomophis henshawi*
SIZE: to 100 cm (3 1/4 ft.)   Snake Eels – Ophichthidae
ID: Variable from mottled red to yellowish or whitish; head flattened except for **an abrupt depression behind eye.** Solitary, bury in sand with only heads protruding to ambush prey. Lagoons and seaward sand patches in 1-25 m.
Indo-Pacific: Arabian Sea to Indonesia, Papua New Guinea, Solomon Is., Hawaii, French Polynesia. – S. Japan to Coral Sea.

CROCODILE SNAKE EEL   *Brachysomophis crocodilinus*
SIZE: to 82 cm (2 3/4 ft.)   Snake Eels – Ophichthidae
ID: Whitish to brown often with dark flecks; head flattened, and not deeply depressed behind eyes, **eyes set far forward on snout;** branched skin flaps along upper lip. Solitary, bury in sand with only heads protruding. Sand patches in 1-15 m.
Indo-Pacific: Madagascar to Andaman Sea, Indonesia, Papua New Guinea and French Polynesia. – S.W. Japan to Australia.

STARGAZER SNAKE EEL   *Brachysomophis cirrocheilos*
SIZE: to 159 cm (5 1/4 ft.)   Snake Eels – Ophichthidae
ID: Light brown with irregular lighter patches on back and sides; fine black spots on head and a line of white spots on back of head; numerous fine teeth protrude. Solitary, bury in sand with only heads protruding to ambush prey. Sand to 10 m.
Indo-Asian Pacific: Red Sea to Indonesia, Papua New Guinea and north to S. Japan.

SHARPSNOUT SNAKE EEL   *Apterichtus klazingai*
SIZE: to 40 cm (16 in.)   Snake Eels – Ophichthidae
ID: White with numerous orange-brown spots; pointed snout, lack pectoral fins. Solitary, bury in sand, seldom in open. Inhabit extensive sand areas of seaward slopes in 2-15 m.
Indo-Pacific: E. Africa to Micronesia and Hawaii. Rarely sighted species, probably more widespread.

BLACK-STRIPED SNAKE EEL   *Callechelys catostoma*
SIZE: to 77 cm (2 1/2 ft.)   Snake Eels – Ophichthidae
ID: Pale tan with **two brown to black body stripes;** dark margin on dorsal fin, overhanging snout with prominent tubular nostrils, no pectoral fins. Solitary, bury in sand. Inhabit sandy patches near reefs in 1-33 m.
Indo-Pacific: Red Sea to Indonesia, Philippines, Micronesia, Solomon Is., Hawaii and French Polynesia. – S. Japan to Australia.

MARBLED SNAKE EEL   *Callechelys marmorata*
SIZE: to 86 cm (2 3/4 ft.)   Snake Eels – Ophichthidae
ID: Yellowish with numerous black spots and blotches; overhanging snout with prominent tubular nostrils, no pectoral fins. Solitary, bury in sand with only head protruding. Inhabit lagoon and seaward sand patches in 1-15 m.
Indo-Pacific: Red Sea and E. Africa to Indonesia, Micronesia, Solomon Is. and French Polynesia. – S. Japan to Australia.

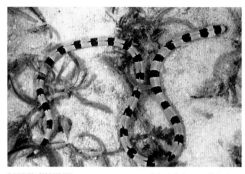

SPOTTED SNAKE EEL   *Myrichthys maculosus*
SIZE: to 100 cm (3 1/4 ft.)   Snake Eels – Ophichthidae
ID: Yellowish to cream with **large dark oval spots;** overhanging snout with long tubular nostrils, small pectoral fins. Solitary, occasionally in open during day, but more commonly out at night. Lagoon and seaward sand in 1-262 m.

Indo-Pacific: Red Sea to Indonesia, Philippines, Micronesia, Solomon Is., Hawaii and French Polynesia. – S. Japan to Australia.

BANDED SNAKE EEL   *Myrichthys colubrinus*
SIZE: to 90 cm (3 ft.)   Snake Eels – Ophichthidae
ID: White with brown to black (usually encircling) bars occasionally with black spots between; overhanging snout, small pectoral fins. Solitary, mimics sea snake. Inhabit sandy patches of lagoons and seaward reefs in 1-30 m.

Indo-Pacific: Red Sea to Indonesia, Micronesia, Papua New Guinea, Solomon Is., French Polynesia. – S.W. Japan to Australia.

CONVICT SNAKE EEL   *Leiuranus versicolor*
SIZE: to 52 cm (21 in.)   Snake Eels – Ophichthidae
ID: **White with broad blackish saddles centered with narrow yellowish bars;** overhanging snout with prominent tubular nostrils, no pectoral fins. Solitary, bury in sand. Sandy patches near reefs to 18 m.

Asian Pacific: Indonesia, Philippines, Micronesia, Papua New Guinea to Great Barrier Reef.

SADDLED SNAKE EEL   *Leiuranus semicinctus*
SIZE: to 65 cm (2 1/4 ft.)   Snake Eels – Ophichthidae
ID: Whitish with wide rounded dark brown saddles that do not quite encircle body on belly. Solitary, bury completely in sand, but occasionally emerge during both day and night to feed. Lagoons and near coastal reefs in 1-10 m.

Indo-Pacific: E. Africa to Indonesia, Philippines, Micronesia, Solomon Is., Hawaii, French Polynesia. – S. Japan to S.E. Australia.

NAPOLEON SNAKE EEL   *Ophichthus bonaparti*
SIZE: to 75 cm (2 1/2 ft.)   Snake Eels – Ophichthidae
ID: Dirty white with dark brown encircling bars and occasional spots between; **head has dark-edged bronze spots and blotches.** Solitary, bury in sand often with heads exposed; occasionally in open at night. Coastal sand slopes in 5-20 m.

Indo-Pacific: S. Africa to Indonesia, Philippines, Micronesia, Papua New Guinea and French Polynesia. – S. Japan to Great Barrier Reef.

Napoleon Snake Eel – Head detail
ID: Easily identified with only head exposed by distinctive **dark-edged bronze spots and blotches.**

BLACK-FINNED SNAKE EEL *Ophichthus altipennis*
SIZE: to 103 cm (3 1/2 ft.) Snake Eels – Ophichthidae
ID: Yellowish brown with white encircling bar behind head, black pectoral fins; overhanging snout with large gray tubular nostrils. Solitary; bury in sand, often with head protruding. Coastal sand patches to 10 m.
Pacific: Indonesia, Philippines, Micronesia, Papua New Guinea, Solomon Is. to French Polynesia. – S. Japan to Australia.

Black-finned Snake Eel – Head Detail
ID: Can be identified when only head exposed by a **pale patch in front of eye.** Similar Blacksaddle Snake Eel [next] lack this feature.

BLACKSADDLE SNAKE EEL *Ophichthus cephalozona*
SIZE: to 115 cm (3 3/4 ft.) Snake Eels – Ophichthidae
ID: Light brown; anterior tip of head brown, remainder of head white with large black saddle behind. Solitary, hide in crevices or bury in sand, often with head protruding; frequent shrimp cleaning stations. Coastal areas in 2-15 m.
Pacific: Indonesia, Micronesia to French Polynesia. – Taiwan to Great Barrier Reef.

LARGE-SPOTTED SNAKE EEL *Ophichthus polyophthalmus*
SIZE: to 50 cm (20 in.) Snake Eels – Ophichthidae
ID: Reddish brown to pale salmon with numerous dark-edged yellow spots; overhanging snout with prominent pale nostrils. Solitary, bury in sand and rubble often with heads protruding. Lagoon and seaward sand patches in 1-20 m.
Indo-Pacific: E. Africa to Indonesia, Philippines, Micronesia and French Polynesia.

LONGFIN SNAKE EEL *Pisodonophis cancrivorus*
SIZE: to 108 cm (3 1/2 ft.) Snake Eels – Ophichthidae
ID: Gray to brown; covered with skin wrinkles highlighted by pale and darker shades, overhanging snout with tubular nostrils. Solitary or in pairs, bury in sand often with heads protruding. Coastal sand and silt patches in 1-25 m.
Indo-West Pacific: Red Sea and E. Africa to Indonesia and Micronesia. – S. Japan to Australia and Samoa.

MOUSTACHE CONGER *Conger cinereus*
SIZE: to 140 cm (4 1/2 ft.) Conger Eels – Congridae
ID: Brown to gray with black well developed pectoral fins; **black streak below eye on upper lip,** may display dark bands. Solitary, occasionally in open at night. Inhabit ledges and crevices of coral reefs in 1-80 m.
Indo-Pacific: Red Sea to Indonesia, Philippines, Micronesia, Solomon Is. and French Polynesia. – S. Japan to Australia.

BIGEYE CONGER   *Ariosoma anagoides*
SIZE: to 30 cm (12 in.)   Conger Eels – Congridae
ID: Grayish shading to whitish on bottom of head and belly; dorsal and anal fin have bluish hue, large eye with white iris. Solitary and cryptic, in open at night. Sand or mud bottoms of sheltered bays in 5-20 m.
Asian Pacific: Indonesia and Philippines to Japan.

SCHEELE'S CONGER   *Ariosoma scheelei*
SIZE: to 20 cm (8 in.)   Conger Eels – Congridae
ID: Gray to brown with **row of pale spots along base of dorsal fin**, all fins translucent; well-developed pectoral fins, large eye with white iris. Solitary and cryptic, occasionally in open at night. Sheltered bays and lagoon reefs in 1-10 m.
Indo-West Pacific: E. Africa to Indonesia, Philippines, Micronesia and Fiji.

BARRED SAND CONGER   *Ariosoma fasciatum*
SIZE: to 60 cm (2 ft.)   Conger Eels – Congridae
ID: White to light brown with numerous brown spots on head and about 12 irregular double brown bars on body; relatively pointed snout and well developed pectoral fins. Solitary. Inhabit sand in lagoons to near seaward reefs in 2-32 m.
Indo-Pacific: Madagascar to Bali and Sulawesi in Indonesia, Philippines, Micronesia, Hawaii and French Polynesia.

ORANGE-BARRED GARDEN EEL
Gorgasia preclara
SIZE: to 33 cm (13 in.)
Garden Eels – Congridae
ID: Prominently marked with alternating white to light gray and orange to brilliant yellow-orange encircling bars. Solitary or form small colonies on sand and rubble bottoms near reefs in 18-75 m.
Indo-West Pacific: Maldives to S.W. Japan, Philippines, Indonesia, Micronesia, Coral Sea and Fiji.

SPAGHETTI GARDEN EEL
Gorgasia maculata
SIZE: to 60 cm (2 ft.)
Garden Eels – Congridae
ID: Gray to light tan covered with small yellow-tan flecks; head and lateral line pores surrounded by white spots, white blotches on head. Form large colonies on sand slopes near coastal and seaward reefs in 15-40 m.
Indo-West Pacific: Islands of W. Indian Ocean to Indonesia, Philippines, Papua New Guinea, Solomon Is. and Fiji.

BARNES GARDEN EEL   *Gorgasia barnesi*
SIZE: to 121 cm (4 ft.)   Garden Eels – Congridae
ID: Longest species in genus; pale gray with numerous small, closely packed brown spots, row of white spots from head down lateral line to tail. Form colonies on sand bottoms in 15-40 m.
Asian Pacific: E. Indonesia, Philippines, Papua New Guinea, Solomon Is. to Vanuatu.

MERCY'S GARDEN EEL   *Heteroconger mercyae*
SIZE: to 68 cm (2 ¼ ft.)   Garden Eels – Congridae
ID: Dark head and body with covered with complex white maze pattern; dorsal fin translucent with alternating black and white spots along outer margin. Large colonies to 50 individuals. Sand and silt bottoms of sheltered bays in 4-10 m.
Asian-Pacific: Indonesia, Philippines and Papua New Guinea.

BANDED GARDEN EEL   *Heteroconger polyzona*
SIZE: to 35 cm (14 in.)   Garden Eels – Congridae
ID: White with numerous close-set narrow dark bars (most do not completely encircle body). Form colonies on sand bottoms near reefs in sheltered bays and on coastal slopes in 1-10 m.
West Pacific: S.W. Japan, Indonesia, Philippines, Papua New Guinea to Vanuatu and Fiji.

TAYLOR'S GARDEN EEL
Heteroconger taylori
SIZE: to 50 cm (20 in.) Garden Eels – Congridae
ID: Whitish to yellowish to tan with covering of small close-set dark spots on head, body and dorsal fin. Solitary or form small groups on silty sand plains adjacent to coastal reefs in 5-15 m.
Asian Pacific: Indonesia, Philippines and Papua New Guinea.

SPOTTED GARDEN EEL   *Heteroconger hassi*
SIZE: to 40 cm (16 in.)   Garden Eels – Congridae
ID: Whitish with dense covering of small black spots; large black blotch over gill opening and another on back about one-third the way down body. Form large colonies on sandy bottoms in lagoons and around seaward reefs in 5-50 m.
Indo-Pacific: Seychelles to Indonesia, Philippines, Micronesia, Solomon Is. and Line Is. – S.W. Japan to Great Barrier Reef and Fiji.

DUSKY GARDEN EEL
Heteroconger enigmaticus
SIZE: to 43 cm (17 in.) Garden Eels – Congridae
ID: Brown to gray with numerous tiny white to yellowish flecks, translucent fins. Solitary or form small groups on sandy bottoms near reefs in sheltered bays and on coastal slopes in 3-25 m.
Localized: Flores and Molucca Is. and West Papua in Indonesia to Papua New Guinea.

MANY-TOOTHED GARDEN EEL   *Heteroconger perissodon*
SIZE: to 54 cm (21 in.)   Garden Eels – Congridae
ID: Mottled brown to gray body flecked with irregular white to bluish gray spots; **white patch on gill opening,** narrow white margin on dorsal fin. Solitary or form colonies on sand or mud bottoms in sheltered bays and coastal slopes in 1-35 m.
Asian Pacific: Andaman Sea., Indonesia and Philippines.

Sharks & Rays

This ID Group consists of fishes whose skeletons are composed of cartilage rather than bone, and are therefore called cartilaginous fishes.

FAMILY: Requiem Sharks – Carcharhinidae
4 Genera – 9 Species Included

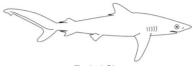

Typical Shape

When thinking of sharks most people imagine the sleek, powerful image of a requiem shark. Members of the large family, represented worldwide by 48 species in 12 genera, have pointed snouts, first dorsal fins positioned in front of ventral fins, round eyes with nictitating membranes and a lower lobe on the tail fin. Family members are responsible for approximately half of all shark attacks and several species should be considered potentially dangerous. Requiem sharks are often difficult to identify. Identification clues include positions of fins, snout shape and color and position of fin markings.

FAMILY: Bamboo & Epaulette Sharks – Hemiscylliidae
2 Genera – 7 Species Included

Typical Shape

These small, slender sharks have small, straight mouths set in front of the eyes, short nasal barbels and rounded anal fins. The bottom-oriented fishes have strong, well-developed paired fins that allow them to "walk" about on the sea floor. The banded patterns of juvenile bamboo sharks, genus *Chiloscyllium,* fade with age. As they mature, the bands on young epaulette sharks, genus *Hemiscyllium,* change into a spotted pattern with two large prominent eyespots.

FAMILY: Hammerhead Sharks – Sphyrnidae
Single Genus – 2 Species Included

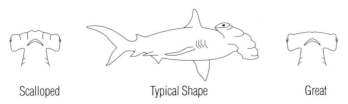

Scalloped   Typical Shape   Great

Hammerheads are the most highly evolved of all sharks. The exact functions of their distinctive T-shaped heads are not completely understood. It is thought that the odd anatomy possibly improves vision, smell, electroreception, as well as helping pin their favorite food, stingrays, to the sea floor while they are being eaten. The shape of the front edge of their wide heads is used for species identification.

SHARK FAMILIES: Other

Wobbegongs – Orectolobidae

Cat Shark – Scyliorhinidae

Whale Shark – Rhincodontidae

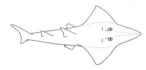

Zebra Sharks – Stegostomatidae

Nurse Sharks – Ginglymostomatidae

Order: Rays – Rajiformes

Wedgefishes – Rhinidae

Guitarfishes – Rhinobatidae

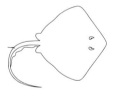

Stingrays – Dasyatidae

Eagle Rays – Myliobatidae

Cownose Rays – Rhinopteridae

Mantas and Devil Rays – Mobulidae

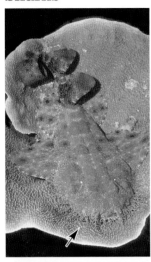

TASSELLED WOBBEGONG

Eucrossorhinus dasypogon

SIZE: to 1.8 m (6 ft.)

Wobbegongs – Orectolobidae

ID: Intricate network of dark-edged spots in shades of brown; broad, flattened head with **continuous fringe of branched tentacles around mouth.** Solitary. Lay on bottom of sheltered coastal and barrier reefs in 1-15 m.

Asian Pacific: N. Australia, Indonesia, Papua New Guinea and Great Barrier Reef.

ORNATE WOBBEGONG   *Orectolobus ornatus*

SIZE: to 2.9 cm (9 1/2 ft.)   Wobbegongs – Orectolobidae

ID: Mottled brown with spots and blotches; wide irregular bar between head and 1st dorsal fin and bars below both dorsal fins and base of tail, flattened head with **clusters of flap-like tentacles.** Solitary. Rest on coral and rocky reefs to 30 m.

Asian Pacific: Indonesia, New Guinea and Australia.

SPOTTED WOBBEGONG   *Orectolobus maculatus*

SIZE: to 3.2 m (10 1/2 ft.)   Wobbegongs – Orectolobidae

ID: Brown with dark saddles and pale irregular circular markings; broad flattened head with mostly **unbranched tentacles in clusters,** pair of longer and wider tentacles near snout tip. Solitary. Rest on coral and rocky reefs to 110 m.

Localized: S. & E. Australia.

NORTHERN WOBBEGONG   *Orectolobus wardi*

SIZE: to 63 cm (2 ft.)   Wobbegongs – Orectolobidae

ID: Brown with saddles or bars and dense reticulum of dark spots; **several widely scattered skin flaps on side of broad flattened head,** pair of nasal barbels longer than skin flap. Solitary and sluggish. Rest on bottom of coastal and lagoon reefs to 5 m.

Localized: Northern half of Australia and Aru Is., Indonesia.

BROWN-BANDED BAMBOO SHARK   *Chiloscyllium punctatum*

SIZE: to 105 cm (3 1/2 ft.)   Bamboo Sharks – Hemiscylliidae

ID: Brown to grayish brown, may retain hint of juvenile banded pattern [next]; long slender body with **pectoral and ventral fins forward of dorsal fins.** Solitary, in open at night. Rest on bottom of coastal reefs and trawling grounds in 1- 85 m.

Indo - Asian Pacific: India to Indonesia, Philippines, Micronesia and Papua New Guinea. – S. Japan to Australia.

Brown-Banded Bamboo Shark – Juvenile

SIZE: to 25 cm (10 in.)

ID: Alternating black or brown and white banding (with maturity white bands become brown); long slender body.

WHITESPOTTED BAMBOO SHARK *Chiloscyllium plagiosum*
SIZE: to 95 cm (3 1/4 ft.)   Bamboo Sharks – Hemiscylliidae
ID: Alternating tan and dark brown encircling bars and scattered white spots; long slender body with pectoral and ventral fins forward of dorsal fins. Solitary, hide in crevices during day and move into open at night. Reefs to 30 m.
Indo-Asian Pacific: India to Thailand, Indonesia, Papua New Guinea and Japan.

PAPUAN EPAULETTE SHARK *Hemiscyllium hallstromi*
SIZE: to 77 cm (2 1/2 ft.)   Bamboo Sharks – Hemiscylliidae
ID: Reddish brown with relatively large; widely spaced brown spots, large white ringed black spot surrounded by smaller spots just behind gill openings, usually no spots on snout. Solitary. Coral reefs to 15 m.
Localized: Gulf of Papua and Port Moresby to Milne Bay Province in Papua New Guinea.

HOODED CARPET SHARK *Hemiscyllium strahani*
SIZE: to 80 cm (2 1/2 ft.)   Bamboo Sharks – Hemiscylliidae
ID: Brown with white spots and dusky dark bands on body and tail; **lower head dark with white band below eye,** incomplete ocelli above pectoral fin; long slender body, pair of nasal barbels. Solitary. Coral reefs in 3-18 m.
Localized: N. New Guinea from Madang to Jayapura region, West Papua in Indonesia.

RAJA EPAULETTE SHARK *Hemiscyllium freycineti*
SIZE: to 72 cm (2 1/2 ft.)   Bamboo Sharks – Hemiscylliidae
ID: Brown with **numerous brown spots on rear head and body,** but sparse on snout; poorly defined ocellus-like marking above pectoral fin, long slender body, pair of short nasal barbels. Solitary. Rest on bottom of sheltered reefs to 12 m.
Localized: Raja Ampat in Indonesia.

EPAULETTE SHARK *Hemiscyllium ocellatum*
SIZE: to 107 cm (3 1/2 ft.)   Bamboo Sharks – Hemiscylliidae
ID: **Tan patches with pale outlines** and numerous scattered small dark spots; large black spot with pale outline above pectoral fin, long slender body, pair of short nasal barbels. Solitary. Rest on bottom of reefs to 10 m, often in stands of staghorn corals.
Localized: Great Barrier Reef.

MILNE BAY EPAULETTE SHARK *Hemiscyllium michaeli*
SIZE: to 70 cm (2 1/4 ft.)   Bamboo Sharks – Hemiscylliidae
ID: White with **brown polygons;** large dark patch above pectoral fin and pair of dark botches on leading edge of each dorsal fin, long slender body, pair of short nasal barbels. Solitary. Rest on the bottom, often under ledges, of coastal and platform reefs in 2-20 m.
Localized: Milne Bay Province in S.E. Papua New Guinea.

Sharks

CORAL CAT SHARK   *Atelomycterus marmoratus*
SIZE: to 70 cm (2¼ ft.)   Cat Sharks – Scyliorhinidae
ID: Variable mixture of spots, stripes and bands in shades of brown to gray to black; **white tips or borders on all fins,** lack barbles. Solitary. Crevices and holes of coastal reefs in 1-15 m.
East Indo-Asian Pacific: Coastal Pakistan and India to Taiwan, Indonesia, Philippines and Papua New Guinea.

BLOTCHY SWELL SHARK   *Cephaloscyllium umbratile*
SIZE: to 1.2 m (4 ft.)   Cat Sharks – Scyliorhinidae
ID: Pale brown with darker saddles and widely scattered dark brown spots; stout body with relatively small dorsal, anal and tail fins, no barbels or flaps on head. Solitary. Rock or sand bottoms of primarily continental shelf waters in 18-220 m.
Asian Pacific: China, Japan and New Guinea.

TAWNY NURSE SHARK   *Nebrius ferrugineus*
SIZE: to 3.2 m (10½ ft.)   Nurse Sharks – Ginglymostomatidae
ID: Shades of gray to brown without markings; pair of short nasal barbels, smallish mouth well in front of eyes, **close-set dorsal fins of nearly same height.** Solitary or small groups. Often rest on bottom of lagoon and seaward reefs in 1-70 m.
Indo-Pacific: Red Sea and E. Africa to Society Is. in French Polynesia. – S. Japan to Australia.

ZEBRA SHARK   *Stegostoma fasciatum*
SIZE: to 3.5 m (11½ ft.)   Zebra Sharks – Stegostomatidae
ID: Tan to gray with numerous leopard-like spots; huge long tail nearly half of total length, pair of nasal barbels, ridges on body. **Juvenile –** Black with white bands. Solitary, often rest on bottom; not considered dangerous. Coastal and offshore reefs to 70 m.
Indo-West Pacific: Red Sea and E. Africa through Asian Pacific. – S. Japan and Taiwan to Australia, New Caledonia and Fiji.

TIGER SHARK   *Galeocerdo cuvier*
SIZE: to 6 m (20 ft.)   Requiem Sharks – Carcharhinidae
ID: Gray with **dusky bars on body;** large head and mouth with short bluntly rounded snout, long slender tail with pointed tip. Solitary and considered dangerous. Coastal and offshore reefs to at least 75 m.
Circumglobal: Tropical and temperate seas.

WHITETIP REEF SHARK   *Triaenodon obesus*
SIZE: to 1.7 m (5½ ft.)   Requiem Sharks – Carcharhinidae
ID: Gray with white underside; **white tips on 1st dorsal fin and upper tail lobe,** occasional dark spots on sides, slender with rounded snout. Solitary or form small groups. Usually rest on bottom of coastal, lagoon and outer reef slopes in 3-122 m.
Indo-Pacific: Red Sea and E. Africa to Hawaii and E. Pacific. – S.W. Japan to Australia.

GRAY REEF SHARK *Carcharhinus amblyrhynchos*
SIZE: to 2.4 m (7 ³/₄ ft.)   Requiem Sharks – Carcharhinidae
ID: Gray with white underside; **broad black tail margin,** 2nd dorsal, anal and underside of pectoral fins usually black. Solitary or form aggregations; occasionally aggressive and dangerous. Outer reef slopes in 1-274 m.
Indo-Pacific: Madagascar and Seychelles to Hawaii and Pitcairn Is. – China to Australia.

DUSKY SHARK *Carcharhinus obscurus*
SIZE: to 4.2 m (14 ft.)   Requiem Sharks – Carcharhinidae
ID: Gray with pale underside; fin tips often dusky, especially underside of large pectoral fins, **dorsal fin pointed.** Solitary or form groups in open water; occasionally aggressive and dangerous. Primarily along continental coastlines to 400 m.
Circumtropical.

SILVERTIP SHARK *Carcharhinus albimarginatus*
SIZE: to 3 m (10 ft.)   Requiem Sharks – Carcharhinidae
ID: Gray with pale underside; **white tips on 1st dorsal, pectoral and tail fin lobes.** Solitary or small groups; considered dangerous. Outer reef slopes, usually below about 20 m.
Indo-Pacific: Red Sea and E. Africa to Society Is. in French Polynesia. – S. Japan to Australia.

BLACKTIP REEF SHARK *Carcharhinus melanopterus*
SIZE: to 2.6 m (8 ¹/₂ ft.)   Requiem Sharks – Carcharhinidae
ID: Brownish gray with white underside; **black tip on 1st and 2nd dorsal, pectoral, anal and lower lobe of tail fins.** Solitary or groups, usually not dangerous. Coastal, lagoon, and outer slopes, usually in 1-20 m.
Indo-Pacific: E. Africa and Red Sea to Hawaii and Pitcairn Is. – S. Japan to Australia.

BULL SHARK *Carcharhinus leucas*
SIZE: to 3.4 m (11 ft.)   Requiem Sharks – Carcharhinidae
ID: Large stout body, very short bluntly rounded snout and small eyes; gray shading to white underside, no markings on fins. Solitary or form groups in open water; considered dangerous. Coastal reefs and estuaries to 152 m.
Circumglobal: Tropical and warm temperate seas.

BLACKTIP SHARK *Carcharhinus limbatus*
SIZE: to 2.6 m (8 ¹/₂ ft.)   Requiem Sharks – Carcharhinidae
ID: Gray with white underside; **anal fin pale to white,** black tips on 2nd dorsal, pectoral and ventral fins and lower tail lobe, silver-white streak on flank. Solitary. Lagoons, inshore waters and reef channels to 50 m.
Circumglobal: Tropical and subtropical seas.

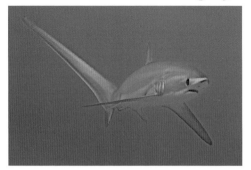

SICKLEFIN LEMON SHARK *Negaprion acutidens*
SIZE: to 2.2 m (7¼ ft.)   Requiem Sharks – Carcharhinidae
ID: Pale yellow-brown with pale underside without distinctive markings; short snout, **pair of widely spaced dorsal fins of nearly equal height.** Solitary and considered dangerous. On or near bottom of bays, estuaries and offshore reefs in 1-30 m.
Indo-Pacific: E. Africa and Red Sea to French Polynesia. – Micronesia to Australia.

PELAGIC THRESHER SHARK *Alopias pelagicus*
SIZE: to 3.3 m (11 ft.)   Thresher Sharks – Alopiidae
ID: Dark blue to bluish gray with silvery burnish and white underside; **long upper lobe of tail almost length of entire body,** long pectoral fins with nearly straight leading edge has a rounded tip. Solitary. Generally oceanic, occasionally near reef dropoffs and seamounts to 150 m.
Circumglobal: Tropical and warm temperate seas.

SCALLOPED HAMMERHEAD *Sphyrna lewini*
SIZE: to 3.7 m (12 ft.)   Hammerhead Sharks – Sphyrnidae
ID: Gray with white underside; head flattened and extended to either side with **prominent central indentation on front edge and pair of lesser indentations on each side.** Solitary or form groups; not generally considered dangerous. Open water of seaward reefs, around seamounts and islands in 1-275 m.
Circumglobal: Tropical and warm temperate seas.

GREAT HAMMERHEAD *Sphyrna mokarran*
SIZE: to 4.2 m (14 ft.)   Hammerhead Sharks – Sphyrnidae
ID: Gray with white underside; head flattened and extended to either side with **front edge slightly curved,** rear edge of ventral fin curved. Similar Scalloped Hammerhead [previous] distinguished by deeply scalloped "hammer". Solitary and considered dangerous. Oceanic, rarely on reefs in 3-75 m.
Circumglobal: Tropical and warm temperate seas.

SHARK RAY *Rhina ancylostoma*
SIZE: to 2.7 m (8¾ ft.)   Wedgefishes – Rhinidae
ID: Large gray shark-like ray; **broad rounded head** with body ridges above eyes and along center line, large broad-based pectoral fins, white spotting on body and fins. Solitary. On or near bottom of coastal seas, occasionally near reefs in 3-90 m.
Indo-Asian Pacific: E. Africa and Red Sea to Papua New Guinea. – S. Japan to Australia.

WHITE-SPOTTED WEDGEFISH *Rhynchobatus australiae*
SIZE: to 2 m (6½ ft.)   Wedgefishes – Rhinidae
ID: Tan to gray to black with white spots and white underside; often white-ringed large black spot above base of pectoral fin, **pointed triangular snout.** Solitary. Often rest on bottom of sandy areas in lagoons and around reefs to 50 m.
Indo-Asian Pacific: Andaman Sea to Philippines, south to Australia.

WHALE SHARK *Rhincodon typus*

SIZE: to 12 m (39 ft.)   Whale Sharks – Rhincodontidae

ID: Huge; dark gray with white underside, numerous white spots scattered on head and arranged in rows and bars on body, broad mouth, ridges on side of body, enormous tail. Solitary. Open clear waters in 0-40 m, often near surface and occasionally around reefs.

Circumtropical.

LONG-SNOUT SHOVELNOSE RAY *Aptychotrema rostrata*

SIZE: to 1.2 m (4 ft.)   Guitarfishes – Rhinobatidae

ID: Shades of gray to brown with **large somewhat indistinct spots;** long pointed triangular snout merges into rounded "wings" below eyes. Solitary. Rest on bottom of sand and seagrass beds of bays, along shorelines and around reefs to 60 m.

Localized: Queensland to Victoria in S.E. Australia.

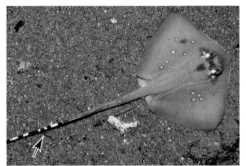

GIANT SHOVELNOSE RAY *Glaucostegus typus*

SIZE: to 2.7 m (8 3/4 ft.)   Guitarfishes – Rhinobatidae

ID: Light brown with **pale snout;** large triangular head merges with rounded pectoral fins, tapering body with pair of relatively small dorsal fins. Solitary. Sand and mud of coastal waters and continental shelf to 100 m.

East Indo-Asian Pacific: India to Indonesia, Papua New Guinea, Solomon Is. and Australia.

BLUE-SPOTTED STINGRAY *Neotrygon kuhlii*

SIZE: disc to 50 cm (20 in.)   Stingrays – Dasyatidae

ID: Brown to olive with blue spots and small black spots; short pointed snout and sharply rounded "wings", tapering tail as long or longer than diameter of disc and **marked with white bars.** Solitary. Rest on sand of lagoons and seaward reefs to 90 m.

Indo-West Pacific: E. Africa to Indonesia, Philippines, Micronesia and Solomon Is. – S. Japan to Fiji.

BLUE-SPOTTED RIBBONTAIL RAY *Taeniura lymma*

SIZE: disc to 35 cm (14 in.)   Stingrays – Dasyatidae

ID: Yellow-brown with numerous blue spots; oval-shaped disc, flattened ribbon-like tail about 1.5 times disc width with 2 spines. Solitary. Rest on sand bottoms, under ledges or in reef holes of coastal, lagoon and outer slopes in 1-20 m.

Indo-West Pacific: Red Sea and E. Africa to Indonesia, Philippines, Papua New Guinea, Solomon Is., Australia to Fiji.

MARBLED STINGRAY *Taeniurops meyeni*

SIZE: disc to 1.85 m (6 ft.)   Stingrays – Dasyatidae

ID: Gray with variable pattern of dense black spots, blotches and mottling; large ovate disc, short tail about same length as disc with single spine. Solitary. rest on sand bottoms of coastal, lagoon and near outer reefs in 2-500 m.

Indo-Pacific: E. Africa and Red Sea to Indonesia, Philippines, Micronesia, Solomon Is. and E. Pacific. – S. Japan to Australia and Fiji.

NARROWTAIL STINGRAY   *Pastinachus gracilicaudus*

SIZE: disc to 75 cm (2 1/2 ft.)   Stingrays – Dasyatidae

ID: Grayish brown, typically with reddish or purplish hue; ventral surface white. Solitary. Rest on sand of estuaries, coastal sand flats and near coral reefs to 60 m.

Asian Pacific: Sarawak and Sabah in Malaysia, Singapore, Java and Kalimantan in Indonesia.

THORNY STINGRAY   *Urogymnus asperrimus*

SIZE: disc to 100 cm (3 1/4 ft.)   Stingrays – Dasyatidae

ID: Pale gray to dark brown with numerous thorns on surface; scattering of white spots, ovate disc with humped central portion, tail about equal to disc length. Solitary, often partially buried in sand. Coastal, lagoon and outer reefs to 130 m.

Indo-West Pacific: Red Sea and E. Africa to Indonesia, Micronesia and Solomon Is. – S. Japan to Australia and Fiji.

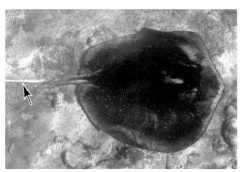

WHITETAIL STINGRAY   *Himantura granulata*

SIZE: disc to 100 cm (3 1/4 ft.)   Stingrays – Dasyatidae

ID: Dark gray to brown usually with numerous white specks, outer edge of disc slate gray; **tail white beyond spines (usually two).** Solitary. On sand bottoms of lagoons, bays and protected areas around shallow reefs to 85 m.

East Indo-Asian Pacific: Maldives to Indonesia, Papua New Guinea, Solomon Is. to Great Barrier Reef and New Caledonia.

TAHITIAN STINGRAY   *Himantura fai*

SIZE: disc to 100 cm (3 1/4 ft.)   Stingrays – Dasyatidae

ID: Light gray to pinkish brown, occasionally blotched or mottled; snout bluntly pointed, rounded "wings" and **long tapering tail** can be nearly 3 times disc diameter. Solitary or form aggregations. Sand and rubble of lagoons to 20 m.

Indo-Pacific: India to Thailand and Society Is. in French Polynesia.

JAVANESE COWNOSE RAY   *Rhinoptera javanica*

SIZE: wing span to 1.5 m (5 ft.)   Cownose Rays – Rhinopteridae

ID: Brown with white underside; nearly triangular disc or "wings" with protruding head and short slender tail about equal to disc length with single spine. Solitary or form groups. Open coastal waters in 1-50 m.

Indo-Asian Pacific: E. Africa to Indonesia and Philippines, north to S. Japan.

ORNATE EAGLE RAY   *Aetomylaeus vespertilio*

SIZE: wing span to 2.5 m (8 ft.)   Eagle Rays – Myliobatidae

ID: Brownish green to goldish brown; reticulated dark lines on forebody, triangular wings with dark rings on wide rear border, protruding head and long slender tail without spine. Solitary. Open water, occasionally over reefs, rubble and muddy bays in 5-50 m.

East Indo-Asian Pacific: Maldives to Malaysia, Philippines, Indonesia and Australia.

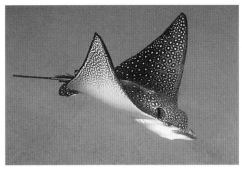

SPOTTED EAGLE RAY *Aetobatus ocellatus*

SIZE: wing span to 3.5 m (11 1/2 ft.) Eagle Rays – Myliobatidae

ID: Gray-brown to nearly black with numerous white spots and white underside; nearly triangular disc or "wings" with protruding head and long slender tail with multiple spines. Solitary. Open water of coastal, lagoon and outer reefs in 1-80 m.

Indo-Pacific: Red Sea to Galapagos and the Gulf of California.

SICKLEFIN DEVIL RAY *Mobula tarapacana*

SIZE: wing span to 3.1 m (10 ft.) Devil Rays – Mobulidae

ID: Mouth under body between eyes; **brown to olive-brown dorsal surface** (all other Devil Rays have dark blue to blue-gray to black dorsal or dark chocolate brown surface); single color dorsal fin; no spine on base of tail; short tail less than disc width.

Circumglobal: Tropical and warm temperate seas.

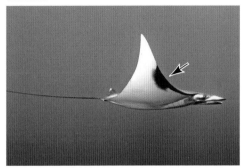

LONGHORN PYGMY DEVIL RAY *Mobula eregoodootenkee*

SIZE: wing span to 1 m (3 1/3 ft.) Devil Rays – Mobulidae

ID: Mouth under body between eyes; **distinct dark shading on the middle front of white ventral surface of pectoral fins;** dorsal surface has brown stripe along the front edge; often white tip on small dorsal fin; no spine on base of tail. Often form large schools.

Indo-West Pacific: Red Sea and E. Africa to Philippines, Indonesia and E. Australia.

BENT FIN DEVIL RAY *Mobula thurstoni*

SIZE: wing span to 1.8 m (6 ft.) Devil Rays – Mobulidae

ID: Mouth under body between eyes; **leading edge of pectoral fins slighty indented near center ;** narrow dark band on top of head behind eyes; ventral surface is silvery with variable amounts of gray shading; white tip on small dorsal fin; no spine on base of tail.

Circumtropical.

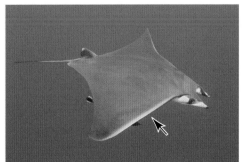

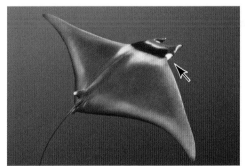

SHORTFIN PYGMY DEVIL RAY *Mobula kuhlii*

SIZE: wing span to 1.2 m (4 ft.) Devil Rays – Mobulidae

ID: Mouth under body between eyes; **dorsal surface has white stripe along leading edge of pectoral fins; black stripe along leading edge of ventral surface;** tail length less than width of pectoral fins; often white tip on small dorsal fin; no spine on base of tail.

Indo-West Pacific: Red Sea and E. Africa to Philippines, Papua New Guinea and N. Australia.

SPINETAIL DEVIL RAY *Mobula mobular*

SIZE: wing span to 3.2 m (10 1/2 ft.) Devil Rays – Mobulidae

ID: Mouth under body between eyes; **dorsal side has wide slightly curving black band between eyes and forebody;** white tip on dorsal fin; spine on base of tail. Often form large schools.

Circumglobal: Tropical and warm temperate seas.

Note: Formerly classified as *Moula japonica.*

Mantas

REEF MANTA RAY   *Mobula alfredi*
SIZE: wing span to 5m (16¹/₂ ft.)   Devil Rays – Mobulidae
ID: Usually white terminal mouth between pair of movable flaps extending from either side; dorsal surface black, usually with **white "shoulder" patches that form the outline of a black "V"**. Solitary, but occasionally form groups over shallow shelf and reef habitats to 20 m.
Indo-Pacific.

Reef Manta Ray – Ventral View
ID: Ventral surface white with various-sized black spots, including between gill slits; often faint wide dusky rear border.

OCEANIC MANTA RAY   *Mobula birostris*
SIZE: wing span to 7m (23 ft.)   Devil Rays – Mobulidae
ID: Usually black terminal mouth between pair of movable flaps extending from sides; dorsal surface black often with **white "shoulder" patches that forms a black "T"**. Solitary or form groups in open water to 24 m.
Circumtropical.

Oceanic Manta Ray – Ventral View
ID: Ventral surface white with various-sized black spots behind the gills and additional dark patches along the trailing edge of individual gill slits; often wide dark rear border.

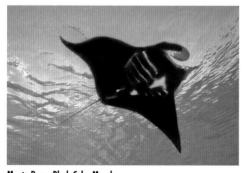

Manta Ray – Black Color Morph
ID: Occasionally both Manta Ray species have mainly black ventral surface. Those with ventral coloration are referred to as Black Color Morph Manta Rays. Usually a white area between the gills and assciated white markings and spots. Identification of species from ventral surface is difficult. Best identifed by mouth color and markings on the dorsal surface.

Manta Ray – Black Color Morph Variation
ID: This variation has more white than typical and illustrates the variability of markings on different individuals.

COMMON NAME INDEX

SCIENTIFIC NAME INDEX

NOTES

  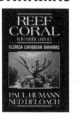